GRAPHIC TECHNOLOGIES, ☞ **W9-BJE-562**

An awareness never before achieved!

The **3-D IMAGE** below
was created for this book by
Graphic Technologies, Inc.

This *amazing* technique can be
used in YOUR publications.
Please turn the page ▶

I M 3 A G E S
3D

As you look at the sample on the preceding page, creative ideas are bound to emerge for ways to use this technology to make your products the most spectacular, innovative and attention-getting on the retailer's shelves.

Give your company the cutting edge over your competition. If this concept in print media catches your attention, you can be certain it will do the same for your customers.

SUGGESTED USES INCLUDE:

Annual Reports	Paperbacks	Labels/Stickers
Books	Newsletters	Maps
Booklets	Software	Newspapers/Tabloids
Catalogs	Manuals	Postcards
Cookbooks	Yearbooks	Stationery
Directories	Calendars	Posters
Galley Copies	Comic Books	Binders/Covers
Journals	Direct Mail Letters	Games/Puzzles
Magazines	Envelopes	Packages
Mass-Market	Greeting Cards	Boxes

R E M E M B E R :

Anything that can be photographed or computer-generated can be produced in 3-D by Graphic Technologies, Inc. Call us at 800-421-5693 for further information about pricing and how Graphic Technologies can benefit you and your company.

GRAPHIC TECHNOLOGIES, INC.
1-800-421-5693

124 Kennedy Road, Ft. Wright, KY 41011   606-341-0259   Fax: 606-341-0456

DIRECTORY
of
PRINTERS

1994-1995 EDITION

M A R I E   K I E F E R

Copyright © 1994 by Evelyn Marie Kiefer

All rights reserved. No part of this book may be reproduced or transmitted in any form or by any means — graphic, electronic, or mechanical, including photocopying, recording, or any information storage or retrieval system — for sale without written permission from the publisher, except for brief quotations included in a review.

Published by: **Ad-Lib Publications**, 51½ West Adams, PO Box 1102, Fairfield IA 52556, 515-472-6617; 800-669-0773; Fax: 515-472-3186.

Publisher's Cataloging in Publication
(Prepared by Quality Books Inc.)

Kiefer, Evelyn Marie.
    Directory of printers, 1994-95 edition / Evelyn Marie Kiefer. --
1994-95 ed.
    p. cm.
    Includes bibliographical references and index.
    ISBN 0-912411-43-0

    1. Printing industry--Directories. 2. Printers--Directories.
I. Title.

Z282.K54 1994   686.2'025
                         QBI94-1393

Printed and bound by Walsworth Publishing of Marceline, MO, USA

ISSN: 0895-139X

Foreword

We've shortened the title of the *Directory* to reflect the fact that those printers listed do much more than just books. While books are our primary focus, we also list those printers who specialize in other items as well.

This *Directory* features over 700 printers of books, catalogs, magazines, and other bound publications. Besides listing the company name, person to contact, address, phone number, fax number, and toll free number (if available), this *Directory* also provides many additional details about each printer. We asked the printers to list their optimum print runs, standard book sizes, binding capabilities, and the other types of items they print besides books. Other information we asked for was a list of the various services each printer offers, including whether or not the printer can do color separations, maintain lists, handle fulfillment, do typesetting, or provide design and layout. Also, in keeping with the times, this edition notes which printers stock acid-free or recycled paper.

This edition also lists the equipment used by the printer, annual sales and number of employees, when it was founded, and whether it is a union shop or not. It also indicates the turnaround times for books, magazines, and catalogs. Finally, for those of you who would like to work with a printer accustomed to working with your kind of organization (whether book publisher, business, non-profit organization, school, or whatever), we have indicated the percentage of their business which comes from various segments of the market.

A new section in this edition is Services/Suppliers. The printer is the final stop when getting either a book, catalog, or other bound publication printed. Other people or companies help put that publication together. Many of those services/suppliers are listed in this new section.

We hope you like the new additions to the book. The information contained in the listings of this *Directory* is from survey forms. Where the printer did not return the survey form to us, we used information available from other published sources, the printer's advertisements, and/or reports received from people who have used their services.

We hope this *Directory* will help you locate the most appropriate printers for your various publication needs. We have worked hard to make this *Directory* the most comprehensive of its kind. Nonetheless, we know there is still room for improvement. Although we have listed all the major book printers as well as the top commercial printers, we know that we've possibly overlooked a good number of reliable printers capable of printing books, catalogs, magazines and other bound publications.

To make this *Directory* even more comprehensive and complete, we need your help. Please send us any further information you discover about the printers we now list. Also, send us the names and addresses of any printers we have overlooked. We'll add them to our PrintBase data files so they'll be included in future editions of the *Directory*. We appreciate your help.

Finally, thanks goes to all who have helped put this *Directory* together: *Barbara Anderson, Jeanette Beasley, Paula Fritchen, John Kremer, Joan Sirdoreus, and Karen Stacy.* Special thanks must go to my husband, Bob, whose support has been unwavering. Thanks.

Marie Kiefer
October 1, 1994

Table of Contents

Chapter 1

Introduction: How to Use This Directory

This *Directory* is designed to save you time and money in finding the best printer for your next book, catalog, journal, magazine, manual, annual report, directory, or other bound publications. It could save you hundreds, even thousands of dollars in printing and binding costs alone. Plus, of course, it can save you time as well as production hassles.

As one user of this *Directory* wrote, "Without your *Directory*, I ended up paying $12,500 for the first printing of one book — and I got plenty of production hassles. With the *Directory*, I paid $4,500 for the identical job — and not one problem."

Two Steps to Savings

How can you get these savings? You need to do two things:
1) Contact the appropriate printers, and
2) Query them in a professional manner.

This *Directory* will help you with both steps.

First, its listings of over 700 printers will enable you to select the appropriate one for your job. If you have a particular project in mind, you can use the indexes to help you target those printers who have the right combination of capabilities and services to meet your needs. Or, if you'd rather, you can browse through the book simply reading the listings until you locate several printers capable of meeting your needs.

Read each listing carefully. Check to see which printers can offer the services you require. Can they print the quantity you want? Do they offer the type of binding you need? What about their other services, credit terms and turnaround times?

Once you have located five or ten printers you think could do the job you require, query them. Always query at least five printers. Don't get hooked on using one printer all the time (except, of course, for a continuing publication such as a magazine where you might contract with a printer for an entire year). As your needs change, you will find that you may have to change printers to find one who can handle the different quantities, sizes, or bindings that you require ... with the quality you want ... at a price you can afford...and with great service and dependability.

To query a printer, develop a standard Request for Quotation (RFQ) form that you can send to all printers. For examples, see the sample book RFQs on pages 21, 28, and 29. For more details about querying printers, read Chapter 3.

Printer Listings — The Details

This 1994-95 edition of the *Directory* has been updated to list over 700 printers capable of printing books, catalogs, magazines, and other bound publications. Where known, the following information is listed about each printer:

- **Main Focus** — This code indicates whether they are primarily a book, catalog, magazine or general commercial printer, or a printing broker.

- **Company Address** — The name of the company, the main person to contact for queries, the address, phone number, and fax number.

- **Print Runs** — Their minimum press run, their normal maximum print run, and their optimum print run for bound publications. Not only will this give you a good idea of the range of their capabilities, but the optimum print run should indicate at what quantity they can offer you their best price.

- **Sizes** — We list whether they are capable of printing the six most common sizes for books as well as whether or not they can handle other odd sizes. The six most common sizes are as follows:
    $4\frac{1}{4}$ x 7 inches — mass market paperback
    $5\frac{1}{2}$ x $8\frac{1}{2}$ inches — common trade paperback
    6 x 9 inches — common hardcover book
    7 x 10 inches — less common size
    $8\frac{1}{2}$ x 11 inches — workbook or magazine
    9 x 12 inches — large art or coffee-table book

- **Bindings** — We indicate whether they offer the following bindings in-house or can subcontract them from outside suppliers:
    Casebound — hardcover, edition binding
    Comb-bound — plastic binding (e.g., cookbooks)
    Looseleaf Binders — for software manuals, etc.
    Otabind — another binding for paperbacks (lays flat when opened)
    Perfectbound — common paperback

Saddlestitched — stapled in center (magazines)
Sidestitched — like *National Geographic*
Spiral Bound — like a notebook
Wire-O Bound — similar to spiral binding

- **Items** — The types of items they regularly print, including annual reports, books, catalogs, cookbooks, directories, journals, magazines, mass-market paperbacks, newsletters, textbooks, software manuals, yearbooks, brochures, calendars, greeting cards, maps, posters, and so forth.

- **Services** — We indicate which of the following services they are capable of providing in-house:

  4-color printing on text pages
  Color separations
  Design and artwork
  Fulfillment and mailing services
  List maintenance
  Opti-Copy prepress system
  Rachwal prepress system
  Acid-free paper stocked
  Recycled paper stocked
  Typesetting
  Typesetting from disks or via modem
  Warehousing

- **Equipment** — We list the kinds of printing presses they use and, when known, the number and sizes of each kind. The five most common ones are:

  Cameron Machinery — cost-effective for quantities between 5000 and 30,000 of trade paperback books.

  Letterpress — While offset presses have replaced letterpresses as the standard printing technology, letterpresses are still commonly used for printing die-cut, embossed, and perforated items. They are also used by printers of limited edition books and other "fine" printing.

  Sheetfed Offset Press — Commonly used for press runs under 10,000 copies. Web presses tend to be more cost-effective for print runs over that quantity (at least for books and other publications).

  Web Offset Press — Cost-effective for print runs over 10,000. Used for print runs as high as 7,000,000.

  Gravure (Rotogravure) — Because of its high make-ready costs, this press is most cost-effective for print runs over half a million.

- **Annual Sales** — This information is listed for those of you who want to know more about the printer you will be working with. Annual sales will indicate whether or not the printer has the cash flow and resources to serve your needs over an extended period of time (especially crucial for magazine and catalog publishers).

- **Number of Employees** – Again, this figure indicates whether the printer has the resources (in this case, the personnel) to handle big jobs or heavy demand.

- **Turnaround Times** – We asked the printers to indicate their typical turnaround time for printing and binding (1) books, (2) catalogs, and (3) magazines. The turnaround time is the number of working days it takes them to complete the job from camera-ready copy to shipment. The listed turnaround times are taken from the printer's reports. We cannot guarantee that the printer will live up to these "typical" turnaround times on every job.

- **Customers** – We also asked the printers to indicate what percentage of their business comes from the following customers:
  Book publishers
  Businesses
  Colleges and universities
  Magazine publishers
  Non-profit groups
  Other organizations
  Schools (elementary and secondary)
  Self-publishers (including authors, genealogists, etc.)
  Others

  This information will be most useful to those of you who would like to work with a printer accustomed to working with similar organizations. While this information may be especially useful for self-publishers, schools, and non-profit groups, it may also be useful to magazine publishers who want to work with a printer accustomed to servicing the special needs of magazine publishers.

- **Union** – We have noted which printers use union workers because we have received a number of requests from unions and non-profit groups who must use union suppliers.

- **Year Established** – In general, the longer a printer has been in business, the more reliable they should be. Use this information only as a guide. Always check a printer's references.

- **Terms** – These are the standard terms offered by the printer. Most printers will offer net 30 terms with approved credit, but some require a deposit when working with a new customer. Note that terms are always negotiable, especially if you have a good credit rating *and* the printer is hungry for business.

- **Statement** – The statements are written by the printers to describe, in their own words, their goals and services. In some cases, these statements have been taken from the printers' advertisements or printed brochures rather than from their statements on our survey form. But in every case, these statements are in their own words.

Other Features

Besides the information contained in the alphabetical listings of the printers, this *Directory* also includes the following information:

- **Short-Run 4-Color Printers**—Chapter 6 features a listing of printers who specialize in printing 4-color catalog sheets, postcards, and other items in short runs (from 500 to 25,000 copies).

- **Foreign Printers**—Chapter 7 features a listing of more than 120 overseas printers and printing brokers. While we do encourage you to use printers in the U.S. and Canada, it is possible to save considerable amounts of money by getting certain items printed overseas (especially 4-color books in short runs).

  The introduction to Chapter 7 describes how to best work with overseas printers. It also describes the advantages and disadvantages to working with such printers.

- **Services/Suppliers** — Chapter 8 features a listing of related publishing services.

- **Glossary**—The glossary provides definitions for about 300 terms commonly used in the printing trade and in this book. If you find a word somewhere in this introduction or in one of the listings that you do not understand, refer to the glossary.

- **Bibliography**—This section provides a list of books we recommend for further reading. Of all these books, the most important is *Getting It Printed* by Mark Beach. This book should be required reading for all printers and all printing buyers. A second bibliography lists books published by Ad-Lib Publications.

- **Indexes**—To make this book as useful as possible, we have provided indexes indicating which printers specialize in the capabilities and services you need. Among these indexes is a special listing of all printers by city and state (or province), so you can locate those printers nearest you.

- **Choosing a Printer**—In the following chapter, we've listed 20 points to consider when selecting a printer. Read these points over carefully. While many smaller publishers tend to select printers solely on price, that can often be a mistake. These points will describe what you should look for in a printer.

  There is an old saying in the printing industry: "Price, speed, and quality—you can have any two." If you want to have top quality, you have to pay the price. If you want speed, you also may have to pay more. No printer can offer you the best of all three (lowest price, fastest turnaround, best quality). Almost all print buying decisions are, therefore, compromises. Know what to look for so you can make the most informed decision.

- **Requests for Quotations** – Chapter 3 describes how to query printers to obtain a reliable pricing quotation for your printing jobs. A well-prepared Request for Quotation (RFQ) will ensure that you get the best price and service from your printer. It also provides the basis for any contract you sign with a printer.

- **Tips on Saving Money** – In Chapter 4, we list some 40 tips that could help you save money on the typesetting and printing of your books and other publications. You should find several points that will help you to streamline your production and save you money, time, or hassles dealing with printers.

- **Printer Data Files** – The information contained in this *Directory* is also available as data files for use with your favorite database program. These data files allow quick and easy selection of all those printers who can match your needs. For example, using these data files and your favorite database program, in a matter of minutes you could print out labels for all printers capable of producing short-run books. These data files can save you time when sending out RFQ's or other requests for information.

Formats include comma-delimited or tab-delimited ASCII (mail-merge) files for IBM or Macintosh compatible computers and DBF. Other formats are also available. Call **800-669-0773** for details.

Chapter 2

Points to Consider When Selecting a Printer

Too many publishers evaluate printers only on the basis of price. This is a mistake. The cheapest printer is not always the best printer. To help you in your evaluation of prospective printers, here are some points you should consider.

- **Experience** — How long have they been printing books? How long have they been in business (what is the history and financial soundness of the firm)? What kinds of books and other printed items are they accustomed to producing?

- **Reputation** — Do they have a reputation for doing quality work and delivering on schedule? Do you know anyone who has used them before? Can they provide references?

- **Quality of Work** — Do they produce good-looking, well-made books? Do their books hold together? Is the printing clear — neither overinked nor underinked? Have you seen samples of their work? Be sure to get a sample of their work before you commit to them — preferably a sample similar in size and binding to the book or other publications you want produced.

- **Price** — How do their prices compare to other printers? If their prices are higher than other printers, what added benefits do they offer? Faster delivery? Better quality? Greater reliability? More services?

- **Service** — Are they willing to work with you to produce the best publication for the price you are able to pay? Do they answer your phone calls, letters, and other queries quickly and courteously? Do they put forth that extra bit of effort that makes working with them a pleasure?

- **Dependability** — Do they deliver books, catalogs, or other publications when promised (are their schedules reasonable)?Do they live up to their promises, both verbal and written? Again, check their references to verify their dependability.

- **Timing** — Do they offer faster delivery than other printers? Most book printers offer a 4 to 6 week lead time for books printed from camera-ready copy. Can this printer offer a faster delivery time and yet produce a quality publication?

- **Terms** — What kind of terms can they offer you? Do they require a large downpayment? Do they offer any discounts for prepayment or quicker payment?

- **Other Services** — Can they provide warehousing and fulfillment services for you? Can they do typesetting and pasteup in-house? Do they have teletypesetting capabilities? Can they maintain your lists?

- **Quantities** — What quantities are they capable of producing? Will they be able to follow up an initial short run with a much larger second run? Will you have to go to another printer to do a higher quantity when your book begins to sell or when you want to roll out a catalog after a test run?

- **Capabilities** — What kind of books or other publications are they accustomed to doing? What sizes and types of bindings? Is your book a special size or binding? Will it economically fit their presses and binding capabilities?

- **Specialization** — Whenever possible, use a printer who specializes in the type of publication you want to publish. Find out which sizes, quantities, bindings, etcetera, they are accustomed to doing. Can they do odd sizes? And, if so, are they set up for easy handling of such odd sizes or bindings?

- **Choices** — Does the printer offer a choice of paper and cover stocks? Do they keep them in stock, or will they have to special order them? Special orders will usually require more time. Does the printer maintain strong relationships with outside vendors?

- **Location** — Where are they located? Location may be important for several reasons: 1) You may be able to get speedier delivery — both in getting copy to them and in getting books from them; 2) It will be easier to make changes in the galleys if necessary; and 3) If you live nearby, you may have a chance to visit their plant and see more samples of their work and meet the people you will be working with.

- **Equipment** — Is the printer's equipment up to date? Can it handle the type of book or publication you want to publish? Is the equipment messy and dirty, or well maintained? (If they do not care how their plant or equipment looks, do you think they will care what your book looks like?) Is the printer's equipment sound, and have they made progressive upgrades that enhance performance? Does the printer employ methods and materials that minimize environmental impact?

- **Working Conditions** — Is a tour of the printer's facility readily available? If you get a chance to visit the printing plant, check out the following points. Is the plant well lighted and ventilated? Does it appear to be a pleasant place to work? Are the floors clean? Is there enough space for the equipment and storage? Is there enough room for the workers to get around efficiently? Does the flow of work seem well thought out?

- **Personnel** — Do the workers seem happy? Are they interested in their work? Do you like the customer service rep you will be working with? Do all the personnel seem responsive to your needs? Do they have the knowledge and experience to do the job? When you become a customer, what level of access to key management can you expect to have?

- **Packing and Shipping** — Will the printer pack the books properly so they are not damaged in shipment? Can they provide shrinkwrapping of books? Can they ship to more than one destination? Can they ship by both UPS and truck? Do they have the capability to handle your list maintenance and fulfillment (for catalogs, magazines, and other periodicals)?

- **Trade Customs** — Does the printer observe trade customs regarding ownership of intermediates, overruns/underruns, and the like?

- **Fitness** — Be sure to select a printer who is right for the job you want done. Don't go to a quick printer for a casebound book, or to a web printer for 100 copies of a short brochure. Use this Directory to help you select those printers who can best serve your needs.

- **Long-Term Relationship** — Never work with a printer you would not be happy to work with again. Aim to develop long-standing relationships with your printers so you can come back to them again and again with confidence. Does the printer conduct random follow-up surveys to gauge their performance?

Chapter 3

How to Request a Printing Quotation

To obtain a reliable quote for the printing of your book or other publication, you should supply all the information the printer will need to make a valid estimate. To ensure that you have included all the necessary information — and to ensure that the same basic data is used by all printers you query — use a Request for Quotation (RFQ) form on your letterhead. See the next page for a copy of the RFQ we sent out when requesting quotations for this *Directory*.

You do not need to copy the form we used; you may adapt it to your own requirements. Regardless of what kind of format you decide to use, you should provide the printer with the following basic information, all of which they need in order to make an accurate estimate:

- **Title of the publication** — If you do not have a title yet, give some reference title or number.

- **Quantity** — State the number of copies you want printed. You may list several options if you are not sure how many copies you want printed, but don't ask for more than two or three (just as a common courtesy). You may also ask for a quote for a preliminary test run and a follow-up roll-out quantity.

  For magazines, you will need to indicate the number of issues per year, plus the projected print runs for each issue through the coming year.

- **Number of Pages** — Include all pages: title page, copyright page, index, blank pages, and so on. Note that most printers can give you a better price if the total number of pages in your book is a multiple of 32 pages (96, 128, 160, 192, 224, ...). If the publication uses a self-cover, the cover should also be included in the page count; otherwise, the cover should be quoted separately.

Ad-Lib Publications
PO Box 1102
Fairfield IA 52556
(515) 472-6617 / Fax: (515) 472-3186

Request for Quotation

Contact: Marie Kiefer
Please quote by March 15, 1994

Quote your best price and turnaround time for the following job:

Specifications:

Book Title: Directory of Printers, 1994-95 Edition

Total Pages: 352 pages

Trim Size: 6 x 9 inches

Text Paper: 55 lb. or 60 lb. offset — Please quote on your best house stock. If possible, quote on a recycled paper grade as well.

Text Ink: Black

Cover Stock:

Cover Ink: 10 pt C1S plus film lamination

4-color, sides 1 and 4, separations will be provided by customer

Binding: Perfectbind and shrinkwrap individually

Material Provided: Camera-ready copy plus some negatives (for about 20 pages of ads). Blue lines required.

Packing: Pack in tightly sealed cartons (275 lb test) not to exceed 40 lb. total weight.

Quote:

Quantity: 3,000 copies — $_____

Delivery: Working days from receipt of camera-ready copy to shipment _____

Terms:

Remarks:

Printer:

Contact:

Address:

Phone:

Thank you. We look forward to working with you on this book.

Illustration 3.1 — Sample RFQ

- **Trim Size** — State the dimensions of the publication, whether 8½ x 11 or 8⅜ x 10⅞, etc. If the dimensions must be exact, note the specification in your RFQ. Otherwise, the printer could well substitute a near fit (for example 5⅜ x 8⅜ for 5½ x 8½ because that size better suits their presses).

  Standard trim sizes (such 5½ x 8½, 6 x 9, or 8½ x 11 for books) will enable you to get a better price from many printers *and* will fit the standard shelving units of libraries and booksellers, but don't rule out an odd size if it is appropriate for the contents of your book. If you want an odd size, be sure to describe its dimensions clearly.

- **Text Paper** — Most books are printed on 50 lb. or 60 lb. white offset or book paper, but if you are publishing a children's or photo book, you will probably want to use a different paper stock. If you want the book to last a hundred years or more, be sure to specify acid-free paper (which is, however, often more expensive). If you intend to supply your own paper, be sure to let the printer know.

- **Text Ink** — If you do not specify a color of ink, it will be black. If you want a four-color book or another accent color, be sure to specify it. In some RFQ's, this may also be referred to as Press Work.

- **Binding** — Do you want your book to be a perfectbound softcover (like this book), or a saddlestitched book (stapled like a magazine), or a smyth-sewn casebound (hardcover), or comb or spiral bound (like many cookbooks)? Perhaps you want to use otabind. You must specify exactly what kind of binding you want for the book.

  If you are publishing a casebound book, you need to specify the grade of binders board and cloth you want to use, plus other specifics. Discuss your options with several printers to get an idea of the specifications you want to use.

- **Cover Stock** — Most trade paperbacks use a 10 pt. C1S (coated one side) cover, though other cover stocks are available. Magazines and catalogs usually use a lighter stock. Again, if you don't know how to specify the cover stock, send the printers samples of the sort of cover stock you want to use. The printer can take it from there.

  For softcover books, we recommend that you ask for a varnish, or better yet, a film lamination or UV coating for the cover (to better protect its surface during shipping and handling; also, it looks better). This additional coating will cost you about 14¢ per copy.

- **Cover Ink** — Will the cover be printed with one or more colors? Any dropouts or screens? Also specify whether you want the cover printed sides 1 and 4 only (the outside) or all sides (including the inside).

- **Copy** — Will you provide camera-ready copy, negatives, press-ready plates, or will you require typesetting and pasteup services? If you require typesetting, you need to provide them with an accurate estimate of the number of words in the book, type size, fonts, and so on.

Will the copy have any photos? Bleeds? Extensive solid areas? If the book is to be printed in full color, will you be providing the color separations or transparencies?

Packing—Do you want your books individually shrinkwrapped or in convenient multiples (5 to 10 books) to protect them during shipment? Do you have any other special packing or shipping requirements?

In general, it is best if you can specify exactly how they should pack your books or other publications and how they should ship them.

For magazines and catalogs, will you provide the packing envelope, labels (in 4-up cheshire format?), and so on, or will the printer be asked to provide many of these fulfillment services as well? In such a case, how often will you be publishing? What kind of mailing schedule will they have to maintain? How many pieces?

Resources for Preparing an RFQ

An RFQ form lays out all the necessary information in a clear, understandable format so that any printer should know exactly what you want. If you do not know how to specify all the details regarding paper stocks, bindings, typesetting, and so on, then do one or more of the following:

- **Read some books on graphic arts and printing.** We highly recommend the following book: *Getting It Printed* by Mark Beach. If every printer and publisher had a copy of this book (and used it), 90% of all printing problems would disappear. It has an excellent glossary, detailed discussion of printing contracts, many superb tips, and much more.

- **Find a local graphic designer** who can help you.

- **Ask questions of the printers** you are considering using. Most will be more than willing to answer your questions if you are seriously considering using them to print your job.

- **Several printers issue publications** to help you understand all the terminology and will send these to you free if you write on your letterhead requesting copies. Here are several of the best:

    Braun-Brumfield's *Book Manufacturing Glossary, Book Paper Samples, Type Sample Book, Camera Copy Preparation Guidelines, and Paper Bulk Chart.*
    Delta Lithograph's *12 Ways to Improve the Turnaround of your Next Publishing Project.*
    Griffin Printing's *Signature* newsletter
    Johnson Printing's *Johnson Journal*
    Malloy Lithographing's *Malloy Quarterly*
    McNaughton & Gunn's *Book Manufacturing Intro Kit*
    Marrakech Express's *Shortruns* newsletter
    Thomson-Shore's superb *Printer's Ink* newsletter

More Notes on RFQ's

- On your RFQ form, besides providing the specifications for your book, you should also require a few other details from the printer. For instance, always ask for their credit terms, normal delivery times, and the approximate delivery charges to your firm (if any).

- Normal terms of credit are net 30 with approved credit. In other words, if your credit is good, you will be expected to pay for the books 30 days after the books are shipped. If your credit is shaky or your business is new, then most printers will require at least ⅓ down, another ⅓ with returned proofs, and the balance on delivery.

- Delivery times can vary from as short as 10 working days (2 weeks) to as long as 8 to 12 weeks for camera-ready copy. If your book requires typesetting or case binding, the time can vary from as short as 4 weeks to as long as 20 or more weeks. If you need fast turnaround, let them know that when you query them.

   If you require delivery by a specific date, be sure to state that on your RFQ when first querying the printer.

- Printing times for magazines and catalogs are usually shorter than books. Weekly magazines, such as *Time* and *Newsweek*, for example, are produced by printers who specialize in large runs at fast speeds and turnarounds. Note also that such magazines are often produced by more than one plant (with regional editions being printed by printers in various parts of the country).

- When choosing a printer, be sure to consider more than just price. How about their turnaround time, their quality, their service, their terms? Read Chapter 2 to make sure you've considered all the facts before committing yourself to a printer.

- Always ask the printer to quote by a specific date (allow 2 - 3 weeks minimum). Printers say they can produce a quote in 1 - 7 days and they should be able to, but our experience is that most printers take much longer. By setting a cutoff date, you ensure that all serious bids will be sent to you in time for you to make a decision.

- Also ask them to state how long the quoted prices are good for. This is vital if you won't be publishing right away (or if there is any chance of a delay).

- Stand by your dates. Make sure you send in your camera-ready copy, negatives, or other artwork when you say you will. You can't expect printers to live up to their commitments unless you live up to yours. If you are going to be delayed in getting materials to them, be sure to let them know.

   If you stand by your dates, get material to them on time, and return your bluelines in a timely manner, then the printer can offer you the kind of service you'd like.

- When you've narrowed your choice to two or three printers, ask them to send you samples of their work. Be sure to inspect the samples carefully.
- Also ask them to give you the names and phone numbers of some of their recent customers. Call these customers and ask them for feedback regarding the printer's service and quality of work. Do this before you send your money or camera-ready copy to a printer. You do not want to spend thousands of dollars for books you would not be proud to sell.
- Be prepared for a significant difference in the response to your queries from different printers. Some printers are very aggressive in marketing their services while others are incredibly blasè. Here, for example, are the responses of a variety of printers to a neophyte book publisher's query:

1) *Their rep was a sweetheart. Spent lots of time talking with me. Sent me books and information up the yingyang. They don't miss a beat ... very professional. Their rep said they were going after the small publishers with fervor. He came down from his original quote twice.*

2) *Never responded.*

3) *Sent paper samples. No letter, no brochure, no hello, never heard from them again.*

4) *They really seemed to want the job. Very personable. Kept in touch by phone every few days. Changed their bid twice.*

5) *Sent a bit of information; nothing to snow you. Never called, just sent a bid.*

6) *Was always consistent on their prices. They never had to budge because they were always the lowest in everything but shipping. They were very professional and helpful. I talked to them quite a bit ... asked every basic question possible.*

7) *Good price, personable people. Didn't kill themselves trying, though. Do send out packets of beautiful and expensive brochures. They spend a lot of money on you.*

8) *My runnerup. Very friendly. Answered all my questions. Took a lot of time with me. Very professional.*

9) *Sent paper samples and a book, but they never followed up. Their bid was good, but I was so busy making friends with other printers that kept calling me that I didn't pursue them.*

10) *They were very helpful and gave me a couple sample books. Promised to send me paper samples, etc. A month passed and nothing sent. I called, and two weeks later they got around to sending me books, samples of paper, brochures, and an estimate. But too late by then.*

11) *Very nice rep, kept in touch, sent me books. Too bad her estimate was high.*

12) *Visited their plant. Very nice people, knocked themselves out for me. But their price was too high.*

13) *Visited them, too. Very nice, but not professional. And their bid was astronomical.*

- Note that printers do vary in their response to your query. If you are seriously interested in working with a printer, call them if they don't call you first.

- If your favorite printers come out significantly higher on a job quote than several other printers, you might want to ask them why. They might offer to come down a bit to match the others or at least come close enough to make the difference negligible when taking into consideration other factors such as service and quality. If they do lower their price, be sure to get a written verification of the change.

- When you are ready to send in your manuscript or camera-ready copy, phone the printer beforehand to confirm prices and the printing schedule. Quoted prices may vary due to changes in paper prices and delivery schedules differ with the printer's workload. Be sure to confirm both.

- When you do send in your copy, enclose a written letter of confirmation (or contract) that describes the agreed-upon price and delivery schedule. This will save you from problems if some question should come up later regarding prices, services, or delivery.

- Always keep a copy of any manuscripts or camera-ready copy you send them. The postal service and printers have both been known to lose or spoil even the best prepared and packaged copy.

> **Note:** Be sure to let printers know you read about them in the *Directory of Book Printers*. It will show them that you are serious about getting good quality work for a reasonable price.

Sample RFQ's

In this book, we have included sample Requests For Quotations which you may use as a pattern to develop your own RFQ for various publications.

The RFQ on page 21 is the one we used to request price quotations for the printing of this *Directory*. It is a simple and straightforward RFQ. Use one like it when the job isn't too complicated. As the job becomes more detailed or complicated, you might want to use an RFQ like the one found on pages 28 and 29 or pages 30 and 31.

The sample RFQ on pages 28 and 29 was taken from *Getting It Printed,* previous edition (reprinted by permission). This sample RFQ is written on a universal RFQ form which can be used to request prices for almost any printing job.

A copy-ready version RFQ can found on pages 30 and 31. Reprinted by permission from the updated edition of *Getting it Printed,* published by North Light Books, Cincinatti, OH. Because the form is so comprehensive and detailed, it should not be used by someone who does not know anything about printing terminology. If you want to use this form, be sure to read the book first.

Getting It Printed accurately defines all the terms and standards you would use to describe any job. It defines, for instance, the differences between basic, good, premium, and showcase quality printing. It also defines all the various grades of paper. If you do not know the differences, you need to read the book. Check your local library, or order the book by calling toll-free **800-669-0773**.

Request for Quotation

Job name __SUMMER CATALOG__   Date __3/2/87__

Contact person __JOHN BRIGHTON__   Date quote needed __3/9/87__

Business name __WILLIAMS MARKETING__   Date job to printer __3/19/87__

Address __919 SECOND AVENUE, ANTELOPE USA 10123__   Date job needed __4/15/87__

Phone __800-282-5800__ Please give ☑ firm quote ☐ rough estimate ☐ verbally ☑ in writing

This is a ☑ new job ☐ exact reprint ☐ reprint with changes _____

Quantity 1) __25,000__   2) __50,000__   3) __40,000__   ☑ additional __1,000__ s

Quality ☐ basic ☐ good ☑ premium ☐ showcase comments __CRITICAL FABRIC MATCH__

Format product description __CATALOG WITH ORDER FORM INSERT__

  flat trim size _____ x _____ folded/bound size __8½__ x __11__

  # of pages __16__ ☐ self cover ☑ plus cover

Design features ☑ bleeds ☑ screen tints # __60__ ☑ reverses # __20__ ☑ comp enclosed

Art ☑ camera-ready ☐ printer to typeset and paste up (manuscript and rough layout attached)

  ☐ plate-ready negatives with proofs to printer's specifications

  trade shop name and contact person _____

Mechanicals color breaks ☐ on acetate overlays ☑ shown on tissues # pieces separate line art __4__

Halftones ☐ halftones # _____ ☐ duotones # _____

Separations ☑ from transparencies # __30__ ☑ from reflective copy # __1__ ☐ provided # _____

  finished sizes of separations __5 @ 3 × 5 ; 8 @ 8½ × 11; 10 @ 4 × 4; 8 @ 5 × 8__

Proofs ☐ galley ☐ page ☑ blueline ☑ loose color ☑ composite color ☐ progressive

Paper	weight	name	color	finish	grade
cover	80#	SUPERCOTE	WHITE	GLOSS	COVER
inside	70#	SNOWLIGHT	WHITE	GLOSS	COATED BOOK
INSERT	70#	RYAN OPAQUE	CREAM	VELLUM	UNCOATED BOOK

☐ send samples of paper ☐ make dummy buy paper from __RIVER PAPER__

Illustration 3.2 — Sample RFQ, Page 1

Request for Quotation (continued)

Printing   ink color(s)/varnish   ink color(s)/varnish

cover   side 1 4-COLOR+ SILVER +VARNISH   side 2 4-COLOR + VARNISH

inside   side 1 4-COLOR+ VARNISH   side 2 4-COLOR + VARNISH

INSERT side 1 BLACK + ONE COLOR   side 2 BLACK + ONE COLOR

_____ side 1 _____ side 2 _____

Ink ☐ special color match ☑ special ink METALLIC SILVER ON COVER ☐ need draw down

coverage is ☐ light ☑ moderate ☐ heavy ☑ see comp attached ☑ need press check

Other printing (die cut, emboss, foil stamp, engrave, thermograph, number, etc.) _____

_____

Bindery

☐ deliver flat press sheets	☐ round corner	☐ pad	☐ Wire-O
☑ trim	☐ punch	☐ paste bind	☐ spiral bind
☐ collate or gather	☐ drill	☑ saddle stitch	☐ perfect bind
☐ plastic coat with _____	☑ score/perforate	☐ side stitch	☐ case bind
☑ fold _____		☐ plastic comb	☐ tip in _____

comments SCORE COVER; PERFORATE INSERT

Packing ☐ rubber band in # ____ s ☐ paper band in # ____ s ☐ shrink/paper wrap in # ____ s

☐ bulk in cartons/maximum weight ____ lbs ☑ skid pack ☐ other _____

Shipping ☐ customer pick up ☑ deliver to QUICK-OUT MAILING SERVICES

_____

☐ quote shipping costs separately ☐ send cheapest way ☐ other _____

Miscellaneous instructions PRINT 500 EXTRA COVERS; SHRINK WRAP

100 CATALOGS IN 10'S; DELIVER EXTRA COVERS AND CATALOGS

TO SUSAN PRESTON WHEN RETURNING MECHANICALS

AND PHOTOS

Illustration 3.2 — Sample RFQ, Page 2

10-5 REQUEST FOR QUOTATION

Overview ————————————————————————————————

Organization name_____

Address _____

Contact person_____ Phone _____ Fax_____

PO #_____ Date_____ Customer #_____

Job name _____ Job # _____

Date quote needed_____ Job to printer_____ Delivery needed _____

☐ new job ☐ exact reprint ☐ reprint with changes   Previous job #_____

Comments

Prepress ————————————————————————————————

Proofs and plate-ready ☐ film ☐ electronic files supplied to printer's specifications

Prepress service_____ Contact_____

Phone _____ Fax_____ Job #_____

Comments

Printing ————————————————————————————————

Quantity 1)_____ 2)_____ 3)_____

Comments

Quality ☐ basic ☐ good ☐ premium ☐ showcase ☐ SWOP ☐ CCUP ☐ PGS ☐ SNAP

Comments

Format Trim size ____x____ Page count ____ ☐ bleeds (see mockup) ☐ plus cover ☐ self-cover

Comments

Ink colors   side one   side two

cover   _____   _____

text   _____   _____

insert   _____   _____

____   _____   _____

____   _____   _____

Comments

Illustration 3.3 — Sample RFQ, Page 1

Coating ☐ varnish ☐ UV ☐ aqueous ☐ spot ☐ flood ☐ dull ☐ gloss ☐ tint_____

Comments

Other printing ☐ die cut ☐ foil stamp ☐ emboss/deboss ☐ other_____

Comments

Paper ——

	weight	brand		color	finish	grade
cover	_____	_____		_____	_____	_____
text	_____	_____		_____	_____	_____
insert	_____	_____		_____	_____	_____
____	_____	_____		_____	_____	_____
____	_____	_____		_____	_____	_____

Show cost of paper separately ☐ no ☐ yes   Suggest alternate stock(s) ☐ no ☐ yes

Comments

Finishing and binding ————————————————————————————————

trim to _____x_____   fold to _____x_____   type of fold_____

☐ see dummy ☐ score ☐ perforate ☐ drill ☐ punch ☐ number ☐ film laminate

☐ plastic comb ☐ spiral plastic ☐ spiral wire ☐ double loop wire ☐ paste ☐ saddle stitch

☐ side stitch ☐ perfect ☐ burst perfect ☐ lay flat ☐ case   binding side_____

Comments

Packing and delivery ————————————————————————————————

Label cartons/pallets_____

☐ bulk pack ☐ band ☐ shrink wrap in bundles of_____   maximum carton weight _____

☐ pallet pack   maximum pallet size/weight_____

☐ will call ☐ deliver ☐ ship via_____

Comments

Illustration 3.3 — Sample RFQ, Page 2

Chapter 4

How to Save Money on Your Book Printing Bill

Here are a few ways you can save money on the design, typesetting, printing, and binding of your books:

- **No rush jobs.** Don't wait until the last minute. Rush jobs only cause headaches, result in errors of omission, and can cost more money in overtime pay and shipping.

- **Present clean typewritten copy** to the typesetter with as few editorial changes as possible. Clean copy allows the typesetter to operate faster, resulting in a lower charge for the typesetter's time.

- **Typeset from disk.** According to the National Composition Association, keyboarding the original input and proofreading make up 53% of the typical costs of regular typesetting. Corrections account for another 13% of typical costs. By providing your own proofread and corrected input, you can save up to 66% of total typesetting costs. Check with your printer or typesetter to see how much you can save by doing your own keyboarding. For instance, one printer will typeset text from disk for $7.00/page versus $12.00/page from manuscript. See the Services Index for printers capable of typesetting from your computer disks or via modem.

- **Use desktop publishing**. If you have a computer and want to have a book that appears professionally typeset, consider using a desktop publishing program.

- Then output via a laser printer (as this book was done) or to a typesetting machine capable of taking input from a desktop publishing program.

- **Seek the help of your printers in cutting costs.** Talk to them early in the planning of the layout and design of your books and other publications. They may be able to suggest minor changes in your specs that will save you money without affecting the quality of your publications.

- **Standardize your format.** Prepare your camera-ready copy so all the pages can be shot using the same camera setting (requiring no special reductions or enlargements).

- **Make all your editorial changes before you send your copy to the typesetter.** Making changes after the copy has been typeset will cost you at least two to three times as much.

- **Use special effects.** You can obtain a two-color effect with only one color by using screens, dropouts, and reverses.

- **Skip the press checks or blue lines.** If you provide complete camera-ready copy to your printer, you may not need to see press proofs. Most quality printers will reproduce your camera-ready copy exactly as you provided it. Hence, you can save the proof charges (which cost anywhere from $.49 to $1.00 per page) and save the time that would otherwise be taken up in sending the proofs back and forth for approval. However, when working with a printer for the first time, it is always safer to require press proofs — and check your press proofs thoroughly.

  This tip, of course, only applies to straight text copy. If your copy involves many halftones, colors, or special effects, you will certainly want to review the proofs before okaying the print run.

- **Avoid close registrations** when doing two or three-color printing. They require extra prep time and can result in a higher reject rate as well.

- **Check with your printer to see how they define camera-ready copy.** With some printers, you will be able to save money by providing your camera-ready copy on single sheets that are the same size as the finished book page. Others require you to use their special layout pages. Check first to find out what they want from you.

- **Avoid special requests** (odd sizes, unusual papers, special effects) unless they contribute to the content of the publication. Extras cost time and money.

- **Design your books to fit the press** (in signatures of 4, 8, 16, or 32 pages). A book of 158 pages will usually cost as much, or more, than a book of 160 pages because of additional labor charges in handling the incomplete signature. Add several empty pages or, better yet, use those extra pages to advertise some of the other books you publish. Or just add a coupon so people can order additional copies of the book for themselves or for friends.

- **Build up your credit rating by paying your bills on time.** Also, take advantage of any discounts for paying early (you can save from 1% to as much as 5%).

- **A lighter weight paper can save you money**, both on the cost of the paper itself and in postage for mailing the book. Many 50 lb. or 55 lb. book papers are now as opaque and as durable as any 60 lb. papers.

- **You can save money by gang-running your books** (that is, by printing several books at the same time). If you can arrange to publish several books in the same size and quantity, you can save on prep costs while getting better quantity discounts on paper costs. You will also save time and money in preparing manuscripts for typesetting and in designing your books if you have a standard format.

  Some printers offer lower prices if you allow your 4-color catalog or brochures to be grouped with others. However, there are some problems to gang-running your 4-color work with others: You have less control over the resulting color match. You must run with certain quantities. You may have to wait longer. Nonetheless, if you want only 25,000 or less copies of a 4-color job, gang-running offers a cost-effective alternative to custom work.

  In this directory, we provide a list of short-run color printers who offer great prices on full-color catalog sheets.

- **Check your quotes.** It is important to check specifications as well as quoted prices when you receive quotes back from printers, since many printers may change one or two specifications to save costs or to fit your book to their capabilities. The most common changes are paper stock and trim size (for example, 5 ⅜ x 8 ⅜ trim size rather than 5 ½ x 8 ½). Be sure that any such changes are acceptable to you.

- **Be sure to get an updated quote** whenever you change any specifications, including the time of printing. Otherwise, you could get a big surprise when the printing bill comes.

- **Book papers are different from offset grades** – they are more opaque, made to a consistent bulk (which is important for accurately fitting covers), usually have less filler and stronger fiber (making them more flexible so they tend to lie flatter), but they also tend to be more expensive. You must make the choice.

- **Paper grades vary for coated stock as well.** If the look or feel of your catalog or other publication is important, be sure to see samples of the paper options offered by the printer or by your paper supplier.

- **Query at least five to ten printers on every book**, especially when the format differs from your standard format or when you're varying the quantity.

- **Print during the off-season.** Some book printers will lower their quotes when work is slow. For instance, we know of one publisher who had three printers call him back to revise their quotes because work was slow in the summer.

- **Plan your printing and publishing schedules.** At the very least, you should have a schedule of what books you will be publishing in the coming year. Planning will save you from the unnatural disasters of rush jobs.

- **You can save money by buying your own paper for your books** (if you know what you are doing). To learn more about how to judge, specify, and buy paper, read *Getting It Printed*.

- **Get everything in writing.** Any modifications in the specifications you agree to over the phone or in a conversation should be put into writing (either included in the contract or attached thereto). If you don't require a signed contract when you work with a printer, you should at least send a letter of confirmation with your manuscript or camera-ready copy. This letter should reconfirm the specifications in the written quote you received from the printer (including final price and delivery date). Attach a copy of your original RFQ as well.

- **Never ask for delivery ASAP** (as soon as possible); always specify the exact date you expect delivery of the completed job. In turn, always stand by your own commitments to get camera-ready materials to the printer on time. If something holds you up, be sure to let the printer know in plenty of time. Otherwise, you might be charged for the printer's downtime.

- **Use a self-cover.** For some of your booklets, manuals, and catalogs, you may be able to use a self-cover (where the cover stock is the same as the text stock). By using a self-cover, you will save the cost of a separate press run.

- **Note shipping costs.** When comparing price quotes, don't forget to include the cost of shipping the books to your warehouse. Shipping costs can make a significant difference. In fact, some West Coast book printers refuse to provide quotes for people on the East Coast because by the time shipping costs are added, they can't be competitive with the Ann Arbor printers.

- **Send everything at the same time.** Provide your printer with all the camera-ready copy, photographs, and instructions at the same time. There is far less room for mistakes to creep into the job if the printer gets everything at once.

- **Try to figure out your true costs for producing books** – including the cost of your time. Value your time. It may be your major expense in producing books and other publications.

- **Match your printing jobs to the printer's capabilities.** You'll save money by using a printer set up to do the kind of job you want. Use this directory to select the most appropriate printers for the kind of books you publish.

- **Look around for inexpensive photo and art sources.** Use stock photo and clip art services rather than hiring freelancers. Or use art students from your local college. You can also get excellent free photos from your local historical society and from many corporate or government PR offices.

- **Edit more carefully.** You can save money on your printing bills by editing your books more thoroughly. Does the book really have to be as long as it is? Can some chapters be trimmed or even deleted without hurting the content, design, and message of the book?

- **Set a firm publishing schedule** and, in turn, a firm production schedule — with plenty of leeway to allow you time to make changes if they are needed. Set a realistic schedule, put it in writing, and stick to it. Don't rush yourself. Rush jobs are sloppy jobs. And sloppy jobs cost you money. Either you must pay to correct the mistakes or, if you don't correct them, you can lose sales because the book is not suitable.

- **Use photographs only when necessary** since each halftone can add from $5 to $25 in prep costs.

- **You can have a number of your photos shot as half-tones at one time**. The photos must have uniform contrasts and be the same size (or be reduced by the same percentage). Then the halftones can be stripped into the production negatives as usual.

- **Always ask for samples of the printer's work.** And be sure to check these samples. Open the book flat, test the binding, check the coating, and so on.

- **In some cases it is possible to save money by using more than one supplier**. For instance, some book binders offer lower prices than are available in-house from a printer.

- **You might try to arrange a long-term contract** with one of your printers to produce a series of books for you. Ask them to quote on the entire job at once. They should be able to give you a lower price because such a long-term contract will allow them to make fuller use of their facilities, save on quantity purchases, and secure a more reliable cash flow.

- **Remember:** The lowest price is not necessarily the best bargain. Don't sacrifice quality, service, or delivery just to save a few dollars.

Printing Trade Customs

Trade Customs have been in general use in the printing industry throughout the United States and Canada for more than 60 years. These may help understand even more about printing in general.

1. QUOTATION: A quotation not accepted within thirty days (30) is subject to review. All prices are based on material costs at the time of quotation.

2. ORDERS: Orders regularly placed, verbal or written, cannot be cancelled except upon terms that will compensate printer against loss incurred in reliance of the order.

3. EXPERIMENTAL WORK: Experimental or preliminary work performed at the customer's request will be charged for at curren trates and may not be used until the printer has been reimbursed in full for the amount of the charges billed.

4. CREATIVE WORK: Creative work, such as sketches, copy, dummikes and all preparatory work developed and furnished by the printer, shall remain his exclusive property and no use of same shall be made, nor any ideas obtained therfrom be used, except upon compensation to be determined by the printer, and not expressly identified and included in the selling price.

5. CONDITION OF COPY: Upon receipt of original copy or manuscript, should it be evident that the condition of the copy differs from that which had been originally described and consequently quoted, the original quotation shall be rendered void and a new quotation issued.

6. PREPARATORY MATERIALS: Working mechanical art, type, negatives, positives, flats, plates and other items when supplied by the printer, shall remain his exclusive property unless otherwise agreed in writing.

7. ALTERATIONS: Alterations represent work performed in addition to the original specifications. Such additional work shall be charged at current rates and be supported with documentation upon request.

8. PRE-PRESS PROOFS: Pre-press proofs shall be submitted with original copy. Corrections are to be made on "master set," returned marked "O.K." or "O.K. with corrections" and signed by customer. If revised proofs are desired, request must be made when proofs are returned. Printer cannot be held responsible for errors under any or all of the following conditions: if the work is printed per customer's OK; if changes are communicated verbally; if customer has not ordered proofs; if the customer has failed to return proofs with indication of changes; or if the customer has instructed printer to proceed without submission of proofs.

9. PRESS PROOFS: Unless specifically provided in printer's quotation, press proofs will be charged for at current rates. An inspection sheet of any form can be submitted for customer approval, at no charge, provided customer is available at the press during the time of make-ready. Lost press time due to customer delay, or customer changes and corrections, will be charged at current rates.

10. COLOR PROOFING: Because of differences in equipment, processing, proofing substrates, paper, inks, pigments, and other conditions between color proofing and production pressroom operations, a reasonable variation in color between color proofs and the completed job shall constitute acceptable delivery.

11. OVER RUNS AND UNDER RUNS: Over runs or under runs not to exceed 10% on quantities ordered, or the percentage agreed upon, shall constitute acceptable delivery. Printer will bill for actual quantity delivered within this tolerance. If customer requires guaranteed exact quantities, the percentage tolerance must be doubled.

12. CUSTOMER'S PROPERTY: The printer will maintain fire, extended coverage, vandalism, malicious mischief and sprinkler leakage insurance on all property belonging to the customer, while such property is in the printer's possession; printer's liability for such property shall not exceed the amount recoverable from such insurance. Customer's property of extraordinary value shall be insured through mutual agreement.

13. DELIVERY: Unless otherwise specified, the price quoted is for a single shipment, without storage, F.O.B. local customer's place of business or F.O.B. printer's platform for out-of-town customers. Proposals are based on continuous and uninterrupted delivery of complete order, unless specifications distinctly state otherwise. Changes related to delivery from customer to printer, or from customer's supplier to printer, are not included in any quotations unless specified. Special priority pickup or delivery service will be provided at current rates upon customer's request. Materials delivered from customer or his suppliers are verified with delivery ticket as to cartons, packages or items shown only. The accuracy of quantities indicated on such tickets cannot be verified and printer cannot accept liability for shortage based on supplier's tickets. Title for finished work shall pass to the customer upon delivery to carrier at shipping point or upon mailing of invoices for finished work, whichever occurs first.

14. PRODUCTION SCHEDULES: Production schedules will be established and adhered to by customer and printer, provided that neither shall incur any liability or penalty for delays due to state of war, riot, civil disorder, fire, labor trouble, strikes, accidents, energy failure, equipment breakdown, delays of suppliers or carriers, action of Government or civil authority and acts of God or other causes beyond the control of customer or printer. Where production schedules are not adhered to by the customer, final delivery date(s) will be subject to renegotiation.

15. CUSTOMER-FURNISHED MATERIALS: Paper stock, inks, camera copy, film, color separations and other customer-furnished material shall be manufactured, packed and delivered to printer's specifications. Additional cost due to delays or impaired production caused by specification deficiencies shall be charged to the customer.

16. TERMS: Payment shall be whatever was set forth in the quotation or invoice unless otherwise provided in writing. Claims for defects, damages or shortages must be made by the customer in writing within a period of fifteen (15) days after delivery of all or any part of the order. Failure to make such claim within the stated period shall constitute irrevocable acceptance and an admission that they fully comply with terms, conditions and specifications.

17. LIABILITY: Printer's liability shall be limited to stated selling price of any defective goods, and shall in no event include special or consequential damages, including profits (or profits lost). As security for payment of any sum due or to become due under terms of any Agreement, printer shall have the right, if necessary, to retain possession of and shall have a lien on all customer property in printer's possession including work in process and finished work. The extension of credit or the acceptance of notes, trade acceptance or guarantee of payment shall not affect such security interest and lien.

18. INDEMNIFICATION: The customer shall indemnify and hold harmless the printer from any and all loss, cost, expense and damages (including court costs and reasonable attorney fees) on account of any and all manner of claims, demands, actions and proceedings that may be instituted against the printer on grounds alleging that the said printer violates any copyrights or any proprietary right of any person, or that it contains any matter that is libelous or obscene or scandalous, or invades any person's right to privacy or other personal rights, except to the extent that the printer contributed to the matter. The customer agrees, at the customer's own expense, to promptly defend and continue the defense of any such claim, demand, action or proceeding that may be brought against the printer, provided that the printer shall promptly notify the customer with respect thereto, and provided further that the printer shall give to the customer such reasonable time as the exigencies of the situation may permit in which to undertake and continue the defense thereof.

Originally promulgated and adopted by the United Typothetae of America, 1922. Updated and adopted by the Graphic Arts Council of North America, 1985.

Chapter 5

How to Read the Printer Listings

The following pages (pages 40 through 189) provide an alphabetical listing of over 650 printers of books, catalogs, magazines, and other bound publications. The details in most of these listings were derived from surveys we sent out in 1994. Other information may have been taken from advertisements, other listings, and articles. A complete listing includes the following details:

Company Name and Address

> **Company Name**
> **Contact Person**
> **Address**
> **Phone Numbers**
> **Parent Company**

Main Focus

At the top right of each listing, we have indicated the main business focus of that printer if they are primarily book printers, catalog printers, magazine printers, or general commercial printers. Where their main focus isn't known, they have been listed as general commercial printers.

> **BK** — Book printers
> **BR** — Printing brokers
> **CA** — Catalog printers
> **GC** — General commercial printers
> **MG** — Magazine printers
> **MS** — Short-run magazine printers
> **SP** — Specialty printer

Print Runs

After the name and address of the company, we list their minimum, maximum, and optimum print runs for publications (where known).

Min: — Minimum print run.
Max: — Maximum print run.
Opt: — Optimum print run (usually the quantity that best fits their equipment and, hence, the quantity for which they can generally offer the best price).

If the print run was not given or listed as an unlimited amount, a "zero " will follow

Sizes

Each listing indicates which of six standard sizes of publications the printers are capable of printing and whether they are also capable of printing other sizes. The six standard sizes are coded by numbers 1 through 6.

1 — 4 1/4 x 7 inches (or similar sizes)
2 — 5 1/2 x 8 1/2 inches
3 — 6 x 9 inches
4 — 7 x 10 inches
5 — 8 1/2 x 11 inches
6 — 9 x 12 inches
X — other sizes, including odd sizes

Bindings

Each listing indicates which bindings the printer is capable of providing in-house and which they can arrange to have done by a regular outside supplier. An "I" after the code (e.g., **HCI**) indicates that they can provide the binding service in-house; an "O" (e.g., **PBO**) indicates they use an outside bindery; an "X" (e.g., **LBX**) means that they did not specify whether they could do the binding in-house or that they use an outside bindery.

HC — Casebinding (hardcovers)
CB — Plastic comb binding
LB — Looseleaf binders
OB — Otabind
PB — Perfect binding (softcovers)
SS — Saddle stitching (stapled in the center)
SD — Side stitching (like *National Geographic*)
SB — Spiral binding (like notebooks)
WB — Wire-O binding

Items

This line in the listing indicates the kinds of publications and other printed items the printer regularly produces. The items are indicated by letter codes as follows:

A — Annual Reports		**P** — Brochures
B — Books		**Q** — Calendars
C — Booklets		**R** — Comic Books
D — Catalogs		**S** — Direct Mail Letters
E — Cookbooks		**T** — Envelopes
F — Directories		**U** — Greeting Cards
G — Galley/Bound Copies		**V** — Labels / Stickers
H — Journals		**W** — Maps
I — Magazines		**X** — Newspapers / Tabloids
J — Massmarket Paperbacks		**Y** — Postcards
K — Newsletters		**Z** — Posters
L — Software Manuals		**1** — Stationery
M — Textbooks		**2** — Binders/Covers/Folders
N — Workbooks		**3** — Games / Puzzles
O — Yearbooks		**4** — Packaging / Boxes

Services

These codes indicate whether the printers offer any of twelve additional services to their customers; everything from color separations through warehousing.

4C — 4-color (full-color process) printing
CS — Color separations
DA — Design, layout, and artwork
FF — Fulfillment and mailing services
LM — List maintenance
OC — Uses the OptiCopy prepress system
RA — Uses the Rachwal prepress system
AF — Stocks acid-free paper
RP — Stocks recycled paper
TY — Typesetting and composition
TD — Typesetting via disk, modem, etc.
WA — Warehouse or store books

Equipment

This line indicates the kind, number, and sizes of printing equipment the printer uses. These details, which can help you decide whether or not the printer has the capabilities to fill your needs, are listed in abbreviated form. For example, **2CB** indicates they have two Cameron belt presses; **4SO(to 24x38)** indicates they have four sheet-fed presses, the largest capable of printing 24 by 38 inch sheets.

CB — Cameron belt press
LP — Letterpress

SO — Sheetfed offset press
WO — Web offset press
GR — Rotogravure press
OP — Other presses

Annual Sales

Our survey asked the printers to list their annual sales figures within a range, e.g. "**1**" indicates that the printer had an annual sales volume under $500,000, while a "**2**" meant their sales volume were between $500,000 and $5,000,000.

1 — under $500,000   **4** — $10,000,000 - $25,000,000
2 — $500,000 - $5,000,000   **5** — $25,000,000 - $50,000,000
3 — $5,000,000 - $10,000,000   **6** — over $50,000,000

Number of Employees

The annual sales figures and number of employees should provide a good indication of the printer's ability to meet your needs, especially if those needs are long term.

Turnaround Times

These codes indicate the normal turnaround times for the printer to produce a job from camera-ready copy to shipment. These times are expressed as a range of working days; e.g., "**B4**" indicates that the printer can produce a perfectbound book within 21 to 30 working days after receipt of camera-ready copy while "**C2**" indicates that they can produce a casebound book within 6 to 10 working days from receipt of camera-ready copy (or negatives). Similarly, an **S2** indicates that they can deliver a finished saddlestitched book in 6 to 10 working days.

Note that these turnaround times are those reported by the printer; we have no way to verify their normal turnaround times.

B1 — 1 to 5 working days.
B2 — 6 to 10 working days.
B3 — 11 to 20 working days.
B4 — 21 to 30 working days.
B5 — 30 to 45 working days.
B6 — over 45 days.

Customers

These figures indicate what percentage of their printing business is done for various types of customers. For example, **BP25** indicates that 25% of their business is for book publishers, while **NP50** indicates that 50% of their business is with non-profit groups. In other words, the number after the code letters indicates the percentage.

This information is most useful to those who want to work with a printer accustomed to serving the needs of clients similar to you. It's useful, for example, if you are a magazine publisher to work with a printer whose major source of business is other magazine publishers, because that printer will be more aware of the potential problems that can arise in producing a magazine on a regular basis.

BP — Book publishers
BU — Businesses, both large and small
CU — Colleges and universities
MP — Magazine publishers
NP — Non-profit groups and organizations
OR — Other organizations and clubs
SC — Schools, elementary and secondary
SP — Self-publishers, authors, genealogists
OT — Other

Union Shop

A **Y** indicates that they are a union shop while an **N** indicates that they employ non-union workers.

Year Established

Where known, we have listed the year in which the printer began business. In a few cases, this date might indicate the year they started printing bound publications rather than the year they began business.

Terms

This line lists the printer's standard terms of payment. Most indicate that they offer net 30 with approved credit. That means if you have good credit, then you may pay your printing bill 30 days after the job is finished. On the other hand, some require that you pay $1/3$ of the bill at the time you give them your camera-ready copy, $1/3$ when you approve the proofs or blues, and the final $1/3$ before they ship the finished job or upon delivery.

Statement

The **statement** consists of the printer's own words describing their goals and services.

Alphabetical Printer Listings

A

A-1 Business Service Inc   GC
Marketing Director
448 N Prior Ave
St Paul MN 55104
612-646-7308; Fax: 612-646-7309
Send RFQ's to: Mark Scanlon

Min: 0 Max: 0 Opt: 500
Sizes: 125
Bindings: SSI-SDI-SBO-WBO
Items: CDFKPSTVY12
Services: DA-RP-TY
Equipment: LP(8.25x14)-5SO(3x5,12x17)
Annual Sales: 1
Number of Employees: 4
Turnaround Times: S2
Customers: BU75-CU5-NP10-OR5-SC5
Union Shop: N
Year Established: 1946
Terms: Net 30

Statement: We are a full service printer-not a book specialist. We also offer bindery work ie: folding, perforating, scoring, padding, numbering, collating, etc.

aBCD   BK
Larry McGoff, Marketing Director
222 Wall St #100
Seattle WA 98121
206-443-1515; Fax: 206-443-4322
Send RFQ's to: Mitch Allcorn
Parent Company: Reischling Network Ltd

Min: 25 Max: 1000 Opt: 500
Sizes: 12345
Bindings: CBX-LBX-OBX-PBX-SSX-SBX
Items: ABCDEFKLMNO
Services: DA-TY-TD
Equipment: 2OP(Ducatech)
Annual Sales: 2
Number of Employees: 10
Turnaround Times: B1C2S1
Customers: BU80-CU15-OT5
Union Shop: N

Year Established: 1994
Terms: 50% down, balance on delivery

Statement: We are one of few US printers who receive original art from remote computers via modem or disk, and cable this art directly into digital presses without any further compromises.

Able Printing Co   GC
Marketing Director
7401 Central Hwy
Pennsauken NJ 08109
609-663-8181; Fax: 609-663-8135

Academy Books   BK
Robert A Sharp, General Manager
10 Cleveland Ave
PO Box 757
Rutland VT 05701
802-773-9194
Send RFQ's to: Robert A Sharp
Parent Company: Sharp Offset Printing

Min: 100 Max: 5000 Opt: 5000
Sizes: 2356X
Bindings: HCI-CBO-PBI-SSI-SBO
Items: ABCDEHINP
Services: AF
Union Shop: N
Year Established: 1946
Terms: 1/3 down, 1/3 proofs, 1/3 done

Statement: Academy has been printing books since 1946.

Accurate Printing   GC
Marketing Director
830 N Spokane St
Post Falls ID 83854
208-773-1103; Fax: 208-773-6287

SERVICES			
4C	4-Color Printing	RA	Rachwal System
CS	Color Separations	AF	Acid-Free Paper
DA	Design / Artwork	RP	Recycled Paper
FF	Fulfillment/Mailing	TY	Typesetting
LM	List Maintenance	TD	Typeset w/ Disk
OC	OptiCopy System	WA	Warehousing

BINDINGS			
HC	Hardcover	SD	Side Stitching
CB	Comb Binding	SB	Spiral Binding
LB	Loose-Leaf Binding	WB	Wire-O Binding
OB	Otabind	I	In-House
PB	Perfect Binding	O	Out of House
SS	Saddle Stitching	X	Unknown

Min: 100 Max: 50000 Opt: 20000
Sizes: 25
Bindings: CBI-PBI-SSI-LBO-SSO-SBO-
    WBO
Items: CEPQTVY
Services: DA-RP-TY-TD
Equipment: 2SO
Annual Sales: 1
Number of Employees: 8
Turnaround Times: B3S3
Customers: BP3-BU70-NP5-OR5-SC2-
    SP10-OT5
Union Shop: N
Year Established: 1991
Terms: Net 30 with credit approval

Accurate Web Inc   **BK**
Marketing Director
32 Windsor PL
Central Islip NY 11722
516-234-3590
Send RFQ's to: Sales Manager

Min: 7000 Max: 200000 Opt: 25000
Sizes: 25
Bindings: PBI
Items: BDF
Equipment: WO

Acme Printing   **GC**
Marketing Director
30 Industrial Way
PO Box 400
Wilmington MA 01887
508-658-0800

Action Printing   **GC**
Tom Carew
N6637 Rolling Meadows Dr
Fond Du Lac WI 54837
414-922-7821; 800-472-7821
Send RFQ's to: Keith Bock

Min: 500 Max: 25000 Opt: 15000
Sizes: 25
Bindings: HCO-CBO-LBO-OBO-PBI-SSI-
    SDO-SBO-WBO
Items: BCDEFHKLMNPQX2
Services: 4C-CS-DA-RP-TY-TD-WA

Equipment: 3SO(20x26)-2WO(22.75
    cutoff)
Annual Sales: 3
Number of Employees: 75
Turnaround Times: B3C4S2
Customers: BP10-BU40-CU20-SC5-OT25
Union Shop: N
Year Established: 1984
Terms: Net 30

Statement: Our goal is to provide quality
books with quick turnaround at a competitive price.

Ad Infinitum Press   **BK**
Marketing Director
7 N MacQuesten Parkway
PO Box 2212
Mount Vernon NY 10550
914-664-5930
Send RFQ's to: William Brandon, Presi-
    dent

Min: 100 Max: 50000 Opt: 0
Bindings: PBI-SSI
Items: BCFGHMN
Services: 4C-CS-DA-TY-WA
Equipment: LP-SO

Statement: We accept print runs of any
quantity that are suitable for sheetfed presses. We offer editorial services, typesetting,
design, pasteup, and galley copies.

Adair Printing Co Inc   **GC**
Marketing Director
18544 W Eight-Mile Rd
Southfield MI 48075
810-569-1122; Fax: 810-569-0951

Adams Press   **BK**
Beverly Freid, Sales Manager
500 N Michigan Ave #1920
Chicago IL 60611
708-676-3426
Send RFQ's to: Beverly Freid

Min: 100 Max: 5000 Opt: 1000
Sizes: 12356
Bindings: HCO-PBI-SSI-SDI-SBI-WBI
Items: BCDEFJMO
Services: 4C-AF-TY-TD-RP

PRINTED ITEMS				
A	Annual Reports	G	Galley Copies	N	Workbooks	U	Greeting Cards
B	Books	H	Journals	O	Yearbooks	V	Labels / Stickers
C	Booklets	I	Magazines	P	Brochures	W	Maps
D	Catalogs	J	Mass-Market Books	Q	Calendars	X	Newspapers
E	Cookbooks	K	Newsletters	R	Comic Books	Y	Postcards
F	Directories	L	Software Manuals	S	Direct Mail Letters	Z	Posters
		M	Textbooks	T	Envelopes	1	Stationery

Equipment: 4SO(17x22)
Annual Sales: 2
Number of Employees: 20
Turnaround Times: B5C6S5
Customers: SP95
Union Shop: N
Year Established: 1942
Terms: 50% deposit, balance with proofs

Statement: Our primary objective is to help self-publishers get their books in print. We supply a professionally printed book at a reasonable cost. We also secure copyrights and Library of Congress catalog card numbers. We pay shipping charges to any address in the USA. Write for our catalog and price list.

Comments: Having advertised in **Writer's Digest** for years, Adams Press is accustomed to working with writers through the mail.

Advanced Data Reproductions BK
Grace Risel, Marketing Director
2012 Northern St
Wichita KS 67216
316-522-5599; Fax: 316-522-5445
Send RFQ's to: Grace Risel

Min: 50 Max: 5000 Opt: 500-1000
Sizes: 12345
Bindings: CBX-PBX-SSX-SDX-SBX-WBX
Items: BCEFLMN
Services: WA
Equipment: 3SO
Annual Sales: 2
Number of Employees: 18
Turnaround Times: C3
Customers: BU95-BP5
Union Shop: N
Year Established: 1978
Terms: Net 30 on approved credit

Advanced Duplicating & Printing BK
Marketing Director
7419 Washington Ave South
Minneapolis MN 55439
612-944-6050; Fax: 612-944-9683
Min: 25 Max: 5000 Opt: 1000
Sizes: 12345
Bindings: CBI-LBI-PBI-SSI-SBI-WBI
Items: BCFLMN
Equipment: SO-WO

Annual Sales: 2
Number of Employees: 35
Turnaround Times: B2
Customers: BP15-BU85
Union Shop: N
Year Established: 1977
Terms: Net 30

Statement: We print one-color books. We have a complete bindery for perfect binding, spiral binding, and wire-o binding.

Adventure Printing GC
Marketing Director
PO Box 29543
Dallas TX 75229
214-638-0588

Advertising Unlimited Inc CA
Marketing Director
PO Box 8000
Sleepy Eye MN 56085
507-794-8000; Fax: 507-794-8100

Adviser Graphics BK
Marketing Director
4757 - 60th St
PO Bag 5012
Red Deer AB Canada T4N 6R4
403-347-8866; Fax: 403-342-2280
Send RFQ's to: Calvin Dallas

Min: 500 Max: 50000 Opt: 10000
Sizes: 123456
Bindings: HCO-LBO-PBO-SSI-SDI-SBO-
    WBI-CBI-OBI
Items: ABCDEFGHIKLMNOQRSTUVW
    XY1234
Services: 4C-DA-FF-LM-AF-TY-TD-WA
Equipment: 2LP(12x18)-3SO(19x25)-
    WO(22.75x35)
Annual Sales: 2
Number of Employees: 21
Turnaround Times: B2C3S3
Customers: BP12-BU60-CU15-NP5-OR7-
    OT1
Union Shop: N
Year Established: 1950
Terms: Deposit with order, net 30 days

SERVICES				BINDINGS			
4C	4-Color Printing	RA	Rachwal System	HC	Hardcover	SD	Side Stitching
CS	Color Separations	AF	Acid-Free Paper	CB	Comb Binding	SB	Spiral Binding
DA	Design / Artwork	RP	Recycled Paper	LB	Loose-Leaf Binding	WB	Wire-O Binding
FF	Fulfillment/Mailing	TY	Typesetting	OB	Otabind	I	In-House
LM	List Maintenance	TD	Typeset w/ Disk	PB	Perfect Binding	O	Out of House
OC	OptiCopy System	WA	Warehousing	SS	Saddle Stitching	X	Unknown

AGI Inc   **GC**
Marketing Director
1950 N Ruby St
Melrose Park IL 60160
708-344-9100; Fax: 708-344-9113

Ainsworth Group Inc   **GC**
Marketing Director
65 Hanson Ave
Kitchener ON Canada N2C 2H6
519-578-0530; Fax: 519-576-4599
800-265-2476

Alan Lithograph Inc   **GC**
Steven Henry, Marketing Director
550 N Oak St
Inglewood CA 90302
310-330-3800; Fax: 310-412-7134
Send RFQ's to: Steven Henry

Min: 10000 Max: 4500000(WO) 50000(SF)
    Opt: 0
Sizes: 123456
Bindings: SSI-SDI-CBO-HCO-LBO-OBO-
    PBO-SBO-WBO
Items: ACDIPSZ
Services: 4C-CS-RP
Equipment: 4SO(2-40,6-40,8-40)-WO(110
    Harris)
Annual Sales: 4
Number of Employees: 160
Turnaround Times: B3C4S1-2
Customers: BU90-MP05-SP05
Union Shop: N
Year Established: 1957
Terms: Net 30

Statement: We are a high-quality printer
whose goal is to deliver high-quality
products, competitively priced, while building a truly great company to work for.

Alcom Printing Group Inc   **GC**
Marketing Director
Lehigh Litho
2285 Ave A
Bethlehem PA 18017
610-691-5050; Fax: 610-861-0565
Send RFQ's to: Mary Lou Gola

Min: 0 Max: 0 Opt: 0 (Varies depending
    on the printed piece)
Sizes: 256
Bindings: SSI-SDI-CBO-HBO-LBO-OBO-
    PBO-SBO-WBO
Items: BCDFKMNPTWXY2
Services: 4C-CS-FF-RA-AF-RP-TY-TD-
    WA
Equipment: 6SO(26,28,35,37)-5WO(11
    to39)
Annual Sales: 4
Number of Employees: 200
Turnaround Times: B2C3S2
Customers: BP25-BU50-CU25
Union Shop: N
Year Established: 1940
Terms: Net 30
Statement: Sell good printing at a
reasonable cost, service the customers
completely, resolve their problems/issues,
and maintain a strategic partnership.

Other Facilities:

Havertown Printing Co
900 Sussex Blvd
Broomall PA 19008
610-544-7000; Fax: 610-544-1677

Tele-Composition Inc
900 Sussex Blvd
Broomall PA 19008
610-544-7000; Fax: 610-544-1677

Alden Press   **CA**
Marketing Director
2000 Arthur Ave
Elk Grove Village IL 60007-6071
708-640-6000; Fax: 708-640-6029
Send RFQ's to: Donald T Carlson, VP
    Sales
Parent Company: World Color Press
Min: 50000 Max: 5000000 Opt: 0
Sizes: 35X
Bindings: SSI
Items: D
Services: 4C
Equipment: 2WO
Annual Sales: 6
Number of Employees: 700
Year Established: 1948

PRINTED ITEMS							
A	Annual Reports	G	Galley Copies	N	Workbooks	U	Greeting Cards
B	Books	H	Journals	O	Yearbooks	V	Labels / Stickers
C	Booklets	I	Magazines	P	Brochures	W	Maps
D	Catalogs	J	Mass-Market Books	Q	Calendars	X	Newspapers
E	Cookbooks	K	Newsletters	R	Comic Books	Y	Postcards
F	Directories	L	Software Manuals	S	Direct Mail Letters	Z	Posters
		M	Textbooks	T	Envelopes	1	Stationery

GC

Alden Press/Wm Feathers Printer
Marketing Director
235 Artino St
Oberlin OH 44074
216-774-1500

Min: 10000 Max: 2000000 Opt: 0
Sizes: 5
Bindings: SSI
Items: CD
Services: 4C
Equipment: WO
Annual Sales: 4
Number of Employees: 80

GC

All-Star Printing
Michael H Duweck, Marketing Director
3101 Grand Oak Dr
Lansing MI 48911
517-393-5200; Fax: 517-393-8301
Send RFQ's to: Cindy Caine

Min: 0 Max: 0 Opt: 0
Sizes: 123456
Bindings: LBI-SSI-SDI-HCO-CSO-OBO-
    PBO-SPO-WBO
Items: ACDIKNPS2
Services: 4C-FF-RP-TY-TD-WA
Equipment: 5SO(to 26x40)
Annual Sales: 3
Number of Employees: 42
Turnaround Times: B3S2
Union Shop: Y
Year Established: 1961
Terms: Net 30 days

Statement: Personal Service, Quality
Product, Timely Delivery

GC

All-Type Printing Inc
Marketing Director
1661 Dixwell Ave
Hamden CT 06514
203-288-7415; Fax: 203-288-3158

MS

Allen Press Inc
Marketing Director
1041 New Hampshire St
PO Box 368
Lawrence KS 66044
913-843-1234

Send RFQ's to: Guy Dresser, VP Financial

Min: 200 Max: 20000 Opt: 7000
Sizes: 23456
Bindings: PBI-SSI
Items: H
Services: 4C-FF-LM-AF-TY-TD-WA
Equipment: 7SO(19x25,28x41)
Customers: CU30-NP70
Union Shop: N
Year Established: 1952
Terms: Cash, net 30

Statement: They specialize in printing
scholarly and scientific journals. They offer
advice on all aspects of journal publishing,
including editing and marketing.

GC

Alpha One Press Inc
Marketing Director
12324 134th Ct NE
Redmond WA 98052
206-823-8510

GC

Alonzo Printing
Marketing Director
3266 Investment Blvd
Hayward CA 94545-3807
510-293-0522; Fax: 510-293-3958

GC

AM Graphics & Printing
Marketing Director
320 S Pacific St
San Marcos CA 92069
619-744-0180; Fax: 619-744-6479
Send RFQ's to: Chris Chadha

Min: 500 Max: 0 Opt: 0
Sizes: 123456
Bindings: SDI-WBO-CBI-LBI-OBI-SSI-
    HCO-PBO-SBO
Items: ABCDFGHIKLMNOPQSTVWZ12
Services: 4C-DA-AF-RP-TY-TD
Equipment: 4SO(27,40,17,15)
Annual Sales: 2
Number of Employees: 14
Turnaround Times: B3C4S2
Customers: BP20-BU55-CU25-SP10
Union Shop: N
Year Established: 1977
Terms: 50% deposit, balance COD

SERVICES			BINDINGS		
4C	4-Color Printing	RA Rachwal System	HC	Hardcover	SD Side Stitching
CS	Color Separations	AF Acid-Free Paper	CB	Comb Binding	SB Spiral Binding
DA	Design / Artwork	RP Recycled Paper	LB	Loose-Leaf Binding	WB Wire-O Binding
FF	Fulfillment/Mailing	TY Typesetting	OB	Otabind	I In-House
LM	List Maintenance	TD Typeset w/ Disk	PB	Perfect Binding	O Out of House
OC	OptiCopy System	WA Warehousing	SS	Saddle Stitching	X Unknown

Statement: A professional printing company that can take your project from start to finish. By offering our clients price lists for 4-color printing, we have taken the "mystery" out of printing.

Ambrose Printing   GC
Marketing Director
210 Cumberland Bend
Box 23028
Nashville TN 37228
615-256-1151

American Litho   GC
Jim Staples, Marketing Director
21062 Forbes St
Hayward CA 94545
510-732-1650; Fax: 510-732-1863
Send RFQ's to: Ted Barstad

Min: 1000 Max: 0 Opt: 500000
Sizes: 123456
Bindings: SDI-SSI-CBO-HCO-LBO-OBO-PBO
Items: ABCDFKPSWYZ12
Services: 4C-CS-DA-RA-RP-TY-TD
Equipment: 6SO(20x28,28x40)-4WO(26.5x17.75)
Annual Sales: 5
Number of Employees: 145
Turnaround Times: B2C4S2
Customers: BU80-MP10-NP5-OR5
Union Shop: Y
Year Established: 1951
Terms: Net 30
Statement: Full service printer

American Press Inc   GC
Paul Grieco, VP Sales
1840 Michael Faraday Dr #120
Reston VA 22090
703-689-4666; Fax: 703-689-4693

American Printers & Litho   GC
Richard D Krebs, Sales Manager
6701 W Oakton St
Niles IL 60714
708-966-6500; Fax: 312-267-6553

Send RFQ's to: Richard D Krebs
Min: 0 Max: 0 Opt: 0
Sizes: 23456
Bindings: HCO-CBO-LBO-PBO-SSI-SDO-SBO-WBO
Items: ABCDEPWZ
Services: 4C
Equipment: 5LP-19SO(11x17,54x77)
Annual Sales: 5
Number of Employees: 260
Turnaround Times: B3C2
Customers: BP8-BU90-MP2
Union Shop: Y
Year Established: 1965
Terms: Net 30

Statement: They specialize in printed premiums; including finger puppets, puzzles, cutouts, punchouts, stickers, mobiles, masks, caps, hats, die-cut menus, and scratch-off promotions.

American Printing Co Inc   GC
Marketing Director
4400 Ave Q (77550)
PO Box 270
Galveston TX 77553
409-763-2412; Fax: 409-763-5854

American Signature Memphis Division   CA
Marketing Director
8649 Hacks Crossroads
Olive Branch MS 38654
601-895-4242; Fax: 601-895-7906

Min: 20000 Max: 3000000 Opt: 0
Bindings: PBI-SSI
Items: ADIPZ
Services: 4C-CS-FF
Equipment: SO-WO
Annual Sales: 6

American Signature Graphics   CA
Marketing Director
6320 Denton Dr
Dallas TX 75235
214-358-1371

PRINTED ITEMS		G	Galley Copies	N	Workbooks	U	Greeting Cards
A	Annual Reports	H	Journals	O	Yearbooks	V	Labels / Stickers
B	Books	I	Magazines	P	Brochures	W	Maps
C	Booklets	J	Mass-Market Books	Q	Calendars	X	Newspapers
D	Catalogs	K	Newsletters	R	Comic Books	Y	Postcards
E	Cookbooks	L	Software Manuals	S	Direct Mail Letters	Z	Posters
F	Directories	M	Textbooks	T	Envelopes	1	Stationery

Send RFQ's to: Bob Schick, National
  Sales Manager

Min: 20000 Max: 3000000 Opt: 0
Sizes: 5X
Bindings: PBI-SSI
Items: ADIPZ
Services: 4C-CS-FF
Equipment: SO-WO
Annual Sales: 6
Number of Employees: 2480

American Signature/Foote   CA
Marketing Director
3101 McCall Dr
Atlanta GA 30340
404-451-4511
Send RFQ's to: Jerry Cotter, VP Sales
Min: 200000 Max: 2000000 Opt: 0
Sizes: 23456X
Bindings: PBI-SSI
Items: ADIP
Services: 4C-CS-FF-TY
Equipment: SO-21WO
Annual Sales: 6
Number of Employees: 2480
Year Established: 1887

American Signature   MG
Marketing Director
PO Box 81608 (68501)
3700 NW 12th St
Lincoln NE 68521
402-474-5825

Min: 50000 Max: 2000000 Opt: 0
Sizes: 2345X
Bindings: PBI-SSI
Items: CDFINPX
Services: 4C-FF
Equipment: 10WO
Annual Sales: 6
Number of Employees: 2480
Year Established: 1887

American Web Inc   MS
Marketing Director
4040 Dahlia St
Denver CO 80216
303-321-2422
Send RFQ's to: Clark Fine, VP Sales

Min: 10000 Max: 200000 Opt: 45000
Sizes: 245X
Bindings: PBI-SSI
Items: HIK
Services: 4C-CS
Equipment: 2WO(22x36)
Annual Sales: 4
Number of Employees: 100
Turnaround Times: S2
Customers: BU5-MP95
Union Shop: N
Year Established: 1981
Terms: 1% 10, net 30

Statement: We are dedicated to serving the
needs of short-run, specialty interest
magazine publishers.

Americomp   BK
Henry Burr
American-Stratford Graphic Service
Putney Rd, Box 8128
Brattleboro VT 05304
802-254-6073; Fax: 802-254-5240
800-451-4328
Send RFQ's to: Henry Burr

Min: 100 Max: 10000 Opt: 1000
Sizes: 23
Bindings: PBI
Items: BGHILN
Services: TY-TD
Equipment: SO
Turnaround Times: B2
Year Established: 1990

Amidon & Associates Inc   GC
Marketing Director
1966 Benson Ave
Saint Paul MN 55116-3299
612-690-2401; 800-328-6502
Send RFQ's to: Stan Amidon, President

Min: 5000 Max: 5000000 Opt: 50000
Sizes: 25
Bindings: SSX
Items: CDKNPRSTY1
Services: 4C-CS
Equipment: 4WO(20x22)
Annual Sales: 3
Number of Employees: 25
Turnaround Times: B3C3
Customers: BU70-MP10-NP10-OR5-SC5
Union Shop: N

SERVICES				BINDINGS			
4C	4-Color Printing	RA	Rachwal System	HC	Hardcover	SD	Side Stitching
CS	Color Separations	AF	Acid-Free Paper	CB	Comb Binding	SB	Spiral Binding
DA	Design / Artwork	RP	Recycled Paper	LB	Loose-Leaf Binding	WB	Wire-O Binding
FF	Fulfillment/Mailing	TY	Typesetting	OB	Otabind	I	In-House
LM	List Maintenance	TD	Typeset w/ Disk	PB	Perfect Binding	O	Out of House
OC	OptiCopy System	WA	Warehousing	SS	Saddle Stitching	X	Unknown

Year Established: 1951
Terms: Net 30

Statement: We specialize in medium and long-run web printing of direct mail pieces, postcards, self-mailers, restaurant placemats, and guest checks—all from one color to full color.

Amos Press Inc   **MS**
Marketing Director
911 Vandemark Rd
PO Box 4129
Sidney OH 45365
513-498-2111; 800-327-1259
Send RFQ's to: Charles Hackett, Sales Manager

Min: 5000 Max: 250000 Opt: 50000
Sizes: 5
Bindings: SSI
Items: K
Annual Sales: 5
Number of Employees: 250

Statement: We provide newsletter publishers with full in-house services from creative design to printing and mailing. In as short as 36 hours, we have typeset, keylined, printed, and mailed 110,000 newsletter packages.

Amsterdam Printing & Litho Corp   **GC**
Marketing Director
55 Wallins Corners Rd
Amsterdam NY 12010
518-842-6000; Fax: 518-843-5204
800-833-6231

Anderson Lithograph Co   **GC**
Ed Binder, Marketing Director
3217 Garfield Ave
Los Angeles CA 90040
213-727-7767; Fax: 213-728-1036

Andrews Printing Co Inc   **GC**
Chuck Schoenberg
2141 Bixby Rd
Lakewood CA 90712
310-426-7123; 800-266-7123

Annex Printing Corporation   **GC**
234 16th St 8th Fl
Jersey City NJ 07310
201-659-8060; Fax: 201-659-7538

Anundsen Publishing Company   **GC**
Erik Anundsen, Sales Manager
108 Washington St
PO Box 230
Decorah IA 52101
319-382-4295; Fax: 319-382-5949
Send RFQ's to: Erik Anundsen

Min: 35 Max: 5000 Opt: 500
Sizes: 12345
Bindings: LBI-PBI-SSI-SDI-HCO-CBO-SBO-WBO
Items: BCEPT
Services: 4C-AF-TY-TD
Equipment: 3LP(10x15,20x26)-7SO(10x15, 20x29)
Annual Sales: 2
Number of Employees: 25
Turnaround Times: B5C6S4
Customers: BP30-OT70
Union Shop: N
Year Established: 1868
Terms: Net 30; genealogies cash only

Statement: We specialize in short-run (35-300 copies) genealogical books for individuals nationwide.

Arandell Corporation   **CA**
James Treis, Marketing Director
N81 W13118 Leon Rd
Menomonee Falls WI 53052
414-255-4400; Fax: 414-253-3162
800-558-8724
Send RFQ's to: James Treis

Min: 75000 Max: 100000 Opt: 0
Sizes: 23456
Bindings: HCO

PRINTED ITEMS						
A	Annual Reports	G	Galley Copies	N	Workbooks	U	Greeting Cards
B	Books	H	Journals	O	Yearbooks	V	Labels / Stickers
C	Booklets	I	Magazines	P	Brochures	W	Maps
D	Catalogs	J	Mass-Market Books	Q	Calendars	X	Newspapers
E	Cookbooks	K	Newsletters	R	Comic Books	Y	Postcards
F	Directories	L	Software Manuals	S	Direct Mail Letters	Z	Posters
		M	Textbooks	T	Envelopes	1	Stationery

Items: ADEPQ
Services: 4C-FF-LM-WA
Equipment: 6WO(38,54)
Annual Sales: 6
Number of Employees: 480
Turnaround Times: B1S1
Customers: BU40-OT60
Union Shop: Y
Year Established: 1980
Terms: Net 30

Statement: Arandell Corporation special-izes in the database marketing, printing, mailing, and distribution for the catalog marketplace.

Arcata Graphics/Fairfield **BK**

Dan Hill, Marketing Director
100 N Miller St
Fairfield PA 17320
717-642-5871; Fax: 717-642-6611
800-356-0603
Send RFQ's to: Regional Sales Office

Min: 1000 Max: 1000000 Opt: 10000-50000
Sizes: 123X
Bindings: HCI-PBI-OBO
Items: BEFLM
Services: AF-RP-TD
Equipment: CB-6WO(17.75, 19⅜ cut off)
Annual Sales: 6
Number of Employees: 600
Turnaround Times: B2C3
Customers: BP100
Union Shop: N
Year Established: 1976
Terms: Net 30

Sales Offices:

Boston Sales Office
1099 Hingham St
Rockland MA 02370
617-871-5300

Brighton Sales Office
7600 W Grand River #220
Brighton MI 48116
313-229-9120

Chicago Sales Office
415 W Gold Rd #16
Arlington Heights IL 60005
708-640-8644

Columbus Sales Office
438 E Wilson Bridge Rd #200
Worthington OH 43085
614-431-8263

Dallas Sales Office
3860 W Northwest Highway #402
Dallas TX 75220
214-351-6168

Los Angeles Sales Office
1960 E Grand Ave #530
El Segundo CA 90245
310-322-4888

Nashville Sales Office
115 Franklin Rd #160
Brentwood TN 37027
615-373-3109

New York Sales Office
1185 Ave of the Americas 27th Fl
New York NY 10036
212-827-2700

Philadelphia Sales Office
GSB Building #311
Belmont and City Ave
Bala Cynwyd PA 19004
610-667-8580

San Francisco Sales Office
Peninsula Office Park
2929 Campus Dr #175
San Mateo CA 94403
415-571-5555

Washington DC Sales Office
4041 Powder Mill Rd #404
Calverton MD 20705
301-595-3450

Argus Litho Inc **MG**

Marketing Director
1031 Broadway
Albany NY 12204
518-432-4411; Fax: 518-433-9216

The Argus Press Inc **CA**

Larry Frerd, Marketing Director
7440 Natchez Ave
Niles IL 60714
708-647-7800; Fax: 708-647-1709
Send RFQ's to: Joe Jensen, Chairman

Min: 1000 Max: 100000 Opt: 50000
Sizes: 123456
Bindings: HCO-CBO-LBO-PBO-SSI-SDI-SBO-WBO-OBO
Items: ABCDEFGHIKLMNPQSUVYZ1

SERVICES				BINDINGS			
4C	4-Color Printing	RA	Rachwal System	HC	Hardcover	SD	Side Stitching
CS	Color Separations	AF	Acid-Free Paper	CB	Comb Binding	SB	Spiral Binding
DA	Design / Artwork	RP	Recycled Paper	LB	Loose-Leaf Binding	WB	Wire-O Binding
FF	Fulfillment/Mailing	TY	Typesetting	OB	Otabind	I	In-House
LM	List Maintenance	TD	Typeset w/ Disk	PB	Perfect Binding	O	Out of House
OC	OptiCopy System	WA	Warehousing	SS	Saddle Stitching	X	Unknown

Services: 4C-CS-FF-OC-AF-RP-TY-TD-WA
Equipment: 8SO(28x41)
Annual Sales: 5
Number of Employees: 130
Turnaround Times: C2S2
Customers: BP5-BU80-CU2-NP2-OR2-OT5
Union Shop: N
Year Established: 1935
Terms: Net 30

Statement: Argus Press believes that total customer satisfaction is our number one objective. We believe that to meet this goal, we must provide the highest level of technical services possible to our clients. Accordingly, we have asssembled people, equipment, and systems to accomplish this task.

Arizona Lithographers Inc **GC**
Marketing Director
351 N Commerce Park Loop
Tucson AZ 85745
602-622-7667
Send RFQ's to: Alison MacDonald, Sales Rep

Min: 5000 Max: 50000 Opt: 20000
Bindings: HCO-PBI-SSI
Items: ADKY (art prints)
Services: 4C
Equipment: SO

Statement: Arizona Litho is a sheet-fed commercial and art print lithographer.

Art Litho Co **GC**
Marketing Director
1500 W Patapsco Ave
Baltimore MD 21230
410-355-3200; Fax: 410-355-3013

Art Print Co **GC**
Marketing Director
Stauffer Industrial Park
Taylor PA 18517
717-562-3060; Fax: 717-562-1573

Artco Printers & Lithographers **GC**
Marketing Director
1015 E Vermont Ave
Anaheim CA 92805
714-535-2169; Fax: 714-533-0660

Artcraft Press **GC**
Marketing Director
The Royle Group
PO Box 7
Waterloo WI 53594
414-478-2176
Send RFQ's to: William Mikalson

Min: 5000 Max: 200000 Opt: 0
Sizes: 2345X
Bindings: PBI-SSI
Items: ACDFINPXYZ
Services: 4C-FF
Equipment: SO-WO

Associated Printers **GC**
Marketing Director
402 Hill Ave
Grafton ND 58237
701-352-0640; Fax: 701-352-1502
Send RFQ's to: Heidi Salwei
Parent Company: Morgan Publishing Inc

Min: 100 Max: 100000 Opt: 30000
Sizes: 345
Bindings: PBX-SSX-HCX
Items: BCDIKPRTX1
Services: 4C-LM-TY-TD-RP
Equipment: 3SO(14x20,25x38,17x25)-2WO(22.75x35)
Annual Sales: 2
Number of Employees: 52
Turnaround Times: B3C5S3
Customers: BU30-CU5-MP40-OR20
Union Shop: N
Year Established: 1940
Terms: 50% deposit with copy; 50% before shipment

Statement: Our company's goal is to establish an atmosphere of trust for small comic book and magazine publishers by providing a commitment of top quality, reasonable prices, timely delivery, and guaranteed service on all printing projects submitted to us.

PRINTED ITEMS					
A	Annual Reports	G	Galley Copies	N	Workbooks	U	Greeting Cards
B	Books	H	Journals	O	Yearbooks	V	Labels / Stickers
C	Booklets	I	Magazines	P	Brochures	W	Maps
D	Catalogs	J	Mass-Market Books	Q	Calendars	X	Newspapers
E	Cookbooks	K	Newsletters	R	Comic Books	Y	Postcards
F	Directories	L	Software Manuals	S	Direct Mail Letters	Z	Posters
		M	Textbooks	T	Envelopes	1	Stationery

Atlantic Co
GC

Marketing Director
701 Business N Hwy 701
Tabor City NC 28463
910-653-3153; Fax: 910-653-9440

Austin Printing Co Inc
GC

George Cella, Marketing Director
130 E Voris St
Akron OH 44313
216-434-6695; Fax: 216-434-0012

Min: 500 Max: 300000 Opt: 50000
Sizes: 123456
Bindings: LBI-SSI-SDI-CBO-HCO-OBO-
PBO-SBO-WBO
Items: ABCDEFIKLPQ
Services: 4C-CS-FF-OC-RP-TY-TD
Equipment: 2LP(28x41)-7SO(28 to 40)
Annual Sales: 4
Number of Employees: 85
Turnaround Times: B2C3S1
Customers: BU90-CU5-NP5
Union Shop: Y
Year Established: 1906
Terms: 1% 15, net 30

Automated Graphic Systems (AGS)
BK

Eileen M Bok, Marketing Director
4590 Graphics Dr
PO Box 188
White Plains MD 20695
301-843-1800; Fax: 301-843-6339
800-678-8760
Send RFQ's to: Mark Edgar

Min: 10 Max: 250000 Opt: 0
Sizes: 123456X
Bindings: CBI-LBI-PBI-SSI-SDI-SBO-
HCO-WBI-OBO
Items: ABCDEFGHIKLMNOP2
Services: 4C-DA-FF-RA-AF-RP-TY-TD-
CS
Equipment: 7SO(26,38,50,54,56)
Annual Sales: 4
Number of Employees: 230
Turnaround Times: B2C3S2
Customers: BP13-BU23-CU3-NP51-OT10
Union Shop: N
Year Established: 1975

Terms: Net 30

Statement: We build books from start to
finish. AGS can assist you with any number
of services including design, typesetting,
desktop publishing, database management,
diskette replication, CD-Rom production,
and fulfillment. We work around the clock
to provide you with quality service and
comittment you can count on.

Comments: They publish a monthly
newsletter called **The AutoGraph**.

Additional Facilities:

Automated Graphic Systems—Ohio
6810 Cochran Rd
Solon OH 44139
800-362-6134

Automated Graphic Imaging/Copy Center
1350 I St NW
Washington DC 20005
202-371-5484

The One-Off CD Shop
4590 Graphics Dr, Box 188
White Plains MD 20695
301-870-0480 or 800-678-8760

B

B & D Litho Inc
GC

Marketing Director
3820 N 38th Ave
Phoenix AZ 85019
602-269-2526; Fax: 602-269-2520

B & W Press Inc
GC

Paul J Beegan, President
401 E Main St
Georgetown MA 01833
508-352-6100; Fax: 508-352-5955
Send RFQ's to: Paul J Beegan

Min: 10000 Max: 100000000 Opt: 100000-
5000000
Items: PST
Services: 4C
Equipment: 6WO(25.5x32)
Annual Sales: 4

SERVICES				BINDINGS			
4C	4-Color Printing	RA	Rachwal System	HC	Hardcover	SD	Side Stitching
CS	Color Separations	AF	Acid-Free Paper	CB	Comb Binding	SB	Spiral Binding
DA	Design / Artwork	RP	Recycled Paper	LB	Loose-Leaf Binding	WB	Wire-O Binding
FF	Fulfillment/Mailing	TY	Typesetting	OB	Otabind	I	In-House
LM	List Maintenance	TD	Typeset w/ Disk	PB	Perfect Binding	O	Out of House
OC	OptiCopy System	WA	Warehousing	SS	Saddle Stitching	X	Unknown

Number of Employees: 120
Turnaround Times: 20-30 days
Customers: BP5-BU60-CU5-MP15-NP5-SC5
Union Shop: N
Year Established: 1965
Terms: 30-60 days no interest

Statement: Our main product line is order form envelopes, inserts, and bind-in-envelopes for catalogs and magazines. B & W Press manufactures more than 7 million orderform envelopes each day on six web presses.

Bagcraft Corp of America SP
Jeff Cortopassi
3900 W 43rd St
Chicago IL 60632
312-254-8000; Fax: 312-254-8204
Send RFQ's to: Jeff Cortopassi
Parent Company: Artra Inc

Min: 30000 Max: 0 Opt: 100000
Items: 4 (Packaging)
Equipment: Flexographic four and six color
Annual Sales: 6
Number of Employees: 1000
Turnaround Times: 60 days
Customers: CU1-OR5-OT90
Union Shop: Y
Terms: 1% 10, net 30

Statement: Bagcraft is a flexible packaging printer producing window bags, carry-out bags, sandwich wraps, Cur-pon(tm) Bags, pocket window bags(tm) made with paper, foil, and film.

Baker Johnson Inc BK
Marketing Director
2810 Baker Rd
Dexter MI 48130
313-426-0200; Fax: 313-426-0301

Min: 100 Max: 25000 Opt: 5000
Sizes: 23456
Bindings: PBI-SSI
Items: BDHMN
Services: 4C-CS-AF-RP
Equipment: SO
Annual Sales: 2
Number of Employees: 35

Turnaround Times: B3S3
Union Shop: N
Year Established: 1989
Terms: Net 30 on approved credit

Statement: By specializing in traditional soft cover bindings, we continue to offer fast turnaround with very competitive pricing.

Baker Press Inc
Marketing Director
8720 Empress Row
Dallas TX 75247
214-630-1700

Bang Printing Company BK
D Hollingsworth, Marketing Director
PO Box 587
Brainerd MN 56401
218-829-2877; Fax: 218-829-7145
800-328-0450
Send RFQ's to: D Hollingsworth

Min: 2000 Max: 1000000 Opt: 50000
Sizes: 12345
Bindings: CBI-PBI-SSI-SBI-HCO-OBO-SDO-WBO
Items: ABCDEFGHKLNPQSTWZ
Services: 4C-CS-DA-FF-OC-RP-TY-TD-WA
Equipment: 6LP(40 Heidelberg)
Number of Employees: 93
Turnaround Times: B3C4S3
Customers: BP45-BU30-CU5-OR20
Union Shop: Y
Year Established: 1984
Terms: Net 30

Baniff Direct Mail Printers GC
Marketing Director
257 Ely Ave
Norwalk CT 06854
203-853-2555

Min: 5000000 Max: 20000000 Opt: 0
Services: 1-6 color
Equipment: WO

PRINTED ITEMS					
A	Annual Reports	G	Galley Copies	N	Workbooks
B	Books	H	Journals	O	Yearbooks
C	Booklets	I	Magazines	P	Brochures
D	Catalogs	J	Mass-Market Books	Q	Calendars
E	Cookbooks	K	Newsletters	R	Comic Books
F	Directories	L	Software Manuals	S	Direct Mail Letters
		M	Textbooks	T	Envelopes

U	Greeting Cards		
V	Labels / Stickers		
W	Maps		
X	Newspapers		
Y	Postcards		
Z	Posters		
1	Stationery		

Banta Company   BK
Perry Rindfleisch, VP Sales/Mktg
Curtis Reed Plaza
PO Box 60
Menasha WI 54952
414-751-7771; Fax: 414-751-7362
Parent Company: Banta Corporation

Min: 2500 Max: 1000000 Opt: 50000
Sizes: 12356X
Bindings: CBX-LBX-PBX-SSX-SBX-WBX
Items: BCDEFJLMN34
Services: 4C-CS-FF-OC-AF-RP-TD-WA
Equipment: 4CB(38,55)-6SO(26,38)-
   13WO(36,38,46)
Annual Sales: 6
Number of Employees: 1000
Turnaround Times: B3C4S3
Customers: BP50-BU20-NP5-OR5-OT20
Union Shop: Y
Year Established: 1901
Terms: Net 30

Statement: Banta Company is an information transfer company in multiple media
including traditional print, CD-ROM, and
other digital vehicles.

Banta Publications Group   MG
Cheryl Cromer, Marketing Director
908 N Elm St #110
Hinsdale IL 60521
708-323-9490; Fax: 708-323-0782
Send RFQ's to: Cheryl Cromer
Parent Company: Banta Corporation

Min: 10000 Max: 350000 Opt: 10000-350000
Sizes: 5
Bindings: PBI-SSI
Items: DHIX
Services: 4C-CS-FF-LM-OC-RP-TD
Equipment: 4SO(28x35)-10WO(22.75x38)
Annual Sales: 6
Number of Employees: 750
Turnaround Times: B3S3
Customers: MP85
Union Shop: Yes and No
Terms: Net 30

Banta Co/Harrisonburg   BK
Marketing Director
3330 Willow Spring Rd
Harrisonburg VA 22801
703-564-3900
Send RFQ's to: Paul Sullivan, Plant
   Manager

Min: 2500 Max: 1000000 Opt: 25000
Sizes: 123456X
Bindings: PBI-SSI
Items: BCDFILN
Services: 4C-CS-FF-OC-AF-RP-WA
Equipment: 3CB-5SO(25x38,28x40)-14WO
Terms: Net 30

Baronet Litho Inc   GC
Marketing Director
307 N Comrie Ave
Johnstown NY 12095
518-762-4627; Fax: 518-762-1302

Barton Press Inc   GC
Frank Matoroti, Marketing Director
55 Lakeside Ave
West Orange NJ 07052
201-736-8730; Fax: 201-736-7929

Min: 3000 Max: 75000 Opt: 25000
Sizes: 123456
Bindings: LBI-SSI-HCO-CBO-OBO-PBO-
   SDO-SPO-WBO
Items: ABCDEKLNOPWZ2
Services: 4C-CS-FF
Equipment: 3LP-SO-WO
Annual Sales: 4
Number of Employees: 175
Customers: BP20-BU80
Union Shop: Y
Year Established: 1991
Terms: Net 30

Bassett Printing Corp   GC
Marketing Director
101 Main St
Bassett VA 24055
703-629-2541; Fax: 703-629-3416

SERVICES				BINDINGS			
4C	4-Color Printing	RA	Rachwal System	HC	Hardcover	SD	Side Stitching
CS	Color Separations	AF	Acid-Free Paper	CB	Comb Binding	SB	Spiral Binding
DA	Design / Artwork	RP	Recycled Paper	LB	Loose-Leaf Binding	WB	Wire-O Binding
FF	Fulfillment/Mailing	TY	Typesetting	OB	Otabind	I	In-House
LM	List Maintenance	TD	Typeset w/ Disk	PB	Perfect Binding	O	Out of House
OC	OptiCopy System	WA	Warehousing	SS	Saddle Stitching	X	Unknown

Baum Printing House Inc   **GC**
Marketing Director
9985 Gantry Rd
Philadelphia PA 19115
215-671-9500; Fax: 215-676-5455

Bawden Printing Inc   **BK**
Cynthia Delashmutt, Marketing Director
400 S 14th Ave
Eldridge IA 52748
319-285-4800; Fax: 319-285-4828
Send RFQ's to: Marketing Coordinator

Min: 3000 Max: 200000 Opt: 15000
Sizes: 1245
Bindings: CBI-PBI-SSI-SBI-WBI-LBI-
HCO-OBO-SDO
Items: BCDFHLN
Services: FF-OC-AF-RP-TD-WA
Equipment: 3SO(19x25)-4WO(22.75x36)
Annual Sales: 5
Number of Employees: 312
Turnaround Times: B3S3
Customers: BP50-BU30-CU10-NP5-SP5
Union Shop: Y
Year Established: 1922
Terms: Net 30 with approved credit

Statement: We are a one and two-color medium to high volume heatset web printing company. We also have full-color sheetfed cover press capability, plus full in-house saddlestiching, perfect binding, labeling, mailing, promotion fulfillment, and distribution.

Bay Port Press   **BK**
Marketing Director
645-D Marsat Ct
Chula Vista CA 91911
619-429-0100; Fax: 619-429-0199
Send RFQ's to: Tina Greeson

Min: 500 Max: 50000 Opt: 10000
Sizes: 12345
Bindings: CBI-LBI-SSI-SDI-HCO-OBO-
PBO-SBO-WBO
Items: BCDEFHIKLMNPQSY2
Services: 4C-DA-AF-TY
Equipment: SO(14x20)-2WO(17.5x22,
36x22.75)
Annual Sales: 2

Number of Employees: 15
Turnaround Times: B3C3S3
Customers: BU50-CU20-NP20-SP10
Union Shop: N
Year Established: 1978
Terms: 1/3 deposit, net 30 w/approval

Statement: We specialize in high quality non-heated web printing in one and two colors. Product specialties are saddle stitched and perfect bound publications.

Bay State Press   **GC**
Ed Gillooly, Marketing Director
2 Watson Pl Bldg 5C
PO BOX 3310
Framingham MA 61701
508-877-0116; Fax: 508-877-7930
Send RFQ's to: John Gleason

Min: 100 Max: 1000000 Opt: 30000
Sizes: 123456X
Bindings: LBX-SSX-SDX
Items: ABCDEFHIKLNOPQSTUWXZ123
Services: 4C-CS-DA-FF-LM-AF-RP-TY-
TD-WA
Equipment: 3WO(28x41,20x26,25x38)-SO-
LP
Annual Sales: 3
Number of Employees: 85
Turnaround Times: B3C4S2
Customers: BP5-BU41-CU8-OR22-NP10-
SC3-SP3-MP3-OT5
Union Shop: N
Year Established: 1988
Terms: Net 30

Statement: Fast turnaround, high quality, two thru six color sheetfed printing.

Beacon Press Inc   **GC**
Marketing Director
4731 Eubank Rd
Richmond VA 23231
804-226-2120
Send RFQ's to: Tom Jenkins

Min: 50000 Max: 2000000 Opt: 0
Sizes: 2345X
Bindings: PBI-SSI-SDI
Items: BCDHIKSX
Services: 4C
Equipment: SO-3WO

PRINTED ITEMS	G	Galley Copies	N	Workbooks	U	Greeting Cards
A Annual Reports	H	Journals	O	Yearbooks	V	Labels / Stickers
B Books	I	Magazines	P	Brochures	W	Maps
C Booklets	J	Mass-Market Books	Q	Calendars	X	Newspapers
D Catalogs	K	Newsletters	R	Comic Books	Y	Postcards
E Cookbooks	L	Software Manuals	S	Direct Mail Letters	Z	Posters
F Directories	M	Textbooks	T	Envelopes	1	Stationery

Beacon Press **GC**
Jay Stevens, Marketing Director
PO Box 1750
Seattle WA 98111-1750
206-624-9699; Fax: 206-762-8394
Send RFQ's to: Chad Ernst
Parent Company: Cheler Corporation

Min: 100 Max: 50000 Opt: 2500
Sizes: 123456
Bindings: HCO-CBI-LBI-PBO-SSI-SDI-
SBI-WBO-OBO
Items: ABCDEFIKLNOPQSTVWYZ124
Services: 4C-FF-LM-AF-RP-TY-TD-WA
Equipment: 3LP(15x20)-5SO(18x21)
Annual Sales: 3
Number of Employees: 25
Turnaround Times: B5C5S3
Customers: BP10-BU50-CU5-MP15-NP5-
OR5-SP10
Union Shop: N
Year Established: 1916
Terms: 1% 10, net 20

Statement: "Quality, Fidelity, Service"

Bedinghaus Business Forms Inc **GC**
Marketing Director
11417 Lippleman Rd
Cincinnati OH 45246
513-772-1900; Fax: 513-772-7500
Send RFQ's to: Gary Hurt, President

Ben-Wal Printing **GC**
Marketing Director
146 Atlantic St
Pomona CA 91768
909-598-9999; Fax: 909-598-9299
Send RFQ's to: Scott Benson

Min: 25 Max: 200000 Opt: 500-10000
Sizes: 1245
Bindings: LBI-PBI-SSI-SDI-CBO-SBO-
WBO
Items: BCDFGKLMNPSY
Services: 4C-CS-DA-AF-RP-TY-TD
Equipment: WO-OP
Annual Sales: 3
Number of Employees: 56
Turnaround Times: B2C4S2
Customers: BP10-BU10-SP5-OT75
Union Shop: N

Year Established: 1991
Terms: Net 30

Bennett Printing Co **GC**
Marketing Director
990 S St Paul St (75201)
PO Box 1971
Dallas TX 75221
214-741-7751; Fax: 214-741-2465

Berryville Graphics **BK**
Marketing Director
Springsbury Rd
PO Box 272
Berryville VA 22611
703-955-2750; Fax: 703-955-2633
Send RFQ's to: Ed Altemose, Marketing
Director
Parent Company: Bertelsmann Printing &
Manufacturing

Min: 2500 Max: 2000000 Opt: 20000
Sizes: 123456
Bindings: HCI-CBO-PBI-SSI-SDI-SBO-
WBO
Items: BCE
Services: 4C-OC-RA-AF-TY-TD-WA
Equipment: 2LP-5SO-3WO
Annual Sales: 6
Number of Employees: 430
Turnaround Times: B5
Customers: BP97-BU1-CU1-NP1
Union Shop: N
Year Established: 1957
Terms: Various, net 30

Bertelsmann Printing & **BK**
Manufacturing
Gregg Aponte
1540 Broadway 23rd Fl
New York NY 10036-4039
212-984-7676; Fax: 212-984-7600
Send RFQ's to: Jerry Allee, VP Marketing

Min: 1500 Max: 1000000 Opt: 20000
Sizes: 123456X
Bindings: HCI-CBO-LBO-PBI-SSI-SDI-
SBO-WBO
Items: BCEFGJLMNPQZ
Services: 4C-CS-OC-RA-AF-RP-TD-WA
Equipment: 57SO-33WO

SERVICES				BINDINGS			
4C	4-Color Printing	RA	Rachwal System	HC	Hardcover	SD	Side Stitching
CS	Color Separations	AF	Acid-Free Paper	CB	Comb Binding	SB	Spiral Binding
DA	Design / Artwork	RP	Recycled Paper	LB	Loose-Leaf Binding	WB	Wire-O Binding
FF	Fulfillment/Mailing	TY	Typesetting	OB	Otabind	I	In-House
LM	List Maintenance	TD	Typeset w/ Disk	PB	Perfect Binding	O	Out of House
OC	OptiCopy System	WA	Warehousing	SS	Saddle Stitching	X	Unknown

Annual Sales: 6
Number of Employees: 1211
Turnaround Times: B3
Customers: BP70-BU10-CU2-NP15-OR1-OT2
Union Shop: Y
Year Established: 1960
Terms: Net 30 from invoice date

Statement: BPMC is a full integrated book manufacturing company consisting of plants in Pennsylvania, Virginia, and California. This is the main corporate office for Berryville Graphics, Delta Lithograph, and Offset Paperback. BPMC is a division of Bertelsmann, a large German publisher which also owns Bantam/Doubleday/Dell.

Best Gagne Book Manufacturers   BK
Guy Lefebvre, Marketing Director
150 Front St East
Toronto ON Canada M5A 1E5
416-362-1872; Fax: 416-362-6081
Send RFQ's to: Guy Lefebvre

Min: 1000 Max: 100000 Opt: 15000
Sizes: 1235X
Bindings: CBI-LBI-OBI-PBI-SDO-SSO-SBO-WBO
Items: BEFJLMN
Services: OC-RA-AF-RP-TD
Equipment: 3SO(20x28,28x41)-4WO(37.5, 35.5,78)
Annual Sales: 4
Number of Employees: 200
Turnaround Times: B3C3S3
Customers: BP90
Union Shop: Y
Year Established: 1970
Terms: Net 30

Best Printing & Duplicating Inc   GC
Marketing Director
11465 Schenk Dr
Maryland Heights MO 63043
314-298-7700; Fax: 314-298-9631

Best Printing Co Inc   GC
Marketing Director
PO Box 1548
Austin TX 78767
512-477-9733; Fax: 512-477-8623

Bethel Park Printing Inc   SP
Marketing Director
5237 Brightwood Rd
Bethel Park PA 15102
412-835-4433

Blake Printery   BK
Marketing Director
2222 Beebee St
San Luis Obispo CA 93401
805-543-6843; 800-234-3320
Send RFQ's to: Richard Blake, President

Min: 1000 Max: 300000 Opt: 5000
Sizes: 123456X
Bindings: HCO-CBI-PBO-SSI
Items: BCDEHIPQUYZ
Services: 4C-DA-TY

Blue Dolphin Press Inc   BK
John Odenkirk, Marketing Director
12380 Nevada City Hwy
Grass Valley CA 95945
916-265-6923; Fax: 916-265-1957
Send RFQ's to: Paul Clemens

Min: 500 Max: 5000 Opt: 3000-5000
Sizes: 125
Bindings: CBI-LBI-PBI-SSI-SDI-SBI-WBI-HCO-OBO
Items: ABCDEFGKNPSTUVYZ12
Services: 4C-CS-DA-FF-OC-AF-RP-TY-TD
Equipment: LP(10x15)-5SO(up to 24 & 38)
Annual Sales: 2
Number of Employees: 13
Turnaround Times: B5C6S5
Customers: BP15-BU60-NP5-SP15
Union Shop: N
Year Established: 1976
Terms: 1/2 down, net 30

Statement: Blue Dolphin specializes in pre-press, service bureau and design in

PRINTED ITEMS							
A	Annual Reports	G	Galley Copies	N	Workbooks	U	Greeting Cards
B	Books	H	Journals	O	Yearbooks	V	Labels / Stickers
C	Booklets	I	Magazines	P	Brochures	W	Maps
D	Catalogs	J	Mass-Market Books	Q	Calendars	X	Newspapers
E	Cookbooks	K	Newsletters	R	Comic Books	Y	Postcards
F	Directories	L	Software Manuals	S	Direct Mail Letters	Z	Posters
		M	Textbooks	T	Envelopes	1	Stationery

both MAC and IBM formats, with six presses. We are especially good at helping a client who does not know how to proceed.

BMP Paper & Printing Inc   **GC**
Marketing Director
4923 W 34th St
Houston TX 77092
713-228-9191; Fax: 713-465-3337

Bockman Printing & Services Co   **GC**
Marketing Director
950 S 25th Ave
Bellwood IL 60104
708-544-4090; Fax: 708-544-7917

Bolger Publications/Creative Printing   **GC**
Dave Borg, Marketing Analysis
3301 Como Ave SE
Minneapolis MN 55414
612-645-6311; Fax: 612-645-1750
Send RFQ's to: Mike Oslund, Sales
   Manager

Min: 1000 Max: 300000 Opt: 50000
Sizes: 123456
Bindings: HCO-CBO-LBO-PBI-SSI-SDO-
   SBO-WBO-OBO
Items: ABCDEFHIKLMNOPQUZ
Services: 4C-CS-DA-FF-LM-OC-AF-RP-
   TY-TD-WA
Equipment: 4SO(28x40)
Annual Sales: 4
Number of Employees: 105
Turnaround Times: B2C4S2
Customers: BU50-CU10-MP10-NP10-
   OR20
Union Shop: N
Year Established: 1934
Terms: Net 30

Statement: Bolger Publications is a full service graphic communications company dedicated to process quality, total customer satisfaction, and continuous improvement.

The Book Press Inc   **BK**
Samuel Ratkewitch, Marketing Director
Putney Rd
Brattleboro VT 05301
802-257-7701; Fax: 802-257-9439
Send RFQ's to: Samuel Ratkewitch

Min: 1000 Max: 0 Opt: 0
Sizes: 23456
Bindings: HCX-PBX
Items: BCDEFLMN2
Services: FF-OC
Equipment: CB-4SO(23x29,54x77)-5WO
   (various)
Annual Sales: 5
Number of Employees: 30
Turnaround Times: B4C4
Customers: PB80-BU10-CU10
Union Shop: Y

Book-Mart Press Inc   **BK**
Michelle T Gluckow, Marketing Director
2001 Forty Second St
North Bergen NJ 07047
201-864-1887; Fax: 201-864-7559
Send RFQ's to: Michelle T Gluckow

Min: 300 Max: 0 Opt: 0
Sizes: 123456X
Bindings: HCI-PBI-SSI-CBO-SBO-WBO
Items: ABCDEFHLMNO
Services: 4C-FF-AF
Equipment: 7SO(41x54,41x56)
Turnaround Times: B3C3S3
Customers: BP-BU-CU-NP-OR-SC-SP
Union Shop: N
Year Established: 1977

Statement: Book-Mart Press specializes in the manufacture of short to medium runs of both softcover and casebound books. All work is done in our New Jersey facilities. We have a reputation of being one of the fastest book manufacturers in the industry. Find out why publishers swear by us and not at us!

BookCrafters Inc/Fredericksburg   **BK**
Marketing Director
3591 Lee Hill Industrial Park
PO Box 892
Fredericksburg VA 22401
703-371-3800; Fax: 703-475-8591

SERVICES				BINDINGS			
4C	4-Color Printing	RA	Rachwal System	HC	Hardcover	SD	Side Stitching
CS	Color Separations	AF	Acid-Free Paper	CB	Comb Binding	SB	Spiral Binding
DA	Design / Artwork	RP	Recycled Paper	LB	Loose-Leaf Binding	WB	Wire-O Binding
FF	Fulfillment/Mailing	TY	Typesetting	OB	Otabind	I	In-House
LM	List Maintenance	TD	Typeset w/ Disk	PB	Perfect Binding	O	Out of House
OC	OptiCopy System	WA	Warehousing	SS	Saddle Stitching	X	Unknown

Min: 300 Max: 100000 Opt: 6000
Sizes: 123456
Bindings: HCX-CBX-LBX-PBX-SSX-
    SBX-WBX
Items: BCDEFHLMN
Services: 4C-FF
Equipment: CB(38)-12SO(29, 38, 40, 50)
Turnaround Times: B4
Customers: BP60-BU3-CU7-NP5-SP15-
    OT10
Union Shop: N
Year Established: 1965
Terms: Net 30; terms to be arranged

Booklet Publishing Company Inc GC

Joe Olcott, President
1902 Elmhurst Rd
Elk Grove Village IL 60007
708-364-1544; Fax: 708-364-0284
Send RFQ's to: Joe Olcott

Min: 500 Max: 50000 Opt: 2500
Sizes: 25
Bindings: PBI-SSI
Items: BCDFKLNP
Equipment: SO
Turnaround Times: B2C2

BookCrafters BK

Pam Eddington, Marketing Director
613 E Industrial Dr
Chelsea MI 48118
313-475-9145; Fax: 313-475-7337
Send RFQ's to: Marketing Department
Parent Company: American Business
    Products

Min: 250 Max: Open Opt: 1000-10000
Sizes: 123456X
Bindings: HCI-CBI-LBI-PBI-SSI-SBI-
    WBI-OBI
Items: BCDEFHLMN
Services: 4C-FF-AF-RP-TD-WA-CS-DA
Equipment: CB(38)-17SO(20.5x29,
    38x50)-WO(6x9)
Turnaround Times: B3C4S3
Customers: BP35-BU8-CU11-OR6-SP35-
    OT5
Union Shop: N
Year Established: 1965
Terms: Net 30; terms to be arranged

It used to be said "You can't get there from here"... but now you can, or at least your books can. In addition to our complete book manufacturing services,

BookCrafters has two full service Distribution Centers.

Call your Account Executive today to learn more.

- *Custom Reporting*
- *Permit Mailings*
- *Polybagging*
- *800 Order Services*
- *Data Communications*
- *Quality Book & Journal Manufacturing*
- *Electronic Prepress Services & Workshops*

≡ BookCrafters

140 Buchanan, Chelsea MI 48118

313-475-9145 FAX 313-475-7337 800-999-BOOK

Statement: Complete book and journal
manufacturers with distribution centers
in Michigan and Virginia. We also offer
electronic prepress services and electric
publishing workshops.

BookMasters Inc BK

Raymond F Sevin, Marketing Director
PO Box 159
Ashland OH 44805
419-289-6051; Fax: 419-281-1731
800-537-6727

PRINTED ITEMS			
A Annual Reports	G Galley Copies	N Workbooks	U Greeting Cards
B Books	H Journals	O Yearbooks	V Labels / Stickers
C Booklets	I Magazines	P Brochures	W Maps
D Catalogs	J Mass-Market Books	Q Calendars	X Newspapers
E Cookbooks	K Newsletters	R Comic Books	Y Postcards
F Directories	L Software Manuals	S Direct Mail Letters	Z Posters
	M Textbooks	T Envelopes	1 Stationery

Send RFQ's to: Sherry Ringler

Min: 100 Max: 100000 Opt: 1000-2000
Sizes: 123456
Bindings: CBI-PBI-SSI-WBI-HCO-LBO-
    OBO-SSO-SBO
Items: BEFHJLMN
Services: 4C-DA-FF-LM-AF-RP-TY-TD-
    WA
Equipment: 3SO(up to 25x38)
Annual Sales: 2
Number of Employees: 110
Turnaround Times: B4C5S4
Customers: BP35-CU5-NP10-SP50
Union Shop: N
Year Established: 1988
Terms: 50% down, 25% with bluelines,
    balance upon completion

Statement: To provide excellent quality and superior service to every customer at a fair and reasonable price.

Boyd Printing Co Inc **BK**
Carl Johnson, Marketing Director
49 Sheridan Ave
Albany NY 12210
800-877-BOYD(2639); Fax: 518-436-7433
Send RFQ's to: Carl Johnson

Min: 1000 Max: 150000 Opt: 10000
Sizes: 2345
Bindings: PBX-SSX
Items: BCHLN
Services: FF-LM-RA-AF-RP-TY-TD-WA
Equipment: 2SO(42x59)-WO(48x84)
Annual Sales: 3
Number of Employees: 100
Turnaround Times: B3S3
Customers: BP60-CU5-NP30-OR5
Union Shop: Y
Year Established: 1889
Terms: Net 30

Statement: Full service from people you can rely on.

Braceland Brothers Inc **GC**
Dan Breitkreutz, Marketing Director
7625 Suffolk Ave
Philadelphia PA 19153-3097
215-492-0200; Fax: 215-492-8538
Send RFQ's to: Debbie Weinberg

Min: 1000 Max: 100000 Opt: 25000

Sizes: 25
Bindings: PBX-SSX-SDX-LBX
Items: BDFLMN
Services: OC-TD
Equipment: 5WO(22.75x36)
Annual Sales: 5
Number of Employees: 250
Turnaround Times: B3S3
Customers: BP30-BU40-CU25-NP5
Union Shop: N
Year Established: 1899
Terms: Net 30

Bradley Printing Company **CA**
Marketing Director
2170 S Mannheim Rd
Des Plaines IL 60018
708-635-8000; Fax: 708-635-7944
Parent Company: World Color Press

Min: 100 Max: 2000000 Opt: 0
Sizes: 25
Bindings: HCO-CBO-LBO-PBI-SSI-SDI-
    SBO-WBO
Items: ACDFKLPQSTUWYZ1
Services: 4C-CS-DA-FF
Equipment: LP-18SO(13x20,28x40)-
    4WO(22x38)
Annual Sales: 6
Number of Employees: 725
Turnaround Times: C2
Customers: BU80-CU5-MP5-NP5-OR5
Union Shop: Y
Year Established: 1964
Terms: Net 30

Braun-Brumfield Inc **BK**
Robert Ryan
100 N Staebler Rd
PO Box 1203
Ann Arbor MI 48103
313-662-3291; Fax: 313-662-1667
Send RFQ's to: Arnold Ziroli, Executive
    President

Min: 100 Max: 10000 Opt: 1000
Sizes: 123456
Bindings: HCI-PBI-SSI-CBO-LBO-SBO-
    WBO
Items: BDEFHMN
Services: FF-OC-RA-AF-RP-TY-TD
Equipment: 10SO(4-50,3-42,1-40,2-29)
Annual Sales: 5

SERVICES				BINDINGS			
4C	4-Color Printing	RA	Rachwal System	HC	Hardcover	SD	Side Stitching
CS	Color Separations	AF	Acid-Free Paper	CB	Comb Binding	SB	Spiral Binding
DA	Design / Artwork	RP	Recycled Paper	LB	Loose-Leaf Binding	WB	Wire-O Binding
FF	Fulfillment/Mailing	TY	Typesetting	OB	Otabind	I	In-House
LM	List Maintenance	TD	Typeset w/ Disk	PB	Perfect Binding	O	Out of House
OC	OptiCopy System	WA	Warehousing	SS	Saddle Stitching	X	Unknown

Number of Employees: 400
Turnaround Times: B3C4
Customers: BP85-CU10-SP5
Union Shop: N
Year Established: 1950
Terms: Net 30

Statement: We are a complete in-house book manufacturer committed to quality and customer satisfaction. To honor this commitment, we utilize the best materials, the most efficient technology available, and that most vital element – dedicated people.

Comments: Write for their excellent resource booklets, **Book Manufacturing Glossary, Book Paper Samples, Type Sample Book, Camera Copy Preparation Guidelines, and Paper Bulk Chart.**

Brennan Printing   **BK**

Robert Brennan, Owner
100 Main St
Deep River IA 52222
515-595-2000; Fax: 515-595-2111
Send RFQ's to: Robert Brennan

Min: 500 Max: 10000 Opt: 2000
Sizes: 123456
Bindings: CBX-LBX-PBX-SSX-SDX-
    WBX
Items: ABCEFKNPST12
Services: TY-WA
Equipment: 3SO(16x23)
Annual Sales: 1
Number of Employees: 8
Turnaround Times: B5S5
Customers: BP10-NP80-SP5-OT5
Union Shop: N
Year Established: 1978
Terms: Net 30

Bro's Lithographic Co   **GC**

Marketing Director
1326 W Washington Blvd
Chicago Il 60607
312-666-0919

Brookshore Lithographers Inc   **GC**

Marketing Director
2075 Busse Rd
Elk Grove Village IL 60007
708-593-1200; Fax: 708-593-0058
800-323-6112
Send RFQ's to: Ray Frick

Min: 50000 Max: 2000000 Opt: 0
Sizes: 123456X
Bindings: SSI
Items: CDPSVYZ
Services: 4C

Statement: Now owned by Banta Corporation, Brookshore prints direct mail formats, including catalogs, mini-catalogs, brochures, gatefolds, inserts, poster pull-outs, and other direct mail packages.

Broughton Printing Co   **GC**

Marketing Director
Marketing Serivces Division
205 W Industrial Blvd
Dalton GA 30720
706-277-1977; Fax: 706-277-3106

Brown Printing   **MS**

Marketing Director
2300 Brown Ave
Waseca MN 56093
507-835-2410; Fax: 507-835-0420
Send RFQ's to: Don Henderson, VP Sales
Parent Company: Grunner & Yahr

Min: 0 Max: 0 Opt: 0
Sizes: 5
Bindings: PBI-SSI-HCO-CBO-LBO-OBO-
    SBO-SDO-WBO
Items: DFHI
Services: 4C-CS-LM-OC-TD
Equipment: 17WO(22.75)-4OP(Roto
    Gravure)
Annual Sales: 6
Number of Employees: 2500
Turnaround Times: B4S3
Customers: MP70-OT30
Union Shop: N
Year Established: 1957
Terms: Net 30

PRINTED ITEMS	G	Galley Copies	N	Workbooks	U	Greeting Cards
A Annual Reports	H	Journals	O	Yearbooks	V	Labels / Stickers
B Books	I	Magazines	P	Brochures	W	Maps
C Booklets	J	Mass-Market Books	Q	Calendars	X	Newspapers
D Catalogs	K	Newsletters	R	Comic Books	Y	Postcards
E Cookbooks	L	Software Manuals	S	Direct Mail Letters	Z	Posters
F Directories	M	Textbooks	T	Envelopes	1	Stationery

Statement: Brown is a full service web offset and rotogravure printer of the highest quality publications, catalogs, and inserts with manufacturing facilities in four states including: Minnesota, Kentucky, Pennsylvania, and California.

Brown Printing/ **MG**
East Greenville Division
William Guthrie, Marketing Director(NY)
668 Gravel Pike
East Greenville PA 18041-2199
215-679-4451; Fax: 215-679-5449
Send RFQ's to: Alan Lowell
Parent Company: Grunner & Yahr

Min: 30000 Max: 1000000 Opt: 100000 - 200000
Sizes: 5
Bindings: PBX-SSX
Items: DHI
Services: 4C-RP-TY-TD-WA
Equipment: 7WO(22.75x38)
Annual Sales: 6
Number of Employees: 750
Turnaround Times: B3S1-3
Customers: BU15-MP85
Union Shop: N
Year Established: 1978
Terms: Net 30

Brown Printing / Franklin **MG**
Marketing Director
300 Brown Rd
PO Box 569
Franklin KY 42134
502-586-9251
Send RFQ's to: Robert Hoffman

Min: 500000 Max: 10000000 Opt: 0
Sizes: 5X
Bindings: PBI-SSI
Items: DIX
Services: 4C-CS-FF
Equipment: 3OP(Gravure)

Bruce Offset Co **GC**
Marketing Director
1099 Greenleaf Ave
Elk Grove Village IL 60007
708-593-5290; Fax: 708-593-0815

Brunswick Publishing Corp **BK**
Walter J Raymond, Marketing Director
Rt 1 Box 1A1
Lawrenceville VA 23868
804-848-3865; Fax: 804-848-0607
Send RFQ's to: Walter J Raymond

Min: 1 Max: 5000 Opt: 500-2000
Sizes: 123456X
Bindings: CBI-LBI-PBI-SSI-SDI-SBO-HCO-OBO-WBO
Items: BCEKMPT1
Services: AF-RP-TY-TD
Equipment: 2SO(12x18,18x24)
Annual Sales: 1
Number of Employees: 9
Turnaround Times: B4C4S4
Customers: BP5-BU20-CU25-SC25-SP20-OT5
Union Shop: N
Year Established: 1968
Terms: 50% deposit, 50% on completion

Statement: As a publishing company, we have the expertise to assist self-publishers and organizations with the planning, typesetting, and printing of their books, including genealogies, school, college, church histories, biographies, etc.

R L Bryan Company **GC**
Marketing Director
301 Greystone Blvd
PO Drawer 368
Columbia SC 29202
803-779-3560; Fax: 803-343-6718
800-476-1844
Send RFQ's to: Jack Whitesides, VP Sales

Min: 1000 Max: 100000 Opt: 40000
Sizes: 123456X
Bindings: HCO-CBO-LBO-PBI-SSI-SDI-SPO-WBO
Items: ABCDEFHIKPQSTVWZ1
Services: 4C-CS-DA-FF-TY-TD-WA
Equipment: 3LP(to 41)-10SO(to 40)
Annual Sales: 3
Number of Employees: 100
Turnaround Times: B6C4S2
Customers: BU65-CU6-NP1-SP22-OT6
Union Shop: N
Year Established: 1915
Terms: Net 30 with approved credit

SERVICES				BINDINGS			
4C	4-Color Printing	RA	Rachwal System	HC	Hardcover	SD	Side Stitching
CS	Color Separations	AF	Acid-Free Paper	CB	Comb Binding	SB	Spiral Binding
DA	Design / Artwork	RP	Recycled Paper	LB	Loose-Leaf Binding	WB	Wire-O Binding
FF	Fulfillment/Mailing	TY	Typesetting	OB	Otabind	I	In-House
LM	List Maintenance	TD	Typeset w/ Disk	PB	Perfect Binding	O	Out of House
OC	OptiCopy System	WA	Warehousing	SS	Saddle Stitching	X	Unknown

BSC Litho **GC**
Marketing Director
3000 Canby St
Harrisburg PA 17103
717-238-9469; Fax: 717-232-4151
Send RFQ's to: S Kleiman, VP Sales

Min: 1000 Max: 2000000 Opt: 0
Sizes: 123456
Bindings: HCO-CBO-LBO-PBO-SSI-SDI-
    SBO-WBO
Items: ABCDEFIKPQSY
Services: 4C-DA-TY-WA
Equipment: 2LP(14x16)-6SO(20x28)-
    2WO(18x23)
Annual Sales: 3
Number of Employees: 100
Turnaround Times: B5C4S5
Customers: CU20-MP20-OR60
Union Shop: N
Year Established: 1921
Terms: Net 30

Buse Printing & Advertising **GC**
Ray Buse, Marketing Director
1616 E Harvard St
Phoenix AZ 85006
602-258-4757; Fax: 602-254-5118
Send RFQ's to: Ray Buse

Min: 1 Max: 1000000000 Opt: 1000000
Sizes: 123456X
Bindings: CBX-LBX-OBX-PBX-SSX-
    SDX-SBX-WBX
Items: ABCDEFGHIJKLMNOPQSTUV
    WXYZ1234
Services: 4C-CS-DA-FF-LM-OC-RA-AF-
    RP-TY-TD-WA
Equipment: LP-SO-WO-OP
Annual Sales: 3
Number of Employees: 110
Turnaround Times: B3S2
Customers: BP10-BU20-CU2-MP5-NP20-
    OR10-SP10-OT23
Union Shop: N
Year Established: 1968
Terms: Per customer

Statement: UV, spot laminating and foil,
special effects, 3-D, all types of specialty
printing including pop-up.

Butler Printing & Laminating Inc **GC**
Marketing Director
250 Hamburg Tpke
Butler NJ 07405
201-838-8550; Fax: 201-838-1767
800-524-0786

Buxton & Skinner Printing Co **GC**
Marketing Director
2419 Glasgow Ave
St Louis MO 63106
314-535-9700; Fax: 314-535-6760
800-524-5500

The William Byrd Press **GC**
Janet Ciaravino, Marketing Director
2901 Byrdhill Rd
Richmond VA 23228
804-264-2711; Fax: 804-262-7759
Send RFQ's to: Sales Office
Parent Company: Cadmus Communica-
    tions

Min: 25000 Max: 150000 Opt: 35000-75000
Sizes: 25X
Bindings: HCO-OBO-PBI-SSI
Items: DFIX
Services: 4C-CS-FF-OC-AF-RP-TD-WA
Equipment: 9WO(2,4,6 unit)
Annual Sales: 5
Number of Employees: 410
Turnaround Times: B3C6S2
Customers: CO10-MP45-NP45
Union Shop: N
Year Established: 1913
Terms: Net 30

Statement: Byrd is a full service provider
of production services for short to medium
run specialty publications.

Sales Offices:
The William Byrd Press Inc
1800 Diagonal Rd #420
Alexandria VA 22314-2480
703-519-8300; Fax: 703-519-8307

The William Byrd Press Inc
11 Penn Plaza #1002
New York NY 10001-2006
212-736-2002; Fax: 212-629-0376

PRINTED ITEMS					
A	Annual Reports	G	Galley Copies	N	Workbooks
B	Books	H	Journals	O	Yearbooks
C	Booklets	I	Magazines	P	Brochures
D	Catalogs	J	Mass-Market Books	Q	Calendars
E	Cookbooks	K	Newsletters	R	Comic Books
F	Directories	L	Software Manuals	S	Direct Mail Letters
		M	Textbooks	T	Envelopes
U	Greeting Cards
V	Labels / Stickers
W	Maps
X	Newspapers
Y	Postcards
Z	Posters
1	Stationery

The William Byrd Press Inc
1600 Broadway #1950
Denver CO 80202-4919
303-832-1488 ext 20; Fax: 303-832-1488

C

C & M Press

BK

Marketing Director
850 East 73rd Ave #5
Denver CO 80229
303-289-4757

Min: 100 Max: 5000 Opt: 1000
Sizes: 123456X
Bindings: HCI-CBI-LBI-PBI-SSI-SBI-WBI
Items: ABCDEFJKLMN
Equipment: 8SO(13x17)
Annual Sales: 3
Number of Employees: 30
Turnaround Times: B3
Customers: BP20-BU60-CU10-SC5-SP5
Union Shop: N
Year Established: 1983
Terms: Net 30

Statement: We are highly specialized and provide no printing services other than the production of books, manuals, and computer documentation.

Cal Central Press

GC

Marketing Director
2629 Fifth St (95818)
PO Box 551
Sacramento CA 95812
916-441-5392; Fax: 916-441-1843
Send RFQ's to: Tom Davis

Min: 200 Max: 2000000 Opt: 125000
Sizes: 123456X
Bindings: CBI-LBI-OBI-PBI-SSI-SDI-SBI-WBI-HCO
Items: ABCDEFGHIJKLMNOPQRWYZ
Services: 4C-CS-FF-OC-RA-RP-TY-TD-WA
Equipment: 14SO(24x41)-4WO(23x36, 17x23)
Annual Sales: 5
Number of Employees: 300
Turnaround Times: B2C4S2

Customers: BP11-BU11-CU11-SC11-MP11-NP11-OR11-SP11-OT11
Union Shop: N
Year Established: 1891
Terms: Net 30 with approved credit

Statement: Provides high quality printing with the best customer service.

Caldwell Printers

GC

Austin Grehan Sr, Marketing Director
215 S First Ave
Arcadia CA 91006
818-447-4601; Fax: 818-447-1370
Send RFQ's to: Austin Grehan Sr

Min: 500 Max: 0 Opt: 0
Sizes: 123456
Bindings: LBI-SSI-SBI-CBO-HCO-OBO-PBO-SDO-WBO
Items: ABCDFIKLNPSTVY1
Services: 4C-DA-RP-TY-TD
Equipment: LP(10x15)-4SO(11x17)
Annual Sales: 5
Number of Employees: 6
Turnaround Times: B3C3S1
Customers: BP5-BU50-CU1-MP5-NP5-SC5-SP5
Union Shop: N
Terms: 1st-COD, 2nd-15days, 3rd-30days

California Offset Printers

MS

Marketing Director
620 W Elk Ave
Glendale CA 91204
213-245-6446; Fax: 213-245-2123
Send RFQ's to: Bill Neckameyer

Min: 10000 Max: 0 Opt: 0
Sizes: 25
Bindings: SSX
Items: CDFHIKLPQRSX
Services: 4C-CS-OC-RP
Equipment: 3WO(2-6 units, 1-10 units)
Annual Sales: 4
Number of Employees: 100
Turnaround Times: B2S2
Customers: BU10-CU5-MP70-NP10-OR5
Union Shop: Y
Year Established: 1962
Terms: Net 30 with approved credit

Statement: Short run magazine printers.

SERVICES				BINDINGS			
4C	4-Color Printing	RA	Rachwal System	HC	Hardcover	SD	Side Stitching
CS	Color Separations	AF	Acid-Free Paper	CB	Comb Binding	SB	Spiral Binding
DA	Design / Artwork	RP	Recycled Paper	LB	Loose-Leaf Binding	WB	Wire-O Binding
FF	Fulfillment/Mailing	TY	Typesetting	OB	Otabind	I	In-House
LM	List Maintenance	TD	Typeset w/ Disk	PB	Perfect Binding	O	Out of House
OC	OptiCopy System	WA	Warehousing	SS	Saddle Stitching	X	Unknown

Calsonic Miura Graphics **GC**
Craig Maron, Marketing Director
9999 Muirlands
Irvine CA 92718
714-583-9000; Fax: 714-855-1542
Send RFQ's to: Craig Maron
Parent Company: Calsonic International
  and Miura Printing (Japan)

Min: 2500 Max: 500000 Opt: 0
Sizes: 123456
Bindings: HCO-CBO-LBO-PBO-SSI-SDI-
  SBO-WBO
Items: ACDIKPQTVWYZ124
Services: 4C-CS-DA-FF-RP-TY-TD-WA
Equipment: 5SO(7-40,6-40,6-26,2-25,2-18)
Annual Sales: 4
Number of Employees: 100
Turnaround Times: B3C4S3
Customers: BU50-OT50
Union Shop: N
Year Established: 1988
Terms: Net 30

Statement: CMG is a complete high quality
commercial printer including prepress
capabilities thru bindery. Product mix includes car brochures, annual reports,
product literature, and all commercial
printing.

Calvert-McBride Printing **GC**
Marketing Director
PO Box 6337
Ft Smith AR 72906
501-646-8311; Fax: 501-646-6036

Camelot Publishing Co **BK**
Donald Spencer, President
39-B Coolidge Ave
Ormond Beach FL 32174
904-672-5672
Send RFQ's to: Donald Spencer

Min: 50 Max: 300 Opt: 150
Sizes: 2345
Bindings: CBI-PBI-SSI
Items: BCDEFLM
Services: DA-AF-TY
Equipment: Xerox Duplicators
Annual Sales: 1
Number of Employees: 5

Turnaround Times: B3S1
Customers: BP40-SP60
Union Shop: N
Year Established: 1972
Terms: Payment in full with order

Statement: We specialize in ultra short-run
books and booklets. Write for FREE
brochure describing services and prices.

Canterbury Press **MS**
Walter Ranger
301 Mill St
PO Box 4299
Rome NY 13440
315-337-5900; Fax: 315-337-4070
800-836-2268
Send RFQ's to: William Kiser, Marketing
  Director

Min: 500 Max: 2500 Opt: 1000
Sizes: 23456X
Bindings: HCO-CBI-PBI-SSI-SBO-WBO
Items: ABCDFHIKNPT
Services: 4C-AF-RP-TY-TD-WA
Equipment: 6SO(19x25,36x49)
Annual Sales: 3
Number of Employees: 82
Turnaround Times: C4S3
Customers: BU10-CU10-OR5
Union Shop: Y
Year Established: 1952
Terms: Net 30

Statement: We primarily produce short-run
paperback books as how-to or field guides.
Additionally, we provide competitive
quality printing with a special expertise in
the medical/pharmaceutical and scientific
fields (short to medium runs).

Capital City Press **BK**
Paul Bozowa, Marketing Director
Airport Dr
PO Box 546
Montpelier VT 05602
802-223-5207; Fax: 802-223-6192
Send RFQ's to: Estimating Department

Min: 1000 Max: 30000 Opt: 3000
Sizes: 123456X
Bindings: HCO-PBI-SSI-CBO-LBO-OBO-
  SDO-SBO-WBO
Items: BDEFHLMN

PRINTED ITEMS					
A	Annual Reports	G	Galley Copies	N	Workbooks	U	Greeting Cards
B	Books	H	Journals	O	Yearbooks	V	Labels / Stickers
C	Booklets	I	Magazines	P	Brochures	W	Maps
D	Catalogs	J	Mass-Market Books	Q	Calendars	X	Newspapers
E	Cookbooks	K	Newsletters	R	Comic Books	Y	Postcards
F	Directories	L	Software Manuals	S	Direct Mail Letters	Z	Posters
		M	Textbooks	T	Envelopes	1	Stationery

Services: 4C-CS-FF-LM-AF-TY-TD-WA
Equipment: 7SO(1-26x40,2-24x36,4-41x56)
Annual Sales: 4
Number of Employees: 230
Turnaround Times: B3C5S3
Customers: BP20-OT80
Union Shop: Y
Year Established: 1960
Terms: Net 30

Statement: We deliver a good product at a great price!

GC
Capital Printing Company Inc
Carolyn A Zodd, Marketing Director
PO Box 17548
Austin TX 78760-7548
512-442-1415; Fax: 512-441-1448

Min: 5000 Max: 100000 Opt: 30000-50000
Sizes: 123456X
Bindings: HCO-CBO-LBI-PBI-SSI-SDI-
    SBO-WBO-OBO
Items: ABCDEFGHIKLMNOPQRSTUV
    WXZ12
Services: 4C-OC-RP
Equipment: 4SO(28x40)-LP
Annual Sales: 3
Number of Employees: 40
Turnaround Times: B3S2
Customers: BP10-BU20-NP10-MP30-OR30
Union Shop: N
Year Established: 1929
Terms: 1% 30, net 30 with approval

Statement: Our goals are to provide only the highest quality product, give total customer satisfaction, and listen and understand the expectations of our customers. We strive to constantly innovate our product and service and honor our commitments whatever the cost or effort. We take pride in our craftsmanship. We are the company.

GC
Cargill's Printing & Stationery
Marketing Director
2910 McKinney St (77003)
PO Box 1608
Houston TX 77251
713-223-0123

MS
Carlson Color Graphics
Marketing Director
3310 SW 7th St
Ocala FL 32674
904-732-7787; Fax: 904-351-5199

Min: 10 Max: 200000 Opt: 50000
Sizes: 5
Bindings: SSI
Items: DI
Services: 4C-DA-FF-TY
Equipment: WO

GC
Carol Printing Corp
Marketing Director
122 W 27th St
New York NY 10001
212-243-5858; Fax: 212-633-8712

GC
Carpenter Lithographic Co
Marketing Director
PO Box 1385
Springfield OH 45501
513-325-1548; Fax: 513-325-4296
Send RFQ's to: Mike Pierce
Parent Company: Central Printing,
    Dayton OH

Min: 2500 Max: 100000 Opt: 50000
Sizes: 123456
Bindings: LBX-PBX-SSX-SDX
Items: BCDEFKLMNO
Services: 4C-FF-TD-WA
Equipment: 5SO(25x38)-2WO(22$^{13}/_{16}$x36)
Annual Sales: 4
Number of Employees: 104
Turnaround Times: B2S1-2
Customers: BP10-BU55-CU10-MP5-OR10-
    OT10
Union Shop: N
Year Established: 1933
Terms: Net 30

Statement: To profitably render a product and service package that will keep our customers coming back.

GC
Carpenter Reserve
Marketing Director
7100 Euclid Ave
Cleveland OH 44103
216-431-0800; Fax: 216-531-5240

SERVICES				BINDINGS			
4C	4-Color Printing	RA	Rachwal System	HC	Hardcover	SD	Side Stitching
CS	Color Separations	AF	Acid-Free Paper	CB	Comb Binding	SB	Spiral Binding
DA	Design / Artwork	RP	Recycled Paper	LB	Loose-Leaf Binding	WB	Wire-O Binding
FF	Fulfillment/Mailing	TY	Typesetting	OB	Otabind	I	In-House
LM	List Maintenance	TD	Typeset w/ Disk	PB	Perfect Binding	O	Out of House
OC	OptiCopy System	WA	Warehousing	SS	Saddle Stitching	X	Unknown

Carqueville TCR Graphics   **GC**
Marketing Director
2200 Estes Ave
Elk Grove Village IL 60007
708-439-8700; Fax: 708-228-3953

Carter Rice   **BK**
Donald Page, Vice President
273 Summer St
Boston MA 02210
617-542-6400; 800-225-6673
Send RFQ's to: Donald Page

Castle Pierce Printing Co   **GC**
Marketing Director
2247 Ryf Rd
Oshkosh WI 54904
414-235-2020; Fax: 414-235-4763

Caxton Printers Ltd   **GC**
Marketing Director
312 Main St
Caldwell ID 83605
208-459-7421; Fax: 208-459-7450

Central Lithograph Co   **GC**
Marketing Director
1278 W 58th St
Cleveland OH 44102
216-281-4884; Fax: 216-281-5109

Central Printing Co   **GC**
Marketing Director
2400 E River Rd
Dayton OH 45439
513-298-4321; Fax: 513-298-6941
800-298-6941

Century Graphics Corp   **GC**
155 Ida St
Omaha NE 68110-2821
402-453-1402; 800-743-7737

Champagne Offset Co Inc   **BK**
Jim Browne, Marketing Director
7 Strathmore Rd
Natick MA 01760
508-651-0400; Fax: 508-651-0402
Send RFQ's to: Dave Ordway

Min: 500 Max: 1000000 Opt: 10000-100000
Sizes: 123456X
Bindings: LBX-PBX-SSX-SDX-SBX-WBX
Items: ACDFGHKLNPQSTUVYZ12
Services: 4C-CS-DA-FF-OC-RA-AF-RP-
   TY-TD-WA
Equipment: 4LP(1-28x40,3-19x25)-
   SO(11x17)
Annual Sales: 3
Number of Employees: 49
Turnaround Times: B3S1-2
Customers: BP10-BU65-CU15-NP10
Union Shop: N
Terms: Net 30

Champion Printing   **GC**
Marketing Director
3250 Spring Grove Ave
Cincinnati OH 45225
513-541-1100; Fax: 513-541-9398
800-543-1957
Send RFQ's to: John C Hassan

Min: 5000 Max: 0 Opt: 150000
Bindings: SSO
Items: CKT
Equipment: 4WO(26)
Annual Sales: 2
Number of Employees: 43
Union Shop: N
Terms: Net 30

Statement: Champion is a specialist in printing direct mail formats, including self-mailers, catalogs, and catalog inserts.

Chernay Printing Inc   **GC**
Marketing Director
7483 S Main St
Coopersburg PA 18036
610-282-3774

PRINTED ITEMS			
A Annual Reports	G Galley Copies	N Workbooks	U Greeting Cards
B Books	H Journals	O Yearbooks	V Labels / Stickers
C Booklets	I Magazines	P Brochures	W Maps
D Catalogs	J Mass-Market Books	Q Calendars	X Newspapers
E Cookbooks	K Newsletters	R Comic Books	Y Postcards
F Directories	L Software Manuals	S Direct Mail Letters	Z Posters
	M Textbooks	T Envelopes	1 Stationery

GC

Cherry Lane Lithography Corp
Marketing Director
1478 Old Country Rd
Plainview NY 11803
718-895-4082

GC

Chicago Press Corp
R C Reeve, Marketing Director
1112 N Homan Ave
Chicago IL 60651
312-276-1500; Fax: 312-276-4595
Send RFQ's to: R C Reeve

Min: 5000 Max: 150000 Opt: 20000-60000
Sizes: 2356
Bindings: SSX
Items: ACDFKLNPS2
Services: 4C-OC-RP-TD-WA
Equipment: 2SO(28x40)
Annual Sales: 3
Number of Employees: 77
Turnaround Times: S2
Customers: BU70-NP30
Union Shop: Y
Year Established: 1927
Terms: Net 30

MS

Citizen Printing
Gregg Davies, Marketing Director
805 Park Ave
PO Box 885
Beaver Dam WI 53916
414-887-0321; Fax: 414-887-0439
Send RFQ's to: Gregg Davies

Min: 20000 Max: 1000000 Opt: 100000-
    300000
Sizes: 25X
Bindings: SSX
Items: DFIX
Services: 4C-CS-LM-OC-RP-TY-TD-WA
Equipment: 42WO(22.75x35)
Annual Sales: 4
Number of Employees: 250
Turnaround Times: S1
Customers: MP85-OT15
Union Shop: N
Terms: Net 30

Statement: State-of-the-art printing facility, providing publishers full electronic prepress services, intergrates "postscript"

imaging with traditional stripping. Complete, including ink jetting, selective binding, tips, blowing, in-line addressing with in-plant postal verification for direct shipment to Chicago BMC.

GC

Clark Printing Co
Marketing Director
3401 Heartland Dr
PO Box 298
Liberty MO 64068
816-792-5300; Fax: 816-792-2031

GC

Clarke Printing Co
Marketing Director
5101 S Zarzamora St
San Antonio TX 78211
210-923-7591; Fax: 210-923-7537

BK

Clarkwood Corporation
Marketing Director
690 Union Blvd
Totowa NJ 07512
201-256-2456; Fax: 201-256-3433
Send RFQ's to: Albert Riecher, President

Min: 2000 Max: 2000000 Opt: 0
Sizes: 5
Bindings: LBI
Items: BL
Services: FF-LM
Equipment: SO-WO

Statement: Clarkwood specializes in the printing and distribution of looseleaf books and subscription services.

BK

Coach House Press
Stan Bevington, President
401 Huron St (Rear)
Toronto ON Canada M5S 2G5
416-979-2217
Send RFQ's to: Stan Bevington

Min: 500 Max: 5000 Opt: 1000
Sizes: 2
Bindings: HCX-PBX
Items: BHKPYZ
Services: 4C-TY
Equipment: LP

SERVICES				BINDINGS			
4C	4-Color Printing	RA	Rachwal System	HC	Hardcover	SD	Side Stitching
CS	Color Separations	AF	Acid-Free Paper	CB	Comb Binding	SB	Spiral Binding
DA	Design / Artwork	RP	Recycled Paper	LB	Loose-Leaf Binding	WB	Wire-O Binding
FF	Fulfillment/Mailing	TY	Typesetting	OB	Otabind	I	In-House
LM	List Maintenance	TD	Typeset w/ Disk	PB	Perfect Binding	O	Out of House
OC	OptiCopy System	WA	Warehousing	SS	Saddle Stitching	X	Unknown

Year Established: 1965

Statement: Coach House is a small Canadian publisher who also does typesetting and printing for other small Canadian publishers. Two telephone numbers available — the other one is 416-979-7374.

Collated Industries **GC**
Marketing Director
716 Pulaski Hwy
Bear DE 19701-1227
302-324-9450; Fax: 302-324-9902

The College Press **GC**
Dale Collins, Assistant Manager
4981 Industrial Dr
PO Box 400
Collegedale TN 37315-0400
615-396-2164; Fax: 615-238-3546
800-277-7377
Send RFQ's to: Al Burdick, Marketing
    Director

Min: 1000 Max: 50000 Opt: 20000
Sizes: 235
Bindings: HCO-CBI-LBO-PBI-SSI-SDI-
    SBO-WBO
Items: ABCDEFIKLMNPT1
Services: 4C-CS-DA-TY-TD
Equipment: 9SO(25,26,38)
Annual Sales: 2
Number of Employees: 30
Turnaround Times: B3C3S3
Customers: BP15-BU20-CU20-MP25-NP20
Union Shop: N
Year Established: 1917
Terms: Net 30

Collins Lithographic Co Inc **GC**
Marketing Director
501 W 23rd St
Baltimore MD 21211
410-889-8686

Colonial Graphics **GC**
John V Turi, President
PO Box 1013
92 Maryland Ave
Paterson NJ 07503-2113
201-345-0600; Fax: 201-345-0083

Send RFQ's to: John V Turi

Min: 1500 Max: 50000 Opt: 25000
Sizes: 25
Bindings: CBX-SSX-SDX-SBX-WBX
Items: ABCDEFHKLPQSWZ
Services: FF-OC-TY-WA
Equipment: 3SO(to 28x40)-WO(22x35)
Annual Sales: 2
Number of Employees: 16
Turnaround Times: B3C3
Union Shop: N
Year Established: 1984
Terms: Per client

Statement: We have provided in-house prep to fulfillment for 33 years. We have furnished our plant with perfecting presses, both sheet and web offset, allowing more product for your dollar.

Color Graphic Press Inc **GC**
Marketing Director
1120 46th Rd
Long Island City NY 11101
718-392-2727

Color World Printers **GC**
Leslie Schultz, Sales Rep
201 E Mendenhall (59715)
PO Box 1088
Bozeman MT 59771-1088
406-587-4508; 800-332-3303

Min: 100 Max: 300000 Opt: 20000
Sizes: 123456X
Bindings: HCO-CBI-LBI-PBI-SSI
Items: ABCDEFHIKLMNPQSUYZ
Services: 4C-DA-FF-AF-TY-TD
Equipment: 4SO(19x25 to 23x29)
Annual Sales: 2
Number of Employees: 35
Turnaround Times: B4C5S4
Customers: BP-BU-CU-MP-SP
Union Shop: N
Year Established: 1967
Terms: New customers: 1/3, 1/3, 1/3

Statement: Your project gets the personal attention of our rep from your first phone call requesting a quote. We want to answer your questions, get your business, and deliver a product the way you want it. We can produce your book from manuscript,

PRINTED ITEMS							
A	Annual Reports	G	Galley Copies	N	Workbooks	U	Greeting Cards
B	Books	H	Journals	O	Yearbooks	V	Labels / Stickers
C	Booklets	I	Magazines	P	Brochures	W	Maps
D	Catalogs	J	Mass-Market Books	Q	Calendars	X	Newspapers
E	Cookbooks	K	Newsletters	R	Comic Books	Y	Postcards
F	Directories	L	Software Manuals	S	Direct Mail Letters	Z	Posters
		M	Textbooks	T	Envelopes	1	Stationery

disc, or camera-ready copy. We are a family-owned business and are committed to quality. Our Montana location keeps our prices very competitive.

ColorCorp Inc
BK

Marketing Director
5575 S Sycamore St #211
Littleton CO 80120
303-795-3020; Fax: 303-795-7203
Send RFQ's to: Martin Pugh

Min: 1000 Max: 100000 Opt: 5000-10000
Sizes: 123456X
Bindings: HCI-PBI-SSI-SBO-WBO
Items: BCDEFQ
Services: 4C-CS-AF
Equipment: 1SSO(28x40)
Annual Sales: 3
Number of Employees: 300
Turnaround Times: B6C6S6
Customers: BP70-CU20-SP10
Union Shop: N
Year Established: 1967
Terms: Net 30

Statement: We focus on shorter run, full color books and calendars for publishing firms throughout the United States.

Colorgraphic Web Offset Printing
GC

Marketing Director
4367 Walden Ave
Lancaster NY 14086
716-684-4300; Fax: 716-684-3659

ColorGraphics Printing Corp
GC

Marketing Director
4129 S 72nd E Ave
Tulsa OK 74145
918-622-3730
Send RFQ's to: Bob Diehl

Min: 0 Max: 0 Opt: 0
Bindings: PBI-SSI
Items: ACDPSZ
Services: 4C-DA-FF-TY
Equipment: 3SO-2WO

Colorlith Corporation
GC

R Celleme, Marketing Director
777 Hartford Ave
Johnston RI 02919
401-521-6000; Fax: 401-751-9436
800-556-7171
Send RFQ's to: J Moura

Min: 500 Max: 0 Opt: 1000000
Sizes: 123456
Bindings: SSX
Items: ABCDEPQSUVW1234
Services: 4C-CS-AF-RP-TY-TD-WA
Equipment: 5SO(19x25,25x38)-WO(17.75x26)
Annual Sales: 2
Number of Employees: 40
Turnaround Times: S3
Customers: BU95-OT5
Union Shop: N
Year Established: 1958
Terms: 1% 10, net 30

Statement: We are full service, short and long run — photography to bindery.

Colortone Press
BK

Marketing Director
1017 Brightseat Rd
Landover MD 20785-3738
301-350-0100; Fax: 301-350-9769
Send RFQ's to: Lloyd Greene, Vice President

Min: 0 Max: 0 Opt: 0
Sizes: 235X
Bindings: HCX-CBX-PBX-SSX-SBX
Items: ABCDEINPQZ
Services: 4C-DA-TY-TD

Statement: Colortone is the printing arm of Acropolis Press. They specialize in printing books requiring high quality color work.

Colotone Riverside Inc
GC

David Besgrove, Marketing Director
4901 Woodall St
Dallas TX 75247
214-631-1150; Fax: 214-951-8035
Send RFQ's to: David Besgrove
Parent Company: Colotone Group

SERVICES		BINDINGS	
4C 4-Color Printing	RA Rachwal System	HC Hardcover	SD Side Stitching
CS Color Separations	AF Acid-Free Paper	CB Comb Binding	SB Spiral Binding
DA Design / Artwork	RP Recycled Paper	LB Loose-Leaf Binding	WB Wire-O Binding
FF Fulfillment/Mailing	TY Typesetting	OB Otabind	I In-House
LM List Maintenance	TD Typeset w/ Disk	PB Perfect Binding	O Out of House
OC OptiCopy System	WA Warehousing	SS Saddle Stitching	X Unknown

Min: 1000 Max: 2000000 Opt: 175000
Sizes: 123456
Bindings: HCO-CBO-LBO-OBO-PBO-
    SDI-SSI-SBO-WBO
Items: ABCDEFIKLPQSTVWYZ124
Services: 4C-CS-FF-OC-RP-TY-TD-WA
Equipment: 3LP(20)-9SO(40 to 60)
Annual Sales: 5
Number of Employees: 225
Turnaround Times: B3C4S2
Customers: BU95-OR5
Union Shop: N
Terms: Net 30

Columbus Bookbinders & Printers   BK
James H Goodlett
1326 Tenth Ave
PO Box 8193
Columbus GA 31908
706-323-9313; Fax: 800-553-2987
800-553-7314
Send RFQ's to: Tom Goodlett

Min: 100 Max: 2000 Opt: 5000
Sizes: 123456
Bindings: CBI-PBI-SSI-SDI-WBI-SBO-
    HCO
Items: BCEFHKLMNPQS
Services: 4C-CS-DA-FF-AF-TY-TD-WA
Equipment: 12SO(11x17,13.75x18,18x15,
    17.5x22.5,19x25)
Annual Sales: 2
Number of Employees: 14
Turnaround Times: B3C6S2
Customers: BP33-BU33-SP33-OT1
Union Shop: N
Year Established: 1983
Terms: 1/3 down, 1/3 proof, 1/3 delivery

Statement: We do books from 100 to
20,000; also, newletters and mailing services. We offer typesetting and typesetting
from disks. Editing, cover design and illustrations are offered.

Combined Communication Services   MS
Marketing Director
1501 Washington St
Mendota IL 61342
815-539-7402

Send RFQ's to: Del Miller, VP Sales
Min: 5000 Max: 500000 Opt: 100000
Sizes: 2345
Bindings: PBI-SSI
Items: DHI
Services: 4C-CS-FF-TY
Equipment: 3SO-2WO
Annual Sales: 6
Number of Employees: 750

Statement: CCS specializes in producing
special interest magazines.

Comfort Printing   GC
Marketing Director
1611 Locust St
St Louis MO 63103
314-241-6991; Fax: 314-241-6994
Send RFQ's to: Mark Gabauer
Parent Company: Comfort Companies Inc

Min: 1000 Max: 1000000 Opt: 0
Sizes: 123456X
Bindings: HCO-CBO-LBO-OBO-PBI-SSI-
    SDI-SBO-WBO
Items: ABCDEFGHIJKLMNOPTVYZ123
Services: 4C-FF-TY-TD-WA
Equipment: LP-SO-WO
Annual Sales: 3
Number of Employees: 48
Turnaround Times: B3C4S2
Customers: BP3-BU25-CU25-NP15-OR6-
    SC3-SP3
Union Shop: Y
Year Established: 1907
Terms: Net 30

Statement: To provide the highest quality
services and products each and every day.
Our vision is to be a moving force in the
St. Louis community. We are committed to
leading the industry by our integrity, innovation and client satisfaction. Our
strength is drawn from an uncompromising
committment to excellence. Our team is
comprised of competent professionals who
deliver a wide variety of quality services.
Our approach is honest, courteous, and
responsive as we continue to strive for excellence.

PRINTED ITEMS	G	Galley Copies	N	Workbooks	U	Greeting Cards
A	Annual Reports	H	Journals	O	Yearbooks	V	Labels / Stickers
B	Books	I	Magazines	P	Brochures	W	Maps
C	Booklets	J	Mass-Market Books	Q	Calendars	X	Newspapers
D	Catalogs	K	Newsletters	R	Comic Books	Y	Postcards
E	Cookbooks	L	Software Manuals	S	Direct Mail Letters	Z	Posters
F	Directories	M	Textbooks	T	Envelopes	1	Stationery

Commercial Lithographing Co **GC**
Marketing Director
1226 Chestnut Ave
Kansas City MO 64127
816-241-2218; Fax: 816-241-6091

Commercial Printing Co **GC**
Marketing Director
222 6th Ave SW (35211)
PO Box 10302
Birmingham AL 35202
205-251-9203; Fax: 205-251-6133

Commercial Documentation Services **BK**
Stephen Rossi, Marketing Director
2661 South Pacific Hwy
Medford OR 97501
503-773-7575; Fax: 503-773-1832
800-388-7575
Send RFQ's to: Stephen Rossi
Parent Company: Commercial Printing
     Company

Min: 5000 Max: 0 Opt: 0Sizes: 2345X
Bindings: PBX-SSX-SBX-WBX
Items: L
Services: 4C-CS-FF-OC
Equipment: 5SO-3WO(23x38,19x32)
Annual Sales: 5
Number of Employees: 250
Turnaround Times: B2S2
Year Established: 1978
Terms: Net 30

Statement: In 1978, CDS became a specialized printer of computer hardware and software documentation. Fully self-contained services include full desktop prepress facility, 3 web presses, 5 color, 5 sheetfed, 6 color with in-line coating and complete bindery and packaging facilities. ISO 9002 certified.

Community Press **BK**
Chuck Revill, Marketing Director
5600 N University Ave
Provo UT 84604
801-225-2299

Send RFQ's to: Chuck Revill
Min: 500 Max: 50000 Opt: 10000
Sizes: 123456X
Bindings: HCI-CBI-PBO-SSI-SBI
Items: ABCDEFHIKMNOPQZ
Services: 4C-DA-TY-WA
Equipment: SO
Terms: Net 30 with approved credit

Complete Book Manufacturing **GC**
   Service
Marketing Director
638 Jefferson St
PO Box 159
Ashland OH 44805
419-289-6052; 800-537-6727

Concord Litho Company Inc **GC**
William P Nourse, Marketing Director
92 Old Turnpike Rd Box 2888
Concord NH 03302-2888
603-225-3328; Fax: 603-225-6120
Send RFQ's to: William P Nourse

Max: 0 Max: 0 Opt: 0
Bindings: HCO-CBO-LBO-OBO-PBO-
     SDO-SBO-WBO-SSO
Items: CDKPQSUZ
Services: 4C-CS-DA-FF-TY-TD-WA
Equipment: 8SO(28x40,54x77)-5WO
     (17x26)
Annual Sales: 5
Number of Employees: 270
Turnaround Times: S2
Customers: BP2-BU45-NP20-SC1-OT32
Union Shop: N
Terms: Net 30

Statement: They specialize in direct response formats, including ruboffs, die-cuts, inserts, and brochures.

Coneco Litho Graphics **BK**
Dennis Brower, Marketing Director
58 Dix Ave
PO Box 3255
Glen Falls NY 12801-3255
518-793-3823; Fax: 518-793-5823
Send RFQ's to: Dennis Brower

SERVICES			
4C	4-Color Printing	RA	Rachwal System
CS	Color Separations	AF	Acid-Free Paper
DA	Design / Artwork	RP	Recycled Paper
FF	Fulfillment/Mailing	TY	Typesetting
LM	List Maintenance	TD	Typeset w/ Disk
OC	OptiCopy System	WA	Warehousing

BINDINGS			
HC	Hardcover	SD	Side Stitching
CB	Comb Binding	SB	Spiral Binding
LB	Loose-Leaf Binding	WB	Wire-O Binding
OB	Otabind	I	In-House
PB	Perfect Binding	O	Out of House
SS	Saddle Stitching	X	Unknown

Min: 1 Max: 15000 Opt: 500-2500
Sizes: 123456X
Bindings: HCX-LBX-OBX-WBX-CBX-
PBX-SSX-SDX-SBX
Items: ABCDEFGHKLMNOPQYZ2
Services: 4C-CS-DA-AF-RP-TY-TD-WA
Equipment: 3SO(19x26)-OP(19x26)
Annual Sales: 2
Number of Employees: 40
Turnaround Times: B3C4S3
Customers: BU30-CU10-NP5-SP30-OT25
Union Shop: N
Year Established: 1984
Terms: 1/3,1/3,1/3 or net 15 days w/ap-
    proval

Statement: To provide top quality products
at competitive prices when our clients need
them.

Consolidated Press Printing Co **CA**
Marketing Director
600 S Spokane St
Seattle WA 98134
206-441-1844; Fax: 206-447-9477

Min: 1000 Max: 150000 Opt: 50000
Sizes: 25
Bindings: PBI-SSI
Items: ABCDFKNPS
Equipment: SO-WO
Year Established: 1948
Terms: Payment in full with order

Statement: Write for their standard price
list for self-cover catalogs and booklets on
a variety of paper stocks.

Consolidated Printers Inc **GC**
Gene Leong
2630 Eighth St
Berkeley CA 94710
510-843-8524; Fax: 510-486-0580
Send RFQ's to: Gene Leong

Min: 1000 Max: 500000 Opt: 150000
Sizes: 12346X
Bindings: HCO-PBI-SSI-WBO
Items: BCDEFLM
Services: OC-RA-AF
Equipment: SO(28x40)-3WO(19⅜x32,
    22.75x36)
Annual Sales: 4

Number of Employees: 70
Turnaround Times: B3C4S3
Customers: BP40-BU40-CU5-SP10-OT5
Union Shop: Y
Year Established: 1969
Terms: Net 30

Statement: Consolidated Printers specializes in printing softcover books for publishers and businesses in Northern California. We provide high quality service and
convenience to our customers.

Continental Web **GC**
Marketing Director
1430 Industrial Dr
Itasca IL 60143
708-773-1903; Fax: 708-773-1909
Send RFQ's to: Kenneth W Field, Presi-
    dent

Min: 25000 Max: 2000000 Opt: 0
Sizes: 245
Bindings: SSI
Items: ADIKPSUWXYZ
Services: 4C-CS-AF-WA
Equipment: 9WO(23x38)
Annual Sales: 6
Number of Employees: 450
Turnaround Times: C3S2
Customers: MP25
Union Shop: N
Terms: Net 30

Cookbook Publishers Inc **BK**
Marketing Director
10800 Lakeview
PO Box 15920
Lenexa KS 66285-5920
913-492-5900; Fax: 913-492-5947
800-227-7282
Send RFQ's to: Victoria Visnosky

Min: 200 Max: 50000 Opt: 500-1000
Sizes: 2356
Bindings: CBI-LBI-SSI-WBI-HCO-PBO
Items: BCEKLN
Services: RP-TY
Equipment: 5SO(11x17)-2WO(9x12)
Annual Sales: 3
Number of Employees: 100
Turnaround Times: B5C6S2
Customers: BP10-BU5-NP80-SP5

PRINTED ITEMS						
A	Annual Reports	G	Galley Copies	N	Workbooks	U	Greeting Cards
B	Books	H	Journals	O	Yearbooks	V	Labels / Stickers
C	Booklets	I	Magazines	P	Brochures	W	Maps
D	Catalogs	J	Mass-Market Books	Q	Calendars	X	Newspapers
E	Cookbooks	K	Newsletters	R	Comic Books	Y	Postcards
F	Directories	L	Software Manuals	S	Direct Mail Letters	Z	Posters
		M	Textbooks	T	Envelopes	1	Stationery

Union Shop: N
Year Established: 1947
Terms: Cookbook plan, no down up to 90
   days for qualified organizations

Statement: We are the premier short-run
cookbook specialists, employee-owned and
operated. We are dedicated to assisting
non-profit organizations with successful
fundraising products. We also specialize in
short-run plastic comb-bound or stitched
booklets and have lamination capabilities
in-house.

BK
Cookbooks by Mom's Press
Tamara Omtvedt, Marketing Director
PO Box 2110
Kearney NE 68848
800-445-6621; Fax: 308-234-3969
Send RFQ's to: Doug Ashley

Min: 200 Max: 50000 Opt: 1000
Sizes: 25
Bindings: HCX-LBX-CBX-PBX-SSX-
   SBX-WBX
Items: ABCDEFHJLMN
Services: 4C-CS-DA-AF-RP-TY
Equipment: SO-WO
Annual Sales: 3
Number of Employees: 75
Turnaround Times: B5C6S3
Customers: NP90-OR10
Union Shop: N
Year Established: 1979
Terms: 50% net 37 days, balance 90 days

Statement: Our specialty is printing personalized cookbooks for organizations to
use for fundraising. By request, we will
send FREE cookbook planning information.

GC
Cooley Printers & Office Supply
Rhonda Counselman, Marketing Director
403 Hudson Ln
Monroe LA 71201
318-325-7541; Fax: 318-387-0034
Send RFQ's to: Rhonda Counselman

Min: 1 Max: 250000 Opt: 0
Sizes: 123456
Bindings: CBX-LBX-SSX-SDX-SBX
Items: ABCDEKPSTVY12

Services: 4C-DA-LM-RP-TY-TD-WA
Equipment: LP-SO
Annual Sales: 2
Number of Employees: 20-25
Turnaround Times: B5C5S2
Customers: BU75-CU10-NP5-SC5-OT5
Union Shop: N
Year Established: 1954
Terms: Net upon invoice, 3% discount on
   prepay

Statement: Woman-owned business.

GC
Coral Graphic Services
Marketing Director
11 Commercial St
Plainview NY 11803
516-935-5900; Fax: 516-935-5902

Min: 0 Max: 0 Opt: 0
Services: 4C-CS

GC
Corley Printing Company
David C Deibel, Marketing Director
3777 Rider Trail South
Earth City MO 63045
314-739-3777; Fax: 314-739-1436
Send RFQ's to: David C Deibel

Min: 300 Max: 250000 Opt: 300-50000
Sizes: 12345
Bindings: PBI-SSI-SDI-LBI-HCO-CBO-
   SBO-WBO
Items: BCDEFKLNW
Services: 4C-RA-AF-RP-TY
Equipment: 2SO(23x35,19x25)-
   2WO(22x36)
Annual Sales: 4
Number of Employees: 64
Turnaround Times: B3C5S3
Customers: BU80-OR10-SP10
Union Shop: N
Year Established: 1929
Terms: 1% 10; net 30 (best terms)

Statement: Our corporate motto, "Better
People — Better Products" says it all.

SERVICES				BINDINGS			
4C	4-Color Printing	RA	Rachwal System	HC	Hardcover	SD	Side Stitching
CS	Color Separations	AF	Acid-Free Paper	CB	Comb Binding	SB	Spiral Binding
DA	Design / Artwork	RP	Recycled Paper	LB	Loose-Leaf Binding	WB	Wire-O Binding
FF	Fulfillment/Mailing	TY	Typesetting	OB	Otabind	I	In-House
LM	List Maintenance	TD	Typeset w/ Disk	PB	Perfect Binding	O	Out of House
OC	OptiCopy System	WA	Warehousing	SS	Saddle Stitching	X	Unknown

Country Press   GC

Carol Ann Vercz, Owner
Route #1, Box 71
Ward Rd
Mohawk NY 13407
315-866-7445
Send RFQ's to: Carol Ann Vercz

Min: 100 Max: 2000000 Opt: 0
Sizes: 123456X
Bindings: HCO-CBI-LBI-PBI-SSI-SDI-
   SBI-WBO
Items: ABCDEFGHIJKLMNPQSTUYZ1
Services: 4C-DA-OC-RP-TY-TD
Equipment: SO(11x17)
Annual Sales: 1
Number of Employees: 6
Turnaround Times: B1C1S1
Customers: BP10-BU40-CU10-NP5-OR20-
   SP5
Union Shop: N
Year Established: 1985
Terms: Payment on delivery

Country Press Inc   BK

Alfred Wolf
1 Commercial Dr
Lakeville MA 02347-0489
508-947-4485; Fax: 508-947-8989
Send RFQ's to: Doug Massey

Min: 11 Max: 3000 Opt: 500
Sizes: 12345X
Bindings: LBX-PBX-SSX-SDX
Items: BEFGMN2
Services: 4C-CS-AF-RP
Equipment: LP-8SO(20x26)
Annual Sales: 2
Number of Employees: 19
Turnaround Times: B2S2
Customers: BP80-CU20
Union Shop: N
Year Established: 1967
Terms: Net 30 days w/approved credit

Statement: Specialize in bound galley and
short run textbooks with laser printed
color covers and fast turnaround time.

Comments: Mailing address: PO Box 489,
Middleboro MA 02346.

Courier Stoughton Inc   BK

Marketing Director
100 Alpine Cir
Stoughton MA 02072
617-341-1800; Fax: 617-341-3973
800-343-5901
Send RFQ's to: James Bailey, VP Sales

Min: 500 Max: 100000 Opt: 15000
Sizes: 123456X
Bindings: HCI-CBI-PBI-SBI-WBI
Items: BFM
Services: RA
Equipment: 6SO-WO(19x22)
Annual Sales: 4
Number of Employees: 250
Customers: BU-CU-OR-SC
Year Established: 1917

Statement: A division of Courier Graphics,
they specialize in producing books for colleges, businesses, and professionals.

Courier Corporation   BK

Marketing Director
165 Jackson St
Lowell MA 01852
508-458-6351; Fax: 508-441-9699

Min: 5000 Max: 2000000 Opt: 0
Sizes: 5
Bindings: PBI
Items: F
Annual Sales: 6
Number of Employees: 1385
Customers: OT(phone directories)

Statement: Courier Corporation is one of
the nation's largest specialty printers.
Courier's book manufacturing services are
provided by its Courier Companies, Inc.
plants in Massachusetts (2) and Indiana
(1); National Publishing Company of
Philadelphia, PA; and Courier International, Ltd of East Kilbride, Glasgow, Scot-
land.

Courier-Kendallville Corp   BK

Nick Smith, Plant Manager
2500 Marion Dr
PO Box 395
Kendallville IN 46755
219-347-3044; Fax: 219-347-3507

PRINTED ITEMS						
A	Annual Reports	G	Galley Copies	N	Workbooks	U	Greeting Cards
B	Books	H	Journals	O	Yearbooks	V	Labels / Stickers
C	Booklets	I	Magazines	P	Brochures	W	Maps
D	Catalogs	J	Mass-Market Books	Q	Calendars	X	Newspapers
E	Cookbooks	K	Newsletters	R	Comic Books	Y	Postcards
F	Directories	L	Software Manuals	S	Direct Mail Letters	Z	Posters
		M	Textbooks	T	Envelopes	1	Stationery

Send RFQ's to: Nick Smith

Min: 1000 Max: 200000 Opt: 20000
Sizes: 2345
Bindings: HCO-PBI-SSI
Items: BEFLMN
Services: AF
Equipment: 6SO(60,77)-llWO(19,38)
Annual Sales: 6
Number of Employees: 400
Turnaround Times: B5
Customers: BP90-BU5-CU5
Union Shop: N
Year Established: 1898
Terms: Net 30

Statement: One of the Courier Companies, Courier Kendallville (formerly The Murray Printing Company) produces quality 1, 2, and 4 color paperback books, workbooks and catalogs within tight manufacturing schedules.

Courier-Westford Inc **BK**

Marketing Director
Pleasant St
Westford MA 01886
508-692-6321; Fax: 508-692-7292

Min: 1000 Max: 100000 Opt: 15000
Sizes: 2345
Bindings: HCI-PBI-SSI
Items: BEFLMN
Services: AF-OC
Equipment: 6SO(60,77)-llWO(19,38)
Turnaround Times: B5
Customers: BP90-BU5-CU5
Union Shop: N
Terms: Net 30

Statement: The largest and most versatile of the Courier Companies' three book manufacturing plants, Courier Westford used to be known as The Murray Printing Company. It's on-site features include conventional and desktop/electronic prepress, paperback and hardbound book production facilities, a short-run printing center, and complete kitting and fulfillment services.

Craftsman Printing Co **GC**

Marketing Director
2700 Westinghouse Blvd (28273)
PO Box 7000
Charlotte NC 28241
704-588-2120; Fax: 704-588-5966

Craftmaster Printers Inc **GC**

Marketing Director
902 Geneva St
Opelika AL 36801
205-749-5611; Fax: 205-745-4653

Craftmasters **GC**

Roger T Craft
309 Lynual St
Sikeston MO 63801
314-472-1872; Fax: 314-471-7202
Send RFQ's to: Rogert T Craft

Min: 250 Max: 12000 Opt: 5000
Sizes: 5X
Bindings: CBX-PBX-SSX-SDX
Items: BCKMNS
Services: DA-WA
Equipment: 6SO(17x22 to 35x45)
Annual Sales: 2
Turnaround Times: B5S4
Customers: BP80-BU5-SP10-OT5
Union Shop: N
Year Established: 1979
Terms: Net 30

Statement: OUR MOTTO: A good business reputation does not spring up overnight, but grows slowly and soundly because it is rooted in the solid ground of customer satisfaction.

The Craftsman Press Inc **CA**

Marketing Director
1155 Valley St
Seattle WA 98109
206-682-8800
Send RFQ's to: George Prue

Min: 5000 Max: 2000000 Opt: 0
Sizes: 25X
Bindings: PBI-SSI-SDI

SERVICES				BINDINGS			
4C	4-Color Printing	RA	Rachwal System	HC	Hardcover	SD	Side Stitching
CS	Color Separations	AF	Acid-Free Paper	CB	Comb Binding	SB	Spiral Binding
DA	Design / Artwork	RP	Recycled Paper	LB	Loose-Leaf Binding	WB	Wire-O Binding
FF	Fulfillment/Mailing	TY	Typesetting	OB	Otabind	I	In-House
LM	List Maintenance	TD	Typeset w/ Disk	PB	Perfect Binding	O	Out of House
OC	OptiCopy System	WA	Warehousing	SS	Saddle Stitching	X	Unknown

Items: DI
Services: 4C-CS-FF-TY
Equipment: SO-5WO(to 22x38)
Annual Sales: 4
Customers: BU60-MP40

Statement: They print about 30 different magazines plus many catalogs.

Crane Duplicating Service

Marketing Director
1611 Main St
West Barnstable MA 02668
508-362-3441; Fax: 508-362-5445
Send RFQ's to: Richard Price, Owner

Min: 10 Max: 5000 Opt: 500
Sizes: 12345
Bindings: CBI-PBI-SSI-SBI
Items: ABCDEFGHJKLMNOP
Services: 4C-CS-DA-AF-RP-TY-TD
Equipment: 7SO(to 13x18)
Annual Sales: 2
Number of Employees: 25
Turnaround Times: B1
Customers: BP60-CU10-SP30
Union Shop: N
Year Established: 1955
Terms: 50% deposit, 2% l0, net 30

Statement: Crane provides bound galleys to the publishing industry. Our goal is to provide the best short-run bound galley and book manufacturing services in the country for runs of 1000 or less.

Crusader Printing Co   GC

Joe W Lewis Jr, Vice President
10th & State St
East Saint Louis IL 62201
618-271-2000; Fax: 618-271-2045
Send RFQ's to: Joe W Lewis Jr

Min: 500 Max: 500000 Opt: 0
Sizes: 123456
Bindings: HCO-CBI-LBO-PBO-SSI-SDI-
    SPO-WBO
Items: ABCDEFIJKLMNOPQSTUXZ1
Services: 4C-DA-FF-LM-RP-TY-TD
Equipment: 3LP-12SO(to 23x35)-WO
Number of Employees: 21
Turnaround Times: B2C2S2
Union Shop: N

Year Established: 1941
Terms: Flexible, net 60

Statement: Very simple: Dedication to quality, technical innovation, and client satisfaction. This Black-owned printing company prints books, catalogs, newspapers, and other items. They are especially supportive of African-American publishers.

Lew A Cummings Company Inc   MS

Marketing Director
215 Canal St
PO Box 4087
Manchester NH 03108
603-625-6901; Fax: 603-623-5132

Min: 1000 Max: 35000 Opt: 20000
Sizes: 25
Bindings: PBI-SSI
Items: HI
Services: FF-TY
Equipment: SO
Number of Employees: 77
Year Established: 1914

Curless Printing Co   GC

Marketing Director
202 E Main St
Blanchester OH 45107
513-783-2403; Fax: 513-783-4690

Cushing-Malloy   BK

Thomas F Weber, VP Sales
1350 N Main St
PO Box 8632
Ann Arbor MI 48107
313-663-8554; Fax: 313-663-5731
Send RFQ's to: Thomas F Weber

Min: 150 Max: 8000 Opt: 2500
Sizes: 123456X
Bindings: HCO-CBO-PBI-SSI-SBO-WBO
Items: BCDEFHJM
Services: 4C-CS-AF
Equipment: 7SO(38x50,25x38,23x36,
    23x29)
Annual Sales: 3
Number of Employees: 100
Turnaround Times: B3C6S3

PRINTED ITEMS			
A Annual Reports	G Galley Copies	N Workbooks	U Greeting Cards
B Books	H Journals	O Yearbooks	V Labels / Stickers
C Booklets	I Magazines	P Brochures	W Maps
D Catalogs	J Mass-Market Books	Q Calendars	X Newspapers
E Cookbooks	K Newsletters	R Comic Books	Y Postcards
F Directories	L Software Manuals	S Direct Mail Letters	Z Posters
	M Textbooks	T Envelopes	1 Stationery

CUSHING-MALLOY, INC.

COMPLETE BOOK MANUFACTURING SINCE 1948

For over forty years, we've been satisfying our family of customers with reliable and honest dealings, fine quality and very competitive prices.

For short and medium-run book production, nobody does it better. Give us the opportunity to convince you.

Call us or write today for more information.

P.O. Box 8632
1350 N. MAIN ST.   (313) 663-8554
ANN ARBOR, MI 48107   FAX (313) 663-5731

Customers: BP40-CU40-SP5-OT5-BU5-
    OR5
Union Shop: N
Year Established: 1948
Terms: Net 30

Statement: Cushing-Malloy's philosophy is simple: To deal honestly, consistently, and professionally with our customers and to produce a product with excellent value and quality, on schedule, time after time.

D

BK
Daamen Printing Company
H Michael Shanahan, Marketing Director
Industrial Park
PO Box 97
West Rutland VT 05701
802-438-5472; Fax: 802-438-5477
Send RFQ's to: Debi Saldi

SERVICES				BINDINGS			
4C	4-Color Printing	RA	Rachwal System	HC	Hardcover	SD	Side Stitching
CS	Color Separations	AF	Acid-Free Paper	CB	Comb Binding	SB	Spiral Binding
DA	Design / Artwork	RP	Recycled Paper	LB	Loose-Leaf Binding	WB	Wire-O Binding
FF	Fulfillment/Mailing	TY	Typesetting	OB	Otabind	I	In-House
LM	List Maintenance	TD	Typeset w/ Disk	PB	Perfect Binding	O	Out of House
OC	OptiCopy System	WA	Warehousing	SS	Saddle Stitching	X	Unknown

Min: 50 Max: 10000 Opt: 1500-2500
Sizes: 12345
Bindings: HCO-CBO-LBO-PBI-SSI-SDI-SBO-WBO
Items: ABCDEFGHLMNO
Services: AF
Equipment: 4SO(29,36,38)
Annual Sales: 2
Number of Employees: 42
Turnaround Times: B3S3
Customers: BP80-CU3-SP5-OT12
Union Shop: N
Year Established: 1978
Terms: Net 30

Statement: To produce a quality product at a competitive price when the customer requires the service.

Daily Journal of Commerce   GC
Marketing Director
2014 NW 24th Ave
PO 10127
Portland OR 97210-0127
503-226-1311; Fax: 503-224-7140

Daily Printing Inc   GC
Marketing Director
2220 Fernbrook Ln N
Minneapolis MN 55447
612-559-0915; Fax: 612-557-8320

Danbury Printing & Litho Inc   CA
William Sherman
1 Prindle Ln
PO Box 2479
Danbury CT 06811
203-792-5500; Fax: 203-744-5633
800-231-8712
Send RFQ's to: Linda Pekrul

Min: 5000 Max: 2000000 Opt: 0
Sizes: 235
Bindings: HCO-PBO-SSI
Items: ACDKPSTZ
Services: 4C-CS-FF-AF-RP-TY-TD
Equipment: 4SO-7WO
Annual Sales: 5
Number of Employees: 300
Turnaround Times: B3C3S2
Union Shop: N
Terms: Net 30

Daniels Printing Co Inc   GC
Marketing Director
40 Commercial St
Everett MA 02149
617-389-7900; Fax: 617-389-5520

Danner Press Corporation   CA
Paul Barrett, Marketing Director
1411 Navarre Rd SW
Canton OH 44706
216-454-5692; Fax: 216-454-4727
Send RFQ's to: Paul Barrett
Parent Company: DB Hess Companies
Min: 50000 Max: 2000000 Opt: 500000
Sizes: 5X
Bindings: PBI-SSI-SBI
Items: DEHIMNOQR
Services: 4C-CS-OC-RA-RP-TD
Equipment: SO(25x38)-5WO(22.75,45.5)
Annual Sales: 5
Number of Employees: 325
Turnaround Times: B3S3
Customers: MP20-SC20-OT60
Union Shop: Y
Year Established: 1945
Terms: 1% 10, net 30

Dartmouth Printing   MS
Stuart V Smith Jr, President
69 Lyme Rd
Hanover NH 03755
603-643-2220; Fax: 603-643-5479

Min: 5000 Max: 2000000 Opt: 0
Sizes: 25X
Bindings: PBI-SSI
Items: DIP
Services: 4C-CS-FF-TY-TD
Equipment: 4SO(25,41)-3WO(23x17,22x34)
Annual Sales: 4
Number of Employees: 250
Year Established: 1843

Statement: Among other magazines, Dartmouth prints **Catalog Age, Folio, Cape Cod Life**, and **U.S. Banker.**

PRINTED ITEMS							
A	Annual Reports	G	Galley Copies	N	Workbooks	U	Greeting Cards
B	Books	H	Journals	O	Yearbooks	V	Labels / Stickers
C	Booklets	I	Magazines	P	Brochures	W	Maps
D	Catalogs	J	Mass-Market Books	Q	Calendars	X	Newspapers
E	Cookbooks	K	Newsletters	R	Comic Books	Y	Postcards
F	Directories	L	Software Manuals	S	Direct Mail Letters	Z	Posters
		M	Textbooks	T	Envelopes	1	Stationery

Data Reproductions Corp **BK**
Marketing Director
1480 North Rochester Hills
Rochester Hills MI 48307
313-426-1229; Fax: 313-426-1239
Send RFQ's to: Nick Janosi

Min: 1000 Max: 20000 Opt: 3000
Sizes: 123456
Bindings: CBI-LBI-PBI-SSI-SDI-SBI-WBI-
    HCO-OBO
Items: BDEFHLMN
Services: 4C-CS-OC-AF-RP-TY
Equipment: 6SO(40)
Annual Sales: 4
Number of Employees: 95
Turnaround Times: B3C4S3
Customers: BP50-BU25-CU10-SP15
Union Shop: N
Year Established: 1967
Terms: Net 30

Statement: Our intimate size, low overhead, committed employees, new equipment, and efficient systems make us a
great partner to print with in the short to
medium-run market! We want to earn your
business with outstanding service and
products. Discover the Data difference!

Dataco A WBC Company **GC**
Marketing Director
1712 Lomas Blvd NE
Albuquerque NM 87106
505-243-2841; Fax: 505-842-9171

Min: 0 Max: 20000 Opt: 5000
Sizes: 123456
Bindings: HCO-CBI-LBI-PBO-SSI-SDI-
    SBO-WBO
Items: ABCDFHIKLPQSTUYZ1
Services: 4C-DA-FF-LM-TY
Equipment: 2SO(19x25)-WO(14x17)
Annual Sales: 3
Number of Employees: 30
Turnaround Times: B3C3
Customers: BP5-BU60-CU5-MP2-NP3-
    OR10-OT15
Union Shop: N
Terms: Net 30

Dataware **GC**
Marketing Director
7570 Renwick Dr
Houston TX 77081
713-432-1023; Fax: 713-432-1381
800-426-4844

De Palma Printing Co **GC**
Marketing Director
1 Teaneck Rd
Ridgefield Park NJ 07660
201-641-6210; Fax: 201-641-8139

Dellas Graphics **GC**
Tom Dellas, Marketing Director
835 Canal St
Syracuse NY 13210
315-474-4641; Fax: 315-474-4650
Send RFQ's to: Tom Dellas

Min: 1000 Max: 1000000 Opt: 25000
Sizes: 123456
Bindings: PBX-SSX-SBX
Items: ABCDEKLPQSTWYZ1
Services: 4C-CS-FF-LM-AF-RP-TD-TY-
    WA
Equipment: 12SO(10x15 to 25x38)
Annual Sales: 3
Number of Employees: 55
Turnaround Times: S2
Customers: BU70-CU10-NP10-OR5-OT5
Union Shop: N
Year Established: 1979
Terms: Net 30

Statement: Printing, mailing, and fulfillment services geared for growth and integrity.

Delmar Printing & Publishing **BK**
Marketing Director
9601 Monroe Rd
PO Box 1013
Charlotte NC 28270
704-847-9801; Fax: 704-945-1218
Send RFQ's to: Commercial Sales Dept
Parent Company: Continental Graphics

Min: 1000 Max: 50000 Opt: 15000

SERVICES				BINDINGS			
4C	4-Color Printing	RA	Rachwal System	HC	Hardcover	SD	Side Stitching
CS	Color Separations	AF	Acid-Free Paper	CB	Comb Binding	SB	Spiral Binding
DA	Design / Artwork	RP	Recycled Paper	LB	Loose-Leaf Binding	WB	Wire-O Binding
FF	Fulfillment/Mailing	TY	Typesetting	OB	Otabind	I	In-House
LM	List Maintenance	TD	Typeset w/ Disk	PB	Perfect Binding	O	Out of House
OC	OptiCopy System	WA	Warehousing	SS	Saddle Stitching	X	Unknown

Sizes: 123456X
Bindings: HCI-CBI-PBI-SSI-SDI-WBI-
LBO-OBO-SBO
Items: ABCDEFIKMNOPQZ
Services: 4C-CS-DA-RP-TY-TD-FF-AF-
WA
Equipment: SO-LP
Annual Sales: 4
Number of Employees: 410
Turnaround Times: B3C4S3
Customers: BP15-SC75-SP5-OT5
Union Shop: N
Year Established: 1951
Terms: Negotiable

Statement: We are a medium-sized company which offers personal attention to our client's needs. We strive to provide a quality product for a fair price with an acceptable turnaround time.

Delprint Inc   GC
Lee Gordon, Marketing Director
2010 S Carboy
Mt Prospect IL 60056
708-364-6000; Fax: 708-364-6012
Send RFQ's to: Lee Gordon

Min: 1000 Max: 0 Opt: 0
Items: V
Equipment: 2LP(10)2SO(40)-OP
Annual Sales: 2
Number of Employees: 30
Customers: BU90-NP10
Union Shop: N
Terms: Net 30

Statement: We print high quality labels and decals. We have 57 products, 8 patents, and 5 printing processes.

Delta Lithograph Company   BK
Jody G Thompson, Marketing Director
28210 North Ave Stanford
Valencia CA 91355-1111
805-257-0584; Fax: 805-257-3867
800-32DELTA (3-3582)
Send RFQ's to: Ara Norwood
Parent Company: Bertelsmann Printing
and Manufacturing Company

Min: 5000 Max: Open Opt: 15000
Sizes: 3X

Bindings: OBI-PBI-SSI-SDO-HCO-CBO-
SBO-WBO-LBO
Items: BDFL
Services: 4C-DA-FF-CS-OC-AF-RP-TD-
WA
Equipment: 2SO(38,40)-3WO(1-30,2-36)
Annual Sales: 5
Number of Employees: 185
Turnaround Times: B3S3
Customers: BP25-BU70-OT5
Union Shop: N
Year Established: 1954
Terms: Net 30 on approved credit

Statement: Delta is dedicated to serving its customers as a media production specialist by providing the best quality printing, manufacturing, and service to all our market segments. We will achieve this by fostering a sense of personal responsibility that empowers each employee to eliminate obstacles that prevent fulfillment of our customers' wants and needs; by promoting teamwork and caring within our organization; by developing customer relationships based upon trust and goodwill; and by using environmentally sound printing processes that are beneficial to society.

Democrat Printing & Litho   MS
Thomas H Whitney, President
114 E 2nd St (72201)
PO Box 191
Little Rock AR 72203
501-374-0271; Fax: 501-374-1610

Min: 5000 Max: 100000 Opt: 0
Sizes: 5
Bindings: SSI
Items: DI
Annual Sales: 4
Number of Employees: 150

Statement: They specialize in short-run to medium-run business and technical magazines as well as catalogs.

Des Plaines Publishing Co   GC
Norm Hirsch, Marketing Director
1000 Executive Way
Des Plaines IL 60018
708-824-1111 Ext 181 or 102;
Fax: 708-824-1112

PRINTED ITEMS					
A	Annual Reports	G	Galley Copies	N	Workbooks	U	Greeting Cards
B	Books	H	Journals	O	Yearbooks	V	Labels / Stickers
C	Booklets	I	Magazines	P	Brochures	W	Maps
D	Catalogs	J	Mass-Market Books	Q	Calendars	X	Newspapers
E	Cookbooks	K	Newsletters	R	Comic Books	Y	Postcards
F	Directories	L	Software Manuals	S	Direct Mail Letters	Z	Posters
		M	Textbooks	T	Envelopes	1	Stationery

Send RFQ's to: Norm Hirsch or
　Jim Forbing

Min: 3000 Max: 3000000 Opt: 0
Sizes: 3456X
Bindings: PBX-SSX-SBX-WBX
Items: ABCDEFHIJKLMNQRXZ
Services: 4C-FF-TY-TD-DA-LM-RA-RP
Equipment: 2WO
Annual Sales: 3
Number of Employees: 130
Turnaround Times: B2S2
Customers: BP7-BU40-CU5-MP20-NP7-
　OR7-SC5-SP1-OT5
Union Shop: Y
Year Established: 1885
Terms: Net due prior to release from our
　dock

Statement: Superior quality and "no un-
pleasant surprises" service with rapid turnaround (we print daily newspapers). Complete typesetting, prepress and mailing in-
house to save time and money.

Desaulniers Printing Company   **GC**
Marketing Director
4905 77th Ave
Milan IL 61264
309-799-7331; Fax: 309-799-3961
Send RFQ's to: Cindy Power

Min: 500 Max: 2000000 Opt: 0
Sizes: 25X
Bindings: CBI-PBI-SSI-SDI-HCO-OBO-
　SBO-WBO
Items: ABCDFHIKNPQSXZ
Services: 4C-CS-FF-LM-TY-WA
Equipment: 4SO(to 28x40)-3WO(to 22x36)
Annual Sales: 4
Number of Employees: 280
Customers: BP-BU-CU-OR
Union Shop: Y
Year Established: 1895
Terms: Net 30 with approved credit

Deven Lithographers   **GC**
Debra Gambella, Marketing Director
15 Huron St
Brooklyn NY 11222
718-383-1700; Fax: 718-349-2615
Send RFQ's to: Debra Gambella

Min: 0 Max: 0 Opt: 0
Sizes: 1245
Bindings: PBO-CBO-SSO-SDO-SBO-
　WBO
Items: ABCDIKNPQRSTWXZ
Services: 4C
Equipment: 3WO(22.75x36,22.75x38,
　23.5x38)
Annual Sales: 4
Number of Employees: 60
Turnaround Times: B2S2
Union Shop: N
Year Established: 1962
Terms: Net 30

Dharma Press   **GC**
Terry Ryder, Marketing Director
1241 21st St
Oakland CA 94607
510-839-3931; Fax: 510-839-0954
Send RFQ's to: Terry Ryder

Min: 500 Max: 60000 Opt: 20000
Sizes: 23456
Bindings: LBX-PBX-SSX-SBX-WBX
Items: ABCDFHIKLNPQSUWYZ
Services: 4C-FF-AF-RP-TY-TD
Equipment: 5SO(29,2-36,40,65)
Annual Sales: 3
Number of Employees: 70
Turnaround Times: B2
Union Shop: N
Year Established: 1970
Terms: Net 30

Statement: We offer fast turnaround, 2 to
4-color, medium to high quality, medium
to long run printing. We have a service in-
tensive, very experienced, stable work
force.

Diamond Graphics Tech Data   **GC**
Patrick Canter, Marketing Director
6324 W Fond du Lac Ave
Milwaukee WI 53218
414-462-2205
Send RFQ's to: Patrick Canter

Min: 1000 Max: 10000 Opt: 3000
Sizes: 123456X
Bindings: HCO-CBI-LBI-PBI-SSI-SBO-
　WBO
Items: ABCDEFIKLMNOPQSTVYZ

SERVICES				BINDINGS			
4C	4-Color Printing	RA	Rachwal System	HC	Hardcover	SD	Side Stitching
CS	Color Separations	AF	Acid-Free Paper	CB	Comb Binding	SB	Spiral Binding
DA	Design / Artwork	RP	Recycled Paper	LB	Loose-Leaf Binding	WB	Wire-O Binding
FF	Fulfillment/Mailing	TY	Typesetting	OB	Otabind	I	In-House
LM	List Maintenance	TD	Typeset w/ Disk	PB	Perfect Binding	O	Out of House
OC	OptiCopy System	WA	Warehousing	SS	Saddle Stitching	X	Unknown

Services: 4C-DA-FF-LM-TY-TD-WA
Equipment: 7SO(19x26)
Annual Sales: 2
Number of Employees: 20
Turnaround Times: B2C2S2
Customers: BP10-BU60-CU5-MP3-NP10-
  SC5-SP7
Union Shop: N
Year Established: 1976
Terms: 50% down, 50% on delivery

Statement: Our motto: Quality, service, and integrity.

Dickinson Press Inc   **BK**
Bob Worcester, Marketing Director
5100 33rd St SE
Grand Rapids MI 49512
616-957-5100; Fax: 616-957-1261
Send RFQ's to: Bob Worcester

Min: 3000 Max: 100000 Opt: 7500
Sizes: 12345X
Bindings: PBI-SSI-HCO-CBO-LBO-OBO-
  SDO-SBO-WBO
Items: BCDEFJMNQ
Services: 4C-RA-AF-TD
Equipment: 2SO(28x40,37x49)-
  3WO(19,17,45)
Annual Sales: 4
Number of Employees: 115
Turnaround Times: B3C4S3
Customers: BP70-NP15-OR15
Union Shop: Y
Year Established: 1884
Terms: 2% 10, net 30

Statement: Dickinson Press Inc is a full service book printer specializing in compact trim sizes, lightweight papers, and educational workbooks. Our facility offers computerized prepress systems direct from customer-supplied disks and complete soft and hardcover book manufacturing capabilities.

Sales Offices

Chicago
Bob Amundson
1128 Amherst Ct
Lake Zurich IL 60047
708-438-1848; Fax: 708-438-1849

East Coast
Larry Ritchie
786 Gravel Pike
Palm PA 18070
215-679-3677; Fax: 215-679-3687

Southeast
Don Latture
49 Ashington Ln
Brentwood TN 37027
615-373-9773; Fax: 615-373-9732

Southwest
Bert Messelink
5225 Silo Ridge
Colorado Springs CO 80917
719-597-7770; Fax: 719-597-7772

The Dingley Press   **CA**
Mary Ann Schwanda, Marketing Director
119 Lisbon St
Lisbon ME 04250-9641
207-353-4151; Fax: 207-353-9886
Send RFQ's to: Mary Ann Schwanda

Min: 150000 Max: 5000000 Opt: 1000000
Sizes: X
Bindings: SSX
Items: D
Services: 4C-OC-RP-WA
Equipment: 3WO(21.5x35,21.5x38,
  19⅜x33)
Annual Sales: 4
Number of Employees: 127
Turnaround Times: S2
Customers: BU100
Union Shop: N
Year Established: 1928
Terms: Net 30

Statement: The Dingley Press is committed to web offset catalog printing with cost saving opportunities for catalogers to print, bind, address and distribute effectively.

Direct Graphics Inc   **GC**
Charles Hackett, Marketing Director
829 Vandemark Rd
Sidney OH 45365
513-498-2194; Fax: 513-492-1100
800-848-4406
Send RFQ's to: Charles Hackett

Min: 25000 Max: 2500000 Opt: 5000000

PRINTED ITEMS					
A	Annual Reports	G	Galley Copies	N	Workbooks
B	Books	H	Journals	O	Yearbooks
C	Booklets	I	Magazines	P	Brochures
D	Catalogs	J	Mass-Market Books	Q	Calendars
E	Cookbooks	K	Newsletters	R	Comic Books
F	Directories	L	Software Manuals	S	Direct Mail Letters
		M	Textbooks	T	Envelopes

U	Greeting Cards
V	Labels / Stickers
W	Maps
X	Newspapers
Y	Postcards
Z	Posters
1	Stationery

Sizes: 24
Bindings: SSX
Items: CDKPSTY
Services: 4C-DA
Equipment: 3WO(17.75x2)
Annual Sales: 4
Number of Employees: 280
Turnaround Times: S3
Customers: BP20-MP60-OT10
Union Shop: N
Terms: 1% 10 days, net 30

Statement: Direct Graphics is a full service printer and mailer providing direct mail processing to the software industry. We combine state of the art technology in our web presses with a high level of craftmanship. Our complete lettershop provides a central mailing location in addition to USPS on-site postal verification. Direct Graphics' ability to receive media through our electronic publishing department, print amd mail a complete package, catalog or selfmailer, has attracted software publishers from throughout the United States.

Direct Press/Modern Litho   CA
Bob Dodson, Marketing Director
386 Oakwood Rd
PO Box 8104
Huntington Station NY 11746
516-271-7000; Fax: 516-271-7008
Send RFQ's to: Bob Dodson

Min: 2500 Max: 0 Opt: 0
Sizes: 123456X
Bindings: HCO-CBO-OBO-LBO-PBO-
    SSI-SDO-SBO-WBO
Items: DPYZ
Services: 4C-CS-RP-TY-TD
Equipment: 3SO(28x40)-WO(17.75x23)
Annual Sales: 4
Number of Employees: 250
Turnaround Times: B6S6
Customers: BU98-CU1-NP1
Union Shop: Y
Year Established: 1917
Terms: 1/2 deposit, balance COD

Statement: A full service printer of color catalog, brochures, direct mail pieces, postcards. Offer a complete package from photography to printed sheets. Sales offices and photo studios throughout the country.

Directories America   MG
Marketing Director
2537 W Montrose
Chicago IL 60618
312-539-6486

Min: 10000 Max: 3000000 Opt: 0
Sizes: 23456X
Bindings: PBI-SSI
Items: BDFHI
Services: 4C-CS-FF-TY-TD
Equipment: SO-WO
Annual Sales: 3
Number of Employees: 30

Statement: We offer complete processing of directories, catalogs, business magazines, and newsstand publications.

Dispatch Printing Inc   GC
Marketing Director
917 Bacon St
Erie PA 16511
814-452-6724; Fax: 814-454-1404

Dittler Brothers   GC
Marketing Director
1375 Seaboard Industrial Blvd
Atlanta GA 30318
404-355-3423; Fax: 404-355-2569
800-927-0777
Send RFQ's to: Stanley Coker

Min: 200000 Max: 5000000 Opt: 0
Sizes: 25X
Bindings: PBI-SSI-SDI
Items: PST
Equipment: SO-WO-OT(Gravure)
Annual Sales: 6
Number of Employees: 1000
Year Established: 1903

Statement: Dittler is a division of Southam Printing of Toronto, Ontario. They print many direct mail formats, including personalized pieces, pop-ups, scratch-offs, holograms, and database printing.

SERVICES				BINDINGS			
4C	4-Color Printing	RA	Rachwal System	HC	Hardcover	SD	Side Stitching
CS	Color Separations	AF	Acid-Free Paper	CB	Comb Binding	SB	Spiral Binding
DA	Design / Artwork	RP	Recycled Paper	LB	Loose-Leaf Binding	WB	Wire-O Binding
FF	Fulfillment/Mailing	TY	Typesetting	OB	Otabind	I	In-House
LM	List Maintenance	TD	Typeset w/ Disk	PB	Perfect Binding	O	Out of House
OC	OptiCopy System	WA	Warehousing	SS	Saddle Stitching	X	Unknown

Diversified Printing   CA

Frank Kubat Jr, General Manager
2632 Saturn St
Brea CA 92621
714-993-4541
Send RFQ's to: Frank Kubat Jr

Min: 2000 Max: 500000 Opt: 50000
Sizes: 235
Bindings: CBI-PBI-SSI-SDI-SBI-WBI
Items: ABCDFLMNR
Services: FF
Equipment: SO-3WO
Turnaround Times: B3C4
Customers: BP50-BU25-CU15-SC10
Union Shop: N
Year Established: 1965
Terms: 2% 10, net 11

Statement: A division of Quebecor Printing, Diversified prints books, manuals, directories, and catalogs.

Dollco Printing   GC

John Westbrook, Marketing Director
2340 St Laurent Blvd
Ottawa ON Canada K1G 3M5
613-738-9181; Fax: 613-738-4655
Send RFQ's to: John Westbrook

Min: 500 Max: 8000000 Opt: 60000
Sizes: 123456
Bindings: HCO-CBO-LBO-OBO-PBI-SSI-
   SDI-SBO-WBO
Items: ABCDEFHIKLMNOPQRSTUVW
   XYZ1234
Services: 4C-CS-DA-OC-RP-AF-TY-TD-
   WA
Equipment: 3LP(10x15,22x28)-10SO(10x15,
   28x40)-2WO(17.75x26,22.75x36)
Annual Sales: 4
Turnaround Times: B1S1
Customers: BP1-BU28-CU3-MP32-NP17-
   OR19
Union Shop: N
Year Established: 1918

Donihe Graphics Inc   CA

Terri English, Marketing Director
PO Box 1788
Kingsport TN 37662
615-246-2800; Fax: 615-246-7297

Send RFQ's to: Terri English
Min: 0 Max: 0 Opt: 0
Sizes: 2356X
Bindings: SSI-SDI
Items: CDKPQR3
Services: 4C-CS-DA-FF-TY-WA-RP-TD
Equipment: WO(17.75x26.5)
Annual Sales: 3
Number of Employees: 75
Turnaround Times: S3-4
Customers: BU50-CU5-OT45
Union Shop: N
Year Established: 1977
Terms: 50% down, balance on delivery

Statement: To provide the best quality printing that money can buy.

R R Donnelley & Sons Company   BK

Marketing Director
1009 Sloan St
Crawfordsville IN 47933-2741
317-364-3693; Fax: 317-364-2559
800-428-0832
Send RFQ's to: Chuck Harpel

Min: 3000 Max: 0 Opt: 10000
Sizes: 123456
Bindings: HCX-CBX-LBX-PBX-SSX-
   SDX-SBX-WBX
Items: ABCDEFGHIJLMNOWX
Services: 4C-CS-DA-FF-LM-OC-RA-AF-
   TY-TD-WA
Equipment: SO-WO
Annual Sales: 6
Number of Employees: 35000
Turnaround Times: B3C4S3
Union Shop: N
Year Established: 1870

Statement: We are committed to ensuring the continuing growth and success of our customer's and our company by providing superior quality products and services.

R R Donnelley & Sons Inc, Harrisonburg   BK

Oliver Philon, Manager
1400 Kratzer Rd
Harrisonburg VA 22801
703-434-8833; Fax: ext. 256
800-428-0832

PRINTED ITEMS					
A	Annual Reports	G	Galley Copies	N	Workbooks
B	Books	H	Journals	O	Yearbooks
C	Booklets	I	Magazines	P	Brochures
D	Catalogs	J	Mass-Market Books	Q	Calendars
E	Cookbooks	K	Newsletters	R	Comic Books
F	Directories	L	Software Manuals	S	Direct Mail Letters
		M	Textbooks	T	Envelopes

U	Greeting Cards		
V	Labels / Stickers		
W	Maps		
X	Newspapers		
Y	Postcards		
Z	Posters		
1	Stationery		

Min: 1000 Max: 2000000 Opt: 25000
Sizes: 12345X
Bindings: HCI-PBI
Items: BCEFLMN
Services: 4C-CS-DA-FF-LM
Equipment: CB-SO-WO
Annual Sales: 6
Number of Employees: 4700
Turnaround Times: B3
Union Shop: N
Year Established: 1864
Terms: Net 30

R R Donnelley & Sons Company, Willard   BK

Gilbert McLondon, Manager
1145 Conwell Ave
Willard OH 44890
419-935-0111; Fax: 419-933-5360
800-428-0832
Send RFQ's to: Gilbert McLondon

Min: 1000 Max: 2000000 Opt: 25000
Sizes: 12345X
Bindings: HCI-PBI
Items: BCEFLMN
Services: 4C-CS-DA-FF-LM
Equipment: SO-WO
Annual Sales: 6
Number of Employees: 4700
Turnaround Times: B3
Union Shop: N
Year Established: 1864
Terms: Net 30

Dragon Press   GC

Leann Folsom, Marketing Director
PO Box 588
Delta Junction AK 99737
907-895-4231; Fax: 907-895-4993
Send RFQ's to: Leann Folsom
Parent Company: Polar Run Inc

Min: 100 Max: 50000 Opt: 0
Sizes: 123456X
Bindings: HCO-CBI-LBO-PBI-SSI-SDI-SBI-WBI
Items: ABCDEFGHIKMNPQSTUV-WYZ124
Services: 4C-DA-AF-RP-TY-TD
Equipment: 4SO(19x25,11x17,12x18)
Annual Sales: 1

Number of Employees: 7
Turnaround Times: B3C5S2
Customers: BP5-BU40-CU5-MP5-NP5-OR5-SC20-SP10
Union Shop: N
Year Established: 1982
Terms: Net 30

Statement: Quality multi-color offset printing and attentive service to our customers — personalized printing.

Dynagraphics   MS

Gerald A Mortimer
6200 Yarrow Dr
Carlsbad CA 92009
619-438-3456; Fax: 619-929-0853

Min: 10000 Max: 500000 Opt: 60000
Sizes: 2345
Bindings: PBI-SSI
Items: CDFIPQ
Services: 4C-CS
Equipment: WO(17x26)-1OT
Annual Sales: 6
Number of Employees: 30
Turnaround Times: C2S2
Customers: BU20-CU10-MP70
Union Shop: N
Year Established: 1977
Terms: Cash

E

E & D Web Inc   CA

Chris Love
Edwards & Deutsch
4633 West 16th St
Cicero IL 60650
708-656-6600; 800-323-5733

Min: 25000 Max: 2000000 Opt: 0
Sizes: 235X
Bindings: SSI
Items: CDKPSTXYZ
Services: 4C
Equipment: 5WO
Annual Sales: 6
Number of Employees: 130
Year Established: 1896

SERVICES				BINDINGS			
4C	4-Color Printing	RA	Rachwal System	HC	Hardcover	SD	Side Stitching
CS	Color Separations	AF	Acid-Free Paper	CB	Comb Binding	SB	Spiral Binding
DA	Design / Artwork	RP	Recycled Paper	LB	Loose-Leaf Binding	WB	Wire-O Binding
FF	Fulfillment/Mailing	TY	Typesetting	OB	Otabind	I	In-House
LM	List Maintenance	TD	Typeset w/ Disk	PB	Perfect Binding	O	Out of House
OC	OptiCopy System	WA	Warehousing	SS	Saddle Stitching	X	Unknown

Turnaround Times: B4C5S3
Customers: BP70-CU20-SP10
Union Shop: N
Year Established: 1944
Terms: Net 30 with approved credit

Statement: We have been printing and binding books for over 50 years. We specialize in personal service. We offer heat set and non-heat set web printing — Smyth sewn casebound and perfect bound books.

Electronic Printing **BK**
Paul Kwiecinski, Marketing Director
81 Emjay Blvd
Brentwood NY 11717
516-434-1277; Fax: 516-434-1299
Send RFQ's to: Estimating Department

Min: 5000 Max: 10000 Opt: 0
Min: (Docutech) 1 Max: (Docutech) 500
Sizes: 123456
Bindings: HCO-CBI-LBO-OBO-PBI-SSI-SDI-WBO
Items: ABCDEFGHKLMNPS
Equipment: 2SO(19x28)
Annual Sales: 2
Number of Employees: 7
Turnaround Times: B2C3S2
Customers: BP80-BU20
Union Shop: N
Year Established: 1960
Terms: Net 30

Emco Printers Inc **GC**
Marketing Director
99 E Elm St
PO Box 192
Everett MA 02149
617-389-0076; Fax: 617-394-9340

Enquire Printing & Publishing Co **GC**
Marketing Director
4715 33rd St
Long Island City NY 11101
718-706-8400; Fax: 718-482-0680

EP Graphics **MG**
Carl H Muselman, President
169 S Jefferson St
Berne IN 46711
219-589-2145; Fax: 219-589-2810

Min: 10000 Max: 2000000 Opt: 250000
Sizes: 123456
Bindings: PBI-SSI-SDI
Items: DEFILNX
Services: 4C-CS-DA-TY-WA
Equipment: 3SO(25x38)-7WO(various)
Annual Sales: 4
Number of Employees: 150
Turnaround Times: B3C3S2
Customers: BP19-MP78-OT3
Union Shop: N
Year Established: 1925
Terms: Net 30

EPI Printers **GC**
Marketing Director
13305 Wayne Rd
Livonia MI 48150
313-261-9400; Fax: 313-261-9538

Epsen Lithographing Co **GC**
Marketing Director
511 N 20th St
Omaha NE 68102
402-342-7000; Fax: 402-342-5593

Etling Printing Co Inc **GC**
Marketing Director
4108 Rexford Dr
St Louis MO 63125
314-544-4400; Fax: 314-631-4577

EU Services **GC**
John C Frantz, Marketing Director
649 N Horners Ln
Rockville MD 20850
301-424-3300; Fax: 301-424-3696
Send RFQ's to: John C Frantz

Min: 500 Max: 5000000 Opt: 100000
Sizes: 12345
Bindings: SSX

PRINTED ITEMS							
A	Annual Reports	G	Galley Copies	N	Workbooks	U	Greeting Cards
B	Books	H	Journals	O	Yearbooks	V	Labels / Stickers
C	Booklets	I	Magazines	P	Brochures	W	Maps
D	Catalogs	J	Mass-Market Books	Q	Calendars	X	Newspapers
E	Cookbooks	K	Newsletters	R	Comic Books	Y	Postcards
F	Directories	L	Software Manuals	S	Direct Mail Letters	Z	Posters
		M	Textbooks	T	Envelopes	1	Stationery

Items: ACDKLNOPQRSTYZ1
Services: 4C-CS-FF-LM-AF-RP-TY-TD-
    WA
Equipment: 6WO(11 to 22)-6SO(11 to 40)
Annual Sales: 4
Number of Employees: 300
Turnaround Times: S2
Customers: BU20-CU10-NP40-OR20-SC10
Union Shop: N
Terms: Net 30

Statement: A full service printing and mailing company specializing in the production
of direct mail packages.

Eureka Printing Co Inc   GC
Marketing Director
106 T St
Eureka CA 95501
707-442-5703; Fax: 707-442-6968
Send RFQ's to: Elletta Tripp

Min: 0 Max: 0 Opt: 0
Sizes: 123456X
Bindings: HCO-CBO-LBI-OBO-PBI-SSI-
    SDI-SBI-WBO
Items: BCDEFKMNOPSTYZ
Services: 4C-DA-AF-RP-TY
Equipment: 2LP(10x15,12x18)-
    2SO(19x25.5,10x15)
Annual Sales: 2
Number of Employees: 9
Turnaround Times: B5S3
Customers: BP10-BU82-CU5-NP2-OR1
Union Shop: N
Year Established: 1973
Terms: Net 30

Eusey Press Inc   BK
Samuel Ratkawitch
27 Nashua St
Leominster MA 01453
508-534-8351; Fax: 508-537-2396

Min: 1000 Max: 0 Opt: 0
Sizes: 4
Bindings: OBX-PBX-SSX
Items: BDEPQ
Services: 4C
Equipment: 14SO(to 54x77)-3WO(22x36)
Annual Sales: 4
Number of Employees: 150
Turnaround Times: B3S3

Customers: BP50-BU50
Union Shop: Y

Eva-Tone Inc   GC
Marketing Director
4801 Ulmerton Rd
Clearwater FL 34622-4194
813-572-7000; Fax: 813-572-7948
800-382-8663
Send RFQ's to: Gary Peters

Min: 1000 Max: 20000 Opt: 10000
Sizes: 123456
Bindings: LBI-OBI-PBI-SSI-WBI
Items: BCDEFGLPNSY2
Services: 4C-CS-DA-FF-LM
Equipment: 5SO(11x17 to 28x40)
Annual Sales: 4
Number of Employees: 230
Turnaround Times: B3S2
Union Shop: N
Year Established: 1990
Terms: Net 30

Statement: To provide high quality perfect
bound books with a specialty in otabind-lay
flat books.

Evangel Press   BK
Jon Stepp, Marketing Director
PO Box 189
Nappanee IN 46550
219-773-3164; Fax: 219-773-5934
Send RFQ's to: Jon Stepp

Min: 300 Max: 10000 Opt: 3000-5000
Sizes: 123456
Bindings: HCO-CBO-OBO-PBI-SSI-SBO-
    WBO
Items: BDEHKNP
Services: AF-TY
Equipment: 3SO(28x40,14x20,11x17)
Annual Sales: 2
Number of Employees: 23
Turnaround Times: B4C5S4
Union Shop: N
Year Established: 1920
Terms: Net 30

SERVICES			BINDINGS		
4C	4-Color Printing	RA Rachwal System	HC Hardcover	SD	Side Stitching
CS	Color Separations	AF Acid-Free Paper	CB Comb Binding	SB	Spiral Binding
DA	Design / Artwork	RP Recycled Paper	LB Loose-Leaf Binding	WB	Wire-O Binding
FF	Fulfillment/Mailing	TY Typesetting	OB Otabind	I	In-House
LM	List Maintenance	TD Typeset w/ Disk	PB Perfect Binding	O	Out of House
OC	OptiCopy System	WA Warehousing	SS Saddle Stitching	X	Unknown

The Everton Publishers Inc **BK**

Marketing Director
PO Box 368
Logan UT 84321
801-752-6022

Min: 100 Max: 5000 Opt: 500
Items: BCF
Customers: OR20-SP80(genealogists)

Statement: We specialize in short runs of family histories and genealogies.

Comments: Physical address: 3233 S Main, Nibley UT 84321.

William Exline Inc **GC**

William B Exline, President
12301 Bennington Ave
Cleveland OH 44135
216-941-0800; 800-321-3062

Min: 300 Max: 50000 Opt: 10000
Sizes: X
Items: CT1

Statement: Among other items, Exline prints bank books and purse books with sewn bindings.

F

Faculty Press Inc **BK**

Walter Heitner, VP Sales
1449 - 37th St
Brooklyn NY 11218
718-851-6666; Fax: 718-853-4151
Send RFQ's to: Art Brezenoff

Min: 500 Max: 0 Opt: 3000000-10000000
Sizes: 123456
Bindings: PBI-SSI-HCO-CBO-LBO-OBO-SDO-SBO-WBO
Items: ABCDEFHIKLNOPQSUWZ1234
Services: 4C-CS-DA-AF-RP
Equipment: 4SO-WO
Annual Sales: 2
Number of Employees: 45
Turnaround Times: B2C3S1

Customers: BP20-BU20-CU20-MP10-NP15-OR10-SC5
Union Shop: Y
Year Established: 1950
Terms: Net 30 with approved credit

Statement: We offer high quality and rapid production turnaround of custom printed items for sales and promotion purposes, non-standard products with specialty printing and finishing.

Fast Print **GC**

Marketing Director
3050 E State Blvd
Fort Wayne IN 46805
219-484-5487; Fax: 219-482-8531

Federated Lithographers **BK**

Daniel Byrne, Marketing Director
369 Prairie Ave
PO Box 158
Providence RI 02901
401-781-8100; Fax: 401-467-8120
Send RFQ's to: Daniel Byrne

Min: 500 Max: 2000000 Opt: 0
Sizes: 123456X
Bindings: HCO-CBO-LBO-PBO-SSI-SDO-SBI-WBI
Items: BCDEFLMNPQTUVWZ
Services: 4C
Equipment: 9SO(up to 55x78)
Annual Sales: 4
Number of Employees: 200
Turnaround Times: B4
Customers: BP50-OT50
Union Shop: Y
Year Established: 1868

Statement: Owned by Quebecor America Book Group.

Fetter Printing **GC**

Robert Gaeta, Vice President
700 Locust Ln (40213)
PO Box 33128
Louisville KY 40232-3128
502-634-4771; Fax: 502-634-3587

Min: 5000 Max: 2000000 Opt: 0

PRINTED ITEMS			
A Annual Reports	G Galley Copies	N Workbooks	U Greeting Cards
B Books	H Journals	O Yearbooks	V Labels / Stickers
C Booklets	I Magazines	P Brochures	W Maps
D Catalogs	J Mass-Market Books	Q Calendars	X Newspapers
E Cookbooks	K Newsletters	R Comic Books	Y Postcards
F Directories	L Software Manuals	S Direct Mail Letters	Z Posters
	M Textbooks	T Envelopes	1 Stationery

Sizes: 2345X
Bindings: SSI-SDI
Items: CDIKPS
Services: 4C-CS-DA-FF-TY-WA
Equipment: 3SO(28x41,43x60)-WO(18x23)
Year Established: 1890

First Impressions Printer & Lithographers Inc **GC**

Marketing Director
700 Touhy Ave
Elk Grove IL 60007
708-439-8600; Fax: 708-439-8679
Send RFQ's to: Vickie Nelson

Min: 0 Max: 0 Opt: 0
Sizes: 123456
Bindings: HCO-CBO-LBI-PBO-SSI-SDI-
SBO-WBO
Items: ACDFHIKLMNPSTVYZ12
Services: 4C-FF-RP-WA
Equipment: 6SO(40x29)-3WO(8.5x11)
Annual Sales: 3
Number of Employees: 91
Turnaround Times: B2S2
Customers: BP40-BU30-CU10-NP20
Union Shop: N
Year Established: 1980
Terms: Net 30

Statement: Full service 1, 2, and 3 color offset printer. We handle disk for IBM and MAC platform, output film print, finish and mail all internally. We are very sensitive to budget and time constraints of each piece produced.

Fisher Printers Inc **MG**

Michael A Morrow, Corporate Sales
2121 N Towne Ln NE
PO Box 1366
Cedar Rapids IA 52406
319-393-5405

Min: 50000 Max: 2000000 Opt: 0
Sizes: 25X
Bindings: SSI
Items: CDFHIKNPSVWXZ
Services: 4C-CS-FF
Equipment: 2WO

Statement: Besides direct mail letters, brochures, inserts, and other promotional materials, they can also print magazines in quantities from 50,000 to 1,000,000 and digests from 100,000 to 2,000,000.

Fisher Printing Co **GC**

Jay Fisher, Marketing Director
405 Grant St
Galion OH 44833
419-460-2190; Fax: 419-468-3701

Min: 2500 Max: 20000000 Opt: 500000
Sizes: 123456
Bindings: SSI-SDI
Items: ACDFPQST
Services: 4C-CS
Equipment: 5SO(28x40)-WO(23)-
2LP(27x40)
Annual Sales: 3
Number of Employees: 70
Turnaround Times: S2
Union Shop: N
Terms: Net 30

Statement: DM color collateral material and envelope manufacture, web. commercial sizes.

Fleming Printing Co **GC**

Mr Val S Stark, Marketing Director
1550 Larkin Williams Rd
St Louis MO 63026
314-343-8900; Fax: 314-343-7165
800-279-2538

Min: 0 Max: 0 Opt: 0
Sizes: 123456
Bindings: HCO-CBO-LBO-OBO-PBO-
SSI-SDO-SBO-WBO
Items: ABCDEFGHIJKMNPQVZ34
Services: 4C-FF-LM-RP-TD-WA
Equipment: 4SO(28x40,14x20)-WO(18x26)-
LP
Annual Sales: 4
Number of Employees: 95
Customers: BU95-CU5
Union Shop: Y
Terms: Net 30

Statement: Fleming is more than a production facility. We are an organization of skilled, creative, seasoned professionals equipped with the latest technology in sheetfed, web, bindery and fullfillment equipment.

SERVICES				BINDINGS			
4C	4-Color Printing	RA	Rachwal System	HC	Hardcover	SD	Side Stitching
CS	Color Separations	AF	Acid-Free Paper	CB	Comb Binding	SB	Spiral Binding
DA	Design / Artwork	RP	Recycled Paper	LB	Loose-Leaf Binding	WB	Wire-O Binding
FF	Fulfillment/Mailing	TY	Typesetting	OB	Otabind	I	In-House
LM	List Maintenance	TD	Typeset w/ Disk	PB	Perfect Binding	O	Out of House
OC	OptiCopy System	WA	Warehousing	SS	Saddle Stitching	X	Unknown

Flower City Printing Inc **GC**
Mark Ashworth, Marketing Director
4800 Dewey Ave
Rochester NY 14612
716-663-9000; Fax: 716-663-4908
Send RFQ's to: Mark Ashworth

Min: 0 Max: 0 Opt: 0
Sizes: 123456
Bindings: SSI-SDI-HCO-CBO-LBO-OBO-
PBO
Items: ABCDEKLPQS24
Services: 4C-CS-DA-FF-RP-TD-WA
Equipment: 5SO(28x40,39x55)
Annual Sales: 4
Number of Employees: 130
Turnaround Times: B4C4S3
Customers: BU85-CU15
Union Shop: N
Year Established: 1971
Terms: Net 30

Focus Direct **GC**
Russell Marino, Executive VP
301 N Frio
PO Box 7789
San Antonio TX 78207-0789
210-227-9185

Min: 0 Max: 0 Opt: 0
Sizes: 123456X
Bindings: HCO-CBO-LBI-PBO-SSI-SDO-
SBO-WBO
Items: ACDFKPQSTVYZ1
Services: 4C-CS-DA-FF-LM-TY-WA
Equipment: LP-SO-WO
Annual Sales: 3
Number of Employees: 30
Turnaround Times: B1C2S3
Customers: BU50-NP30-OR15-OT5
Union Shop: N
Year Established: 1936
Terms: COD, credit with approval limit

Statement: We specialize in producing
direct marketing services and materials.

The Forms Man Inc **BK**
Marketing Director
35 Jefryn Blvd
Deer Park NY 11729
516-242-0009; Fax: 516-242-7748

800-877-9455
Min: 3000 Max: 50000 Opt: 5000-10000
Sizes: 123456
Bindings: LBX-PBX-SSX
Items: BCDFPQSTWZ
Services: 4C-CS-DA-RP-WA
Equipment: SO-WO
Turnaround Times: B4S3
Customers: SP80
Union Shop: N
Year Established: 1954
Terms: Net 30

Fort Orange Press Inc **BK**
William Dorsman, Marketing Director
31 Sand Creek Rd
Albany NY 12203
518-489-3233; Fax: 518-489-1638
800-777-3233
Send RFQ's to: William Dorsman

Min: 2500 Max: 100000 Opt: 50000
Sizes: 2356
Bindings: CBX-PBX-SSX-SDX
Items: ACDEFHKLNPQWZ12
Services: 4C-DA-TY-TD-WA-RP
Equipment: LP-SO
Annual Sales: 2
Number of Employees: 55
Turnaround Times: B2S1
Union Shop: Y
Year Established: 1905
Terms: Depending upon receipt of credit
application

Foster Printing Service **GC**
Beth Gomez, Vice President Sales
4295 S Ohio St
PO Box 2089
Michigan City IN 46360
219-879-8366; 800-382-0808

Min: 100 Max: 25000 Opt: 0
Sizes: 123456
Bindings: PBI-SSI
Items: ACGKPZ
Services: 4C-TY
Equipment: LP-1OSO(10x15,25x35)
Annual Sales: 3
Number of Employees: 30
Customers: MP70-OT30
Union Shop: Y

PRINTED ITEMS	G	Galley Copies	N	Workbooks	U	Greeting Cards
A Annual Reports	H	Journals	O	Yearbooks	V	Labels / Stickers
B Books	I	Magazines	P	Brochures	W	Maps
C Booklets	J	Mass-Market Books	Q	Calendars	X	Newspapers
D Catalogs	K	Newsletters	R	Comic Books	Y	Postcards
E Cookbooks	L	Software Manuals	S	Direct Mail Letters	Z	Posters
F Directories	M	Textbooks	T	Envelopes	1	Stationery

Terms: Net 30

Statement: They specialize in promotional printing, including ad reprints, brochures, spec sheets, statement stuffers, and catalog sheets.

Frederic Printing Inc
GC

Marketing Director
14701 E 38th Ave
Aurora CO 80011
303-371-7990

Friesen Printers
BK

Don Penner, Marketing Director
PO Box 720
Altona MB Canada R0G 0B0
204-324-6401; Fax: 204-324-1333
Send RFQ's to: Tim Fast

Min: 3000 Max: 50000 Opt: 10000
Sizes: 3456
Bindings: HCI-CBI-PBI-SSI-SBI-WBI
Items: BEHMNOPQ
Services: 4C-CS-FF-CO-AF-RP-WA
Equipment: 10SO(up to 38x50)
Annual Sales: 5
Number of Employees: 350
Turnaround Times: B4
Customers: BP75-SC25
Union Shop: N
Year Established: 1907
Terms: Net 30

Statement: Their U.S. address is: PO Drawer B, Neche ND 58265. We are a sheet-fed printer specializing in 4-color work. We believe that we are the lowest-cost producer of color books this side of Hong Kong.

Fry Communications
GC

Marketing Director
800 West Church Rd
Mechanicsburg PA 17055
717-766-0211; Fax: 717-691-0341
800-334-1429
Send RFQ's to: Gary Shugart, VP Sales

Frye & Smith
CA

John Udink, Sales Manager
150 Baker St E
Costa Mesa CA 92626
714-540-7005; Fax: 714-979-1496

Min: 20000 Max: 3000000 Opt: 0
Sizes: 5X
Bindings: PBI-SSI
Items: ADIPZ
Services: 4C-CS-FF
Equipment: SO-WO
Annual Sales: 6
Number of Employees: 2480

Ft Dearborn Lithograph Co
GC

Marketing Director
6035 W Gross Point Rd
Niles IL 60714
312-774-4321; Fax: 312-774-1091

Fundcraft Publishing
BK

Marketing Director
410 Highway 72 West
PO Box 340
Collierville TN 38017
901-853-7070; 800-351-7822

Min: 500 Max: 25000 Opt: 2000
Sizes: 23
Bindings: CBI
Items: E
Services: 4C-DA-TY
Customers: OR100
Terms: No downpayment

Statement: They specialize in printing cookbooks for fundraising organizations.

Futura Printing Inc
GC

Bob Steinmetz
3050 SW 14th Pl #18
Boynton Beach FL 33426
407-734-0825; Fax: 407-734-0862
Send RFQ's to: Bob Steinmetz

Min: 25 Max: 0 Opt: 0
Sizes: 123456
Bindings: HCX-CBX-LBX-PBX-SSX-SDX
Items: ABCDEFIKOPQSTVWXYZ1

SERVICES		BINDINGS		
4C 4-Color Printing	RA Rachwal System	HC Hardcover	SD Side Stitching	
CS Color Separations	AF Acid-Free Paper	CB Comb Binding	SB Spiral Binding	
DA Design / Artwork	RP Recycled Paper	LB Loose-Leaf Binding	WB Wire-O Binding	
FF Fulfillment/Mailing	TY Typesetting	OB Otabind	I In-House	
LM List Maintenance	TD Typeset w/ Disk	PB Perfect Binding	O Out of House	
OC OptiCopy System	WA Warehousing	SS Saddle Stitching	X Unknown	

Services: 4C-TY
Equipment: 2SO(11x17,18x24.5)
Annual Sales: 1
Number of Employees: 6
Turnaround Times: B3C4S2
Customers: BP20-CU10-NP10-SC40-SP20
Union Shop: N
Year Established: 1977
Terms: Net 30

Statement: To fill the gap for those who want to self-publish without hype of our sales or circulation. We act strictly as printers with the customer owning all rights.

G

Gamse Lithographing Co   GC
Marketing Director
7413 Pulaski Hwy
Baltimore MD 21237
410-866-4700; Fax: 410-866-5672

Gateway Press Inc   GC
Darrell Embry, Marketing Director
4500 Robards Ln
Louisville KY 40218
502-454-0431; Fax: 502-459-7930
Send RFQ's to: Darrell Embry

Min: 10000 Max: 10000000 Opt: 100000-
   200000
Sizes: 123456X
Bindings: HCO-CBO-OBO-LBO-PBI-SSI-
   SDI-SBO-WBO
Items: ABCDEFHIKLMNPWZ
Services: 4C-CS-DA-FF-OC-AF-RP-TY-
   TD-WA
Equipment: 3SO(28x40)-4WO(22.75x36)
Annual Sales: 5
Number of Employees: 285
Turnaround Times: B3S3
Customers: BU50-OR10-MP10-OT30
Union Shop: N
Year Established: 1950
Terms: Net 30 with approved credit

Statement: To produce the best product possible at an affordable price and to treat each customer like they are the only one we have.

Gateway Press   BR
Marketing Director
1001 N Calvert St
Baltimore MD 21202
410-837-8271

Customers: SP(genealogists)

Comments: They broker printing services for genealogists.

Gaylord Printing   GC
David Duncan
15555 Woodrow Wilson St
Detroit MI 48238
313-883-7800; Fax: 313-883-2917
800-486-7746

Min: 5000 Max: 2000000 Opt: 0
Sizes: 2345X
Bindings: SSI
Items: ACDKNPSWXZ
Services: 4C-CS
Equipment: SO-WO
Annual Sales: 4
Number of Employees: 190

Geiger Brothers   BK
Eugene Geiger, President
PO Box 1609 (04241)
Lewiston ME 04240
207-783-2001; Fax: 207-777-7083
Send RFQ's to: Eugene Geiger

Min: 0 Max: 0 Opt: 0
Sizes: 25X
Bindings: LBI-SSI-SBI-Smythe Sewn
Items: BQ
Equipment: LP-SO-WO

Statement: Specialize in printing looseleaf time management systems planning diaries and address books.

General Printing Inc   GC
Marketing Director
13163 NW 42 Ave
Miami FL 33054
305-685-7900; Fax: 305-769-9198
800-282-0214
Send RFQ's to: Christopher E Rose

PRINTED ITEMS			
A Annual Reports	G Galley Copies	N Workbooks	U Greeting Cards
B Books	H Journals	O Yearbooks	V Labels / Stickers
C Booklets	I Magazines	P Brochures	W Maps
D Catalogs	J Mass-Market Books	Q Calendars	X Newspapers
E Cookbooks	K Newsletters	R Comic Books	Y Postcards
F Directories	L Software Manuals	S Direct Mail Letters	Z Posters
	M Textbooks	T Envelopes	1 Stationery

Statement: Short-run printing, typesetting, and graphic design; 19 years experience.

George Lithograph **GC**
Mark Lindsay, Marketing Director
650 2nd St
San Francisco CA 94107
415-397-2400; Fax: 415-267-4626
Send RFQ's to: Jim Brownlow

Min: 10 Max: 20000 Opt: 200-5000
Sizes: 2345X
Bindings: CBX-LBX-PBX-SSX-SDX-
    WBX
Items: ABCDEFKLP
Services: 4C-CS-FF-OC-RP-TD-WA
Equipment: SO-WO
Annual Sales: 5
Number of Employees: 260
Turnaround Times: B2S2
Customers: BU100
Union Shop: Yes and No
Year Established: 1925
Terms: Net 30

Statement: George Lithograph is among the largest commercial printers in Northern California. Located in San Francisco since 1925, the company offers a full range of promotional printing services to customers representing a variety of industries throughout the Bay Area. The company's responsiveness to market changes and the needs of its clients has enabled the company to grow and thrive while many long standing printers have faded from the San Francisco market.

Another plant located at:

George Lithograph
460 Valley Dr
Brisbane CA 94005
415-467-0600; Fax: 415-715-2107

Geryon Press Limited **GC**
Stuart McCarty, Owner
PO Box 70
Tunnel NY 13848
607-693-1572
Send RFQ's to: Stuart McCarty

Min: 10 Max: 1000 Opt: 400

Sizes: 23
Bindings: HCO-PBI-SSI
Items: BCDEPYZ
Services: DA-TY
Equipment: LP
Union Shop: N
Terms: To be arranged

Statement: Geryon is a small letterpress shop specializing in typesetting and printing poetry books, postcards, and broadsides.

Gilliland Printing Inc **BK**
Floyd L Ferris, Marketing Director
215 N Summit Ave
Arkansas City KS 67005
316-442-0500; Fax: 316-442-8504
800-332-8200
Send RFQ's to: Sales Department

Min: 500 Max: 20000 Opt: 2000-10000
Sizes: 23456X
Bindings: PBX-SSX
Items: BCDFNP
Services: OC-AF-RP-TD-CS-WA
Equipment: 3SO(25x38,26x40,19.5x25.5)
Annual Sales: 3
Number of Employees: 73
Turnaround Times: B3S3
Customers: BP70-CU30
Union Shop: N
Year Established: 1947
Terms: 100% unless credit app is approved — otherwise net 30

Statement: To be absolutely unequaled in speed of turnaround time for competitively priced high quality soft bound books.

Glundal Color Service **CA**
Greg Tino, Marketing Director
6700 Joy Rd
East Syracuse NY 13057
315-437-1391

Min: 15000 Max: 3000000 Opt: 500000
Sizes: 25
Bindings: CBO-LBO-PBO-SSI-SDO-SBO-
    WBO
Items: ACDFIKPQZ
Services: 4C-CS-DA
Equipment: 4SO(25x38)-WO(17x23)

SERVICES				BINDINGS			
4C	4-Color Printing	RA	Rachwal System	HC	Hardcover	SD	Side Stitching
CS	Color Separations	AF	Acid-Free Paper	CB	Comb Binding	SB	Spiral Binding
DA	Design / Artwork	RP	Recycled Paper	LB	Loose-Leaf Binding	WB	Wire-O Binding
FF	Fulfillment/Mailing	TD	Typeset w/ Disk	OB	Otabind	I	In-House
LM	List Maintenance	TY	Typesetting	PB	Perfect Binding	O	Out of House
OC	OptiCopy System	WA	Warehousing	SS	Saddle Stitching	X	Unknown

Annual Sales: 3
Number of Employees: 100
Turnaround Times: C3S4
Customers: BP10-BU70-CU10-OR10
Union Shop: N
Terms: Net 30

Golden Belt Printing Inc **GC**
Marketing Director
1125 281 Bypass
PO Box 997
Great Bend KS 67530-0997
800-299-6351; Fax: 316-792-5322
Send RFQ's to: Roy Myers

Min: 500 Max: 10000 Opt: 2000-4000
Sizes: 123456
Bindings: CBI-PBI-SSI-SDI-HCO-LBO-
  OBO-SBO-WBO
Items: BCDEFKNPQTY12
Services: 4C-FF-RP-TY-TD-WA
Equipment: 4SO(17,18,25)
Annual Sales: 2
Number of Employees: 9
Turnaround Times: B3C4S2
Customers: BP20-BU50-CU10-NP10-SC5-
  SP5
Union Shop: N
Year Established: 1980
Terms: 1% 10, net 30 w/approval

Statement: Our mission as a small full-service printer: To honor God in all our business and personal dealings; to understand our customer's needs and exceed their expectations; to provide friendly individual attention, dependable service, and fair pricing.

Golden Horn Press **BK**
Marketing Director
2120 Dwight Way
Berkeley CA 94704
510-845-4355

Min: 300 Max: 10000 Opt: 0
Sizes: 12
Bindings: PBX-SSX
Items: BJ

Statement: They specialize in printing smaller books (4 x 7 and 5½ x 8½) in quantities from 300 to 10,000.

Goodway Graphics of Virginia **GC**
Marketing Director
6628 Electronic Dr
Springfield VA 22151
703-941-1160; Fax: 703-658-9511

Min: 500 Max: 25000 Opt: 15000
Sizes: 35
Bindings: PBI-CBI-SSI
Items: BCDEFHLN
Equipment: 4SO(35x45)-2WO(26x36)-
  2(Miller)
Number of Employees: 250

Statement: We are primarily a black and white book, catalog, and directory printer. Our pricing is best on 8½ x 11 and 6 x 9 high volume work.

Gorham Printing **GC**
Marketing Director
334 Harris Rd
Rochester WA 98579
206-273-0970
Send RFQ's to: Kurt Gorham

Min: 100 Max: 3000 Opt: 1000
Sizes: 235
Bindings: CBI-PBI-SSI-HCO-LBO-OBO-
  SDO-SBO-WBO
Items: BCDEFMN
Services: TY-TD
Equipment: 2SO(19x26)
Annual Sales: 1
Number of Employees: 3
Turnaround Times: B4C5S4
Customers: BP10-SC10-SP80
Union Shop: N
Year Established: 1976
Terms: 50% down, balance on delivery

Statement: We are a small operation and take pleasure in working with our customers, helping make their books just the way they want them. FREE catalog upon request.

Gowe Printing Company **GC**
Gary Hartman, Marketing Director
620 E Smith Rd
Medina OH 44256
216-725-4161; Fax: 216-225-4531

PRINTED ITEMS	G Galley Copies	N Workbooks	U Greeting Cards
A Annual Reports	H Journals	O Yearbooks	V Labels / Stickers
B Books	I Magazines	P Brochures	W Maps
C Booklets	J Mass-Market Books	Q Calendars	X Newspapers
D Catalogs	K Newsletters	R Comic Books	Y Postcards
E Cookbooks	L Software Manuals	S Direct Mail Letters	Z Posters
F Directories	M Textbooks	T Envelopes	1 Stationery

800-837-4693
Send RFQ's to: Gary Hartman

Min: 50000 Max: 3000000 Opt: 500000
Sizes: 245X
Bindings: SSI-OBO-PBO
Items: BCDFIPRX
Services: 4C-CS-FF-LM-OC-AF-RP-TY-TD-WA
Equipment: 6WO(22.75x35,21x40,22.75x36)
Annual Sales: 5
Number of Employees: 300
Turnaround Times: B3S2
Customers: BU40-CU10-MP10-NP5-SC5-SP20-OT10
Union Shop: N
Year Established: 1954
Terms: Net 30 with approval

MS

Graftek Press
A E Skip Sorg, VP Sales
11595 McConnell Rd
PO Box 1149
Woodstock IL 60098
815-338-6750

Min: 50000 Max: 350000 Opt: 150000
Sizes: 4X
Bindings: HCO-PBI-SSI-SBO
Items: CDFHIX
Services: 4C-CS
Equipment: 3SO(18x25,28x40)-9WO(23x36)
Annual Sales: 5
Number of Employees: 545
Turnaround Times: S2
Customers: BU2-CU1-MP95-NP2
Union Shop: N
Year Established: 1970
Terms: Net 30

Statement: Printers of medium-run special interest and trade publications, they have two printing facilities: Graftek Press in Crystal Lake, Illinois, and Elkhorn Web Press in Elkhorn, Wisconsin.

CA

Graphic Arts Center
J D Droge, Sales
2000 NW Wilson St
Portland OR 97209
503-224-7777; Fax: 503-222-0735

Min: 2000 Max: 2000000 Opt: 0
Sizes: 23456X
Bindings: HCI-PBI-SSI-SDI-SBI-WBI
Items: ADPS
Services: 4C-CS-FF-TY
Equipment: 5SO(to 26x40)-4WO(to 23x36)
Annual Sales: 6
Number of Employees: 650

Statement: GAC specializes in printing catalogs of all shapes and sizes. They also publish their own calendars and coffee-table books.

GC

Graphic Arts Publishing
Marketing Director
5325 Kendall St
Boise ID 83706
208-375-1010

Min: 1000 Max: 1000000 Opt: 0
Sizes: 235
Bindings: SSI
Items: D
Equipment: SO-WO

Statement: Will print as few as 1000 catalogs or as many as 1,000,000.

BK

Graphic Litho Corporation
Jesse Kamien, Marketing Director
130 Shepard St
Lawrence MA 01843
508-683-2766; Fax: 508-681-7588
Send RFQ's to: Jesse Kamien

Min: 1000 Max: 200000 Opt: 20000-100000
Sizes: 123456X
Bindings: CBI-PBI-SSI-SDI-SBO-WBO
Items: ACDFIKNPQRSUVWYZ4
Services: 4C-CS-OC-AF-RP-TD
Equipment: 4SO(2-77,54,25)
Annual Sales: 2
Number of Employees: 29
Turnaround Times: B3C5S2
Customers: OT80
Union Shop: N
Year Established: 1960
Terms: 1% 10, net 30

Statement: Graphic Litho specializes in large format (to 77") printing and prepress services. Multicolor and process specialists with full digital capability. Art posters and books, maps, labels, and book covers.

SERVICES			
4C	4-Color Printing	RA	Rachwal System
CS	Color Separations	AF	Acid-Free Paper
DA	Design / Artwork	RP	Recycled Paper
FF	Fulfillment/Mailing	TY	Typesetting
LM	List Maintenance	TD	Typeset w/ Disk
OC	OptiCopy System	WA	Warehousing

BINDINGS			
HC	Hardcover	SD	Side Stitching
CB	Comb Binding	SB	Spiral Binding
LB	Loose-Leaf Binding	WB	Wire-O Binding
OB	Otabind	I	In-House
PB	Perfect Binding	O	Out of House
SS	Saddle Stitching	X	Unknown

Graphic Printing **BK**

Marketing Director
310 N Clay St
New Carlisle OH 45344
513-845-3757
Parent Company: McGregor & Werner

Min: 100 Max: 20000 Opt: 5000
Sizes: 123456X
Bindings: HCO-CBI-LBI-PBI-SSI-SDI-SBI-WBI
Items: BCDFGHKLMNPS
Services: 4C
Equipment: 7SO(29 to 55)
Annual Sales: 4
Number of Employees: 125
Turnaround Times: B3
Customers: BP20-BU15-CU10-NP40-SC10-SP5
Union Shop: N
Year Established: 1950
Terms: 1% 10, net 30 with approved credit

Graphic Ways Inc **GC**

Mike Yeager, Marketing Director
8332 Commonwealth Ave
Buena Park CA 90621-2591
714-521-5920; Fax: 714-228-0403
Send RFQ's to: Mike Yeager

Min: 5000 Max: 100000 Opt: 25000
Sizes: 123456
Bindings: HCO-CBO-LBO-OBO-SSI-SDO-SBO-WBI
Items: ABCDEIKPQSTYZ2
Services: 4C-CS-DA
Equipment: 4SO(40)
Annual Sales: 3
Number of Employees: 50
Turnaround Times: B4C5S3
Customers: BP25-BU75
Union Shop: N
Year Established: 1970
Terms: Net 30

Statement: Complete printing capability-from digital photography and electronic pre-press through 6 color 40" presses. Twenty five years experience in production of photographic books and art catalogs.

Graphics Ltd **GC**

Vona W Lauman, Vice President
401 N College Ave
Indianapolis IN 46202-3605
317-263-3456

Min: 300 Max: 30000 Opt: 15000
Sizes: 123456
Bindings: HCO-CBO-LBI-PBO-SSI-SDI-SBO-WBO
Items: ABCDEFGHIKLPT
Services: 4C-CS-TY-TD
Equipment: 7SO(11x17,28x41)
Annual Sales: 3
Number of Employees: 25
Turnaround Times: B4C3S2
Customers: BU50-NP40-OR10
Union Shop: Y
Year Established: 1920
Terms: Net 30 with approved credit

Gray Printing **MS**

Robert A Gray, President
401 E North St
PO Box 840
Fostoria OH 44830
419-435-6638; Fax: 419-435-9410
800-458-4721

Min: 8000 Max: 200000 Opt: 50000
Sizes: 25X
Bindings: CBI-PBI-SSI-WBI
Items: ADFHIKLNPXZ
Services: 4C-DA-TY-TD
Equipment: 6SO(to 28x40)-3WO(23x38)
Annual Sales: 4
Number of Employees: 200
Turnaround Times: B4C3S2
Customers: BU25-CU10-MP60-NP5
Union Shop: Y
Year Established: 1888
Terms: Net 30

Graytor Printing **GC**

Marketing Director
149 Park Ave
PO Box 187
Lyndhurst NJ 07071
201-933-0100

PRINTED ITEMS			
A Annual Reports	G Galley Copies	N Workbooks	U Greeting Cards
B Books	H Journals	O Yearbooks	V Labels / Stickers
C Booklets	I Magazines	P Brochures	W Maps
D Catalogs	J Mass-Market Books	Q Calendars	X Newspapers
E Cookbooks	K Newsletters	R Comic Books	Y Postcards
F Directories	L Software Manuals	S Direct Mail Letters	Z Posters
	M Textbooks	T Envelopes	1 Stationery

GC
Great Eastern Color Lithograpic
Marketing Director
40 Violet Ave
Poughkeepsie NY 12601
914-454-7420; Fax: 914-454-7507

GC
Great Eastern Printing Co Inc
Marketing Director
7 Aerial Way
Syosset NY 11791
516-931-3900; Fax: 516-931-3757

BK
Great Impressions Printing & Graphics
Marketing Director
444 W Mockingbird
Dallas TX 75247
214-631-BOOK

Min: 0 Max: 0 Opt: 0
Sizes: 12345
Bindings: HCO-CBI-LBO-PBI-SSI-SDI-
    SBI-WBI-OBO
Items: BCDEFKLNPQ4
Services: 4C-FF-AF-RP-TY-TD-WA
Equipment: 5SO-WO
Annual Sales: 2
Number of Employees: 34
Turnaround Times: B3C5S2
Union Shop: N
Year Established: 1986
Terms: 2% 10, net 30

GC
Great Lakes Lithograph Co
Tom Schultz, Marketing Director
4005 Clark Ave
Cleveland OH 44109-1186
216-651-1500; Fax: 216-651-8311
Send RFQ's to: Tom Schultz

Min: 0 Max: 0 Opt: 0
Sizes: 123456
Bindings: LBI-SSI-SDI-HCO-CBO-OBO-
    PBO-SBO-WBO
Items: ABCDEFHIKLMNOPQSUWZ12
Services: 4C-CS-FF-LM-OC-AF-RP-TD-
    WA
Equipment: 4SO(3-28x40,20x26)-
    WO(17.75x26.5)

Annual Sales: 4
Number of Employees: 100
Turnaround Times: B3C4S1
Customers: BU80-CU10-NP5-OT5
Union Shop: N
Year Established: 1932
Terms: Net 30

Statement: Great Lakes Lithograph is an
ongoing source of counsel, information,
coordination, image reproduction, and distribution services to assist organizations to
achieve their respective promotional and
sales objectives. Services include projects
and programs both simple and complex,
with above-average risks and consequences
for our customers.

GC
Great Northern/Design Printing
Tony Albert, President
5401 Fargo Ave
Skokie IL 60077
708-674-4740
Send RFQ's to: Tony Albert

Min: 0 Max: 0 Opt: 0
Items: ADPQZ
Services: 4C-CS

MS
Greenfield Printing & Publishing
David L Moon
1025 N Washington St
Greenfield OH 45123
513-981-2161; 800-543-3881
Send RFQ's to: David L Moon

Min: 2000 Max: 250000 Opt: 0
Sizes: 2345X
Bindings: PBI-SSI
Items: I
Services: 4C-CS-FF
Equipment: 2SO-2WO

GC
Gregath Publishing Company
Fredrea Cook, Marketing Director
PO Box 505
Wyandotte OK 74370-1045
800-955-5253; Fax: 918-542-4148
Send RFQ's to: Fredrea or Carrie Cook

Min: 50 Max: 0 Opt: 0

SERVICES		BINDINGS	
4C 4-Color Printing	RA Rachwal System	HC Hardcover	SD Side Stitching
CS Color Separations	AF Acid-Free Paper	CB Comb Binding	SB Spiral Binding
DA Design / Artwork	RP Recycled Paper	LB Loose-Leaf Binding	WB Wire-O Binding
FF Fulfillment/Mailing	TY Typesetting	OB Otabind	I In-House
LM List Maintenance	TD Typeset w/ Disk	PB Perfect Binding	O Out of House
OC OptiCopy System	WA Warehousing	SS Saddle Stitching	X Unknown

Sizes: 123456X
Bindings: LBI-SSI-SDI-HCO-CBO-OBO-
   PBO-SBO-WBO
Items: ABCDEFHIKLMNOPQSTUVWX
   YZ12
Services: 4C-DA-LM-AF-TY-RP-TD-WA
Equipment: 5SO-WO
Annual Sales: 1
Number of Employees: 9
Turnaround Times: B3C3S2
Customers: BU2-NP1-OT1-SP90-OR1
Union Shop: N
Year Established: 1970
Terms: Cash

Statement: To serve our clients with the
best quality at the best price. We are proud
to have been selected to represent
America in International Book Exhibitions.

Griffin Printing & Lithograph **BK**
Marketing Director
544 West Colorado St
Glendale CA 91204
213-245-3671; Fax: 818-242-1172
800-826-4049
Send RFQ's to: Larry Davis

Min: 1000 Max: 250000 Opt: 0
Sizes: 1245
Bindings: HCX-PBX-SBX-WBX-SSX-
   SDX
Items: ABCDEFHKLNPQRUX
Services: FF-OC-RP
Equipment: 3SO-4WO
Annual Sales: 4
Number of Employees: 135
Turnaround Times: B3C4S3
Customers: BP58-CU7-NP3-SC2-OT5-
   BU25
Union Shop: N
Year Established: 1928
Terms: COD and net 30

Statement: We provide a quality service to
our customers and clients customers.

Griffin Printing and Lithograph **BK**
Company
Jerry Travers, Plant Manager
1801 Ninth St
Sacramento CA 95814
916-448-3511; Fax: 916-448-3597

800-448-3511

Min: 1000 Max: 25000 Opt: 5000
Sizes: 25
Bindings: HCO-CBO-LBO-PBI-SSI-SDO-
   SBO-WBO
Items: BCDEFKLMN
Services: OC
Equipment: SO-WO
Annual Sales: 2
Number of Employees: 30
Turnaround Times: B3
Customers: BP70-BU30
Union Shop: N
Year Established: 1924
Terms: Net 30 with approved credit

Statement: We offer personalized service
and 3 to 4 week turnarounds. Formerly
Spilman Printing, this company has merged
with Griffin.

Grinnell Lithographic Co Inc **GC**
Marketing Director
185 Grant Ave
Islip NY 11751
516-581-3300; Fax: 516-581-3348

GRIT Commercial Printing **GC**
Services
Bill Ott, General Manager
208 W Third St
PO Box 965
Williamsport PA 17701
717-326-1771; Fax: 717-326-6940

Min: 1000 Max: 50000 Opt: 5000
Sizes: 123456X
Bindings: SSI-SDI-HCO-CBO-LBO-OBO-
   PBO-SBO-WBO
Items: ACDKIPQSXZ
Services: 4C-DA-TY-TD-WA
Equipment: 7SO(14x20,25x38,19x25)-WO
Annual Sales: 2
Number of Employees: 65
Turnaround Times: B3C4S2
Customers: BU10-BP65-CU10-SC5-SP10
Union Shop: Y
Year Established: 1920
Terms: Net 30

PRINTED ITEMS							
A	Annual Reports	G	Galley Copies	N	Workbooks	U	Greeting Cards
B	Books	H	Journals	O	Yearbooks	V	Labels / Stickers
C	Booklets	I	Magazines	P	Brochures	W	Maps
D	Catalogs	J	Mass-Market Books	Q	Calendars	X	Newspapers
E	Cookbooks	K	Newsletters	R	Comic Books	Y	Postcards
F	Directories	L	Software Manuals	S	Direct Mail Letters	Z	Posters
		M	Textbooks	T	Envelopes	1	Stationery

Statement: Our goal is to produce the highest quality printing, direct mail service, and customer service.

GRT Book Printing **BK**
Marketing Director
3960 East 14th St
Oakland CA 94601
510-534-5032; Fax: 510-534-1873
800-643-0300 (in California)
Send RFQ's to: Mark Ross

Min: 200 Max: 50000 Opt: 0
Sizes: 12345X
Bindings: CBI-LBI-SSI-SDI-PBO-SBI-WBO
Items: BCEFHLMN
Services: RP
Equipment: SO-WO
Turnaround Times: B3S3
Customers: BP25-BU30-CU10-NP10-SP25
Union Shop: N
Year Established: 1973
Terms: 50 down, balance COD

Statement: GRT specializes in printing softcover books for customers located in California only.

GTE Directories Printing Corp **BK**
Marketing Director
1221 Business Center Dr
Mt Prospect IL 60056
708-391-5100

Min: 5000 Max: 2000000 Opt: 0
Sizes: 23
Bindings: PBI
Items: BF
Equipment: WO
Customers: BU-OT

Statement: GTE is a directory printer. We print millions of telephone books.

GTE Directories Corp **GC**
Marketing Director
111 Rawls Rd
Des Plaines IL 60018
708-635-1200

Min: 5000 Max: 200000 Opt: 0

Sizes: 23
Bindings: PBI
Items: BF
Equipment: WO
Customers: BU-OT

Statement: GTE is a directory printer. We print millions of telephone books.

Gulf Printing **GC**
Barbara Shumacher, Sales Manager
2210 W Dallas Ave
Houston TX 77019
713-529-4201; 800-423-9537

Min: 2000 Max: 2000000 Opt: 0
Sizes: 2345X
Bindings: PBI-SSI-SDI
Items: ABCDFIKPSZ
Services: 4C-CS-FF
Equipment: 5SO(20x28,43x60)-2WO
Annual Sales: 6
Number of Employees: 529

Statement: A subsidiary of Southwestern Bell Corporation.

Gulf South Printing **GC**
Marketing Director
1129 E Vermilion St
Lafayette LA 70501
318-235-5231; Fax: 318-233-0526

Guynes Printing Company **GC**
George W Andresen, President
2709 Girard Blvd NE
Albuquerque NM 87107
505-884-8882; Fax: 505-889-3148

Min: 500 Max: 20000 Opt: 5000
Sizes: 12345X
Bindings: HCO-CBI-LBI-PBI-SSI-SDI-SBI-WBI
Items: ABCDEFIJKLPQSUYZ
Services: 4C-CS-FF-AF-RP-TY-TD
Equipment: 4SO(19x25,26x40)
Annual Sales: 3
Number of Employees: 84
Turnaround Times: B3C2
Customers: BP15-BU15-CU2-SP10
Union Shop: N
Year Established: 1932
Terms: 2% 10, net 30

SERVICES				BINDINGS		SD	Side Stitching
4C	4-Color Printing	RA	Rachwal System	HC	Hardcover	SB	Spiral Binding
CS	Color Separations	AF	Acid-Free Paper	CB	Comb Binding	WB	Wire-O Binding
DA	Design / Artwork	RP	Recycled Paper	LB	Loose-Leaf Binding	I	In-House
FF	Fulfillment/Mailing	TY	Typesetting	OB	Otabind	O	Out of House
LM	List Maintenance	TD	Typeset w/ Disk	PB	Perfect Binding	X	Unknown
OC	OptiCopy System	WA	Warehousing	SS	Saddle Stitching		

H

Haddon Craftsmen **BK**

Dan Williams, Marketing Director
1001 Wyoming Ave
Scranton PA 18509
717-348-9211; Fax: 717-348-9266
Parent Company: RR Donnelley & Sons

Min: 1000 Max: 0 Opt: 500000
Sizes: 12356
Bindings: HCI-OBI-PBI
Items: BEFGLMN
Services: FF-LM-RA-RP-TY-TD-WA
Equipment: 6WO(2-17.75,4-19⅜
Annual Sales: 5
Number of Employees: 1000
Turnaround Times: B3C3
Customers: PB5CU5
Union Shop: Y
Year Established: 1900
Terms: 1% - 15 days after EOM

Statement: To provide quick-turn manufacturing to the book publishing and book club markets.

Hamilton Printing Inc **GC**

Marketing Director
1703 S Brook St
Louisville KY 40208
502-635-7465

Hammer Lithograph Corp **GC**

Marketing Director
330 Metro Pk
Rochester NY 14623
716-424-3880; Fax: 716-424-3886

Harlo Printing Co **BK**

Marketing Director
50 Victor St
Detroit MI 48203
313-883-3600; Fax: 313-883-0072
Send RFQ's to: Kitty Russo

Min: 250 Max: 10000 Opt: 3000
Sizes: 123456
Bindings: HCO-CBI-LBO-PBI-SSI-SDI-SBI-WBI
Items: ABCDEFKLMNOPQSTYZ1
Services: 4C-DA-FF-AF-TY-TD-WA
Equipment: LB-SO-WO
Annual Sales: 2
Number of Employees: 25
Turnaround Times: B3C3
Customers: NP20-OR10-SC10-SP60
Union Shop: N
Year Established: 1946
Terms: 2% 10, net 30

Statement: Harlo has advertised in **Writer's Digest** for years and, therefore, is quite accustomed to working with writers and self-publishers.

Harmony Printing **GC**

Marketing Director
1200 E 210 Highway
PO Box 377
Liberty MO 64068-0377
800-827-1155; Fax: 816-781-0856
Send RFQ's to: Randall "Hutch" Hutchinson

Min: 5000 Max: 2000000 Opt: 40000-700000
Sizes: 123456
Bindings: PBI-SSI-HCO-CBO-SBO-WBO
Items: ABCDEFHIJKLPQWX
Services: 4C-CS-DA-FF-TY-TD
Equipment: SO(40)-2WO(Full)
Annual Sales: 4
Number of Employees: 80
Turnaround Times: B3C4S2
Customers: BP2-BU70-CU4-MP15-NP5-OR2
Union Shop: N
Year Established: 1970
Terms: Net 30 upon approval

Statement: We provide very good quality and service at competitive prices. A great fit for anyone who wants fast even distribution on your publication. We can do it all!

Hart Graphics **MG**

Marketing Director
800 SE Main St
PO Box 879
Simpsonville SC 29681
803-967-7821

PRINTED ITEMS			
A Annual Reports	G Galley Copies	N Workbooks	U Greeting Cards
B Books	H Journals	O Yearbooks	V Labels / Stickers
C Booklets	I Magazines	P Brochures	W Maps
D Catalogs	J Mass-Market Books	Q Calendars	X Newspapers
E Cookbooks	K Newsletters	R Comic Books	Y Postcards
F Directories	L Software Manuals	S Direct Mail Letters	Z Posters
	M Textbooks	T Envelopes	1 Stationery

Min: 100000 Max: 2000000 Opt: 0
Sizes: X
Bindings: PBI
Items: CI
Services: 4C-FF
Equipment: WO
Number of Employees: 100

Statement: This plant is designed to produce only perfectbound digests. They print **TV Guide** as well as other digest-sized books and publications.

Hart Graphics　　　GC
David E Hart, President
8000 Shoal Creek Blvd (78757)
PO Box 968
Austin TX 78767-0968
512-454-4761; 800-531-5471

Min: 50000 Max: 2000000 Opt: 0
Sizes: 25X
Bindings: PBI-SSI
Items: BDPS
Services: 4C-CS-FF-TY
Equipment: LP-SO-WO
Annual Sales: 6
Number of Employees: 750

Statement: Their speciality is full-color magazine inserts and envelopes. They can provide inserts with perforations, rub-offs, scratch-n-sniff, and other special effects. They were one of the first companies in the country to use the Otabind system.

Hart Press　　　MS
Marketing Director
Banta Publications Group
100 Banta Rd
Long Prairie MN 56347-1903
612-732-2121
Parent Company: Banta Corporation

Min: 10000 Max: 500000 Opt: 250000
Sizes: 245X
Bindings: PBI-SSI
Items: DIX
Services: 4C-CS-FF-LM-TD
Equipment: 3S0(25x38)-8WO(23)
Customers: CU10-MP70-NP20
Union Shop: Y
Year Established: 1927

Terms: Net 30

Statement: We provide efficient, quality service for 220 short to long-run publications.

Hatco Printing Corp　　　GC
Marketing Director
75 Oser Ave
Hauppauge NY 11788
516-231-1234; Fax: 516-434-9556

Havertown Printing Co　　　GC
Marketing Director
900 Sussex Blvd
Broomall PA 19008
215-544-7000; Fax: 215-544-1677

Hawkes Publishing　　　BK
John D Hawkes, President
5947 S 350 West
Salt Lake City UT 84107
801-266-5555; Fax: 801-266-5599
Send RFQ's to: John D Hawkes

Min: 20 Max: 100000 Opt: 5000-10000
Sizes: 25
Bindings: CBX-PBX-SSX-SDX
Items: BCEHIKLMNP
Services: TY-TD
Equipment: SO-WO(11x17)
Annual Sales: 1
Number of Employees: 3
Turnaround Times: B2S2
Customers: BP80-SP20
Union Shop: N
Year Established: 1965
Terms: 50% down, balance on delivery

Statement: To print and publish positive, helpful books.

Heartland Press Inc　　　MS
Tom McKay, Marketing Director
520 Second Ave East
Spencer IA 51301
800-932-9675; Fax: 712-262-6353
Send RFQ's to: Ken Grams

SERVICES			BINDINGS				
4C	4-Color Printing	RA	Rachwal System		SD	Side Stitching	
CS	Color Separations	AF	Acid-Free Paper	HC	Hardcover	SB	Spiral Binding
DA	Design / Artwork	RP	Recycled Paper	CB	Comb Binding	WB	Wire-O Binding
FF	Fulfillment/Mailing	TY	Typesetting	LB	Loose-Leaf Binding	I	In-House
LM	List Maintenance	TD	Typeset w/ Disk	OB	Otabind	O	Out of House
OC	OptiCopy System	WA	Warehousing	PB	Perfect Binding	X	Unknown
				SS	Saddle Stitching		

Min: 10000 Max: 200000 Opt: 30000
Sizes: 25X
Bindings: PBX-SSX
Items: ADEFHIPQ
Services: 4C-FF
Equipment: 2WO(6 unit, 5 unit)
Annual Sales: 4
Number of Employees: 120
Turnaround Times: B3S3
Customers: MP70-OT30
Union Shop: N
Year Established: 1974
Terms: 2% 10, net 30

Statement: At Heartland Press we will
develop the skills and utilize the highest
technology available to position ourselves
as the RR Donnelly of short run publications printing by the year 2000.

Heath Printers Inc **GC**
Marketing Director
1617 Boylston Ave
Seattle WA 98122
206-323-3577; Fax: 206-325-1636

Hederman Brothers Printing Co **GC**
Marketing Director
PO Box 491
Jackson MS 39205
601-961-7300; Fax: 601-961-7335

Henington Publishing Company **BK**
Marc Wensel, Marketing Director
112 W Main St
Wolfe City TX 75496
903-496-2226; Fax: 214-248-0100
Parent Company: Henington Industries Inc

Min: 100 Max: 5000 Opt: 0
Sizes: 2356
Bindings: CBX-HCX-PBX-SSX-SDX-
SBX-WBX-LBX
Items: BCEFOPQT
Services: 4C-AF
Equipment: 4SO(17x22,25x38)-LP
Annual Sales: 2
Turnaround Times: B6C6S6
Customers: BP50-SC50
Union Shop: N

Year Established: 1949
Terms: 1/3 down, 1/3 blues, 1/3 finish

Hennegan Company **CA**
Kevin D Ott
1001 Plum St
Cincinnati OH 45202
513-621-7300; Fax: 513-421-6995
Send RFQ's to: Nancy J Jacobs

Min: 0 Max: 0 Opt: 0
Sizes: 12345
Bindings: PBX-SSX-WBX
Items: ABCDFHIMNPYZ
Services: 4C-CS
Equipment: SO(44x60)-5SO(28x40)-
2WO(22x38)
Annual Sales: 5
Number of Employees: 325
Customers: BU100
Union Shop: Y
Year Established: 1886
Terms: Net 30

Heritage Printers Inc **BK**
Bill Loftin Sr, Marketing Director
PO Box 669407
Charlotte NC 28266-9407
704-392-1032; Fax: 704-393-3376
Send RFQ's to: Hazel Hathcock

Min: 50 Max: 7500 Opt: 2000
Sizes: 123456X
Bindings: HCO-CBO-OBO-PBO-SSO-
SBO-WBO
Items: BEH
Services: 4C-TY
Equipment: 2LP(28x41)
Annual Sales: 2
Number of Employees: 11
Turnaround Times: B5C6
Customers: BP25-CU75
Union Shop: N
Year Established: 1956
Terms: Net 30

Statement: Fine book manufacturer using
traditional linotype/monotype composition
and letterhead printing. 1700 fonts type.
Type matching for reissue titles set in
metal. Reproduction proofs for offset
painting.

PRINTED ITEMS					
A	Annual Reports	G	Galley Copies	N	Workbooks
B	Books	H	Journals	O	Yearbooks
C	Booklets	I	Magazines	P	Brochures
D	Catalogs	J	Mass-Market Books	Q	Calendars
E	Cookbooks	K	Newsletters	R	Comic Books
F	Directories	L	Software Manuals	S	Direct Mail Letters
		M	Textbooks	T	Envelopes
U	Greeting Cards
V	Labels / Stickers
W	Maps
X	Newspapers
Y	Postcards
Z	Posters
1	Stationery

BK

D B Hess Company
Edward J Webler, VP Sales
1150 McConnell Rd
Woodstock IL 60098
815-338-6900

Min: 3000 Max: 250000 Opt: 30000
Sizes: 5
Bindings: PBI-SSI-SBI
Items: BDFMN
Services: 4C-FF
Equipment: 2SO(28x40)-4WO(22)
Annual Sales: 4
Number of Employees: 75
Turnaround Times: B4C2S2
Customers: BP60-BU40
Union Shop: N
Year Established: 1978
Terms: Net 30

Statement: They have two divisions: educational (for workbooks, lab manuals, and test booklets) and commercial (for catalogs, annual reports, and training manuals). Both divisions specialize in one standard format, 8½ x 11 inches, in one, two, or four colors.

BK

Hignell Printing
Kevin Polley, Marketing Director
488 Burnell St
Winnepeg MB Canada R3G 2B4
204-783-7237; Fax: 204-774-4053
Send RFQ's to: Kevin Polley

Min: 500 Max: 10000 Opt: 3000
Sizes: 1235
Bindings: HCI-CBO-LBO-PBI-SSI-SDO-
        SBO-WBO
Items: BDFHKMNV
Services: 4C-AF-RP-TY-TD
Equipment: 3SO(29,38,40)
Annual Sales: 3
Number of Employees: 50
Turnaround Times: B3C3
Customers: BP70-BU5-CU15-OT10
Union Shop: N
Year Established: 1940
Terms: New accounts: 1/3, 1/3, 1/3

Statement: We are short-run book specialists with excellent turnaround time and top quality.

GC

Hillsboro Printing Co
Marketing Director
2442 W Mississippi Ave
Tampa FL 33629
813-251-2401; Fax: 813-251-0831

GC

Hiney Printing Co Inc
Marketing Director
1034 Home Ave
Akron OH 44310
216-535-1566; Fax: 216-535-5454

GC

Hinz Lithographing Company
Lloyd Shin, President
1750 W Central Rd
Mt Prospect IL 60056
708-253-2020; Fax: 708-253-6758

Min: 5000 Max: 25000000 Opt: 0
Sizes: 1235
Bindings: HCO-PBO-SSI-SBO
Items: ACDIKPQSWZ
Services: 4C-CS-TY
Equipment: 2SO(25x38)-2WO(18x26)
Annual Sales: 3
Number of Employees: 100
Turnaround Times: B3C3S2
Customers: BU98-SC2
Union Shop: N
Year Established: 1947
Terms: Net 30

GC

A B Hirschfeld Press
Trent Cunningham, Sales Manager
5200 Smith Rd
Denver CO 80216
303-320-8500; Fax: 303-329-3111
Send RFQ's to: Trent Cunningham

Min: 10000 Max: 1500000 Opt: 75000
Sizes: 123456X
Bindings: HCI-CBI-LBI-PBI-SSI-SDI-SBI-
        WBI
Items: ABCDEHIKLPQSTUWYZ
Services: 4C-CS-DA-FF-RP-TY-TD-WA
Equipment: 11SO-4WO(22x36,31x46)
Annual Sales: 5
Number of Employees: 250
Turnaround Times: B3C2S2

SERVICES			BINDINGS			
4C	4-Color Printing	RA Rachwal System	HC	Hardcover	SD	Side Stitching
CS	Color Separations	AF Acid-Free Paper	CB	Comb Binding	SB	Spiral Binding
DA	Design / Artwork	RP Recycled Paper	LB	Loose-Leaf Binding	WB	Wire-O Binding
FF	Fulfillment/Mailing	TY Typesetting	OB	Otabind	I	In-House
LM	List Maintenance	TD Typeset w/ Disk	PB	Perfect Binding	O	Out of House
OC	OptiCopy System	WA Warehousing	SS	Saddle Stitching	X	Unknown

Customers: BP5-BU70-CU3-MP3-SC5-
SP3-OT11
Union Shop: Y
Year Established: 1920
Terms: 50% down, balance prior to ship-
ping

Statement: We are proud to have the
reputation of quality and on-time delivery
in an industry full of empty promises.
Please call for a competitive bid.

Histacount Corp **GC**
Marketing Director
965 Walt Whitman Rd
Melville NY 11747
516-351-4900; Fax: 516-673-4794
800-645-5220

Hoechstetter Printing **GC**
Marketing Director
218 N Braddock Ave
Pittsburgh PA 15208
412-241-8200; Fax: 412-242-3835
Parent Company: Graphics Industries

Min: 0 Max: 0 Min: 0
Bindings: PBI-SSI
Items: BCDIN
Services: 4C
Eqiupment: SO

Holladay-Tyler Printing Corp **MS**
Howard Sullivan, President/CEO
7100 Holladay-Tyler Rd
Glenn Dale MD 20769-0459
301-464-9100; Fax: 301-464-6047
Parent Company: Southam Graphics Inc

Min: 20000 Max: 2000000 Opt: 0
Bindings: HCX-PBX-SSX-SDX
Items: BCDFHIN
Services: 4C
Equipment: WO
Annual Sales: 6
Number of Employees: 530

Rae Horowitz Book Manufac-turers **BK**
Alan Horowitz, VP Marketing
300 Fairfield Rd
PO Box 1308
Fairfield NJ 07004
201-575-7070; Fax: 201-575-4565
Send RFQ's to: Alan Horowitz

Min: 1000 Max: 500000 Opt: 10000
Sizes: 123456X
Bindings: HCI-PBI-SSI-SDI-SBO-WBO
Items: BEF
Services: 4C-CS-FF-AF-RP-WA
Equipment: 7SO(20x28,54x77)-WO(22x36)
Annual Sales: 5
Number of Employees: 400
Customers: BP90-CU5-OR5
Union Shop: N
Year Established: 1920
Terms: Net 30

Statement: We specialize in manufacturing
high quality 1, 2, and 4 color books and
components in our "one-stop" shopping
plant.

Horticultural Printers Co **GC**
Marketing Director
3638 Executive Blvd
Mesquite TX 75149
214-289-0705; Fax: 214-285-4881

House of Printing **GC**
Marketing Director
2544 Leghorn St
Mountain View CA 94043
415-964-9701; Fax: 415-969-8481

I

Imperial Litho/Graphics Inc **GC**
Lee F Zierten, Marketing Director
210 S 4th Ave
Phoenix AZ 85003
602-257-8500; Fax: 602-495-2544

PRINTED ITEMS	G	Galley Copies	N	Workbooks	U	Greeting Cards
A	Annual Reports	H	Journals	O	Yearbooks	V	Labels / Stickers
B	Books	I	Magazines	P	Brochures	W	Maps
C	Booklets	J	Mass-Market Books	Q	Calendars	X	Newspapers
D	Catalogs	K	Newsletters	R	Comic Books	Y	Postcards
E	Cookbooks	L	Software Manuals	S	Direct Mail Letters	Z	Posters
F	Directories	M	Textbooks	T	Envelopes	1	Stationery

Send RFQ's to: Lee F Zierten

Min: 1000 Max: 100000 Opt: 25000
Sizes: 25
Bindings: PBX-SSX-SBX-WBX
Items: ACDFKLP
Services: 4C-CS-DA-FF-AF-RP-WA
Equipment: 6S0(28x40)-2WO
   (17.75x36,17.75x20)
Annual Sales: 4
Number of Employees: 210
Turnaround Times: B3C4S2
Customers: BP5-BU75-CU5-MP5-NP10
Union Shop: N
Year Established: 1959
Terms: Net 30 with approved credit

Independent Printing Company **GC**

Marketing Director
215 E 42nd St
New York NY 10017
212-689-5100

Min: 25 Max: 200 Opt: 1000
Sizes: 235
Bindings: CBI-LBI-PBI-SSI-WBI
Items: BGHP
Services: SO-WO(mini)
Terms: Net 30 with credit approval

Statement: Specialize in printing and ultra-short runs from 25 to 2500 copies.

Independent Publishing Company **BK**

Ned Burke, Marketing Director
PO Box 15126
Sarasota FL 34277
813-366-0608; Fax: 813-366-0608
Send RFQ's to: Ned Burke

Min: 100 Max: 2500 Opt: 500
Sizes: 25
Bindings: PBI-SSI
Items: BCFHIKPT1
Services: 4C-TY-TD
Equipment: 2SO
Annual Sales: 1
Number of Employees: 5
Turnaround Times: B3S2
Customers: BU10-MP10-SP70-BP10
Union Shop: N
Year Established: 1973
Terms: Payable in advance

Statement: Printing small publications is our specialty — not a sideline. We are not a "quick" shop. We, as desktop publishers, also know the time and care you put into your product. We will print it with the same care. Low cost and high quality.

Industrial Printing Co **GC**

Marketing Director
1635 Coining Dr
Toledo OH 43612
419-476-9101; Fax: 419-476-8692

Inland Press **GC**

Marketing Director
2001 W Lafayette Blvd
Detroit MI 48216
313-961-6000; Fax: 313-961-7817

Inland Printing Co Inc **GC**

Marketing Director
2009 West Ave S
PO Box 1268
La Crosse WI 54602-1268
608-788-5800; Fax: 608-787-5870
800-657-4413

Inland Printing Company **BR**

Steven Spaeth, Marketing Director
PO Box 414
Syosset NY 11791
516-367-4700; Fax: 516-367-4700
Send RFQ's to: Steven Spaeth

Min: 5000 Max: 500000 Opt: 0
Sizes: 2356
Items: ACDFIMX
Annual Sales: 3
Terms: Net 30

Statement: Inland Printing Co is a sales agency representing a number of diversified printers. Our full-service firm offers the highest quality printing and service with price savings.

SERVICES		BINDINGS			
4C	4-Color Printing	RA	Rachwal System	HC	Hardcover	SD	Side Stitching
CS	Color Separations	AF	Acid-Free Paper	CB	Comb Binding	SB	Spiral Binding
DA	Design / Artwork	RP	Recycled Paper	LB	Loose-Leaf Binding	WB	Wire-O Binding
FF	Fulfillment/Mailing	TY	Typesetting	OB	Otabind	I	In-House
LM	List Maintenance	TD	Typeset w/ Disk	PB	Perfect Binding	O	Out of House
OC	OptiCopy System	WA	Warehousing	SS	Saddle Stitching	X	Unknown

Insert Color Press **MG**

Marketing Director
90 Air Park Dr
Ronkonkoma NY 11779
516-981-5300; 800-356-3943

Min: 0 Max: 0 Opt: 0
Sizes: 5X
Bindings: SSI
Items: IX

Intelligencer Printing **CA**

Stephen J Brody, Sales Manager
330 Eden Rd
PO Box 17603
Lancaster PA 17601
717-291-3100; Fax: 717-569-2752
800-233-0107
Send RFQ's to: Stephen J Brody

Min: 25000 Max: 2000000 Opt: 0
Sizes: 23456X
Bindings: PBI-SSI-SDI
Items: ADIPQUZ
Services: 4C-CS
Equipment: SO-WO
Annual Sales: 5
Number of Employees: 225
Year Established: 1794

Statement: Write for a copy of their **Intellectual** newsletter.

Interform Corporation **GC**

Tom Wisinski, Marketing Director
1901 Mayview Rd
Bridgeville PA 15017
412-221-3300; Fax: 412-221-8937
Send RFQ's to: Andy Booth
Parent Company: Guaranty Reassurance
    Corporation

Min: 0 Max: 0 Opt: 0
Sizes: 123456
Bindings: HCO-CBI-LBO-OBO-PBO-SSI-
    SDI-SBO-WBO
Items: ABCDEFGHIKLNPQSTYZ12
Services: 4C-CS-DA-FF-LM-AF-RP-TY-
    TD-WA
Equipment: 3LP(10x15,11x17)-
    3SO(19x25,14x20,28x40)
Annual Sales: 5

Number of Employees: 325
Turnaround Times: B4C5S3
Union Shop: Y
Year Established: 1964
Terms: 2% 15, net 30

Statement: For over 30 years, the name Interform has become synonymous with quality and service. Throughout those years, independent distributors have come to know that sending a job to Interform means you are assured it will be done right and on time.

International Lithographing **GC**

Marketing Director
11631 Caroline Rd
Philadelphia PA 19154
215-677-9000

Interprint Inc **GC**

Marketing Director
12350 US Hwy 19N
Clearwater FL 34624
813-531-8957; Fax: 813-536-0647

Interstate Printing Company **CA**

Stanley Erickson, Sales Director
2002 N 16th St
PO Box 3667
Omaha NE 68103
402-341-8028; Fax: 402-341-6168
800-788-4177

Min: 3000 Max: 2000000 Opt: 0
Sizes: 25X
Bindings: SSX
Items: ABCDKNPQRSWXYZ
Services: 4C-FF-TY-TD-WA
Equipment: CB-LP-SO-WO
Annual Sales: 2
Number of Employees: 125
Turnaround Times: B5C2S2
Customers: BU75-CU15-NP5-SC5
Union Shop: Y
Year Established: 1917
Terms: Net 10 following month

Statement: They specialize in gang printing and mailing of insurance company letters, magazines with print runs between 20,000

PRINTED ITEMS	G	Galley Copies	N	Workbooks	U	Greeting Cards
A Annual Reports	H	Journals	O	Yearbooks	V	Labels / Stickers
B Books	I	Magazines	P	Brochures	W	Maps
C Booklets	J	Mass-Market Books	Q	Calendars	X	Newspapers
D Catalogs	K	Newsletters	R	Comic Books	Y	Postcards
E Cookbooks	L	Software Manuals	S	Direct Mail Letters	Z	Posters
F Directories	M	Textbooks	T	Envelopes	1	Stationery

and 500,000, and direct mail catalogs and inserts with runs between 50,000 and 1,000,000.

IPC Publishing Services **GC**
501 Colonial Dr
PO Box 26
Saint Joseph MI 49085
616-983-7105; Fax: 616-983-5736
Send RFQ's to: Greg Forbes, President
Parent Company: Journal Communications

Min: 2500 Max: 70000 Opt: 7-35000
Sizes: 2345X
Bindings: CBX-LBX-OBX-PBX-SSX-SBX-WBX
Items: BDEFHILMN
Services: 4C-CS-DA-FF-OC-TY-TD-WA
Equipment: 11SO(40)-4WO(23⁹⁄₁₆, 19 ³⁄₁₆ cutoff
Annual Sales: 6
Number of Employees: 560
Turnaround Times: B1S2
Customers: Technical/medical 50%, software 50
Union Shop: N
Year Established: 1953
Terms: 1% 10, net 30

Statement: IPC Publishing believes people are the foundation of all professional relationships and the essence of these relationships is integrity, reliability, and honesty. IPC's mission is to provide its cusotmers with quality products and service at competitive prices with a faster turnaround than is normal in the industry. To achieve this goal, IPC continually invests in state-of-the-art equipment to provide its customers with modern manufacturing efficiencies. However, the heart and soul of the company is a staff of truly dedicated people who feel close, personal communication with their customers is the key to achieving IPC's objective of continuous quality improvements.

I/P/D Printing **GC**
Marketing Director
5800 Peachtree Rd
Chamblee GA 30341
404-458-6351; Fax: 404-454-6326

Ivy Hill Packaging **GC**
Sue Goodlett, General Manager
PO Box 18640
4325 Old Shepherdsville Rd
Louisville KY 40218
502-458-5303; Fax: 502-458-2125

Min: 0 Max: 0 Opt: 0
Items: PV
Equipment: SO

Statement: Mainly music labels and packaging.

J

Jacob North Printing Co Inc **GC**
Marketing Director
2615 O St
Lincoln NE 68510
402-475-5335; Fax: 402-475-0381

Japs-Olson Company **GC**
Michael Beddor, Vice President
30 N 31st Ave
Minneapolis MN 55411
612-522-4461; Fax: 612-522-3837
800-548-2897

Min: 0 Max: 0 Opt: 0
Sizes: 23456
Bindings: SSI
Items: CDKPSTV
Services: 4C-FF-TY
Equipment: 8SO(to 26x40)-7WO(to 26x23)
Annual Sales: 5
Number of Employees: 225
Turnaround Times: C2S2
Customers: BU65-MP20-NP10-SP5
Union Shop: N
Terms: Net 30

Statement: We are complete printers for the direct marketing industry: order envelopes, web and sheetfed catalogs, service bureau, and mailing.

SERVICES		BINDINGS			
4C	4-Color Printing	RA Rachwal System	HC Hardcover	SD	Side Stitching
CS	Color Separations	AF Acid-Free Paper	CB Comb Binding	SB	Spiral Binding
DA	Design / Artwork	RP Recycled Paper	LB Loose-Leaf Binding	WB	Wire-O Binding
FF	Fulfillment/Mailing	TY Typesetting	OB Otabind	I	In-House
LM	List Maintenance	TD Typeset w/ Disk	PB Perfect Binding	O	Out of House
OC	OptiCopy System	WA Warehousing	SS Saddle Stitching	X	Unknown

Jefferson/Keeler Printing Co **GC**
Marketing Director
1234 S Kingshighway Blvd
St Louis MO 63110
314-533-8087; Fax: 314-533-2369

Jersey Printing Company **MS**
Arthur Shlossman, CEO
111 Linnet St
PO Box 79
Bayonne NJ 07002
201-436-4200; Fax: 201-436-0116
Send RFQ's to: Arthur Shlossman

Min: 100 Max: 50000 Opt: 10000
Sizes: 123456
Bindings: PBI-SSI
Items: ABCDFHIKPQYZ
Services: 4C-CS-DA-FF-TY-TD-WA
Equipment: 9LP-5SO(20x28,25x38)
Annual Sales: 4
Number of Employees: 125
Turnaround Times: B5C4S2
Customers: BP3-BU36-CU5-MP38-NP5-
    OT13
Union Shop: Y
Year Established: 1910
Terms: Net 30

Statement: Jersey prints museum and gallery quality posters, catalogs of art reproductions, annual reports, house organs, publications, and advertising brochures.

Jet LithoColor Inc **GC**
Marketing Director
1500 Centre Cir Dr
Downers Grove IL 60515
708-932-9000; Fax: 708-932-9101

JK Creative Printers **BK**
Mike Nobis, Marketing Director
2029 Hollister Whitney Pkwy
Quincy IL 62301
217-222-5145; Fax: 217-222-5149
Send RFQ's to: Mike Nobis

Min: 500 Max: 500000 Opt: 10000-50000

Sizes: 123456
Bindings: PBX-SSX-SDX-SBX
Items: ABCDIKLMNPSTVYZ12
Services: 4C-DA-OC-AF-RP-TY-TD
Equipment: 4LP(18x20)-10SO(10x15,
    24x32)
Annual Sales: 2
Number of Employees: 34
Turnaround Times: B3S2
Customers: BP30-BU40-CU20-OR10
Union Shop: Y
Year Established: 1914
Terms: Net 30

Statement: To provide top quality color and black and white printing on tight schedule and at reasonable prices.

The Job Shop **GC**
David Shephard, Owner
3 Water St
PO Box 305
Woods Hole MA 02543
508-548-9600

Min: 100 Max: 0 Opt: 1000
Sizes: 12345
Bindings: HCO-CBI-LBI-PBI-SSI-SDI
Items: BCDFKLNPSTVY1
Services: DA-TY-TD
Equipment: 2SO(11x17,14x18)
Annual Sales: 1
Number of Employees: 5
Turnaround Times: B4C3
Customers: BP10-BU25-CU5-NP15-OR20-
    SP5-OT2
Union Shop: N
Year Established: 1967
Terms: 30% down, 30% delivery, net 30

Statement: Our goal is to be as adaptable as possible to the needs of short-run publishers of books and newsletters.

JohnsByrne Company **MS**
Marketing Director
7350 Croname Rd
Niles IL 60648
708-647-7227; Fax: 708-647-2238

Min: 0 Max: 0 Opt: 0
Bindings: SSI
Items: DIKQZ

PRINTED ITEMS			
A Annual Reports	G Galley Copies	N Workbooks	U Greeting Cards
B Books	H Journals	O Yearbooks	V Labels / Stickers
C Booklets	I Magazines	P Brochures	W Maps
D Catalogs	J Mass-Market Books	Q Calendars	X Newspapers
E Cookbooks	K Newsletters	R Comic Books	Y Postcards
F Directories	L Software Manuals	S Direct Mail Letters	Z Posters
	M Textbooks	T Envelopes	1 Stationery

Services: 4C-CS
Equipment: SO

MS
The Johnson & Hardin Company
Marketing Director
3600 Red Bank Rd
Cincinnati OH 45227
513-271-8834; Fax: 513-271-3603
Send RFQ's to: Andrew M Jamison

Min: 5000 Max: 2000000 Opt: 100000
Sizes: 2345X
Bindings: HCX-PBX-SBX
Items: ABCDEFHIKLMNP
Services: 4C-TY
Equipment: LP-SO-WO
Year Established: 1902

GC
Johnson Graphics
Marketing Director
120 Frentress Lake Rd
East Dubuque IL 61025
815-747-6511

Min: 2000 Max: 50000 Opt: 0
Sizes: 2345X
Bindings: SSI
Items: ACDFIKNP
Services: FF
Equipment: SO

GC
The Johnson Press Inc
Marketing Director
800 North Ct St
Pontiac IL 61764-0110
815-844-5161; Fax: 815-842-1349

GC
Johnson Printing & Packaging Co
Marketing Director
40 77th Ave NE
Fridley MN 55432
612-574-1700; Fax: 612-574-0191

GC
Jostens Printing & Publishing
Marketing Director
2505 Empire Dr
Winston-Salem NC 27113
919-765-0070

Min: 1000 Max: 1000000 Opt: 30000
Sizes: 123456
Bindings: HCI-CBO-LBO-PBI-SSI-SDI-
    SBO-WBO
Items: ABCDEFGHIKLMNOPQXYZ
Services: 4C-CS-DA-TY-TD-WA
Equipment: 8SO(25x38,28x40)
Annual Sales: 4
Number of Employees: 375
Turnaround Times: B5C3S3
Customers: BP5-BU40-CU50-MP5
Union Shop: N
Terms: Net 30

BK
Jostens Printing & Publishing
Marketing Director
1312 Dickson Highway
PO Box 923
Clarksville TN 37041
615-647-5211

Min: 100 Max: 100000 Opt: 10000
Sizes: 123456X
Bindings: HCI-CBI-PBI-SSI-SBI
Items: BCDEINOQZ
Services: 4C-CS-DA-TY-TD
Equipment: SO(25x38)(many)
Annual Sales: 6
Number of Employees: 8000
Union Shop: N
Year Established: 1950
Terms: Net 30 with approved credit

Statement: A yearbook publisher with divisions in Tennessee, California, Kansas, and Pennsylvania.

BK
Jostens Printing & Publishing
Marketing Director
401 Science Park Rd
PO Box 297
State College PA 16801
814-237-5771

Min: 100 Max: 100000 Opt: 10000
Sizes: 123456X
Bindings: HCI-CBI-PBI-SSI-SBI
Items: BCDEINOQZ
Services: 4C-CS-DA-TY-TD
Equipment: SO(25x38)(many)
Annual Sales: 6
Number of Employees: 8000

SERVICES				BINDINGS		SD	Side Stitching
4C	4-Color Printing	RA	Rachwal System	HC	Hardcover	SB	Spiral Binding
CS	Color Separations	AF	Acid-Free Paper	CB	Comb Binding	WB	Wire-O Binding
DA	Design / Artwork	RP	Recycled Paper	LB	Loose-Leaf Binding	I	In-House
FF	Fulfillment/Mailing	TY	Typesetting	OB	Otabind	O	Out of House
LM	List Maintenance	TD	Typeset w/ Disk	PB	Perfect Binding	X	Unknown
OC	OptiCopy System	WA	Warehousing	SS	Saddle Stitching		

Union Shop: N
Year Established: 1950
Terms: Net 30 with approved credit

Statement: A yearbook printer with divisions in Pennsylvania, Tennessee, Kansas, and California.

Jostens Printing & Publishing **BK**
Marketing Director
4000 S Adams St
Topeka KS 66609
913-266-3300

Min: 100 Max: 100000 Opt: 10000
Sizes: 123456X
Bindings: HCI-CBI-PBI-SSI-SBI
Items: BCDEINOQZ
Services: 4C-CS-DA-TY-TD
Equipment: SO(25x38)(many)
Annual Sales: 6
Number of Employees: 8000
Union Shop: N
Year Established: 1950
Terms: Net 30 with approved credit

Statement: A yearbook printer with divisions in Pennsylvania, Tennessee, Kansas, and California.

Jostens Printing & Publishing **BK**
Marketing Director
29625 Rd 84
PO Box 991
Visalia CA 93291
209-651-3300

Min: 100 Max: 100000 Opt: 10000
Sizes: 123456X
Bindings: HCI-CBO-LBO-PBI-SSI-SDI-SBO-WBO
Items: BCDEFHILNOPQSYZ
Services: 4C-CS-DA-TY
Equipment: 7SO(25x38)
Annual Sales: 6
Number of Employees: 8000
Turnaround Times: B4C4S4
Customers: BP5-BU30-MP35-OR8-SC10-SP8-OT4
Union Shop: N
Year Established: 1950
Terms: 35% down, 35% proofs, 30% net 30

Statement: A yearbook printer with divisions in Pennsylvania, Tennessee, Kansas, and California.

Jostens /Printing & Publishing **BK**
Patrick M Hickey, Marketing Director
5501 Norman Center Dr
Minneapolis MN 55437
612-830-3300; Fax: 612-830-0818

Min: 200 Max: 12000 Opt: 1000
Sizes: 3456
Bindings: HCI-CBI-OBI-PBI-SSI-SDI-SBI-WBI
Items: ABCDEFHIMNOPQZ1
Services: 4C-CS-DA-AF-RP-TY-TD-WA
Equipment: 42SO(38 to 40)
Annual Sales: 6
Number of Employees: 3800
Turnaround Times: B3C4S2
Customers: BU20-SC80
Union Shop: Y
Year Established: 1959
Terms: Net 10

Joyce Printing Inc **GC**
Marketing Director
600 Cleveland Ave
Albany CA 94710
510-525-3401; Fax: 510-525-7142

Judd's Incorporated **MS**
Marketing Director
1500 Eckington Pl NE
Washington DC 20002
202-636-9283

Min: 0 Max: 0 Opt: 0
Bindings: PBI-SSI
Annual Sales: 6
Number of Employees: 1200

Judd's Incorporated **MS**
Marketing Director
Box 777 Route 55E
Strasburg VA 22657
703-465-3731; Fax: 703-465-3737

Min: 0 Max: 0 Opt: 0

PRINTED ITEMS			
A Annual Reports	G Galley Copies	N Workbooks	U Greeting Cards
B Books	H Journals	O Yearbooks	V Labels / Stickers
C Booklets	I Magazines	P Brochures	W Maps
D Catalogs	J Mass-Market Books	Q Calendars	X Newspapers
E Cookbooks	K Newsletters	R Comic Books	Y Postcards
F Directories	L Software Manuals	S Direct Mail Letters	Z Posters
	M Textbooks	T Envelopes	1 Stationery

Bindings: PBI-SSI
Annual Sales: 6
Number of Employees: 1200

Statement: NY: 212-921-9180; Atlanta: 404-980-6711.

Julin Printing **GC**
Ruth Julin, President
225 S Locust St
Monticello IA 52310
319-465-3558
Send RFQ's to: Ruth Julin

Min: 2000 Max: 2000000 Opt: 10000
Sizes: 123456
Bindings: HCO-CBI-LBI-PBI-SSI-SDI-SBI-WBI
Items: ABCDEFIKPQSTUWYZ1
Services: 4C-CS-DA-AF
Equipment: 6SO(17 to 61)
Annual Sales: 2
Number of Employees: 20
Turnaround Times: B3C3S3
Customers: BP10-BU60-CU20-MP10
Union Shop: N
Year Established: 1954
Terms: Net 30

K

K & H Printers & Lithographers **GC**
Marketing Director
1611 Broadway
Everett WA 98201
206-252-2145
Send RFQ's to: Stacey Coley

K-B Offset Printing **CA**
Marketing Director
1006 W College Ave
State College PA 16801
814-238-8445

Min: 0 Max: 0 Opt: 0
Sizes: 2345X
Bindings: SSI
Items: ACDIKP
Services: 4C

Statement: 4-color brochures at prices you've been looking for. Also catalogs reports, and periodicals.

Kaufman Press Printing **GC**
Marketing Director
PO Box 68
Syracuse NY 13207
315-471-1817
Send RFQ's to: Gary J Dedell

Min: 1000 Max: 20000 Opt: 0
Sizes: 123456X
Bindings: HCO-CBI-LBI-PBO-SSI-SDI-SBO-WBO
Items: ACDHIKLPQSTVWXYZ1
Services: 4C-CS-DA-FF-TY-WA
Equipment: 14LP-11SO-5WO
Annual Sales: 2
Number of Employees: 50
Turnaround Times: B3C2S2
Customers: BP5-BU60-CU5-MP5-NP10-OR8-SC5
Union Shop: Y
Year Established: 1910
Terms: 1% 10, net 20

Keys Printing Co Inc **GC**
Dick Flinn, Marketing Director
PO Box 8
Greenville SC 29602
803-288-6560; Fax: 803-297-1661
Send RFQ's to: Dick Flinn

Min: 1000 Max: 50000 Opt: 20000
Sizes: 23456
Bindings: LBX-SSX-SDX-SBX
Items: CDEKLPTWZ14
Services: 4C-CS-FF-LM-OC-AF-RP-TY-TD-WA
Equipment: 13SO(40)
Annual Sales: 4
Number of Employees: 106
Turnaround Times: S2
Union Shop: N
Year Established: 1896
Terms: Net 30

Statement: QUALITY! QUALITY! QUALITY!

SERVICES			BINDINGS		
4C	4-Color Printing	RA Rachwal System	HC Hardcover	SD	Side Stitching
CS	Color Separations	AF Acid-Free Paper	CB Comb Binding	SB	Spiral Binding
DA	Design / Artwork	RP Recycled Paper	LB Loose-Leaf Binding	WB	Wire-O Binding
FF	Fulfillment/Mailing	TY Typesetting	OB Otabind	I	In-House
LM	List Maintenance	TD Typeset w/ Disk	PB Perfect Binding	O	Out of House
OC	OptiCopy System	WA Warehousing	SS Saddle Stitching	X	Unknown

Kimberly Press   BK
Bill McNally, President
5390 Overpass Rd
Santa Barbara CA 93111-2008
805-964-6469
Send RFQ's to: Bill McNally

Min: 500 Max: 5000 Opt: 2000
Sizes: 2345
Bindings: HCI-PBI-SSI
Items: BCGHI
Services: 4C-FF

Statement: We are serious local printers in a high overhead resort area. Our prices will never be as low as printers in Michigan. Many of Santa Barbara's 100+ publishers choose to have their printing done here. We specialize in producing scholarly books and journals.

Kimm Printing   GC
Marketing Director
428 Washington Ave N
Minneapolis MN 55401
612-332-3311; Fax: 612-332-0809
Parent Co: Kimm Company

Kinney Printing Co   GC
Marketing Director
3375 W Columbus Ave
Chicago IL 60652
312-436-2525

Kirby Lithographic Co Inc   BK
Thomas Mackey III
2900 South Eads St
Arlington VA 22202
703-684-7600; Fax: 703-683-5918
800-932-3594
Send RFQ's to: Thomas Mackey III

Min: 250 Max: 20000 Opt: 2500
Sizes: 123456X
Bindings: HCO-CBO-LBO-PBI-SSI-SDI-
  SBO-WBO
Items: BCFHLNP2
Services: OC-AF-TD
Equipment: 5SO(44x64,41x55,25x38,19x25)

Annual Sales: 3
Number of Employees: 50
Turnaround Times: B2C3S2
Customers: BP75-CU5-NP10-OR10
Union Shop: N
Year Established: 1927
Terms: Net 30

Statement: Kirby provides extraordinary quality and service often forgotten today. We have over 60 years of experience producing books and journals of distinction.

KNI Incorporated   BK
Peggy Bryant, Marketing Director
1261 S State College Pkwy
Anaheim CA 92806
714-956-7300; Fax: 714-635-1744
Send RFQ's to: Jerry Bernstein

Min: 500 Max: 25000 Opt: 5000-10000
Sizes: 12345X
Bindings: HCO-CBI-LBI-PBI-SSI-SDO-
  SBO-WBO
Items: BCDEFHLMN
Services: 4C-OC-RP
Equipment: 2SO(20x29,14x20)-
  4WO(23$\frac{9}{16}$x38,21.5x30,22x17.5)
Annual Sales: 3
Number of Employees: 60
Turnaround Times: B4S4
Customers: BP50-BU25-CU5-SC5-SP5-
  OR5-OT5
Union Shop: N
Year Established: 1970
Terms: Various

Statement: Short to medium-run manufacturer of books, manuals, directories, and catalogs.

Kolor View Press   GC
Shelly Pichler, Marketing Director
112 W Olive
Aurora MO 65605
417-678-2135; Fax: 417-678-3626
800-225-6567
Parent Company: MWM Dexter Inc

Min: 500 Max: 1000000 Opt: 10000
Sizes: 123456
Bindings: SSX-SBX-WBX-SDX

PRINTED ITEMS					
A	Annual Reports	G	Galley Copies	N	Workbooks
B	Books	H	Journals	O	Yearbooks
C	Booklets	I	Magazines	P	Brochures
D	Catalogs	J	Mass-Market Books	Q	Calendars
E	Cookbooks	K	Newsletters	R	Comic Books
F	Directories	L	Software Manuals	S	Direct Mail Letters
		M	Textbooks	T	Envelopes

U	Greeting Cards
V	Labels / Stickers
W	Maps
X	Newspapers
Y	Postcards
Z	Posters
1	Stationery

Items: ABCDEFKPQUWZY1
Services: 4C-CS-TY-DA-FF-LM-OC-RP-
TD-WA-AF
Equipment: LP(11x17)-3SO(40)
Annual Sales: 4
Number of Employees: 150
Turnaround Times: S4
Customers: BU90-NP10
Union Shop: N
Year Established: 1922
Terms: 50% down, balance COD

Statement: Provide quality full-color printing, competitive prices and guaranteed
turnaround. 5, 10, 15 day KolorKwik Service or your order is FREE! Economical
five week service is also available.

Kordet Graphics Inc   CA
Marketing Director
15 Neil Ct
Oceanside NY 11572
516-766-4114

Min: 0 Max: 0 Opt: 0
Sizes: 5
Bindings: SSI
Items: DP
Services: 4C-CS-DA-TY
Equipment: SO
Annual Sales: 5
Number of Employees: 375

The C J Krehbiel Company   GC
R D Hastings, Marketing Director
3962 Virginia Ave
Cincinnati OH 45227
513-271-6035; Fax: 513-271-6082
Send RFQ's to: R D Hastings

Min: 5000 Max: 1000000+ Opt: 750000
Sizes: 123456
Bindings: SDI-SBI-WBI-HCI-OBI-PBI-
SSI-CBO-LBO
Items: ABCDEFGHIJKLMNOPQWZ
Services: 4C-DA-FF-OC-AF-RP-WA
Equipment: 4SO-4WO
Annual Sales: 4
Number of Employees: 175
Turnaround Times: B2C3S2
Union Shop: N
Year Established: 1872

Terms: Net 30, will consider 2% 10

Statement: The C J Krehbiel Company is a
graphic arts company supplying print
media buyers, on a national scale, with the
highest quality products and services at a
fair price.

L

La Crosse Graphics   BK
W Krause, Marketing Director
3025 E Ave South (54601)
PO Box 249
LaCrosse WI 54602
608-788-2500; Fax: 608-788-2660
800-832-2503
Send RFQ's to: Tom Morgan

Min: 1000 Max: 50000 Opt: 10000
Sizes: 123456
Bindings: HCI-CBO-LBO-PBI-SSI-SDI-
SBO-WBO
Items: ABCDEFHIKPQSTWYZ1
Services: 4C-DA-AF-RP-TY-TD
Equipment: 4LP(25x38)-2SO(25x36)
Annual Sales: 2
Number of Employees: 56
Turnaround Times: B4C3S4
Customers: BP10-BU53-CU20-MP2-NP15
Union Shop: N
Year Established: 1987
Terms: Net 30

Statement: We offer top quality with on
time delivery.

Lakeland Litho-Plate Inc   GC
Marketing Director
2863 E Grand Blvd
Detroit MI 48202
313-871-1535; Fax: 313-871-4462

Lakeway Printers Inc   GC
Marketing Director
1609 W 1st North St
Morristown TN 37814
615-581-5630; Fax: 615-581-3061

SERVICES		BINDINGS		
4C	4-Color Printing	RA	Rachwal System	HC	Hardcover	SD	Side Stitching
CS	Color Separations	AF	Acid-Free Paper	CB	Comb Binding	SB	Spiral Binding
DA	Design / Artwork	RP	Recycled Paper	LB	Loose-Leaf Binding	WB	Wire-O Binding
FF	Fulfillment/Mailing	TY	Typesetting	OB	Otabind	I	In-House
LM	List Maintenance	TD	Typeset w/ Disk	PB	Perfect Binding	O	Out of House
OC	OptiCopy System	WA	Warehousing	SS	Saddle Stitching	X	Unknown

MS

Lancaster Press

Marketing Director
3575 Hempland Rd
Lancaster PA 17603
717-285-9095; Fax: 717-285-7261

Min: 5000 Max: 95000 Opt: 0
Sizes: 235
Bindings: PBI-SSI
Items: BDHI
Services: 4C-CS-FF-TY-TD
Equipment: SO-WO
Annual Sales: 4
Number of Employees: 225

MS

The Lane Press

Tom Drumheller, VP Marketing
PO Box 130
Burlington VT 05402
802-863-5555; Fax: 802-865-1714
800-733-3740
Send RFQ's to: Tom Drumheller

Min: 10000 Max: 200000 Opt: 45000
Sizes: 5X
Bindings: PBI-SSI
Items: HI
Services: 4C-CS-RP-TD-WA
Equipment: 3SO(25x35)-26WO(22.75x36)
Annual Sales: 5
Number of Employees: 325
Turnaround Times: B3S3
Customers: CU20-MP45-NP15
Union Shop: N
Year Established: 1904
Terms: Net 30

Statement: A tradition of unparalleled excellence in the printing of magazines, catalogs, and tabloids since 1904.

CA

Lasky Company

Ronald H Barnhard, President
67 E Willow St
Milburn NJ 07041
201-376-9200; Fax: 201-376-3832

Min: 5000 Max: 2000000 Opt: 0
Sizes: 2345X
Bindings: PBO-SSO
Items: ACDIKPXZ
Services: 4C-CS-TY

Equipment: 4SO(25x39,28x40)-4WO(to 23x38)
Annual Sales: 6
Number of Employees: 430
Year Established: 1917

Statement: We love to print jobs that other printers say are too tough or impossible to do, because that's where we always show our true colors.

GC

LBA Custom Printing Inc

Marketing Director
207 Arco Dr
Toledo OH 43607
419-535-3151; Fax: 419-535-9538

GC

Lebanon Valley Offset Inc

Marketing Director
1325 E Main St
Annville PA 17003
717-867-4601; Fax: 717-867-4156
800-444-4586

GC

Lehigh Press/Cadillac

Mr Bob Friedman, Marketing Director
25th & Lexington St
Broadview IL 60153
708-681-3612; Fax: 708-681-3673
Send RFQ's to: Mr Bob Friedman

Min: 10000 Max: 15000000 Opt: 1500000
Items: SW
Services: 4C
Equipment: 10WO(26 to 50)(17.75 to 45.67)
Annual Sales: 6
Number of Employees: 200
Customers: BU100
Union Shop: Y
Terms: Net 30

GC

Lehigh Press Inc

Marketing Director
51 Haddonfield Rd
Cherry Hill NJ 08002
609-665-5200

Min: 20000 Max: 2000000 Opt: 0

PRINTED ITEMS		G	Galley Copies	N	Workbooks	U	Greeting Cards
A	Annual Reports	H	Journals	O	Yearbooks	V	Labels / Stickers
B	Books	I	Magazines	P	Brochures	W	Maps
C	Booklets	J	Mass-Market Books	Q	Calendars	X	Newspapers
D	Catalogs	K	Newsletters	R	Comic Books	Y	Postcards
E	Cookbooks	L	Software Manuals	S	Direct Mail Letters	Z	Posters
F	Directories	M	Textbooks	T	Envelopes	1	Stationery

Sizes: 23456X
Bindings: SSI
Items: ADIPSTWYZ
Services: 4C-CS-LM
Equipment: WO
Annual Sales: 6
Number of Employees: 1060
Year Established: 1924

Statement: Lehigh prints magazines, catalogs, direct mail components, inserts, book covers, and book jackets.

BK
Les Editions Marquis
Thomas Hussey, Marketing Manager
305 E Tache Blvd
Montmagny PQ Canada G5V 1C7
418-248-0737

Min: 300 Max: 10000 Opt: 2000
Sizes: 24
Bindings: HCO-CBO-LBO-PBI-SSO-
 SDO-SBO-WBO
Items: ABCDEFLMNOPWZ
Services: 4C-CS-AF-TY-TD
Equipment: 6SO(19x25,28x40)
Annual Sales: 2
Number of Employees: 35
Turnaround Times: B3C3
Customers: BP70-BU5-CU15-SP5-OT5
Union Shop: Y
Year Established: 1937
Terms: Net 30

Statement: We print top quality paperbacks and color publications and are competitive for low runs.

GC
Lighthouse Press Inc
Marketing Director
Lighthouse Group
177 E Industrial Park Dr
Manchester NH 03109
603-668-7760

Min: 0 Max: 0 Opt: 0
Sizes: 123456X
Bindings: CBI-SDI-SSI-WBI-HCO-LBO-
 OBO-PBO-SBO
Items: ABCDFHIKLNPQSXYZ1
Services: 4C-RP-TD-WA
Equipment: 4SO(28 to 40)-8WO(36)
Annual Sales: 3

Number of Employees: 120
Turnaround Times: B2S2
Union Shop: N
Year Established: 1980

GC
Litho Industries Inc
Marketing Director
PO Box 14106
Durham NC 27709
919-596-7000; Fax: 919-596-7555

GC
Litho Productions Inc
Marketing Director
2800 Perry St
PO Box 9423
Madison WI 53715
608-271-8100; Fax: 608-271-8237

Min: 0 Max: 0 Opt: 0
Sizes: 123456
Bindings: SSX
Items: ABCDFPQST1
Services: 4C-CS-AF-RP-WA
Equipment: 4SO(26x40)
Annual Sales: 3
Number of Employees: 60
Turnaround Times: S2
Union Shop: Y
Year Established: 1949
Terms: Net 30

Statement: Litho Productions shall continually strive to achieve excellence in our "Niche" as a high quality, multi-color, sheetfed offset lithographer. Our purpose is to contribute to the success of our customers while conforming to their expectations and providing unparalleled quality and service.

GC
Litho Sales Inc
Marketing Director
110 S Adams St
Glendale CA 91205
818-246-5501; Fax: 818-246-5576

SERVICES				BINDINGS			
4C	4-Color Printing	RA	Rachwal System	HC	Hardcover	SD	Side Stitching
CS	Color Separations	AF	Acid-Free Paper	CB	Comb Binding	SB	Spiral Binding
DA	Design / Artwork	RP	Recycled Paper	LB	Loose-Leaf Binding	WB	Wire-O Binding
FF	Fulfillment/Mailing	TY	Typesetting	OB	Otabind	I	In-House
LM	List Maintenance	TD	Typeset w/ Disk	PB	Perfect Binding	O	Out of House
OC	OptiCopy System	WA	Warehousing	SS	Saddle Stitching	X	Unknown

Litho Specialties   GC
Nancy Bjornson, Marketing Director
1280 Energy Park Dr
St Paul MN 55108
612-644-3000; Fax: 612-644-4839
Send RFQ's to: Nancy Bjornson

Min: 5000 Max: 1000000 Opt: 250000-
   500000
Sizes: 123456
Bindings: HCO-CBO-OBO-PBO-SDO-
   SBO-WBO-LBO-SSI
Items: ABCDEKPZ1
Services: 4C-CS
Equipment: 2SO(28x40)-WO(22x38)
Annual Sales: 4
Number of Employees: 135
Turnaround Times: B3C4S1
Customers: BP5-BU85-NP5-OR5
Union Shop: N
Year Established: 1990
Terms: Net 30

Statement: Litho Specialties produces
multi-color, high quality prepress through
finishing and distribution.

Lithocolor Press   BK
Marketing Director
9825 W Roosevelt Rd
Westchester IL 60154
708-345-5530; Fax: 708-345-1283
Send RFQ's to: John Matheson

Min: 3000 Max: 50000 Opt: 15000
Sizes: 12356
Bindings: HCO-CBO-PBI-SSI-SBO-WBO
Items: BCDFIJLN
Services: OC
Equipment: 2SO(25x38)-2WO(22.75x38.75)
Annual Sales: 3
Number of Employees: 60
Turnaround Times: B4C5S3
Customers: BP73-OR5-MP12-NP30-OT8
Union Shop: N
Year Established: 1945
Terms: Net 30

Statement: Lithocolor's primary corporate
purpose is servicing our customer's needs,
to be service oriented with efficient
flexability, to cater to our customer's
needs, to produce a quality product consistent with our customer's needs and intended applications.

Lithograph Printing Company   BK
Russ Gordon, Marketing Director
4222 Pilot Dr
Memphis TN 38118
901-365-7100; Fax: 901-795-7815
Send RFQ's to: Russ Gordon

Min: 5000 Max: 500000 Opt: 0
Sizes: 12356X
Bindings: HCO-CBO-LBO-PBO-SSI-
   SDO-SBO-WBO
Items: ABCDEKNPQSYZ
Services: 4C-CS-FF-RP-TY-TD-WA
Equipment: 3SO(28x40)-2WO(17x26)
Annual Sales: 4
Number of Employees: 130
Turnaround Times: B3C5S2
Union Shop: N
Year Established: 1985
Terms: 1% 10, net 30

Lithographic Industries Inc   GC
Marketing Director
2445 Gardner Rd
Maywood IL 60153
708-865-1018; Fax: 708-865-0738

Lithographix Inc   GC
Marketing Director
13500 S Figueroa St
Los Angeles CA 90061
213-770-1000; Fax: 213-770-1731

Lithoid Printing Corporation   BK
Arthur W Beer
19 Cotters Ln
East Brunswick NJ 08816
908-238-4000; Fax: 908-238-9628
Send RFQ's to: Arthur W Beer

Min: 100 Max: 50000 Opt: 5000
Sizes: 123456
Bindings: HCO-CBO-LBO-SBO-WBO-
   PBI-SSI-SDI
Items: ABCDEFGHIKLNPQSX

PRINTED ITEMS			
A Annual Reports	G Galley Copies	N Workbooks	U Greeting Cards
B Books	H Journals	O Yearbooks	V Labels / Stickers
C Booklets	I Magazines	P Brochures	W Maps
D Catalogs	J Mass-Market Books	Q Calendars	X Newspapers
E Cookbooks	K Newsletters	R Comic Books	Y Postcards
F Directories	L Software Manuals	S Direct Mail Letters	Z Posters
	M Textbooks	T Envelopes	1 Stationery

Services: 4C-CS-DA-FF-LM-OC-AF-RP-
TY-TD-WA
Equipment: SO-WO
Annual Sales: 3
Number of Employees: 72
Turnaround Times: B3C3S3
Union Shop: N
Year Established: 1951
Terms: Net 30 with approved credit

Statement: When Lithoid Printing Corporation began in 1951, so did our company philosophy—serve the customer with
quality products, competitive prices, dependable delivery and personal service
dedicated with a strong commitment to the
success of our customers.

Lithotone Inc **GC**
Marketing Director
1313 W Hively Ave
Elkhart IN 46517
219-294-5521; Fax: 219-294-6851

Lithotype Co **GC**
Marketing Director
333 Point San Bruno Blvd
South San Francisco CA 94080
415-871-1750; Fax: 415-871-0714

Little Falls Color Print **GC**
Marketing Director
Riverside Industrial Park
Little Falls NY 13365
315-823-0550; Fax: 315-823-2884

The Little River Press Inc **BK**
Richard Nettina, President
55 NE 73rd St
Miami FL 33138
305-757-7504

Min: 1000 Max: 500000 Opt: 15000
Sizes: 125
Bindings: HCO-PBI-SSI
Items: ABCDFHIJKNP
Services: 4C
Terms: 1% 10, net 30

Long Island Web Printing **CA**
Janice Kerner, Sales Manager
26 Jericho Turnpike
Jericho NY 11753
516-997-7000; Fax: 516-334-4055
Send RFQ's to: Janice Kerner

Min: 5000 Max: 500000 Opt: 250000
Sizes: 45X
Bindings: SSI
Items: CDFX
Services: DA-FF-RP-TY
Equipment: CB(22x35)-LP(22x35)-WO
Annual Sales: 2
Number of Employees: 40
Turnaround Times: S2
Customers: BP10-BU40-CU10-SP40
Union Shop: N
Year Established: 1967
Terms: Net 30

Statement: We are a division of Marks-
Roiland Communications and specialize in
printing on lightweight paper stocks, such
as newsprint, directory, 35# groundwood
and 50# offset. Cheshire mailing available.

John D Lucas Printing Company **BK**
Bob Jones, VP Sales
1820 Portal St
Baltimore MD 21224
410-633-4200; Fax: 410-633-1202

Min: 1000 Max: 500000 Opt: 0
Sizes: 23456X
Bindings: HCI-PBI-SSI
Items: ABCDFHIKMNPQSUYZ
Services: 4C-CS-TY
Equipment: 6SO(to 58)-2WO(22x36)
Annual Sales: 5
Number of Employees: 320
Terms: 1/3 down, 1/3 proofs, 1/3 done

Statement: This commercial printer
produces lottery tickets, game boards,
cards, calendars, posters, catalogs, annual
reports, books, directories, technical journals, and textbooks.

SERVICES				BINDINGS			
4C	4-Color Printing	RA	Rachwal System	HC	Hardcover	SD	Side Stitching
CS	Color Separations	AF	Acid-Free Paper	CB	Comb Binding	SB	Spiral Binding
DA	Design / Artwork	RP	Recycled Paper	LB	Loose-Leaf Binding	WB	Wire-O Binding
FF	Fulfillment/Mailing	TY	Typesetting	OB	Otabind	I	In-House
LM	List Maintenance	TD	Typeset w/ Disk	PB	Perfect Binding	O	Out of House
OC	OptiCopy System	WA	Warehousing	SS	Saddle Stitching	X	Unknown

M

Mack Printing Group Inc **MS**

Thomas N Plath, VP Sales
1991 Northhampton St
Easton PA 18042
610-258-9111

Min: 10000 Max: 500000 Opt: 80000
Sizes: 56X
Bindings: PBI-SSI
Items: DFHILX
Services: 4C-DA-FF-LM-TY-TD-WA
Equipment: 7SO(18x25,25x38)-
    6WO(23x38)
Annual Sales: 6
Number of Employees: 621
Turnaround Times: C3S3
Customers: MP40-OR60
Union Shop: N
Year Established: 1907
Terms: Net 30

Mackintosh Typography Inc **BK**

Lynne Stark, Business Manager
319 Anacapa St
Santa Barbara CA 93101
805-962-9915
Send RFQ's to: Lynne Stark

Min: 300 Max: 3000 Opt: 2000
Sizes: 123456
Bindings: HCO-PBI-SSI
Items: BCDHIKPVYZ
Services: 4C-CS-DA-TY-TD
Equipment: LP(10x15)-SO(18x25)
Annual Sales: 0
Number of Employees: 10
Customers: BP60-CU10-SP15-OT15
Union Shop: N
Terms: 1/2 down, 1/2 on completion

Statement: We do fine book production
for literary and academic publishers.

The Mad Printers **BK**

Marketing Director
800 Wickham Ave
Mattituck NY 11952
516-298-5100

Min: 300 Max: 10000 Opt: 0
Sizes: 2345
Bindings: PBI
Items: BH
Services: TY-TD
Equipment: SO

Mail-O-Graph **CA**

William H Harper, President
206 West 4th St
PO Box 407
Kewanee IL 61443
309-852-2602

Min: 10000 Max: 2000000 Opt: 50000
Sizes: 245
Bindings: SSI-SDI
Items: CDFHSTX
Services: 4C-TY-WA
Equipment: 6WO
Annual Sales: 2
Number of Employees: 30
Turnaround Times: B3C3S3
Customers: BU90-CU2-MP2-NP2-OR4
Union Shop: Y
Year Established: 1933
Terms: Net 30

Malloy Lithographing **BK**

Joe Upton, Marketing Director
5411 Jackson Rd
PO Box 1124
Ann Arbor MI 48106-1124
313-665-6113; Fax: 313-665-2326
800-722-3231
Send RFQ's to: Tim Scarbrough

Min: 500 Max: 50000 Opt: 5000
Sizes: 123456X
Bindings: HCI-CBI-LBI-PBI-SSI-SDO-
    SBO-WBO-OBI
Items: BCDEFHLMNO
Services: OC-RA-AF-RP-FF
Equipment: 5SO(20x29,41x54)-5WO
    (38 to 45)
Annual Sales: 5
Number of Employees: 400
Turnaround Times: B3C4S3
Customers: BP98-SP2
Union Shop: N
Year Established: 1960
Terms: Net 30

PRINTED ITEMS			
A Annual Reports	G Galley Copies	N Workbooks	U Greeting Cards
B Books	H Journals	O Yearbooks	V Labels / Stickers
C Booklets	I Magazines	P Brochures	W Maps
D Catalogs	J Mass-Market Books	Q Calendars	X Newspapers
E Cookbooks	K Newsletters	R Comic Books	Y Postcards
F Directories	L Software Manuals	S Direct Mail Letters	Z Posters
	M Textbooks	T Envelopes	1 Stationery

Statement: It is Malloy's desire to target and cultivate customers who value service above all other things, and with whom we can develop long-lasting relationships. We will strive to build and strengthen these relationships by continually providing such superior service as to give our customers cause to boast about us.

Mandarin Offset **GC**
Timothy C Linn, Marketing Manager
1 Madison Ave 25th Fl
New York NY 10010
212-481-7170; Fax: 212-683-0882
Send RFQ's to: Timothy C Linn
Parent Company: Octopus Books Inc

Maple-Vail Book Mfg Group **BK**
Bill Long, Marketing Director
Willow Springs Ln
PO Box 2695
York PA 17405
717-764-5911; Fax: 717-764-4702
Send RFQ's to: Rick Weidner
Parent Company: Maple Press Company

Min: 500 Max: 500000 Opt: 2500-10000
Sizes: 23456
Bindings: HCI-PBI
Items: BELMN
Services: FF-OC-AF-RP-TY-TD-WA
Equipment: 13SO(54x77)-5WO(19x23)
Annual Sales: 6
Number of Employees: 1100
Turnaround Times: B4C4
Customers: BP85-CU15
Union Shop: Y
Year Established: 1903
Terms: Net 30

Sales Offices:

Baltimore MD
410-323-3244; Fax: 410-435-8722

Bordentown NJ
609-298-4013; Fax: 609-298-1418

Boston MA
617-770-0016; Fax: 617-770-0317

Chicago IL
708-799-7510; Fax: 708-799-0834

Gatlinburg TN
615-430-5900; Fax: 615-436-0420

New York NY
212-213-1115; Fax: 212-689-1410

San Fancisco CA
510-934-1440; Fax: 510-934-7020

Maquoketa Web Printing **GC**
Robert B Melvold, Marketing Director
1287 East Maple St
Maquoketa IA 52060
319-652-4971; Fax: 319-652-4666
Send RFQ's to: Robert B Melvold

Min: 2000 Max: 2000000 Opt: 100000
Sizes: 1245X
Bindings: LBI-PBI-SSI-CBO-OBO-SDO-SBO-WBO
Items: BCDFHIKNX
Services: 4C-DA-TY-TD
Equipment: 12SO(22.75x35)-12WO (22.75x35)
Annual Sales: 3
Number of Employees: 60
Turnaround Times: B3S2
Customers: BP5-BU40-CU15-MP10-OT30
Union Shop: N
Year Established: 1976
Terms: Net 30

Marathon Communications Group **GC**
Debi Traeder, Marketing Director
2001 Second St
PO Box 1908
Wausau WI 54402-1908
715-845-4231; Fax: 715-845-9476

Min: 0 Max: 0 Opt: 10000-30000
Sizes: 123456X
Bindings: HCO-CBI-LBI-PBO-SSI-SDI-SBO-WBI
Items: ACDEIKLPQSTWZ12
Services: 4C-CS-DA-FF-LM-AF-RP-TY-TD-WA
Equipment: 5CB(28x41,13x20,20x26,10x15)-13SO
Annual Sales: 4
Number of Employees: 168
Turnaround Times: B3C3S2
Customers: BU85-CU5-MP5-OR5
Union Shop: N
Year Established: 1926

SERVICES				BINDINGS		SD	Side Stitching
4C	4-Color Printing	RA	Rachwal System	HC	Hardcover	SB	Spiral Binding
CS	Color Separations	AF	Acid-Free Paper	CB	Comb Binding	WB	Wire-O Binding
DA	Design / Artwork	RP	Recycled Paper	LB	Loose-Leaf Binding	I	In-House
FF	Fulfillment/Mailing	TY	Typesetting	OB	Otabind	O	Out of House
LM	List Maintenance	TD	Typeset w/ Disk	PB	Perfect Binding	X	Unknown
OC	OptiCopy System	WA	Warehousing	SS	Saddle Stitching		

Terms: Net 30

Sales Office:

Marathon Communications Group
8 Skokie Highway #201
Lake Bluff IL 60044

BK

Mark IV Press

Robert Rinere, President
325 Rabro Dr
Hauppauge NY 11788
516-348-5252

Min: 500 Max: 20000 Opt: 0
Sizes: 2345
Bindings: CBI-LBI-PBI-SBI
Items: BCDEFGJL
Services: 4C-CS-DA-FF-TY-TD
Equipment: SO

BK

Marrakech Express

S Sigmund, Marketing Director
500 Anclote Rd
Tarpon Springs FL 34689-6701
813-942-2218; Fax: 813-937-4758
800-940-6566
Send RFQ's to: Shirley Copperman

Min: 250 Max: 10000 Opt: 2500
Sizes: 123456X
Bindings: HCO-CBI-PBI-SSI-SBI-WBI
Items: ABCDEFHIKLMNOPRS
Services: 4C-CS-FF-LM-AF-RP-TD-WA
Equipment: 6SO(40)
Annual Sales: 2
Number of Employees: 17
Turnaround Times: B3S1
Customers: BP31-BU8-CU11-MP38-NP5-
   OR2-SP4-OT1
Union Shop: N
Year Established: 1976
Terms: To be arranged

Statement: We give personal attention to
the needs of small book and magazine
publishers with quality workmanship and
the lowest possible pricing. For us, the
challenge is always to finish up with a publication you can proudly distribute and sell.
Write for a FREE copy of their **Shortruns**
newsletter.

GC

Mars Graphic Services

Thomas Black, President
1 Deadline Dr
Westville NJ 08093
609-456-8666; Fax: 609-456-1528

Bindings: SSI
Items: CIPST
Services: 4C
Equipment: 5WO
Annual Sales: 4
Number of Employees: 140

BR

Maverick Publications Inc

Gary Asher, Marketing Director
PO Box 5007
Bend OR 97708
503-382-6978; Fax: 503-382-4831
Send RFQ's to: Gary Asher
Min: 250 Max: 20000 Opt: 2000
Sizes: 123456
Bindings: CBI-LBI-PBI-SSI-SDI-SBI-
   HCO-WBO
Items: ABCDEFKLMNO
Services: 4C-CS-DA-AF-RP-TY-TD
Equipment: 2SO(19x25)
Annual Sales: 2
Number of Employees: 12
Turnaround Times: B5C5S4
Customers: BP10-SP80-CU5-OR5
Union Shop: N
Year Established: 1968
Terms: 1/2 down, balance with proofs

Statement: Maverick Publications specializes in working with the first-time author
or self-publisher. We take the time to ex-
plain every step of the book production
process, including pre-production cost
flexability and marketing.

GC

May Printing Co

Marketing Director
4110 Clearwater Rd
St Cloud MN 56301
612-656-5000; Fax: 612-656-5163

PRINTED ITEMS			
A Annual Reports	G Galley Copies	N Workbooks	U Greeting Cards
B Books	H Journals	O Yearbooks	V Labels / Stickers
C Booklets	I Magazines	P Brochures	W Maps
D Catalogs	J Mass-Market Books	Q Calendars	X Newspapers
E Cookbooks	K Newsletters	R Comic Books	Y Postcards
F Directories	L Software Manuals	S Direct Mail Letters	Z Posters
	M Textbooks	T Envelopes	1 Stationery

The Mazer Corporation GC
Beth A Miller, Marketing Director
2501 Neff Rd
Dayton OH 45414
513-276-6181; Fax: 513-276-3197
Send RFQ's to: Robin Deis

Min: 50 Max: 0 Opt: 5000
Sizes: 12345
Bindings: CBX-LBX-PBX-SSX-SDX-SBX-
   WBX
Items: BCFGHJKLMNOPSTVWZ124
Services: 4C-DA-FF-RP-TY-TD-WA-OC
Equipment: 4SO(28 to 40)-2WO(36)
Number of Employees: 200
Turnaround Times: B3S3
Customers: BP70-BU5-CU15-OR10
Union Shop: N
Year Established: 1963
Terms: Net 30

Statement: The Mazer Corporation is a
publishing services company engaged in
the creative design, manufacture and distribution of our customer's learning
products for the education and training
markets. We bring value to our customers
through our knowledge of education, packaging, and communication technologies.

Plants:

3100 Industrial Dr
PO Box 5520 EKS
Skyline Industrial Park
Johnson City TN 37601
615-928-7101; Fax: 615-928-7786

704 Rolling Hills Dr
Johnson City TN 37604
615-928-7200; Fax: 615-928-4581

Sales Offices:

6913 Jackman Blvd
Winter Park FL 32792
407-657-1060; Fax: 407-672-1363

193 Wilton Rd West
Ridgefield CT 06877
203-438-9779; Fax: 203-438-1319

2340 Darby Ct
Bel Air MD 21015
410-569-8895; Fax: 410-569-8942

1575 Beacon St #2R
Brookline MA 02146
617-232-6096; Fax: 617-232-2334

226 Bay Colony Dr
Naperville IL 60565
708-305-0389; Fax: 708-305-0537

2800A Lafayette St #106
Portsmouth NH 03801
603-436-7197; Fax: 603-436-0814

MBP (Multi Business BK
Press)Lithographics
Stan Theissen, Marketing Director
135 North Main
Hillsboro KS 67063
316-947-3966; Fax: 316-947-3392
800-844-1655
Send RFQ's to: Austin Havens
Parent Company: Lee Company

Min: 500 Max: 30000 Opt: 10000
Sizes: 123456
Bindings: CBX-PBX-SSX-SDX-SBX
Items: ABCDEFGHIJKNPSTVY12
Services: 4C-FF-LM-AF-TY-TD-WA
Equipment: 2LP(20)-5SO(26,38,48,25)
Turnaround Times: B3S3
Customers: BP33-MP27-OR9-SP19-OT12
Union Shop: N
Year Established: 1914
Terms: Net 30

Statement: Our goal is to help the author
become successfully published and financially rewarded through services that enhance the distinction of the work in a way
that others find valuable.

McArdle Printing Co BK
G William Teare Jr, Marketing Director
800 Commerce Dr
Upper Marlboro MD 20772
301-390-8500; Fax: 301-390-8525
Send RFQ's to: G William Teare Jr

Min: 1000 Max: 200000 Opt: 20000
Sizes: 235
Bindings: SSI-SDI-HCO-CBO-LBO-OBO-
   PBO-SBO-WBO
Items: BCDEFHIKMP
Services: OC-AF-RP-TD-WA
Equipment: 3SO(28x40)-4WO(17.5,31,36)-
   4OP
Annual Sales: 5
Number of Employees: 200

SERVICES				BINDINGS		SD	Side Stitching
4C	4-Color Printing	RA	Rachwal System	HC	Hardcover	SB	Spiral Binding
CS	Color Separations	AF	Acid-Free Paper	CB	Comb Binding	WB	Wire-O Binding
DA	Design / Artwork	RP	Recycled Paper	LB	Loose-Leaf Binding	I	In-House
FF	Fulfillment/Mailing	TY	Typesetting	OB	Otabind	O	Out of House
LM	List Maintenance	TD	Typeset w/ Disk	PB	Perfect Binding	X	Unknown
OC	OptiCopy System	WA	Warehousing	SS	Saddle Stitching		

Turnaround Times: B3C3S2
Customers: BP60-BU10-NP25-OR5
Union Shop: Y
Year Established: 1947
Terms: Net 30

Statement: The mission of the McArdle Printing Company is to expeditiously manufacture and distribute customer time-sensitive informational products, serving publishers, trade associations, professional societies, other non-profit organizations, financial institutions and governmental organizations and in so doing make our customers successful in the fulfillment of their missions.

McClain Printing Company   GC
Ken Smith, Marketing Director
PO Box 403
Parsons WV 26287
304-478-2881; Fax: 304-478-4658
800-654-7179
Send RFQ's to: George A Smith

Min: 250 Max: 10000 Opt: 3000-5000
Sizes: 12345
Bindings: HCO-CBI-LBO-PBI-SSI-SDO-
    SBO-WBO
Items: ABCDEFHIJKMNOPQSXYZ
Services: 4C-CS-FF-AF-TY-TD-WA
Equipment: 5SO(19x25,20x29)
Annual Sales: 2
Number of Employees: 20
Turnaround Times: B5C4S4
Customers: BU60-CU5-NP5-SP30
Union Shop: N
Year Established: 1958
Terms: Net 30

Statement: We are the right choice for profitable results.

John M McCoy Printers Inc   GC
Marketing Director
2907 Supply Ave
Los Angeles CA 90040
213-723-5148; Fax: 213-721-3060

McDowell Publications   BK
Marketing Director
11129 Pleasant Ridge Rd
Utica KY 42376
502-275-4075

Min: 100 Max: 5000 Opt: 500
Bindings: HCX-PBX-SSX
Items: BCF
Customers: OR20-SP80(genealogists)

Statement: They specialize in printing genealogies and local histories.

The McFarland Press   MS
Marketing Director
Crescent and Mulberry Sts
PO Box 3645
Harrisburg PA 17105
717-234-6235; Fax: 717-234-1587
Min: 3000 Max: 20000 Opt: 10000
Sizes: 25
Bindings: CBO-PBI-SSI-SPO-WBO
Items: ACDFHIKLMPQUVWZ
Services: 4C
Equipment: 5SO(25x38-70)
Annual Sales: 2
Number of Employees: 125
Turnaround Times: B4C3S2
Customers: BP10-BU30-MP30-NP30
Union Shop: Y
Year Established: 1886
Terms: Net 30

McGill Jensen   CA
Joyce Larson-Stoefen, Marketing Manager
641 Fairview Ave N
St Paul MN 55104-1792
612-645-0751; Fax: 612-645-7537

Min: 5000 Max: 3000000 Opt: 0
Sizes: 23456X
Bindings: SSI
Items: ACDFIKNPSWXZ
Services: 4C-CS-DA-FF-TY
Equipment: LP-5SO-8WO
Annual Sales: 6
Number of Employees: 3800

PRINTED ITEMS			
A Annual Reports	G Galley Copies	N Workbooks	U Greeting Cards
B Books	H Journals	O Yearbooks	V Labels / Stickers
C Booklets	I Magazines	P Brochures	W Maps
D Catalogs	J Mass-Market Books	Q Calendars	X Newspapers
E Cookbooks	K Newsletters	R Comic Books	Y Postcards
F Directories	L Software Manuals	S Direct Mail Letters	Z Posters
	M Textbooks	T Envelopes	1 Stationery

McGraphics Inc GC
Marketing Director
2310 Millar Ave
Saskatoon SK Canada S7K 2C4
306-665-3560

McKay Printing Services GC
Cecilia Beckley, Partner
208 N Wabash St
Michigan City IN 46360
708-841-7300; Fax: 219-872-8531
800-227-1432

Min: 0 Max: 0 Opt: 0
Bindings: SSI-PBI
Items: CDIKNP
Services: FF
Equipment: SO-WO
Customers: BP-BU-SC

Statement: McKay is a graphic arts company specializing in quality printing and mailing/fulfillment, with special emphasis on direct mail advertising.

McLaren, Morris & Todd GC
John Morris, Group President
Consumer Printed Products Group
2270 American Dr
Mississauga ON Canada L4V 1B5
416-677-3592; Fax: 416-677-3675
Send RFQ's to: Bill Poutney, General
    Manager & VP
Parent Company: Southam Graphics Inc

McNaughton & Gunn Inc BK
Ronald Mazzola, Sales Director
960 Woodland Dr
Saline MI 48176
313-429-5411; Fax: 800-677-BOOK
Send RFQ's to: Ronald A Mazzola

Min: 250 Max: 50000 Opt: 3000-10000
Sizes: 23456
Bindings: HCO-OBO-SBO-WBO-PBI-SSI
Items: BDEFHLMN
Services: RA-AF-RP-TD-CS
Equipment: 7SO(24x32,41x55)-
    WO(41x45)

Annual Sales: 5
Number of Employees: 280
Turnaround Times: B3C6S4
Customers: BP40-CU10-SP50
Union Shop: N
Year Established: 1975
Terms: Net 30 with prior credit approval

Statement: Complete book manufacturers, servicing sheetfed and web production. All standard trim sizes. Case binding, perfect binding and electric pre-press services.

McNaughton & Gunn, Inc.

A Book Manufacturing Service Company

"Competitive Web and Sheetfed Schedules"

*Call one of our offices today,
for an estimate on your title!
Or Fax your specifications to
1-800-677-BOOK*

Ronald A. Mazzola
*Executive Director,
Marketing/Sales*
P.O. Box 10
Saline, MI   48176
Ph: 313 429 5411
extention 206
Fax: 1 800 677 BOOK

West Coast Office
Frank Gaynor
Ph: 707 939 9343
Fax: 707 939 9346

Mid Atlantic Office
Jack Richardson
Ph: 301 762 7000
Fax: 301 217 0316

Midwest Office
Davis L. Scott
Ph: 708 537 0169
Fax: 708 537 0183

East Coast Office
Glenn Fillman
Ph: 212 979 6590
Fax: 212 979 6597

McNaughton & Gunn, Inc.

SERVICES				BINDINGS			
4C	4-Color Printing	RA	Rachwal System	HC	Hardcover	SD	Side Stitching
CS	Color Separations	AF	Acid-Free Paper	CB	Comb Binding	SB	Spiral Binding
DA	Design / Artwork	RP	Recycled Paper	LB	Loose-Leaf Binding	WB	Wire-O Binding
FF	Fulfillment/Mailing	TY	Typesetting	OB	Otabind	I	In-House
LM	List Maintenance	TD	Typeset w/ Disk	PB	Perfect Binding	O	Out of House
OC	OptiCopy System	WA	Warehousing	SS	Saddle Stitching	X	Unknown

McQuiddy Printing Company **GC**
Hal Rehorn, Marketing Director
711 Spence Ln
Nashville TN 37217
615-366-6565; Fax: 615-367-0923
Send RFQ's to: Hal Rehorn

Min: 5000 Max: 5000000 Opt: 0
Sizes: 123456
Bindings: HCO-CBO-LBO-OBO-PBO-
   SSI-SDI-SBO-WBO
Items: ACDFKPQZ2
Services: 4C-CS-OC-RP-TY-TD
Equipment: 3SO(28,40)-2WO(M1001,
   M300)
Annual Sales: 4
Number of Employees: 110
Turnaround Times: B3C5S1
Customers: BU70-MP20-OR10
Union Shop: N
Year Established: 1908
Terms: Net 30 to 60

Meaker the Printer **GC**
Thom Meaker, Vice President
802 W Jefferson St
Phoenix AZ 85007
602-254-2171

Min: 300 Max: 2000000 Opt: 7000
Sizes: 2345X
Bindings: HCO-CBI-PBO-SSI
Items: ABCDEFHIKNPSYZ1
Services: 4C-DA-TY-TD
Terms: Net 10 with approved credit

Media Printing **CA**
John Checchia, Vice President
8050 NW 74th Ave
Miami FL 33166
305-888-1300; Fax: 305-888-8542
800-544-WEBS

Min: 50000 Max: 2000000 Opt: 0
Sizes: 5X
Bindings: SSI
Items: DP
Services: 4C-FF
Equipment: 3WO(19x33,23x38)
Year Established: 1971

Statement: We are your one source for
fine quality catalogs, brochures, flyers,
coupons, business reply cards, and more.

Media Publications **GC**
Marketing Director
3050 Coronado Dr
Santa Clara CA 95054
408-492-0400; Fax: 408-492-0475
Send RFQ's to: Bud Meacham

Min: 300 Max: 10000 Opt: 0
Sizes: 123456
Bindings: CBI-LBI-OBI-PBI-SBI-SSI-SDI-
   WBI
Items: BCDEFLMNP
Services: 4C
Equipment: 7SO-WO-OP
Annual Sales: 3
Number of Employees: 52
Turnaround Times: B1-2S1-2
Customers: 15BP-75BU-5CU-5SP
Union Shop: N
Year Established: 1981
Terms: 1% 10, net 30

Statement: Docutech min: 50, max: 300.
Specializing in printing books and manuals
(photo direct) 10 to 12 working days.
Xerox Docutech books and manuals in 5 to
10 working days.

Meehan-Tooker & Company **GC**
Tom Murphy, VP Sales
55 Madison Cir Dr
East Rutherford NJ 07073
201-933-9600; Fax: 201-933-8322

Min: 5000 Max: 2000000 Opt: 0
Sizes: 235X
Bindings: PBO-SSO
Items: ADPSWXZ
Services: 4C-CS
Equipment: SO(28x40)-7WO(22x38)
Annual Sales: 6
Number of Employees: 380

Statement: Meehan-Tooker   is   an
employee-owned company specializing in
printing inserts, catalogs, coupons, annual
reports, and other direct response promotions.   They   can   provide   die-cutting,
microfragrances, perfs, rub-offs, and other
special effects.

PRINTED ITEMS					
A	Annual Reports	G	Galley Copies	N	Workbooks	U	Greeting Cards
B	Books	H	Journals	O	Yearbooks	V	Labels / Stickers
C	Booklets	I	Magazines	P	Brochures	W	Maps
D	Catalogs	J	Mass-Market Books	Q	Calendars	X	Newspapers
E	Cookbooks	K	Newsletters	R	Comic Books	Y	Postcards
F	Directories	L	Software Manuals	S	Direct Mail Letters	Z	Posters
		M	Textbooks	T	Envelopes	1	Stationery

Merced Color Press GC
Marketing Director
2201 Cooper Ave (95340)
PO Box 3139
Merced CA 95344
209-384-0444; Fax: 209-722-7504
Parent Company: World Color Press

Mercury Printing Company GC
Raymond Kirkley, Marketing Director
2929 Convair Rd
Memphis TN 38132
901-345-8480
Parent Company: Graphics Industries of
  Atlanta

Min: 500 Max: 200000 Opt: 20000
Sizes: 123456X
Bindings: HCO-CBI-LBO-PBI-SSI-SDI-
  SBO-WBO
Items: ABCDEFIKLMNOPQRTVWYZ1
Services: 4C-DA-FF-AF-TY-TD-WA
Equipment: LP-5SO(19x25,28x41)
Turnaround Times: B2C3S2
Customers: BU90-CU3-MP6-OT1
Union Shop: N
Year Established: 1961
Terms: Net 30

Merrick Printing Co Inc GC
Marketing Director
808 E Liberty St
Louisville KY 40204
502-584-6258; Fax: 502-589-3512

Merrill Corporation BK
Marketing Director
1 Merrill Cir
St Paul MN 55108
612-646-4501

Statement: Sales offices in Los Angeles,
CA; Dallas, TX; and Chicago, IL; see each
listing in the Directory.

Merrill Corporation BK
Bill Rakow, Sales Manager
1926 East 14th
Los Angeles CA 90021-2891
213-231-4133
Send RFQ's to: Linda Wexler

Min: 3000 Max: 100000 Opt: 20000
Sizes: 12345
Bindings: HCX-CBX-PBX-SSX-SBX-
  WBX
Items: BCDFHJLMN
Services: 4C-CS-FF-AF-WA
Equipment: 3SO(40)-WO(38)
Number of Employees: 35
Turnaround Times: B3C3S3
Customers: BP20-BU45-CU5-NP5-SC5-
  SP20
Union Shop: Y
Year Established: 1938
Terms: 2% 10, net 30

Statement: Our company has specialized
for 50 years in the production of books,
catalogs, manuals, and directories for commercial and trade customers.

Merrill Corporation BK
Marketing Director
333 N Stemmons Freeway
Dallas TX 75207
214-698-9777

Merrill Corporation BK
Marketing Director
650 W Washington
Chicago IL 60661
312-930-2700

Metromail Corporation GC
Joe Rowland, Marketing Director
901 W Bond St
Lincoln NE 68521
402-475-4591; Fax: 402-473-4811
Send RFQ's to: Joe Rowland

Min: 175 Max: 5000 Opt: 2000
Sizes: 24
Bindings: HCX-LBX-PBX

SERVICES				BINDINGS			
4C	4-Color Printing	RA	Rachwal System	HC	Hardcover	SD	Side Stitching
CS	Color Separations	AF	Acid-Free Paper	CB	Comb Binding	SB	Spiral Binding
DA	Design / Artwork	RP	Recycled Paper	LB	Loose-Leaf Binding	WB	Wire-O Binding
FF	Fulfillment/Mailing	TY	Typesetting	OB	Otabind	I	In-House
LM	List Maintenance	TD	Typeset w/ Disk	PB	Perfect Binding	O	Out of House
OC	OptiCopy System	WA	Warehousing	SS	Saddle Stitching	X	Unknown

Items: FLN
Services: RA
Equipment: 3SO(19.5x25)-3WO(24^{13}/$_{16}$x38)
Annual Sales: 3
Number of Employees: 110
Turnaround Times: B3C4
Customers: BP30-BP70
Union Shop: N
Year Established: 1978
Terms: Net 30

MS

Metroweb
Stephen Waldner, Marketing Director
PO Box 18760
Erlanger KY 41018
606-525-1168; Fax: 606-525-8219
Send RFQ's to: Stephen Waldner

Min: 15000 Max: 0 Opt: 0
Sizes: X
Bindings: SSX
Items: DFHI
Services: 4C-OC-RP-TD
Equipment: 2WO(17x26,22.75x36)
Annual Sales: 4
Number of Employees: 160
Turnaround Times: B2S2
Customers: CU10-MP65-NP25
Union Shop: N
Year Established: 1977
Terms: Net 30 upon credit approval

Statement: Metroweb is a company which prints magazines in partnership with its employees and customers. Our mission is to continually improve the conditions and status of all elements of the partnership so that customers will be continually delighted by the results of all our efforts.

GC

Midwest Litho Arts
Marketing Director
125 E Oakton
Des Plaines IL 60018
708-296-2000; Fax: 708-296-2785
Parent Company: World Color Press

GC

Miller Printing Inc
Marketing Director
2357 Ventura Dr #100
PO Box 25190
Woodbury MN 55125
612-738-2677; Fax: 612-738-1898

GC

Mitchell Press
Jack Mellor, Sales Manager
PO Box 6000
1706 W First Ave
Vancouver BC Canada V6B 4B9
604-731-5211

Min: 2000 Max: 2000000 Opt: 0
Sizes: 2345X
Bindings: PBI-SSI
Items: ABCDFHIKNP
Services: 4C-CS-DA-TY
Equipment: SO-WO

BK

MMI Press
Mountain Missionary Institute
Marketing Director
PO Box 279
Harrisville NH 03450
603-827-3111

Min: 3000 Max: 100000 Opt: 0
Sizes: 2345X
Bindings: HCI-PBI-SSI
Items: BCDFJMN
Services: 4C
Equipment: CB

Statement: Besides printing their own publications, they sometimes take on outside work.

GC

Modern Printing & Lithography Inc
Marketing Director
1000 Federal Rd
Brookfield CT 06804
203-775-6291; Fax: 203-775-6809

PRINTED ITEMS	G	Galley Copies	N	Workbooks	U	Greeting Cards
A Annual Reports	H	Journals	O	Yearbooks	V	Labels / Stickers
B Books	I	Magazines	P	Brochures	W	Maps
C Booklets	J	Mass-Market Books	Q	Calendars	X	Newspapers
D Catalogs	K	Newsletters	R	Comic Books	Y	Postcards
E Cookbooks	L	Software Manuals	S	Direct Mail Letters	Z	Posters
F Directories	M	Textbooks	T	Envelopes	1	Stationery

Moebius Printing   **CA**

Ron Hughes, VP Marketing
300 N Jefferson St
PO Box 302
Milwaukee WI 53202
414-276-5311; Fax: 414-276-8725

Min: 5000 Max: 5000000 Opt: 0
Sizes: 23456X
Bindings: SSI
Items: ACDJIPQSUYZ
Services: 4C-CS-FF
Equipment: 3SO-5WO
Annual Sales: 6
Number of Employees: 380
Year Established: 1911

Monroe Litho Inc   **GC**

Marketing Director
39 Delevan St
Rochester NY 14605
716-454-3290; Fax: 716-325-4478

Monument Printers &   **BK**
Lithographer Inc

Marketing Director
6th St & Madalyn Ave
PO Box 629
Verplanck NY 10596
914-737-0992; Fax: 914-737-0783
800-227-2081
Send RFQ's to: Norman McGowan

Min: 100 Max: 20000 Opt: 5000
Sizes: 123456
Bindings: CBI-PBI-SSI-SDI-SBO
Items: BCDFGH
Services: TY-TD
Equipment: 6SO(25,38)
Annual Sales: 2
Number of Employees: 50
Turnaround Times: B3S3
Customers: BP70-BU5-CU10-OT15
Union Shop: N
Year Established: 1950
Terms: Net 30; new accounts 50/50

Moran Printing Company   **GC**

Chuck Fry, Marketing Director
9125 Bachman Rd
Orlando FL 32824
407-859-2030; Fax: 407-826-5284
Send RFQ's to: Chuck Fry

Min: 1000 Max: 500000 Opt: 100000
Sizes: 6
Bindings: HCO-CBI-LBI-SSI-SBO-WBO
Items: ABCDEFHIKLMNOPQSTZ4
Services: 4C-CS-FF-AF-RP-TY-WA
Equipment: 3LP(28x40)-6SO(28x40)
Annual Sales: 4
Number of Employees: 140
Turnaround Times: B4C3S2
Customers: BP13-BU57-CU10-MP15-NP5
Union Shop: N
Year Established: 1983
Terms: Net 30 with approved application

Statement: For 6th year awarded "Best Printer In Florida" by virtue of our winning the statewide contest. Printing Association of Florida. We produce 400 L.S. waterless printing for Walt Disney & other critical buyers.

Moran Printing Inc   **GC**

George Choquette
5425 Florida Blvd
PO Box 66538
Baton Rouge LA 70806
504-923-2550; Fax: 504-923-1078
Send RFQ's to: George Choquette

Min: 2000 Max: 200000 Opt: 20000
Sizes: 123456
Bindings: CBI-LBI-PBI-SSI-SDI-HCO-
    SBO-WBO
Items: ABCDEFHKLPZ24
Services: 4C-FF-RP-TY-TD
Equipment: 6LP(to 50)-6SO(to 40)
Annual Sales: 3
Number of Employees: 65
Turnaround Times: B2C6S2
Customers: BU60-CU2-MP1-NP1-SC2-
    SP1-OT40
Union Shop: Y
Year Established: 1881
Terms: Net 30

Statement: In addition to high quality multi-color commercial printing, we offer

SERVICES		BINDINGS			
4C	4-Color Printing	RA	Rachwal System	HC	Hardcover	SD	Side Stitching
CS	Color Separations	AF	Acid-Free Paper	CB	Comb Binding	SB	Spiral Binding
DA	Design / Artwork	RP	Recycled Paper	LB	Loose-Leaf Binding	WB	Wire-O Binding
FF	Fulfillment/Mailing	TY	Typesetting	OB	Otabind	I	In-House
LM	List Maintenance	TD	Typeset w/ Disk	PB	Perfect Binding	O	Out of House
OC	OptiCopy System	WA	Warehousing	SS	Saddle Stitching	X	Unknown

two unique services: a software developer's service that helps with everything from duplicating the disks to producing the packaging.

GC
Morgan Press
Lloyd Morgan, President
145 Palisade St
Dobbs Ferry NY 10522
914-693-0023

Min: 1000 Max: 50000 Opt: 0
Bindings: SSI
Items: ABCDHKPQTUYZ1
Services: 4C-DA-TY-WA
Equipment: 2LP(10x15,22x30)-5SO(11x17, 38x50)
Annual Sales: 2
Number of Employees: 30
Turnaround Times: B6C4
Customers: BP25-BU65-NP10
Union Shop: N
Year Established: 1965
Terms: Net 30 with proper credit

Statement: We do short-run commercial printing. We are publishers of greeting cards and technical photographic books under the Morgan and Morgan name.

BK
Morgan Printing
Mark Hillis, Marketing Director
900 Old Koenig Ln #135
Austin TX 78756
512-459-5194; Fax: 512-451-0755
Send RFQ's to: Mark Hillis

Min: 50 Max: 3000 Opt: 800
Sizes: 23456
Bindings: HCO-SBO-WBO-CBO-PBI-SSI
Items: BCDEFHKLMNP
Services: 4C-AF-TY-TD-RP
Equipment: 4SO(11x17,2-14x20,15x20)
Annual Sales: 2
Number of Employees: 17
Turnaround Times: B3C5S2
Customers: BP10-BU20-CU30-NP5-OT5-SP30
Union Shop: N
Year Established: 1979
Terms: 50% down, 50% COD

Statement: To enjoy each working day

and provide well crafted books and newsletters with superior customer service.

GC
Morningrise Printing
Jane McLaughlin, Trustee
1525 W MacArthur Blvd #1
Costa Mesa CA 92626
714-957-8494; Fax: 714-549-1241
Send RFQ's to: Jane McLaughlin

Min: 200 Max: 5000 Opt: 2000
Sizes: 1245
Bindings: HCO-CBO-PBO-SSI-SDI-SBO-WBO
Items: BCDKLPT
Services: DA-TY
Equipment: 2SO(12x18)
Annual Sales: 1
Number of Employees: 5
Turnaround Times: B4C2
Customers: BU80-SP15-OT5
Union Shop: N
Year Established: 1978
Terms: 50% deposit, 50% on delivery

Statement: We operate under the premise that every job is a portrait of the person who did it. If our customers aren't happy, we're not happy, and we want to be happy.

CA
Motheral Printing
Marketing Director
510 S Main St (76102)
PO Box 629
Fort Worth TX 76101-0629
817-335-1481; Fax: 817-332-4672

Min: 5000 Max: 2000000 Opt: 0
Sizes: 2345X
Bindings: SSI
Items: ACDHIPZ
Services: 4C
Equipment: SO-WO
Year Established: 1934

GC
William H Muller Printing Co
Bob McCluskey, VP Sales
3550 Thomas Rd
PO Box 698
Santa Clara CA 95070
408-988-8400; Fax: 408-988-2519

PRINTED ITEMS					
A	Annual Reports	G	Galley Copies	N	Workbooks
B	Books	H	Journals	O	Yearbooks
C	Booklets	I	Magazines	P	Brochures
D	Catalogs	J	Mass-Market Books	Q	Calendars
E	Cookbooks	K	Newsletters	R	Comic Books
F	Directories	L	Software Manuals	S	Direct Mail Letters
		M	Textbooks	T	Envelopes

U	Greeting Cards
V	Labels / Stickers
W	Maps
X	Newspapers
Y	Postcards
Z	Posters
1	Stationery

Min: 5000 Max: 350000 Opt: 175000
Sizes: 25
Bindings: HCO-CBO-LBI-PBO-SSI-SDO-
  SBO-WBO
Items: ACDHILPQSZ
Services: 4C-CS
Equipment: 4SO(19x25,28x40)-WO
Annual Sales: 6
Number of Employees: 125
Turnaround Times: B2C3S3
Customers: BU90-MP10
Union Shop: N
Year Established: 1956
Terms: Net 30

Multiple Business Forms Inc   GC
Doug Emo, General Manager
7765 Transmere Dr
Mississauga ON Canada L5S 1V5
416-678-2351; Fax: 416-675-7590

Statement: A division of Southam
Graphics, Inc.

Multiple Business Forms Inc   GC
Marcel Bayard, President
Guthrie Ave
Dorval QB Canada H9P 2V2
514-636-6163; Fax: 514-631-5329

Statement: A division of Southam
Graphics, Inc.

MultiPrint Co Inc   CS
Monica Pratl, Marketing Director
5555 W Howard St
Skokie IL 60077
800-858-9999; Fax: 708-677-7544
Send RFQ's to: Sales Department
Parent Company: Great Lakes Graphics
  Inc

Min: 500 Max: 250000 Opt: 50000
Sizes: 123456
Bindings: PBO-SBO-WBO-HCO-CBO-
  LBO-OBO-SDO-SSI
Items: ABCDEFKPQSTUVYZ124
Services: 4C-CS-DA-RA-TY-TD-WA
Equipment: 4SO(29x40)-OP
Annual Sales: 4
Number of Employees: 150

Turnaround Times: B4C5S2
Customers: BU75-CU10-NP5-OR5-SC5
Union Shop: N
Year Established: 1932
Terms: 50% deposit, 1%-10, net 30

Statement: We also offer a "price list" for
full color printing of brochures, catalogs,
sales sheets, postcards and posters as well
as a "free idea sampler" kit.

Murphy's Printing Company   GC
Bob Murphy, Owner
1731 Dell Ave
Campbell CA 95008
408-866-8070

Min: 100 Max: 0 Opt: 2000
Sizes: 123456X
Bindings: PBO-SSI
Items: ABCDKPSTVYZ1
Services: 4C

N

National Graphics Corp   GC
William Watkins, Marketing Director
724 E Woodrow Ave
Columbus OH 43207
614-445-3211; Fax: 614-445-3243
Send RFQ's to: William Watkins

Min: 2500 Max: 2000000 Opt: 100000
Sizes: 2345
Bindings: HCO-CBI-LBO-PBI-SSI-SDO-
  SBO-WBO
Items: CDEHIKPQTVXZ23
Services: 4C-TY-TD-WA
Equipment: 7LP(27x41)-7SO(28.75x41)-
  3WO(23⁹⁄₁₆x38)
Annual Sales: 4
Number of Employees: 110
Turnaround Times: B3C5S3
Customers: BU50-MP5-OR5-OT40
Union Shop: Y
Year Established: 1842
Terms: Net 30

Statement: National Graphics Corporation
is a customer-oriented company committed
to long-term, mutually beneficial relation-

SERVICES				BINDINGS		SD	Side Stitching
4C	4-Color Printing	RA	Rachwal System	HC	Hardcover	SB	Spiral Binding
CS	Color Separations	AF	Acid-Free Paper	CB	Comb Binding	WB	Wire-O Binding
DA	Design / Artwork	RP	Recycled Paper	LB	Loose-Leaf Binding	I	In-House
FF	Fulfillment/Mailing	TY	Typesetting	OB	Otabind	O	Out of House
LM	List Maintenance	TD	Typeset w/ Disk	PB	Perfect Binding	X	Unknown
OC	OptiCopy System	WA	Warehousing	SS	Saddle Stitching		

to long-term, mutually beneficial relationships with our clients, vendors, and employees. We endeavor to provide quality printing at a competitive price and to achieve profitability that will provide for continuous growth and allow us to invest in new equipment and technology to better serve our customers.

BK
National Lithographers Inc
Jon Leavy, Marketing Director
7700 NW 37th Ave
Miami FL 33147
800-446-4753; Fax: 305-633-7157
Send RFQ's to: Jon Leavy

Min: 1000 Max: 1000000 Opt: 50000-250000
Sizes: 123456
Bindings: HCO-OBO-PBO-SSI-SBO-WBO
Items: ACDFIKPQSTUWYZ2
Services: 4C-CS-DA-FF-AF-RP-WA
Equipment: SO(20x26,23x31,28x40)
Annual Sales: 3
Number of Employees: 75
Turnaround Times: B3C5S2
Customers: BU30-CU10-MP5-NP5-OR5-OT45
Union Shop: N
Year Established: 1954
Terms: Net 30

Statement: To provide high quality commercial printing; efficiently, on time, and cost effective to our clients.

BK
National Publishing Company
Marketing Director
PO Box 8386
Philadelphia PA 19101-8386
215-732-1863; Fax: 215-732-1314
Parent Company: Courier Corporation

Min: 0 Max: 500000 Opt: 80000
Bindings: HCI-PBI-SDI-SSI
Items: BFM
Services: 4C-FF
Equipment: LP-WO
Customers: CU-NP-OR-OT
Year Established: 1863

Statement: Being fully equipped makes us fully capable of handling your every publishing need. From manuscripts to printing to binding to cartoning for domestic and overseas shipments, National Publishing Company's wide range of specialized and state-of-the-art equipment ensures that each assignment is completed with quality, economy, and personal attention.

BK
National Reproductions Corp
Marketing Director
29400 Stephenson Highway
Madison Heights MI 48071
313-398-7900; Fax: 313-398-7081
800-628-2299
Send RFQ's to: Brian Kennedy, Account Executive

Min: 50 Max: 3000 Opt: 1000
Sizes: 2356
Bindings: CBI-PBI-SSI-SDI
Items: BCDEFGHLMNPS
Services: 4C-CS
Equipment: SO
Annual Sales: 3
Number of Employees: 30
Turnaround Times: B3C3
Customers: NP-OR-SP
Union Shop: N
Year Established: 1953
Terms: Net 30 with approved credit

GC
Nationwide Printing
Hubert C Lewis, President
5906 N Jefferson St
Burlington KY 41005
606-586-9005

Min: 1000 Max: 100000 Opt: 25000
Sizes: 123456
Bindings: SSI
Items: ACDHKLMNPQSTYZ1
Services: 4C
Equipment: 5SO(19x25)
Annual Sales: 2
Number of Employees: 10
Turnaround Times: C2S2
Customers: BU65-NP15-SC20
Union Shop: N
Year Established: 1984
Terms: Net 30

PRINTED ITEMS						
A	Annual Reports	G	Galley Copies	N	Workbooks	U	Greeting Cards
B	Books	H	Journals	O	Yearbooks	V	Labels / Stickers
C	Booklets	I	Magazines	P	Brochures	W	Maps
D	Catalogs	J	Mass-Market Books	Q	Calendars	X	Newspapers
E	Cookbooks	K	Newsletters	R	Comic Books	Y	Postcards
F	Directories	L	Software Manuals	S	Direct Mail Letters	Z	Posters
		M	Textbooks	T	Envelopes	1	Stationery

Statement: Nationwide specializes in direct response envelopes as well as all types of direct mail printing.

Naturegraph Publishers **BK**
Barbara Brown, Marketing Director
PO Box 1075
Happy Camp CA 96039
916-493-5353; Fax: 916-493-5240
Send RFQ's to: Barbara Brown

Min: 500 Max: 15000 Opt: 5000
Sizes: 12356
Bindings: PBX
Items: B
Services: 4C
Equipment: 2SO(18x24,19x26)
Annual Sales: 1
Number of Employees: 5
Turnaround Times: B5
Customers: BP20-MP30-SP50
Union Shop: N
Year Established: 1960
Terms: 1/2 down, balance on completion

Statement: We are publishers of 100 titles of nature and Native American topics. When our presses have any extra time, we use our equipment to print outside jobs.

Louis Neibauer Co Inc **GC**
Nathan Neibauer, President
20 Industrial Dr
Warminster PA 18974
215-322-6200
Send RFQ's to: Nathan Neibauer

Min: 1000 Max: 50000 Opt: 5000
Sizes: 23456
Bindings: PBX-SSX
Items: ABCDEFGHIKLMNOPQRSZ124
Services: 4C-DA-FF-RP-TY-TD-WA
Equipment: 5SO(up to 40)
Annual Sales: 2
Number of Employees: 32
Turnaround Times: B4C5S3
Union Shop: N
Year Established: 1967
Terms: 50% w/order, balance net 20

Statement: High quality short-run printing on coated up to 8 colors. No job too complicated.

Network Graphics **BK**
Dan Crider
906 NW 14th
Portland OR 97209
503-223-5226; Fax: 503-294-0228

Min: 1000 Max: 15000 Opt: 0
Items: B

Statement: Pacific Northwest leader in book manufacturing. 1,000 to 15,000 run lengths. Fast quotes.

Network Studios Inc **GC**
Bill Poutney, General Mgr & VP
1030 Islington Ave
Etobicoke ON Canada M8Z 6A4

Statement: A division of Southam Graphics, Inc.

Newsfoto Publishing Company **BK**
Caryn Truitt, Sales Manager
PO Box 1392
2027 Industrial Blvd
San Angelo TX 76902
915-949-3776

Bindings: HCI
Items: BFO
Services: 4C-CS-TY
Equipment: SO

Statement: A subsidiary of Taylors.

Neyenesch Printers **MS**
Marketing Director
2750 Kettner Blvd
San Diego CA 92101
619-297-2281; Fax: 619-299-7250

Min: 5000 Max: 25000 Opt: 15000
Sizes: 2345X
Bindings: PBI-SSI-SDI
Items: ABCDFHIKNSWZ
Services: 4C-CS-TY
Equipment: LP-SO-WO

SERVICES				BINDINGS			
4C	4-Color Printing	RA	Rachwal System	HC	Hardcover	SD	Side Stitching
CS	Color Separations	AF	Acid-Free Paper	CB	Comb Binding	SB	Spiral Binding
DA	Design / Artwork	RP	Recycled Paper	LB	Loose-Leaf Binding	WB	Wire-O Binding
FF	Fulfillment/Mailing	TY	Typesetting	OB	Otabind	I	In-House
LM	List Maintenance	TD	Typeset w/ Disk	PB	Perfect Binding	O	Out of House
OC	OptiCopy System	WA	Warehousing	SS	Saddle Stitching	X	Unknown

The Nielsen Lithographing Co **GC**
Steven C Finzer
3731 Eastern Hills Ln
Cincinnati OH 45209
513-321-5200; Fax: 513-321-5315

Min: 1000 Max: 0 Opt: 200000
Sizes: 123456
Bindings: PBX-SSX
Items: ACDPV
Services: 4C-CS-DA-TY
Equipment: SO-WO
Annual Sales: 4
Number of Employees: 204
Turnaround Times: B1S1
Customers: OT100
Union Shop: N
Year Established: 1924
Terms: Net 30

Nies/Artcraft Printing Companies **GC**
Jim Poneta, Marketing Director
5900 Berthold Ave
Saint Louis MO 63110
314-647-3400

Min: 0 Max: 0 Opt: 0
Items: AI
Services: 4C
Equipment: SO-WO
Customers: BU100
Year Established: 1902

Statement: They specialize in corporate publications, from annual reports to company magazines.

Nimrod Press Inc **GC**
Seth Tower, President
170 Brookline Ave
Boston MA 02215
617-437-7900

Min: 300 Max: 50000 Opt: 3000
Sizes: 235X
Bindings: HCX-CBI-PBI-SSI-SBI
Items: ABCDEFHILMNP
Services: 4C-TY-TD
Equipment: SO
Annual Sales: 4

Number of Employees: 185
Terms: Net 30

Statement: A financial and general commercial printer, Nimrod also prints book jackets and some books.

Nittany Valley Offset **GC**
Marketing Director
1015 Benner Pike
PO Box 920
State College PA 16804-0920
814-238-3071

Noll Printing Company **CA**
Frederick Scheiber, VP Sales
100 Noll Plaza
Huntington IN 46750
219-356-2020; Fax: 219-356-4584
800-348-2886

Min: 75000 Max: 1000000 Opt: 250000
Sizes: 1245X
Bindings: PBI-SSI
Items: CDFHIJXZ
Services: 4C-CS-FF-LM-AF-TY-TD
Equipment: 5WO(22x36,22x38)
Annual Sales: 5
Number of Employees: 391
Turnaround Times: B3C3S3
Customers: BU60-MP15-NP10-OR5-SC10
Union Shop: Y
Year Established: 1930
Terms: Net 30 with approved credit

Statement: Two plants: another plant at 41 E Park Dr, Huntington IN, 219-356-3690. A subsidiary of Our Sunday Visitor Inc, we are experts in printing lightweight stocks to save postage.

Northeast Graphics **MG**
Marketing Director
1290 Ave of Americas #1480
New York NY 10104
212-924-7551; Fax: 212-245-5283
Send RFQ's to: Alan Bortner, Chairman

Min: 20000 Max: 2000000 Opt: 0
Bindings: SSI
Items: DIX

PRINTED ITEMS						
A	Annual Reports	G	Galley Copies	N	Workbooks	U	Greeting Cards
B	Books	H	Journals	O	Yearbooks	V	Labels / Stickers
C	Booklets	I	Magazines	P	Brochures	W	Maps
D	Catalogs	J	Mass-Market Books	Q	Calendars	X	Newspapers
E	Cookbooks	K	Newsletters	R	Comic Books	Y	Postcards
F	Directories	L	Software Manuals	S	Direct Mail Letters	Z	Posters
		M	Textbooks	T	Envelopes	1	Stationery

Services: 4C
Equipment: 4SO-6WO
Annual Sales: 6
Number of Employees: 350
Year Established: 1948

Statement: They print publications and newspaper inserts.

Northeast Offset Inc **GC**
Marketing Director
11 Alpha Rd
Chelmsford MA 01824
508-256-9939; Fax: 508-250-1249

Northeast Web Printing Inc **GC**
Robert Andrews, Marketing Director
425 Smith St
Farmingdale NY 11735
516-753-9000; Fax: 516-753-0128
Send RFQ's to: Robert Andrews
Parent Company: This Week Publications
     Inc

Min: 10000 Max: 200000 Opt: 100000
Sizes: 5
Items: CX
Services: 4C
Equipment: 19WO(22.75x35)
Annual Sales: 2
Number of Employees: 50
Turnaround Times: S3
Customers: BU5-OT95
Union Shop: N
Year Established: 1978
Terms: Net 30

Statement: Our primary business is as a printer of weekly newspapers. Average press run of 48 pages, 15,000.

Northlight Studio Press **GC**
Lew Bell, Marketing Director
Route 14
PO Box 568
Barre VT 05641-0568
802-479-0565; Fax: 802-479-5245
Send RFQ's to: Lew Bell

Min: 100 Max: 5000 Opt: 1500
Sizes: 2356
Bindings: HCO-CBI-LBO-PBI-SSI-SDO-
     SBO-WBO

Items: ABCDEFGKPT1
Services: 4C-RP-TY-TD
Equipment: 3SO(19x25)
Annual Sales: 2
Number of Employees: 16
Turnaround Times: B3C3
Customers: BP20-BU45-CU30-OT5
Union Shop: N
Year Established: 1974
Terms: Net 30

Statement: We strive to provide the best possible product to our customers.

Northprint International Inc **MS**
Marketing Director
1321 SE 8th St
Grand Rapids MN 55744
218-326-9407; 800-662-5784

Min: 25000 Max: 500000 Opt: 0
Bindings: SSI
Items: DI
Equipment: WO

Northwest Web **CA**
Marketing Director
3592 West 5th Ave
Eugene OR 97402
503-345-0552; Fax: 503-484-0226

Min: 25000 Max: 3000000 Opt: 500000
Sizes: 1346X
Bindings: SSI-HCO-CBO-LBO-OBO-
     PBO-SDO-SBO-WBO
Items: CDIKL
Services: 4C
Equipment: 2WO(6, 8 Units, 22x39)
Annual Sales: 4
Number of Employees: 95
Turnaround Times: S2
Customers: BU95-CU1-NP5-SP35
Union Shop: N
Year Established: 1972
Terms: Net 30

Statement: QUALITY, SERVICE, PRICE

SERVICES				BINDINGS			
4C	4-Color Printing	RA	Rachwal System	HC	Hardcover	SD	Side Stitching
CS	Color Separations	AF	Acid-Free Paper	CB	Comb Binding	SB	Spiral Binding
DA	Design / Artwork	RP	Recycled Paper	LB	Loose-Leaf Binding	WB	Wire-O Binding
FF	Fulfillment/Mailing	TY	Typesetting	OB	Otabind	I	In-House
LM	List Maintenance	TD	Typeset w/ Disk	PB	Perfect Binding	O	Out of House
OC	OptiCopy System	WA	Warehousing	SS	Saddle Stitching	X	Unknown

Nystrom Publishing Company **GC**

Lynn Reemtsma, Marketing Director
9100 Cottonwood Ln
Osseo MN 55369
612-425-7900; Fax: 612-425-0898
Send RFQ's to: Jerry Nystrom, President

Min: 1000 Max: 80000 Opt: 5000
Sizes: 23456
Bindings: HCO-PBO-SSI-SBO-WBO
Items: ABCDFHIKPW
Services: 4C-FF-RP-TY-TD-WA
Equipment: 3SO(18x24,25x38)
Annual Sales: 2
Number of Employees: 13
Turnaround Times: B4C3S3
Customers: BP5-BU25-MP10-NP30-SC25-OT5
Union Shop: N
Year Established: 1978
Terms: Net 30

Statement: We offer quality printing at reasonable prices with a quick turnaround.

O

Oaks Printing Company **GC**

Donald Frace, Marketing Director
195 Nazereth Pike
Bethlehem PA 18017
610-759-8511; Fax: 610-759-8770
Send RFQ's to: Bonnie Levy

Min: 1000 Max: 0 Opt: 0
Sizes: 123456
Bindings: SSX
Items: ABCDEFHIKLMNPQSZ
Services: 4C-CS-DA-FF-RP-TD-WA
Equipment: 6SO(19x25,28x40)
Annual Sales: 3
Number of Employees: 49
Turnaround Times: B3C4S3
Union Shop: Y
Year Established: 1974
Terms: Net 30

Odyssey Press Inc **BK**

Charles Parker, Marketing Director
113 Crosby Rd #15
Dover NH 03820
603-749-4433; Fax: 603-749-1425
Send RFQ's to: Doug Stone

Min: 10 Max: 2000 Opt: 0
Sizes: 12345
Bindings: CBX-LBX-PBX-SSX-WBX
Items: BCEFGHJKLN
Services: FF-AF-RP
Equipment: 4SO(12x18)
Annual Sales: 2
Number of Employees: 27
Turnaround Times: B2S2
Customers: BP32-CU5-NP5-SP5-OR53
Union Shop: N
Year Established: 1989
Terms: Net 30

Statement: We produce short-run quantities of single and two color, time sensitive materials. Two manufacturing facilities guarantee quick turnaround of journals, books, reprints, and bound galleys. The second plant is located at 1414 Key Highway, Baltimore MD 21230.

Offset By Craftsman **GC**

Marketing Director
1107 W Elizabeth Ave
Linden NJ 07036
908-862-8811; Fax: 908-862-6840

Offset House **GC**

Marketing Director
89 Sandhill Rd
Essex Junction VT 05452
802-878-4440; Fax: 802-879-4865

Offset Paperback Manufacturers **BK**

Robert O'Connor, VP Sales
Route 309
PO Box N
Dallas PA 18612
717-675-5261; Fax: 717-675-8714
Send RFQ's to: Robert O'Connor
Parent Company: Bertelsmann Printing

PRINTED ITEMS					
A	Annual Reports	G	Galley Copies	N	Workbooks
B	Books	H	Journals	O	Yearbooks
C	Booklets	I	Magazines	P	Brochures
D	Catalogs	J	Mass-Market Books	Q	Calendars
E	Cookbooks	K	Newsletters	R	Comic Books
F	Directories	L	Software Manuals	S	Direct Mail Letters
		M	Textbooks	T	Envelopes
U	Greeting Cards				
V	Labels / Stickers				
W	Maps				
X	Newspapers				
Y	Postcards				
Z	Posters				
1	Stationery				

Min: 5000 Max: 2000000 Opt: 100000
Sizes: 12X
Bindings: PBI
Items: BJ
Services: 4C-CS-FF-WA
Equipment: 2SO(25x38)-9WO(22x36,
   29x60)
Annual Sales: 5
Number of Employees: 500
Turnaround Times: B4
Customers: BP100
Union Shop: Y
Year Established: 1972
Terms: Net 30, but negotiable

Ohio Valley Litho Color   CS

Marketing Director
Econocolor Catalog
7405 Industrial Rd
Florence KY 41042
606-525-7405; Fax: 606-525-7654
800-877-7405

Min: 5000 Max: 25000 Opt: 20000
Items: DPZ
Services: 4C-CS

Omaha Printing Company   GC

Lanny Worley, VP Sales
4700 F St
Omaha NE 68117
402-734-4400

Min: 1000 Max: 2000000 Opt: 0
Sizes: 123456X
Bindings: LBO-PBO-SSI-SDI-SBO-WBO
Items: ACDEIKPQSTVYZ1
Services: 4C-CS-TY-TD-WA
Equipment: 4LP(13x19)-9SO(to 25x38)-
   3WO-OP
Annual Sales: 4
Number of Employees: 150
Turnaround Times: B4C4S2
Customers: BU80
Union Shop: N
Year Established: 1858
Terms: Net 30

Statement: We specialize in direct mail and
related pieces.

OMNIPRESS   BK

Steve Harrell
2600 Anderson St
PO Box 7214
Madison WI 53707-7214
608-246-2600; Fax: 608-246-4237
800-828-0305
Send RFQ's to: Derrick Carter, Sales
   Manager

Min: 50 Max: 2000 Opt: 500
Sizes: 235
Bindings: CBX-LBX-PBX-SDX-WBX
Items: BFGHLMN
Services: FF-RP-TD
Turnaround Times: B2
Union Shop: N
Year Established: 1977
Terms: No credit extended

Statement: OMNIPRESS specializes in
quick and dependable service on the
reproduction of high page count, standard
size books and manuals from 50 to 2000
copies.

Optic Graphics Inc   BK

David Termotto, Marketing Director
101 Dover Rd
Glen Burnie MD 21060
410-768-3000; Fax: 410-760-4082
800-638-7107
Send RFQ's to: Attn: Estimating Dept

Min: 500 Max: 0 Opt: 5000-50000
Sizes: 1235X
Bindings: HCO-CBI-LBI-PBI-SSI-SDI-
   WBI-OBO-SBO
Items: BCEFHLMNZ234
Services: 4C-CS-LM-OC-TY-TD
Equipment: 2SO(28x41)-2WO(19⅜,23.75)-
   2OP
Annual Sales: 5
Number of Employees: 300
Turnaround Times: B3C4S3
Customers: BP50-BU30-NP20
Union Shop: N
Year Established: 1980
Terms: Net 30

Statement: One of the nation's leading
manufacturers of packaging products in
custom looseleaf systems, specialty publishing materials, and multi-media kits.

SERVICES				BINDINGS			
4C	4-Color Printing	RA	Rachwal System	HC	Hardcover	SD	Side Stitching
CS	Color Separations	AF	Acid-Free Paper	CB	Comb Binding	SB	Spiral Binding
DA	Design / Artwork	RP	Recycled Paper	LB	Loose-Leaf Binding	WB	Wire-O Binding
FF	Fulfillment/Mailing	TY	Typesetting	OB	Otabind	I	In-House
LM	List Maintenance	TD	Typeset w/ Disk	PB	Perfect Binding	O	Out of House
OC	OptiCopy System	WA	Warehousing	SS	Saddle Stitching	X	Unknown

GC

Original Copy Centers
Marketing Director
3333 Chester Ave
Cleveland OH 44114
216-881-3500

Min: 100 Max: 2000 Opt: 500
Sizes: 5
Bindings: LBX-SSX
Items: BCDFLN
Services: Demand printing

GC

Original Smith Printing Co Inc
Marketing Director
2 Hardman Dr
Bloomington IL 61701
309-663-0325; Fax: 309-662-6566

GC

Outstanding Graphics
Garry Moore
1417 50th St
Kenosha WI 53140
414-658-8990; Fax: 414-658-4049
Send RFQ's to: Gail Anderson

Min: 500 Max: 250000 Opt: 3000
Sizes: 2345
Bindings: SSI-SDI-HCO-LBO-PBO-SBO-
    WBO
Items: BCDEFIKLNOPSTVY12
Services: 4C-RP-TY
Equipment: 4SO(10x15,23x29)
Annual Sales: 2
Number of Employees: 12
Turnaround Times: B3C4S2
Customers: BP5-BU20-CU50-MP20-SC5
Union Shop: N
Year Established: 1969
Terms: Net 10

Statement: We specialize in printing saddle
stitched books, runs from 500 to 5,000, one
color body, with one, two, or four color
covers from 8 pages to 72 pages.

MS

The Ovid Bell Press Inc
Tim Steinbeck, Marketing Director
1201-05 Bluff St
Fulton MO 65251
314-642-2256; Fax: 314-642-8467

Send RFQ's to: Tim Steinbeck
Min: 500 Max: 0 Opt: 10000
Sizes: 2356
Bindings: PBI-SSI-HCO-CBO-OBO-SBO-
    WBO
Items: ABCDEFHIKLMN
Services: 4C-FF-OC-RA-AF-RP-TY-TD-
    WA
Equipment: 7SO(25x38)
Annual Sales: 4
Number of Employees: 165
Turnaround Times: B3C6S3
Customers: BP12-CU3-MP65-OT5
Union Shop: N
Year Established: 1927
Terms: Net 30

Statement: Our objective is to meet the
customer's needs, paying particular attention to all production details and working
with the customer to assure that the final
product is one in which both of us can take
pride.

BK

The Oxford Group
Bruce Little, Marketing Director
2 Bridge St
Norway ME 04268
207-743-8958; Fax: 207-743-2256
Send RFQ's to: Bruce Little

Min: 100 Max: 500000 Opt: 100000
Sizes: 2356X
Bindings: SSI-CBO-HCO-LBO-OBO-
    PBO-SDO-SBO-WBO
Items: ABCDEF-
    HIKLMNPQRSUVWXYZ1
Services: 4C-CS-DA-FF-TY-TD-WA
Equipment: 3SO(to 28)-WO(35)
Annual Sales: 2
Number of Employees: 30
Turnaround Times: B4C5S4
Union Shop: N
Year Established: 1836
Terms: 50% down, 50% COD

PRINTED ITEMS							
A	Annual Reports	G	Galley Copies	N	Workbooks	U	Greeting Cards
B	Books	H	Journals	O	Yearbooks	V	Labels / Stickers
C	Booklets	I	Magazines	P	Brochures	W	Maps
D	Catalogs	J	Mass-Market Books	Q	Calendars	X	Newspapers
E	Cookbooks	K	Newsletters	R	Comic Books	Y	Postcards
F	Directories	L	Software Manuals	S	Direct Mail Letters	Z	Posters
		M	Textbooks	T	Envelopes	1	Stationery

P

Pace Lithographers Inc   GC
Marketing Director
18030 Cortney Ct
La Puente CA 91748
818-961-5416; Fax: 818-961-1028

Padgett Printing Corporation   GC
John Mauldin, Marketing Director
1313 N Industrial Blvd
Dallas TX 75207
214-742-4261; Fax: 214-748-8740
Send RFQ's to: John Mauldin
Min: 2500 Max: 3000000 Opt: 75000
Sizes: 245
Bindings: HCO-CBO-LBI-OBO-PBO-SSI-
SDO-SBO-WBO
Items: ACDFIKLMNPQSTUVWYZ134
Services: 4C-CS-FF-TD-WA
Equipment: 2LP(12x18,13x20)-4SO(13x20)
-WO(17.5x26.5)
Annual Sales: 4
Number of Employees: 102
Turnaround Times: B3C4S2
Customers: BP5-BU88-CU5-NP2-SC2-SP8
Union Shop: N
Year Established: 1903
Terms: 2% 10, net 30

Page Litho Inc   CA
Frank Pecherski, Marketing Director
6445 E Vernor Hwy
Detroit MI 48207
313-921-6880; Fax: 313-921-6771
Send RFQ's to: Frank Pecherski

Min: 1000 Max: 100000 Opt: 25000
Sizes: 245
Bindings: LBX-PBX-SSX-SDX
Items: DFL2
Services: 4C-RP-TD
Equipment: 2SO(20x26)-2WO(23x75)
Annual Sales: 3
Turnaround Times: B3S3
Customers: BP10-BU90
Union Shop: N
Year Established: 1981
Terms: Net 30

Panel Prints Inc   GC
Karl Schoettle, Marketing Director
1001 Moosic Rd
Old Forge PA 18518
717-457-8334; Fax: 717-457-6440
Send RFQ's to: Elaine Evans

Min: 1000 Max: 1000000 Opt: 5000-500000
Items: QUWZ4
Services: 4C-CS
Equipment: 17SO
Annual Sales: 5
Number of Employees: 350
Customers: BP5-BU90-OR5
Union Shop: N
Terms: Net 30

Statement: The goal of Panel Prints is to provide total customer satisfaction for users of large-sheet, color lithography and related services.

Panorama Press Inc   GC
Mark Callahan, Marketing Director
1 Entin Rd
Clifton NJ 07014
800-969-8901; Fax: 201-472-5270
Send RFQ's to: Bill Breakstone, Sales
Manager

Min: 2500 Max: 100000 Opt: 50000
Sizes: 123456
Bindings: PBX-SSX-SDX-SBX-WBX
Items: ACDFKPQSYZ2
Services: 4C-CS-DA-FF-OC-RP-TD-WA
Equipment: 6SO(28x41)
Annual Sales: 4
Number of Employees: 150
Turnaround Times: B3C4S2
Customers: BU80-MP5-NP5-OR10
Union Shop: N
Year Established: 1952
Terms: Net 30

Statement: High quality 4 to 8 color collatoral sheetfed lithography and 250 line screen waterless driography—in-house electronic prepress to diecutting. Pioneers in environmentally responsible printing through our Second Nature Division.

SERVICES			BINDINGS		SD	Side Stitching
4C	4-Color Printing	RA Rachwal System	HC	Hardcover	SB	Spiral Binding
CS	Color Separations	AF Acid-Free Paper	CB	Comb Binding	WB	Wire-O Binding
DA	Design / Artwork	RP Recycled Paper	LB	Loose-Leaf Binding	I	In-House
FF	Fulfillment/Mailing	TY Typesetting	OB	Otabind	O	Out of House
LM	List Maintenance	TD Typeset w/ Disk	PB	Perfect Binding	X	Unknown
OC	OptiCopy System	WA Warehousing	SS	Saddle Stitching		

Pantagraph Printing   **BK**
Fred Dolan, President
PO Box 1406
217 W Jefferson St
Bloomington IL 61701
309-829-1071

Min: 0 Max: 0 Opt: 0
Bindings: HCX-PBX
Items: BFHMN
Services: TY
Equipment: SO-WO

Paraclete Press   **BK**
Marketing Director
Southern Eagle Cartway
RR 6A
Brewster MA 02631
800-451-5006; Fax: 508-255-5705
Send RFQ's to: Loretta Jack

Min: 0 Max: 40000 Opt: 10000-20000
Sizes: 123456
Bindings: PBI-SSI-HCO-CBO-SBO-WBO
Items: ABCDEFHIJKMNPQWYZ
Services: 4C-DA-AF-RP-TY
Equipment: CB(20x28)-LP(40x28)-
    SO(12x18)
Number of Employees: 50
Turnaround Times: B3S3
Customers: BP15-BU50-MP25-SP10
Union Shop: N
Year Established: 1981
Terms: 1/2 down, net 30, first time

Statement: Paraclete is a full service
printer specializing in top quality 4 color
and book printing.

Park Press   **GC**
Daniel Kallemeyn, Marketing Director
930 East 162nd St
South Holland IL 60473
708-331-6352; Fax: 708-596-9688
Send RFQ's to: Donald Woo

Min: 5000 Max: 3000000 Opt: 50000
Sizes: X
Bindings: SSX
Items: CDFHIKLNX
Services: 4C-RP
Equipment: 4WO(22.75x35)

Annual Sales: 2
Number of Employees: 25
Turnaround Times: S2
Customers: BU20-CU10-MP5-NP5-SC10-
    SP50
Union Shop: N
Year Established: 1950
Terms: Net 30 w/approval

Statement: Park Press provides affordable,
effective ways for our customers to com-
municate with their customers. We currently sell high volume printing on inexpensive papers offering lower costs on
catalogs and newspapers.

Parker Graphics   **GC**
James J Parker, President
712 N Main St
PO Box 159
Fuquay-Varina NC 27526-0159
919-552-9676

Min: 500 Max: 50000 Opt: 10000
Sizes: 123456
Bindings: CBI-LBI-PBI-SSI-SDI
Items: CDEFKNPTV1
Services: 4C-TY
Equipment: 6SO(10x15,20x28)-WO(23x38)
Annual Sales: 1
Number of Employees: 5
Turnaround Times: C6S5
Customers: BU90-NP5-SP5
Union Shop: N
Year Established: 1969
Terms: Net on receipt

Parker Printing & Color Co   **GC**
Marketing Director
812 Prospect St
Trenton NJ 08618
609-394-8151; Fax: 609-394-8199

Patterson Printing Company   **BK**
Leroy Patterson, Marketing Director
1550 Territorial Rd
Benton Harbor MI 49022
616-925-2177; Fax: 616-925-6057
Send RFQ's to: Leroy Patterson

PRINTED ITEMS					
A	Annual Reports	G	Galley Copies	N	Workbooks
B	Books	H	Journals	O	Yearbooks
C	Booklets	I	Magazines	P	Brochures
D	Catalogs	J	Mass-Market Books	Q	Calendars
E	Cookbooks	K	Newsletters	R	Comic Books
F	Directories	L	Software Manuals	S	Direct Mail Letters
		M	Textbooks	T	Envelopes

U	Greeting Cards
V	Labels / Stickers
W	Maps
X	Newspapers
Y	Postcards
Z	Posters
1	Stationery

Min: 500 Max: 100000 Opt: 5000
Sizes: 12345
Bindings: HCX-CBX-LBX-PBX-SSX-
  WBX
Items: BCDEFHLN
Services: 4C-FF-RA-AF-RP-TD-WA
Equipment: 2WO(22x36)
Annual Sales: 4
Number of Employees: 110
Turnaround Times: B3C4S3
Customers: BP-CU
Union Shop: N
Year Established: 1950
Terms: 2% 10, net 30

Statement: Patterson specializes in serving publishers of information materials (educational textbooks, student workbooks, technical manuals, spirit masters, transparencies, and educational kits).

Paust Inc   GC

Roland Paust, Sales Manager
14 N 10th St
PO Box 1326
Richmond IN 47375
317-962-1507

Min: 10 Max: 2000000 Opt: 1000
Sizes: 123456
Bindings: HCO-CBO-LBI-PBO-SSI-SDI-
  SBI-WBO
Items: ABCDFIKLNPQSTUVWYZ1
Services: 4C-CS-DA-LM-AF-TY
Equipment: 10SO(10x15,23x29)
Annual Sales: 3
Number of Employees: 35
Turnaround Times: B3C2S2
Customers: BP10-BU70-CU10-NP10
Union Shop: N
Year Established: 1945
Terms: Payment in full with order

Peake Printers Inc   GC

Marketing Director
2500 Schuster Dr
Hyattsville MD 20781
301-341-4600; Fax: 301-341-1162

Pearl-Pressman-Liberty Communications Group   BK

Elliot Schindler, Marketing Director
912 N 5th St
Philadelphia PA 19123
215-925-4900; Fax: 215-925-2302
Send RFQ's to: Elliot Schindler

Min: 0 Max: 0 Opt: 0
Sizes: 123456X
Bindings: SSX
Items: ABCDEKPQSUXYZ1
Services: 4C-CS-DA-RP-TY-TD-WA
Equipment: 4SO(28x40)
Annual Sales: 4
Number of Employees: 105
Union Shop: Y
Terms: Net 30

Peconic Companies   BK

Marketing Director
800 Wickham Ave
Mattituck NY 11952
516-298-9478; Fax: 516-298-9478
Send RFQ's to: William Green

Min: 100 Max: 2000 Opt: 700
Sizes: 25
Bindings: HCX-PBX-SSX
Items: ABCDEFGHLMPSTYZ
Services: 4C-RA-AF-TD
Equipment: 7SO(11.25x17.5,13.5x18.25,
  17.5x22)
Annual Sales: 2
Number of Employees: 12
Turnaround Times: B3C4S3
Customers: BP80-BU3-CU2-NP3-OR3-
  SC2-SP5-OT2
Union Shop: N
Year Established: 1973
Terms: 1/3 down, 1/3 on proofs, 1/3 on
  completion

Statement: We are a small family based company, fulfilling the Ultra Short-Run printing needs of the industry.

Pend Oreille Printers Inc   GC

Marketing Director
310 Church St
Sandpoint ID 83864
208-263-9534; Fax: 208-263-9091

SERVICES				BINDINGS			
4C	4-Color Printing	RA	Rachwal System	HC	Hardcover	SD	Side Stitching
CS	Color Separations	AF	Acid-Free Paper	CB	Comb Binding	SB	Spiral Binding
DA	Design / Artwork	RP	Recycled Paper	LB	Loose-Leaf Binding	WB	Wire-O Binding
FF	Fulfillment/Mailing	TY	Typesetting	OB	Otabind	I	In-House
LM	List Maintenance	TD	Typeset w/ Disk	PB	Perfect Binding	O	Out of House
OC	OptiCopy System	WA	Warehousing	SS	Saddle Stitching	X	Unknown

Pendell Printing   **MS**
David Moore, VP Marketing/Sales
1700 James Savage Rd
Midland MI 48642
517-496-3333; Fax: 517-496-9165
800-448-4288 ext 166

Min: 5000 Max: 150000 Opt: 30000
Sizes: 5
Bindings: PBI-SSI
Items: CFHI
Services: 4C-CS-DA-LM-OC-AF-TY-TD-WA
Equipment: SO-4WO
Annual Sales: 5
Number of Employees: 400
Turnaround Times: B2
Customers: MP85
Union Shop: N
Year Established: 1953
Terms: Net 30

Statement: Pendell specializes in shortrun publications and offers 2 complete color electronic desktop divisions; sheetfed, halfweb and fullweb presses, a complete prepress service, a complete bindery service; uv-coating, polybagging, mail list management and in house loading and distribution services.

Penn Lithographics Inc   **GC**
Marketing Director
16221 Author St
Cerritos CA 90701
310-926-0455; Fax: 310-926-8955

PennWell Printing   **MS**
Terrence P Stillin, Marketing Director
1421 S Sheridan Ave
PO Box 1260
Tulsa OK 74112
918-832-9338; Fax: 918-831-9477
Send RFQ's to: Terrence P Stillin

Min: 5000 Max: 75000 Opt: 35000
Sizes: 5X
Bindings: PBX-SSX
Items: DHIKPQX1
Services: 4C-CS-DA-LM-AF-RP-TY-TD-WA

Equipment: SO(26x40)-2WO(23⁹⁄₁₆x36)
Annual Sales: 4
Number of Employees: 200
Turnaround Times: B2S2
Customers: OR30-MP70
Union Shop: Y
Year Established: 1910
Terms: 2% 10, net 30

Statement: PennWell is a full service printer specializing in short to medium run special interest publications.

Pentagram Press   **BK**
Michael Tarachow
4925 S Nicollet Ave
Minneapolis MN 55409
612-824-4576

Min: 1 Max: 1000 Opt: 0
Sizes: 12345
Bindings: HCO-CBO
Items: BCEPSTU1
Services: DA-AF
Equipment: 3LP(10x15,15x20)
Year Established: 1974

Statement: Pentagram Press uses only the letterpress process to print books, chapbooks, stationery suites, wedding invitations, cards, etc.

Penton Press Publishing   **MG**
Marketing Director
680 N Rocky River Rd
Berea OH 44017
216-243-5700

Min: 1000 Max: 500000 Opt: 0
Sizes: 2345X
Bindings: PBI-SSI
Items: CDIPX
Services: 4C
Equipment: SO-WO

Perfect Plastic Printing Corp   **GC**
Marketing Director
345 Kautz Rd
St Charles IL 60174
708-584-1600; Fax: 708-584-0648

PRINTED ITEMS					
A	Annual Reports	G	Galley Copies	N	Workbooks
B	Books	H	Journals	O	Yearbooks
C	Booklets	I	Magazines	P	Brochures
D	Catalogs	J	Mass-Market Books	Q	Calendars
E	Cookbooks	K	Newsletters	R	Comic Books
F	Directories	L	Software Manuals	S	Direct Mail Letters
		M	Textbooks	T	Envelopes

U	Greeting Cards		
V	Labels / Stickers		
W	Maps		
X	Newspapers		
Y	Postcards		
Z	Posters		
1	Stationery		

The Perlmuter Printing Co   GC

Ken Allen, Marketing Director
4437 E 49th St
Cleveland OH 44125
216-271-5300; Fax: 216-271-7650
Send RFQ's to: Ken Allen

Min: 5000 Max: 3000000 Opt: 450000
Sizes: 123456
Bindings: PBX-SSX
Items: ADIPWZ
Services: 4C-CS-DA-FF-OC-RP-TY-TD-
WA
Equipment: 3SO(28x40)-3WO(22.75x38)
Annual Sales: 5
Number of Employees: 350
Turnaround Times: B3S3
Customers: CU2-MP5-OT93
Union Shop: N
Year Established: 1917
Terms: Net 30

Statement: We are a full service national web and sheetfed printer, specializing in 4-color commercial printer.

Perry Printing Corp   CA

Timothy Smith
575 W Madison St
PO Box 97
Waterloo WI 53594
414-478-3551; Fax: 414-478-1500

Min: 100000 Max: 5000000 Opt: 0
Sizes: 1245
Bindings: PBI-SSI
Items: ACDFHINPVXZ
Services: 4C-FF-TD-WA
Equipment: 14WO
Annual Sales: 6
Number of Employees: 1600
Turnaround Times: C1S1
Customers: MP40-OR60
Union Shop: Y
Year Established: 1956
Terms: Net 30

Statement: A subsidiary of Journal Communications, they print many catalogs.

Perry/Baraboo   GC

Marketing Director
1300 Sauk Ave
Baraboo WI 53913
608-356-7787
Parent Company: Perry Printing

Min: 100000 Max: 5000000 Opt: 0
Sizes: 1245
Bindings: PBI-SSI
Items: ACDFHINPVXZ
Services: 4C-FF-TD-WA
Equipment: 14WO
Annual Sales: 6
Number of Employees: 1600
Turnaround Times: C1S1
Customers: MP40-OR60
Union Shop: Y
Year Established: 1956
Terms: Net 30

Petersen Graphics Group   GC

Marketing Director
2920 W Sample St
South Bend IN 46619
219-233-6171; Fax: 219-288-4521

Petty Company Inc   GC

Marketing Director
420 W Industrial Ave
Effingham IL 62401
217-347-7721

Min: 50000 Max: 5000000 Opt: 0
Sizes: 25X
Bindings: SSI
Items: DPST
Services: 4C-FF-TY
Equipment: SO(24x36),7WO(to 22x38)
Annual Sales: 6
Number of Employees: 450

Statement: Petty specializes in producing various direct mail formats, including catalogs, self-mailers, inserts, game cards, coupons and bind-ins.

SERVICES		BINDINGS			
4C	4-Color Printing	RA	Rachwal System	HC	Hardcover
CS	Color Separations	AF	Acid-Free Paper	CB	Comb Binding
DA	Design / Artwork	RP	Recycled Paper	LB	Loose-Leaf Binding
FF	Fulfillment/Mailing	TY	Typesetting	OB	Otabind
LM	List Maintenance	TD	Typeset w/ Disk	PB	Perfect Binding
OC	OptiCopy System	WA	Warehousing	SS	Saddle Stitching

SD	Side Stitching
SB	Spiral Binding
WB	Wire-O Binding
I	In-House
O	Out of House
X	Unknown

PFP Printing Corporation **BK**

Frank Wood, General Manager
8041 Cessna Ave #132
Gaithersburg MD 20879-4118
301-258-8353; Fax: 301-670-4147
Send RFQ's to: Frank Wood

Min: 200 Max: 6000 Opt: 750
Sizes: 12345
Bindings: PBI-SSI-SDI-HCO-CBO-LBO-
SBO-WBO
Items: BCDEFGHIKLMNPST1
Services: 4C-FF-AF-TY-TD
Equipment: 4SO(up to 14x20)-WO
(14x18)
Annual Sales: 2
Number of Employees: 9
Turnaround Times: B2C4S2
Customers: BP50-BU15-MP25-SP10
Union Shop: N
Year Established: 1987
Terms: 50% deposit, net 30 — 1st time

Statement: PFP specializes in production of softbound books in short and medium run lengths, principally single-color text with single or multiple-color covers. We strive to achieve customer satisfaction regarding quality, schedule, cost, and headache-free service.

Philipp Lithographing Co **GC**

Marketing Director
1960 Wisconsin Ave
Grafton WI 53024
414-377-1100; Fax: 414-377-6660

Phillips Brothers Printers Inc **BK**

Marketing Director
1555 W Jefferson
PO Box 580
Springfield IL 62705
217-787-3014; Fax: 217-787-9624
800-637-9327

Min: 0 Max: 0 Opt: 0
Bindings: PBI-SDI-CBI
Items: BDLP2
Services: 4C
Equipment: SO-2WO
Customers: CU-BP-BU-OT

Union Shop: Y
Year Established: 1883

Statement: Above all, Phillips Brothers Printers is friendly because our number one goal is satisfying our customers. For over 110 years, we have been achieving our goal. Regardless of who you are, we will treat you with respect and your work with pride. See for yourself how PBP places its customers first.

Pine Hill Press Inc **GC**

Joe Mierau
Highway 81 South
PO Box 340
Freeman SD 57029-0340
605-925-4228; Fax: 605-925-4756
800-676-4228
Send RFQ's to: Joe Mierau

Min: 50 Max: 20000 Opt: 1000
Sizes: 123456X
Bindings: HCX-CBX-LBX-PBX-SSX-
SDX-SBX-WBX
Items: ABCDEFHKLMNPSTZ
Services: 4C-DA-AF-RP-TY-TD
Equipment: LP(10x14)-6S0(20x28)
Annual Sales: 2
Number of Employees: 18
Turnaround Times: B4C5S4
Customers: BP20-BU10-CU7-OR3-SP50-
OT10
Union Shop: N
Year Established: 1940
Terms: Books: 50% down, 50% 30

Statement: We provide complete book printing services for publishing companies, self-publishers, individuals, and organizations. We also provide complete commercial printing services.

Pinecliffe Printers **CS**

Marketing Director
1815 N Harrison
Shawnee OK 74801
405-275-7351

Min: 0 Max: 0 Opt: 0
Sizes: 5
Services: 4C-CS
Equipment: SO

PRINTED ITEMS			
A Annual Reports	G Galley Copies	N Workbooks	U Greeting Cards
B Books	H Journals	O Yearbooks	V Labels / Stickers
C Booklets	I Magazines	P Brochures	W Maps
D Catalogs	J Mass-Market Books	Q Calendars	X Newspapers
E Cookbooks	K Newsletters	R Comic Books	Y Postcards
F Directories	L Software Manuals	S Direct Mail Letters	Z Posters
	M Textbooks	T Envelopes	1 Stationery

Pioneer Press GC
Marketing Director
1232 Central Ave
Wilmette IL 60091
708-251-4300

Min: 5000 Max: 1000000 Opt: 0
Sizes: 2345X
Bindings: SSI
Items: BCDFIKPS
Services: 4C-TY
Equipment: SO-WO

Pioneer Printing & Stationery Co GC
Marketing Director
514 W 19th St
Cheyenne WY 82001
307-635-4114; Fax: 307-632-7810

Plain Talk Printing Company GC
Marketing Director
511 E Sixth St
Des Moines IA 50309
515-282-0485; Fax: 515-282-5169

Min: 0 Max: 0 Opt: 0
Sizes: 123456
Items: ACDEFHIKLPQSTVXYZ1
Services: 4C
Equipment: 3LP(to 21x26)-6SO(to 26x40)
Annual Sales: 3
Number of Employees: 125
Turnaround Times: B4C3S3
Customers: BP2-BU65-CU5-MP2-NP5-
    OR16-SP5
Union Shop: Y
Year Established: 1869
Terms: Net 30

Plus Communications BK
Marketing Director
2828 Brannon Ave
Saint Louis MO 63139
314-776-1110

Min: 1000 Max: 100000 Opt: 20000
Sizes: 2356
Bindings: CBI-LBI-PBI-SSI-SDI
Items: BCDFHKLMNPQ

Services: 4C-DA-FF-TY-TD-WA
Equipment: 3SO(26,40),WO(35)
Annual Sales: 3
Number of Employees: 30
Turnaround Times: B4C4
Customers: BP40-BU15-CU25-NP10-SP10
Union Shop: N
Year Established: 1979
Terms: Net 30

Statement: We are short to medium-run
book printers. We provide high quality
covers with lamination and good one or
two-color text.

Plymouth Printing Co Inc GC
Marketing Director
450 North Ave E
Cranford NJ 07016
908-276-8100; Fax: 908-276-6566

Port City Press Inc BK
Kenneth J Tornvall
1323 Greenwood Rd
Baltimore MD 21208
410-486-3000; Fax: 410-486-0706
800-858-PORT
Send RFQ's to: Stephanie Williams
Parent Company: Judd's Incorporated

Min: 500 Max: 50000 Opt: 5000
Sizes: 123456X
Bindings: LBI-PBI-SSI-SDI-HCO-CBO-
    OBO-SBO-WBO
Items: BCDEFHLMN
Services: 4C-FF-LM-OC-RA-AF-RP-TY-
    WA-TD
Equipment: SO(54x77)-3WO(22x35)
Annual Sales: 5
Number of Employees: 280
Turnaround Times: B3C4S3
Customers: BP40-CU5-NP40-OT15
Union Shop: N
Year Established: 1961
Terms: 30 days

Statement: To produce our customer's
publications accurately, on time, and
profitably.

SERVICES				BINDINGS		SD	Side Stitching
4C	4-Color Printing	RA	Rachwal System	HC	Hardcover	SB	Spiral Binding
CS	Color Separations	AF	Acid-Free Paper	CB	Comb Binding	WB	Wire-O Binding
DA	Design / Artwork	RP	Recycled Paper	LB	Loose-Leaf Binding	I	In-House
FF	Fulfillment/Mailing	TY	Typesetting	OB	Otabind	O	Out of House
LM	List Maintenance	TD	Typeset w/ Disk	PB	Perfect Binding	X	Unknown
OC	OptiCopy System	WA	Warehousing	SS	Saddle Stitching		

Send RFQ's to: Murray Tuchman

Min: 500 Max: 5000000 Opt: 50000-750000
Sizes: 2356
Bindings: CBI-LBI-PBI-SSI-SDI-SBI-
    HCO-OBO-WBO
Items: ABCDEFHIKLMNOPQSTWYZ23
Services: 4C-CS-FF-RP-WA
Equipment: SO-WO
Annual Sales: 3
Number of Employees: 45
Turnaround Times: B3C5S2
Customers: BP15-BU65-MP5-NP5-SC2-
    OT1-CU5-OR1-SP1
Union Shop: N
Year Established: 1970
Terms: 50% down, balance COD

Statement: Service and value to established
growth oriented organizations.

The Printing House Inc   GC
Marketing Director
RR 6 Box 2000
Quincy FL 32351
904-875-1500

Printing Service Co   GC
Marketing Director
630-52 S Main St
Dayton OH 45402
513-461-4580; Fax: 513-223-2013

Printing: Campanella & Rome   BR
Mary Campanella
3319 Tennyson
San Diego CA 92106
619-223-2929; Fax: 619-223-2470
800-554-9880

Printloid Corp   GC
Marketing Director
10-10 44th Ave
Long Island City NY 11101
718-361-1770; Fax: 718-361-2706

Printpack Inc   GC
Marketing Director
3510 Asheville Hwy
Hendersonville NC 28739
704-693-1723; Fax: 704-692-1716
800-438-4442

Printright International   BK
Marketing Director
Unit 204 Building 12
200 E Ave R
Palmdale CA 93550
805-274-9272; Fax: 805-274-4338
Send RFQ's to: Ted Jaffe
Parent Company: Printright of Hong Kong

Min: 1000 Max: 100000 Opt: 5000-35000
Sizes: 123456X
Bindings: CBI-HCI-PBI-SSI-SDI-LBO-
    OBO-SBI-WBI
Items: ABCDEFNOPQUYZ24
Services: 4C-CS-AF-RP
Annual Sales: 3
Number of Employees: 47
Turnaround Times: B5C6S5
Customers: BP30-BU20-SP50
Union Shop: N
Year Established: 1979
Terms: 20% down, 30% proof, 30% print,
    20% on shipping

Printwest   GC
Dallas Robinson, Marketing Director
1150 11th Ave
Regina SK Canada S4R 1C9
306-525-2304; Fax: 306-757-2439
Send RFQ's to: Art Zimmerman
Parent Company: Directwest/Saskwheat
    Pool

Min: 100 Max: 1000000 Opt: 300000
Sizes: 12345
Bindings: LBX-PBX-SSX-SDX-SBX
Items: ABCDEFHIKPQSX2
Services: 4C-CS-DA-FF-OC-RP-TY-TD-
    WA
Equipment: LP-SO-WO
Annual Sales: 4
Number of Employees: 150
Turnaround Times: B4C5S3

PRINTED ITEMS			
A Annual Reports	G Galley Copies	N Workbooks	U Greeting Cards
B Books	H Journals	O Yearbooks	V Labels / Stickers
C Booklets	I Magazines	P Brochures	W Maps
D Catalogs	J Mass-Market Books	Q Calendars	X Newspapers
E Cookbooks	K Newsletters	R Comic Books	Y Postcards
F Directories	L Software Manuals	S Direct Mail Letters	Z Posters
	M Textbooks	T Envelopes	1 Stationery

Customers: BP10-BU25-CU5-MP25-NP5-
   OR10-SP5
Union Shop: Y
Year Established: 1950
Terms: Net 30

Statement: "Customer satisfaction."

Process Displays Company GC
Marketing Director
7108 31st Ave N
Minneapolis MN 55427
612-546-1133; Fax: 612-546-0188

Professional Press GC
Marketing Director
304 Warren Way
PO Box 4371
Chapel Hill NC 27515-4371
919-942-8020; Fax: 919-942-3094

Progress Printing GC
Craig Shaffer, Marketing Director
3523 Waterlick Rd
Lynchburg VA 24502
804-239-9213; Fax: 804-237-1618
800-572-7804
Send RFQ's to: Daniel Thornton

Min: 1000 Max: 5000000 Opt: 200000
Sizes: 123456
Bindings: HCO-CBI-LBI-PBI-SSI-SDI-
   SBO-WBO-OBO
Items: ABCDEFHIKLNPQSTUVWYZ12
Services: 4C-CS-DA-FF-OC-AF-TY-TD-
   WA-RP
Equipment: 3LP(28x14)-15SO-
   2WO(17x23)
Annual Sales: 5
Number of Employees: 282
Turnaround Times: B3C4S2
Customers: BP10-BU40-CU20-MP10-NP2-
   OT18
Union Shop: N
Year Established: 1962
Terms: Net 30

Statement: Progress Printing is the largest
privately owned commercial printer in Virginia with more than five acres under one
roof. Progress specializes in a wide range
of professional services for publishers, including electronic prepress, sheetfed and
web offset, and full-service finishing. We
are a customer-driven shop committed to
exquisite quality, exceptional service, and
competitive value.

Progressive Typographers Inc BK
Suzanne Schiding
York County Industrial Park
PO Box 278
Emigsville PA 17318
717-764-5908; Fax: 717-767-4092
800-673-2500
Send RFQ's to: Marta Dunker

Min: 0 Max: 0 Opt: 0
Services: CS-DA-OC-TY-TD
Annual Sales: 4
Number of Employees: 200
Union Shop: N
Year Established: 1950
Terms: Net 30

ProLitho GC
Ned Warner, President
201 E Bay Blvd
Provo UT 84606
801-373-7335

Min: 0 Max: 0 Opt: 0
Items: ABCDFKLNPSTYZ1
Services: 4C
Equipment: 4SO-WO
Number of Employees: 100

Statement: A general commercial printer,
ProLitho also produces computer software
manuals and support materials.

Promotional Printing Corporation CA
Marketing Director
6902 Palestine Rd
Houston TX 77020
713-673-1005

Min: 50000 Max: 2000000 Opt: 0
Sizes: 5X
Bindings: SSI
Items: DIPS

SERVICES				BINDINGS		SD	Side Stitching
4C	4-Color Printing	RA	Rachwal System	HC	Hardcover	SB	Spiral Binding
CS	Color Separations	AF	Acid-Free Paper	CB	Comb Binding	WB	Wire-O Binding
DA	Design / Artwork	RP	Recycled Paper	LB	Loose-Leaf Binding	I	In-House
FF	Fulfillment/Mailing	TY	Typesetting	OB	Otabind	O	Out of House
LM	List Maintenance	TD	Typeset w/ Disk	PB	Perfect Binding	X	Unknown
OC	OptiCopy System	WA	Warehousing	SS	Saddle Stitching		

Services: 4C-CS-TY
Equipment: 2WO(21x36)

Statement: This subsidiary of the Houston Chronicle Company specializes in printing catalogs.

Publishers Choice Reprints **MS**
Richard Wright, Marketing Director
9391 Grogan's Mill Rd
The Woodlands TX 77380
800-256-8271; Fax: 713-367-8708
Parent Company: Woodlands Printing

Min: 0 Max: 0 Opt: 0
Sizes: 6
Bindings: SSX
Items: I
Services: 4C-DA-TY-TD
Equipment: SO
Turnaround Times: S2
Customers: MP99-OT1
Union Shop: N

Publishers Express Press **BK**
Jerry Elling, Marketing Director
200 West 5th St South
Ladysmith WI 54848
715-532-5300; Fax: 715-532-7458
Send RFQ's to: Jerry Elling
Parent Company: Flambeau Litho Corporation

Min: 500 Max: 25000 Opt: 10000
Sizes: 123456X
Bindings: LBI-PBI-SSI-SDI-SBI-HCO-OBO-WBO-CBO
Items: BCDEFKLMNPZ
Services: 4C-CS-FF-LM-WA
Equipment: 3SO(12.5x18,2-25x36)
Annual Sales: 2
Number of Employees: 21
Turnaround Times: B3S3
Customers: BP60-BU15-CU6-OT11-SP8
Union Shop: N
Year Established: 1970
Terms: Net 30, if an open account

Statement: To work with Mac platform publishers in taking E-files direct to the imposed film for the production of books. We provide outstanding service at an economical price while producing a quality product.

Publishers Press **BK**
Brent McPhie, Marketing Director
1900 West 2300 South
PO Box 27408
Salt Lake City UT 84119
801-972-6600; Fax: 801-972-6601
800-456-6600
Send RFQ's to: Brent McPhie

Min: 1000 Max: 500000 Opt: 10000-50000
Sizes: 123456X
Bindings: HCX-CBX-LBX-PBX-SSX-SBX-WBX-OBX-SDX
Items: ABCDEFGHIJKLMNO
Services: 4C-FF-OC-AF
Equipment: 5SO(36,41)-2WO(17x26,19x30)
Annual Sales: 5
Number of Employees: 115
Turnaround Times: B4C4
Customers: BP50-BU20-CU5-SP25
Union Shop: N
Year Established: 1958
Terms: Net 30 upon credit approval

Statement: Our goal is to meet the needs of our clients, whatever those needs. We take great pride in the number of repeat customers we have.

Publishers Press **BK**
Marketing Director
PO Box 37500
Louisville KY 40233
502-955-6526; Fax: 502-955-5586
800-627-5801
Send RFQ's to: Michael J Simon

Publishers Printing Co **MS**
Marketing Director
100 Frank E Simon Ave
Shepherdsville KY 40165
502-543-2251; Fax: 502-543-8808
Send RFQ's to: Dan Weber

Min: 10000 Max: 300000 Opt: 40000
Sizes: 25
Bindings: PBX-SSX
Items: I
Services: 4C-CS-OC-RP-TY-TD-WA
Equipment: 14SO(36,18)
Annual Sales: 6

PRINTED ITEMS					
A	Annual Reports	G	Galley Copies	N	Workbooks
B	Books	H	Journals	O	Yearbooks
C	Booklets	I	Magazines	P	Brochures
D	Catalogs	J	Mass-Market Books	Q	Calendars
E	Cookbooks	K	Newsletters	R	Comic Books
F	Directories	L	Software Manuals	S	Direct Mail Letters
		M	Textbooks	T	Envelopes

U	Greeting Cards	
V	Labels / Stickers	
W	Maps	
X	Newspapers	
Y	Postcards	
Z	Posters	
1	Stationery	

Number of Employees: 1500
Turnaround Times: B3S3
Customers: MP100
Union Shop: N
Year Established: 1866

Statement: A family owned, high quality, "special interest" magazine printer.

Q

Q & Q Printing Co   **GC**
Marketing Director
40 Hague St
Detroit MI 48202
313-872-5151; Fax: 313-872-8313

Quad Graphics - Thomaston Press   **MG**
Marketing Director
100 McIntosh Parkway
PO Box 552
Thomaston GA 30286
706-647-4000

Min: 50000 Max: 2000000 Opt: 0
Sizes: 25X
Bindings: PBI-SSI
Items: JIX
Services: 4C-FF-TY
Equipment: 8WO
Annual Sales: 6
Number of Employees: 4771
Turnaround Times: S1

Quad/Graphics, Lomira   **MG**
Marketing Director
952 Badger Rd
Lomira WI 53048
414-269-4700

Min: 100000 Max: 7000000 Opt: 0
Sizes: 2345X
Bindings: PBI-SSI
Items: ACDIPW
Services: 4C-CS-FF-TY
Equipment: WO-3OP(Gravure)
Annual Sales: 6
Number of Employees: 4771

Turnaround Times: S1
Customers: BU25-MP75
Year Established: 1972

Quad/Graphics, Pewaukee   **MG**
Marketing Director
W224 N3322 Duplainville Rd
Pewaukee WI 53072-4195
414-246-2800; Fax: 414-246-5165

Min: 100000 Max: 7000000 Opt: 0
Sizes: 2345X
Bindings: PBI-SSI
Items: ACDIPW
Services: 4C-CS-FF-TY
Equipment: 24WO
Annual Sales: 6
Number of Employees: 4771
Turnaround Times: S1
Customers: BU25-MP75
Year Established: 1972

Statement: Quad Graphics prints many of the largest circulation magazines and catalogs in the United States.

Quad/Graphics Inc   **MG**
Marketing Director
56 DuPlainville Rd
Saratoga Springs NY 12866-9050
518-581-4000

Min: 100000 Max: 7000000 Opt: 0
Sizes: 2345X
Bindings: PBI-SSI
Items: ACDIPW
Services: 4C-CS-FF-TY
Equipment: 24WO
Annual Sales: 6
Number of Employees: 4771
Turnaround Times: S1
Customers: BU25-MP75
Year Established: 1972

Quad/Graphics, Sussex   **CA**
Marketing Director
N63 W23075 Highway 74
Sussex WI 53089
414-246-9200

Min: 100000 Max: 7000000 Opt: 0

SERVICES				BINDINGS			
4C	4-Color Printing	RA	Rachwal System	HC	Hardcover	SD	Side Stitching
CS	Color Separations	AF	Acid-Free Paper	CB	Comb Binding	SB	Spiral Binding
DA	Design / Artwork	RP	Recycled Paper	LB	Loose-Leaf Binding	WB	Wire-O Binding
FF	Fulfillment/Mailing	TY	Typesetting	OB	Otabind	I	In-House
LM	List Maintenance	TD	Typeset w/ Disk	PB	Perfect Binding	O	Out of House
OC	OptiCopy System	WA	Warehousing	SS	Saddle Stitching	X	Unknown

Sizes: 2345X
Bindings: PBI-SSI
Items: ACDIPW
Services: 4C-CS-FF-TY
Equipment: 24WO
Annual Sales: 6
Number of Employees: 4771
Turnaround Times: S1
Customers: BU25-MP75
Year Established: 1972

BK
Quebecor Printing (USA)Corp
Mark Anderson, Marketing Director
125 High St
Boston MA 02110
800-678-6299; Fax: 612-690-7438
Send RFQ's to: Mark Anderson
Parent Company: Quebecor Printing Inc

Min: 0 Max: 0 Opt: 0
Sizes: 123456X
Bindings: HCX-CBX-LBX-PBX-OBX-
    SSX-SDX-SBX-WBX
Items: ABCDEFGHIJKLMNOPQR2
Services: 4C-CS-FF-LM-OC-AF-RP-TD-
    WA
Equipment: CB-LP-SO-WO-OP(Gravure)
Annual Sales: 6
Number of Employees: 10000
Union Shop: Y
Year Established: 1880

Statement: Quebecor Printing (USA)
Corp. is a leading book printer, as well as a
leader in the printing of newspaper inserts,
magazines, catalogs and other commercial
printing products. Superior service, high
quality and competitive pricing — all
designed around meeting and exceeding
customer expectations — have made
Quebecor Printing the new force in the
printing industry. Its 26 printing and related services plants located in 17 U.S.
states are supported by a nationwide network of 21 sales offices. The company also
is part of Quebecor Printing Inc., which
owns 65 printing and related services
facilities in the U.S., Canada, France and
Mexico.

BK
Quebecor Book Group
S Ratkewitch, Sales Director
9 East 38th St
New York NY 10016
212-779-2772; Fax: 212-779-2779
Send RFQ's to: S Ratkewitch
Parent Company: Quebecor Printing Inc

Min: 2000 Max: 0 Opt: 25000-250000
Sizes: 2356
Bindings: HCX-CBX-OBX-PBX-SSX-
    SDX-SBX-WBX
Items: ABDEFIJLMNQSWYZ4
Services: 4C-CS-OC-AF-FP-TD-WA
Equipment: 12SO(20x29,54x78)-
    13WO(17x23),CB
Annual Sales: 6
Turnaround Times: B2C3S2
Customers: BP60-BU20-CU10
Union Shop: Y
Terms: Net 30

BK
Quebecor/Semline
Marketing Director
180 Wood Rd
Braintree MA 02184
617-848-2380; Fax: 617-843-3643

Min: 1000 Max: 0 Opt: 0
Sizes: 23456X
Bindings: HCX-CBX-LBX-OBX-PBX-
    SSX-SDX-SBX-WBX
Items: BCDEFLMNOPQWYZ234
Services: 4C-OC
Equipment: CB-18SO(to 55x78)-10WO(to
    22x38)
Annual Sales: 6
Number of Employees: 600
Turnaround Times: B4C5S4
Customers: BP60-BU30-CU10
Union Shop: Y
Year Established: 1886
Terms: Net 30

CA
Quebecor Printing/St Cloud
Tom Poupard, Production Service
    Manager
660 Mayhew Lake Rd NE
PO Box 1007
Saint Cloud MN 56302
612-654-2400; Fax: 612-252-0062

PRINTED ITEMS			
A Annual Reports	G Galley Copies	N Workbooks	U Greeting Cards
B Books	H Journals	O Yearbooks	V Labels / Stickers
C Booklets	I Magazines	P Brochures	W Maps
D Catalogs	J Mass-Market Books	Q Calendars	X Newspapers
E Cookbooks	K Newsletters	R Comic Books	Y Postcards
F Directories	L Software Manuals	S Direct Mail Letters	Z Posters
	M Textbooks	T Envelopes	1 Stationery

800-678-6299

Min: 10000 Max: 2000000 Opt: 0
Sizes: 5
Bindings: PBI-SSI
Items: FHI
Services: 4C-CS-DA-FF-TY
Equipment: SO-16WO

Quebecor Printing Inc   CA
Marketing Director
10481 Yonge St
Richmond Hill ON L4C 3C6 Canada
416-884-9121

Min: 75000 Max: 4000000 Opt: 0
Sizes: 5
Bindings: HCO-CBO-LBO-PBI-SSI-SDO-
SBO-WBO
Items: ABCDEFHIJKLPRSXZ
Services: 4C-FF-WA
Equipment: 27SO-135WO
Turnaround Times: C2S2
Customers: BU65-MP35
Union Shop: Y
Year Established: 1890
Terms: Credit to be arranged

Quebecor Printing Inc   CA
Dennis Meyer, Marketing Manager
1999 Shepard Rd
Saint Paul MN 55116
612-690-7200; Fax: 612-690-7357

Min: 30000 Max: 5000000 Opt: 0
Sizes: 5
Bindings: PBI-SSI
Items: CDHIP
Services: 4C-CS-DA-FF-TY
Equipment: SO-16WO
Customers: BU-CU-MP-NP-OR
Year Established: 1884

Quebecor /Dallas   CA
Marketing Director
4800 Spring Valley Rd #704
PO Box 809067
Dallas TX 75234
214-233-3400

Min: 250000 Max: 2000000 Opt: 0

Sizes: 235X
Bindings: PBI-SSI
Items: DX
Services: 4C-CS-FF
Equipment: 8WO-13OP(Gravure)

Other Quebecor Plants:
9 S 1st Ave
Mount Vernon NY 10550
914-667-4400

11 S 2nd Ave
St Charles IL 60174
708-377-2299

1900 Glades Rd
Boca Raton FL 33431
407-394-6021

451 Arcata Blvd
Clarksville TN 37040
615-553-4400

201 Spear St
San Francisco CA 94105
415-541-9652

Buffalo Inc
T C Industrial Park
Depew NY 14043
716-686-2500

99 River St
Providence RI 02904
401-331-1771

6300 Hazelton National D
Orlando FL 32822
407-851-3681

62 Southfield Ave
Stamford CT 06902
203-973-0040

PO Box 686
Dickson TN 37056
615-446-6700

2325 Carillon Pt
Kirkland WA 98033
206-889-1248

15821 Ventura Blvd
Encino CA 91436
818-501-2080

PO Box 330
1800 Parkway Pl
Marietta GA 30067
404-590-7800

SERVICES				BINDINGS			
4C	4-Color Printing	RA	Rachwal System	HC	Hardcover	SD	Side Stitching
CS	Color Separations	AF	Acid-Free Paper	CB	Comb Binding	SB	Spiral Binding
DA	Design / Artwork	RP	Recycled Paper	LB	Loose-Leaf Binding	WB	Wire-O Binding
FF	Fulfillment/Mailing	TY	Typesetting	OB	Otabind	I	In-House
LM	List Maintenance	TD	Typeset w/ Disk	PB	Perfect Binding	O	Out of House
OC	OptiCopy System	WA	Warehousing	SS	Saddle Stitching	X	Unknown

733 3rd Ave
New York NY 10017
212-697-7000

1 Dedham Pl
Dedham MA 02026
617-326-1070

5101 Industrial Blvd NE
Minneapolis MN 55421
612-572-5800

PO Box 16037
Memphis TN 38186
901-332-6391

404 N Wesley Ave
Morris IL 61054
815-734-4121

7400 Impala Dr
Richmond VA 23228
804-264-3800

PO Box 6363
San Jose CA 95150
408-435-2300

7364 Baltimore Annapolis
Glen Burnie MD 21061
410-761-0440

Rt 372
Atglen PA 19310
610-593-5173

4931 Birch St
Newport Beach CA 92660
714-474-5525

8690 Wolff Ct #200
Westminster CO 80030
303-657-9676

59 W Steegers Rd
Arlington Heights IL 60005
708-593-2878

Queen Beach Printers Inc **GC**
Marketing Director
937 Pine Ave
Long Beach CA 90813
310-436-8201; Fax: 310-435-2209

Quigley Printing Co Inc **GC**
Marketing Director
790 Potts Ave
Green Bay WI 54304
414-494-7441

Quinn-Woodbine Inc **BK**
Lisa Newsome, Marketing Director
Oceanview Rd
Woodbine NJ 08270
609-861-5352; Fax: 609-861-5355
Send RFQ's to: Peg Davis

Min: 1000 Max: 10000 Opt: 0
Sizes: 123456
Bindings: HCI-PBI-SSO-SBO-WBI
Items: BDEFH
Services: FF-OC-AF
Equipment: 6SO(54,42,38)
Annual Sales: 2
Number of Employees: 80
Turnaround Times: B4C4S4
Customers: BP80-CU10-SP5-OT5
Union Shop: N
Year Established: 1977
Terms: Net 30 with approved credit

Statement: We specialize in short run book manufacturing from repro copy to perfect and smyth sewn casebinding. We have built our reputation by providing timely, high quality, cost efficient book manufacturing.

R

Racanelli Printing Co **GC**
Marketing Director
640 Dell Rd
Carlstadt NJ 07072
201-933-8000; Fax: 201-933-8758

Rand McNally Media Services Co **BK**
Bill LeMonds
Book Manufacturing Division
US Bypass 60
Versailles KY 40383-1496
606-873-3181

Statement: This book and map publisher also prints books for other companies.

PRINTED ITEMS	G	Galley Copies	N	Workbooks	U	Greeting Cards
A Annual Reports	H	Journals	O	Yearbooks	V	Labels / Stickers
B Books	I	Magazines	P	Brochures	W	Maps
C Booklets	J	Mass-Market Books	Q	Calendars	X	Newspapers
D Catalogs	K	Newsletters	R	Comic Books	Y	Postcards
E Cookbooks	L	Software Manuals	S	Direct Mail Letters	Z	Posters
F Directories	M	Textbooks	T	Envelopes	1	Stationery

RBW Graphics

CA

Ken Cruickshank, VP Marketing
1749 Twentieth St East
Owen Sound ON Canada N4K 5R2
519-376-8330; Fax: 519-376-1164

Min: 50000 Max: 6000000 Opt: 250000
Sizes: 25X
Bindings: HCI-PBI-SSI
Items: BDFINQ
Services: 4C-FF-TY-TD
Equipment: LP-4SO(to 28x41)-
    7WO(38x47)
Annual Sales: 5
Number of Employees: 600
Turnaround Times: B5C5S3
Customers: BP17-BU43-CU5-MP30-NP5
Union Shop: N
Year Established: 1853
Terms: Net 30

Statement: A division of Southam, RBW is a major catalog, magazine, and book printer in Canada.

Relief Printing East

GC

Marketing Director
112 Commerce Dr
Madison FL 32340
904-973-2290; Fax: 904-973-6950

Repro-Tech

GC

Marketing Director
1275 Bloomfield Ave Bldg 6-36
Fairfield CT 07004
201-785-0011

Min: 0 Max: 5000 Opt: 3000
Sizes: 235
Bindings: HCX-CBX-LBX-PBX-SSX-
    SDX-SBX-WBX
Items: ABCDEFGKLNPSTV1
Services: DA-LM-TY-TD-WA
Equipment: 5SO(11x17,14x18)
Annual Sales: 2
Number of Employees: 75
Turnaround Times: B2C2
Union Shop: N

Statement: We are a full-service commercial printing company offering typesetting, electronic publishing, graphic design, offset printing, and binding. We specialize in runs between 500 and 5000.

George Rice & Sons

GC

Marketing Director
2001 N Soto St
Los Angeles CA 90032
213-223-2020
Parent Company: World Color Press

Min: 0 Max: 0 Opt: 0
Bindings: SSI
Items: AI
Annual Sales: 6
Number of Employees: 540

Rich Printing Company Inc

BK

Tom Hutchins, Marketing Director
7131 Centennial Blvd
Nashville TN 37209
615-350-7300; Fax: 615-350-7303
Send RFQ's to: Tom Hutchins

Min: 5000 Max: 200000 Opt: 0
Sizes: 2345
Bindings: PBX-SSX
Items: BDFIMNX
Services: RP
Equipment: 2SO(25x36)-WO(22x36)
Annual Sales: 3
Turnaround Times: B4S4
Customers: BP2-BU45-CU30-MP5
Union Shop: N
Year Established: 1936
Terms: Net 30

Ringier America

GC

Lawrence Vodak, Marketing Director
One Pierce Pl #800
Itasca IL 60143-1272
708-285-6000; Fax: 708-285-6584
Send RFQ's to: Ron Covelli, VP, Book
    Sales

Min: 25000 Max: 40000000 Opt: 200000
Sizes: 123456X
Bindings: HCI-CBI-PBI-SSI-SDI
Items: BDEFHIJM
Services: 4C-CS-FF
Equipment: SO-WO-OP(Gravure and
    flexographic)

SERVICES				BINDINGS		SD	Side Stitching
4C	4-Color Printing	RA	Rachwal System	HC	Hardcover	SB	Spiral Binding
CS	Color Separations	AF	Acid-Free Paper	CB	Comb Binding	WB	Wire-O Binding
DA	Design / Artwork	RP	Recycled Paper	LB	Loose-Leaf Binding	I	In-House
FF	Fulfillment/Mailing	TY	Typesetting	OB	Otabind	O	Out of House
LM	List Maintenance	TD	Typeset w/ Disk	PB	Perfect Binding	X	Unknown
OC	OptiCopy System	WA	Warehousing	SS	Saddle Stitching		

Annual Sales: 6
Number of Employees: 5700
Turnaround Times: B4C4S4
Customers: BP20-BU40-MP30-OR10
Union Shop: N
Terms: Net 30

Statement: We are the nation's largest privately held printer. Nationwide, we bring the finest pre-press, bindery, and distribution technology available today—and the first opportunity to use the technology of tomorrow.

MG
Ringier America - Brookfield Wisconsin Division
Marketing Director
12821 W Bluemound Rd
Brookfield WI 53005-8098
414-827-2300; Fax: 414-827-2309

Min: 10000 Max: 2000000 Opt: 0
Sizes: 2345
Bindings: PBI-SSI
Items: ABDIMPQ
Services: 4C-CS-TY-WA
Equipment: 39WO
Annual Sales: 6
Number of Employees: 5800

MG
Ringier America Inc
Marketing Director
Corinth Division
1 Golden Dr
PO Box 1555
Corinth MS 38834
601-287-3744; Fax: 601-286-1224

Min: 200000 Max: 2000000 Opt: 0
Sizes: 5X
Bindings: SSI-SDI
Items: DIWXZ
Services: 4C-FF
Equipment: WO-OP(Gravure)
Annual Sales: 6
Number of Employees: 5800

BK
Ringier America - Dresden
Norman Hartgan, Plant Manager
Mass Market Book Division
2073 Evergreen St
Dresden TN 38225
901-364-4100

Min: 30000 Max: 2000000 Opt: 100000
Sizes: 1
Bindings: PBI
Items: J
Annual Sales: 6
Number of Employees: 5800

CA
Ringier America - Jonesboro
Marketing Director
4708 Krueger Dr
Jonesboro AR 72401-9198
501-935-7000

Min: 10000 Max: 2000000 Opt: 0
Sizes: 2345
Bindings: SSI
Items: ADIMPQ
Services: 4C-CS-TY-WA
Equipment: 39WO
Annual Sales: 6
Number of Employees: 5800

CA
Ringier America - Phoenix
Marketing Director
2802 West Palm Ln
Phoenix AZ 85009-2599
602-272-3221

Min: 10000 Max: 2000000 Opt: 0
Sizes: 2345
Bindings: SSI
Items: ADIMPQ
Services: 4C-CS-TY-WA
Equipment: 39WO
Annual Sales: 6
Number of Employees: 5800

CA
John Roberts Company
Bill Culbert, Sales Manager
9687 East River Rd
Minneapolis MN 55433
612-755-5500

PRINTED ITEMS						
A	Annual Reports	G	Galley Copies	N	Workbooks	U	Greeting Cards
B	Books	H	Journals	O	Yearbooks	V	Labels / Stickers
C	Booklets	I	Magazines	P	Brochures	W	Maps
D	Catalogs	J	Mass-Market Books	Q	Calendars	X	Newspapers
E	Cookbooks	K	Newsletters	R	Comic Books	Y	Postcards
F	Directories	L	Software Manuals	S	Direct Mail Letters	Z	Posters
		M	Textbooks	T	Envelopes	1	Stationery

Min: 10000 Max: 750000 Opt: 0
Sizes: 25X
Bindings: SSI
Items: ABCDNPZ
Services: 4C-CS
Equipment: 6LP(to 22x32)-14SO(to
    28x40)-4WO
Annual Sales: 4
Number of Employees: 180
Year Established: 1951

GC

Rodgers RF Lithograph Co
Marketing Director
2211 S Jackson Ave
Tulsa OK 74107
918-587-8165; Fax: 918-587-6921

CA

Ronalds Printing
Marketing Director
8000 Blaise Pascal
Montreal PQ Canada H1E 2S7
514-648-1880
Parent Company: Quebecor Printing

Min: 75000 Max: 4000000 Opt: 0
Sizes: 5
Bindings: HCO-CBO-OBI-LBO-PBI-SSI-
    SDO-SBO-WBO
Items: ABCDEFHIJKLPRSXZ
Services: 4C-FF-WA
Equipment: 27SO-135WO
Turnaround Times: C2S2
Customers: BU65-MP35
Union Shop: Y
Year Established: 1890
Terms: Credit to be arranged

BK

Rose Printing Company
Charles Rosenberg, President
Rose Industrial Park
2503 Jackson Bluff Rd
Tallahassee FL 32304
904-576-4151; Fax: 904-576-4153
Send RFQ's to: Charles Rosenberg

Min: 250 Max: 100000 Opt: 1800
Sizes: 123456X
Bindings: HCX-CBX-OBX-PBX-SSX-
    SBX-WBX
Items: BDEFHLMNOP2
Services: 4C-FF-LM-OC-AF-RP-TD-WA

ROSE
PRINTING COMPANY, INC.

2503 Jackson Bluff Road
Tallahassee, FL 32304

Complete In House
Book Manufacturing Since 1932

• **Books**   • **Directories**
• **Journals**   • **Manuals**
• **Softcover**   • **Hardcover**
• **RepKover Layflat Binding**

7 Regional Offices to Serve You.

Contact: Charles Rosenberg

904-576-4151
800-227-3725
Fax 904-576-4153

Equipment: 6SO(23x29,38x50)-
    3WO(21x40,23x38)
Annual Sales: 4
Number of Employees: 240
Turnaround Times: B3C4S3
Customers: BP65-MP10-CU15-SP5-OR5
Union Shop: N
Year Established: 1932
Terms: Net 30 with approved credit

Statement: Meeting customer needs in
the short run (200-5000) book markets
and mini-book markets. Quality, fast
turnaround at competitive costs have contributed to a steady growth.

SERVICES				BINDINGS			
4C	4-Color Printing	RA	Rachwal System	HC	Hardcover	SD	Side Stitching
CS	Color Separations	AF	Acid-Free Paper	CB	Comb Binding	SB	Spiral Binding
DA	Design / Artwork	RP	Recycled Paper	LB	Loose-Leaf Binding	WB	Wire-O Binding
FF	Fulfillment/Mailing	TY	Typesetting	OB	Otabind	I	In-House
LM	List Maintenance	TD	Typeset w/ Disk	PB	Perfect Binding	O	Out of House
OC	OptiCopy System	WA	Warehousing	SS	Saddle Stitching	X	Unknown

Ronalds Printing **GC**

Don Brown, Publications Director
1070 SE Marine Dr
Vancouver BC Canada V5X 2V4
604-321-2231
Parent Company: Quebecor Printing

Min: 10000 Max: 275000 Opt: 150000
Sizes: 12345
Bindings: HCO-CBO-LBO-PBI-SSI-SDO-
    SBO-WBO
Items: BCDFHILMNPQSWZ
Services: 4C-FF-TY-TD
Equipment: LP-4SO(to 59)-5WO(23x38)
Turnaround Times: B3C2S2
Customers: BP3-BU47-MP25-OT25
Union Shop: Y
Year Established: 1950
Terms: Net 30

S Rosenthal & Company Inc **MS**

Ron Magnan, Sales Director
9933 Alliance Rd
Cincinnati OH 45242
513-984-0710; Fax: 513-984-5643
800-325-7200
Send RFQ's to: Stu Stormer

Min: 20000 Max: 2000000 Opt: 200000
Sizes: 245X
Bindings: PBI-SSI-HCO-CBO-OBO-SBO-
    WBO
Items: DFHIJNPQXZ
Services: 4C-DA-FF-TY-TD-RP-OC-AF
Equipment: 2SO(25x38)-5WO(22x38,
    31x46)
Annual Sales: 5
Number of Employees: 300
Turnaround Times: B2C5S2
Customers: BU30-BP5-MP65
Union Shop: Y
Year Established: 1868
Terms: Net 30

Statement: S Rosenthal & Company Inc, a
graphic communications company, is a
quality-minded, service oriented, customer
responsive, national printer of publications, catalogs and commercial printing. Its
purpose is to produce a product to satisfy
its customers' needs, to provide an environmentally responsible workplace for its
employees and community; and, to accom

-plish high productivity through training,
professional planning, organization and
total employee and management teamwork, at the necessary level of profit to
keep abreast with emerging technology in
our industry.

Rotary Press Inc **GC**

Marketing Director
18221 Andover Pk W
Seattle WA 98188
206-575-0144; Fax: 206-575-1650

Rothchild Printing Co Inc **GC**

Marketing Director
7900 Barnwell Ave
Flushing NY 11373
718-899-6000; Fax: 718-397-1921

Roxbury Publishing Company **BK**

Claude Teweles, Executive Editor
537 N Orlando Ave
PO Box 491044
Los Angeles CA 90049
213-653-1068
Send RFQ's to: Claude Teweles

Min: 100 Max: 2000000 Opt: 0
Sizes: 123456X
Bindings: HCO-CBI-LBO-PBI-SSI-SDI-
    SBO-WBO
Items: BEFJMN
Services: 4C-CS-DA-FF-AF-TY-TD-WA
Equipment: 5SO(to 41x54)-3WO(to 40x45)
Annual Sales: 1
Number of Employees: 75
Turnaround Times: B5
Customers: BP50-SP50
Union Shop: Y
Year Established: 1979
Terms: 1/2 down, 1/2 with blues

Statement: Roxbury, a textbook and trade
publishing company, does typesetting and
printing of books for outside customers,
including self-publishers.

PRINTED ITEMS					
A	Annual Reports	G	Galley Copies	N	Workbooks
B	Books	H	Journals	O	Yearbooks
C	Booklets	I	Magazines	P	Brochures
D	Catalogs	J	Mass-Market Books	Q	Calendars
E	Cookbooks	K	Newsletters	R	Comic Books
F	Directories	L	Software Manuals	S	Direct Mail Letters
		M	Textbooks	T	Envelopes

U	Greeting Cards
V	Labels / Stickers
W	Maps
X	Newspapers
Y	Postcards
Z	Posters
1	Stationery

Royle Printing
GC

Bruce Rutz, Marketing Director
Royle Communications Group
112 Market St
Sun Prairie WI 53590
608-837-5161; Fax: 608-837-3946
Parent Company: Royle Communications
    Group

Min: 10000 Max: 1500000 Opt: 300000
Sizes: 12356
Bindings: SSI-SDI-HCO-PBO-CBO-LBO-
    OBO-WBO-SBO
Items: ABCDHIKNQWXZ2
Services: 4C-CS-LM-OC-RP-TD-WA
Equipment: 4SO(25,36,40)-4WO
    (20x23⅛, 22.75x38)
Annual Sales: 5
Number of Employees: 350
Turnaround Times: B2S1
Customers: BU80-CU10-MP10
Union Shop: N
Year Established: 1982
Terms: Net 30

Statement: Our mission is to provide superior quality graphic communication products and services that exceed the expectations of our customers now and in the future.

RPP Enterprises Inc
GC

Jeffrey R Pauly, President
1920 Enterprise Ct
Libertyville IL 60048
708-680-9700

Min: 5000 Max: 2000000 Opt: 0
Sizes: 256
Bindings: SSI-SDI
Items: ACDFKPQ
Services: 4C-DA-WA
Equipment: 3LP-5SO(to 25x38)-
    WO(18x23)
Annual Sales: 3
Number of Employees: 125
Turnaround Times: C3
Customers: BU90-OR10
Year Established: 1927
Terms: Net 30

S

Sakurai USA Inc
GC

Marketing Director
Oliver Offset Presses
501 Lively Blvd
Elk Grove Village IL 60007
708-640-7400; 800-458-4720

Saltzman Printers Inc
GC

Marketing Director
50 Madison St
Maywood IL 60153
708-344-4500; Fax: 708-344-9423
800-952-2800

Sanders Printing Company
BK

Bob Sanders, Marketing Director
508 Main St
PO Box 160
Garretson SD 57030
605-594-3427; Fax: 605-594-3840
Send RFQ's to: Bob Sanders

Min: 100 Max: 5000 Opt: 1000
Sizes: 123456
Bindings: LBI-SSI-SDI-CBO-HCO-OBO-
    SBO-WBO-PBO
Items: ABCDEFGHIKLMNPSTVXYZ1
Services: 4C-CS-DA-RP-TY-TD-WA
Equipment: LP(10x15)-2SO(12x18,
    18x25)
Annual Sales: 2
Number of Employees: 8
Turnaround Times: B5C5S4
Customers: BP2-BU85-NP2-SC4-SP7
Union Shop: N
Year Established: 1983
Terms: Net 30 commercial, 1/2 with proof
    for books

The Saratoga Printing Co
BK

Mark Sharadin, Customer Service
120 Henry St
Saratoga Springs NY 12866
518-584-2054; Fax: 518-587-9077
Send RFQ's to: Robert Black

SERVICES				BINDINGS			
4C	4-Color Printing	RA	Rachwal System	HC	Hardcover	SD	Side Stitching
CS	Color Separations	AF	Acid-Free Paper	CB	Comb Binding	SB	Spiral Binding
DA	Design / Artwork	RP	Recycled Paper	LB	Loose-Leaf Binding	WB	Wire-O Binding
FF	Fulfillment/Mailing	TY	Typesetting	OB	Otabind	I	In-House
LM	List Maintenance	TD	Typeset w/ Disk	PB	Perfect Binding	O	Out of House
OC	OptiCopy System	WA	Warehousing	SS	Saddle Stitching	X	Unknown

Min: 500 Max: 50000 Opt: 10000
Sizes: 23456
Bindings: CBI-LBI-SSI-SDI-PBO-SBO-
WBO
Items: ABCDEFIJKMNOPQSTVWYZ12
Services: 4C-AF-RP-WA
Equipment: LP(14x20)-5SO(4x5 to 28x41)
Annual Sales: 2
Number of Employees: 12
Turnaround Times: B3S2
Customers: BP20-BU20-CU20-NP20-SC20
Union Shop: N
Year Established: 1941
Terms: Net 30 with approved credit

Statement: We produce books, signatures, covers, dustjackets, maps, posters and more. Fair pricing of quality work and impeccable service.

Sayers Printing Co **GC**
Marketing Director
9600 Manchester Rd
St Louis MO 63119
314-968-5400; Fax: 314-968-1927

Schelin Printing Co **GC**
Bill Winters, Marketing Director
45 East Lockwood
St Louis MO 63119
314-968-2910; Fax: 314-968-2904
Send RFQ's to: Matt Rvorden

Min: 0 Max: 0 Opt: 0
Sizes: 123456
Bindings: SSX-SDX
Items: ACDIKLNPQSTYZ12
Services: 4C-CS-FF-RP-TD-WA
Equipment: 5SO(18,40)
Annual Sales: 3
Number of Employees: 85
Turnaround Times: B3S2
Customers: BU90-CU5-NP2-OR3
Union Shop: N
Terms: Net 30

Schiff Printing **GC**
Scott Barthelmes, Marketing Director
1107 Washington Blvd
Pittsburgh PA 15206
412-441-5760; Fax: 412-441-0133

Send RFQ's to: Scott Barthelmes
Min: 500 Max: 1000000 Opt: 20000
Sizes: 123456X
Bindings: SDI-SSI-HCO-CBO-LBO-PBO-
SBO-WBO
Items: ABCDEFKNPQSYZ
Services: 4C-CS-TY-TD-DA-FF
Equipment: 3SO(18x25,27x40,12.5x17.5)
Annual Sales: 4
Number of Employees: 30
Turnaround Times: B2C4S2
Customers: BU20-CU40-NP20-OT20
Union Shop: N
Year Established: 1954
Terms: Net 30

Statement: To provide high quality sheetfed printing, Schiff offers a complete package ranging from typesetting, electronic file output and color separations, to printing and binding.

Schmidt Printing Inc **GC**
Marketing Director
1101 Frontage Rd NW
Byron MN 55920-1386
507-775-6400; Fax: 507-775-6655

Schoenfeld Ludlow Printing **GC**
Marketing Director
3809 43rd Ave
Long Island City NY 11101
718-392-9600; Fax: 718-786-9405

Schumann Printers **MS**
Marketing Director
701 S Main St
PO Box 128
Fall River WI 53932
414-484-3348; Fax: 414-484-3661
Send RFQ's to: Jack G Schumann
Min: 10000 Max: 250000 Opt: 50000
Sizes: 5
Bindings: SSI-PBI
Items: HI
Services: 4C-CS-FF
Equipment: WO
Year Established: 1963

PRINTED ITEMS					
A	Annual Reports	G	Galley Copies	N	Workbooks	U	Greeting Cards
B	Books	H	Journals	O	Yearbooks	V	Labels / Stickers
C	Booklets	I	Magazines	P	Brochures	W	Maps
D	Catalogs	J	Mass-Market Books	Q	Calendars	X	Newspapers
E	Cookbooks	K	Newsletters	R	Comic Books	Y	Postcards
F	Directories	L	Software Manuals	S	Direct Mail Letters	Z	Posters
		M	Textbooks	T	Envelopes	1	Stationery

BK
Science Press
Volker Kruhoeffer, VP Commercial Sales
300 W Chestnut St
PO Box 497
Ephrata PA 17522-0497
717-738-9300; Fax: 717-738-9368
Send RFQ's to: Volker Kruhoeffer
Parent Company: The Mack Printing
    Group

Min: 1000 Max: 100000 Opt: 5000-25000
Sizes: 123456X
Bindings: HCX-LBX-PBX-SSX-SBX-
    WBX
Items: ABCDEFHIKLMNPQ
Services: 4C-CS-FF-OC-RA-AF-RP-TY-
    TD-WA
Equipment: 6SO(25x38,41x50)-
    WO(23.75x36)
Annual Sales: 5
Number of Employees: 360
Turnaround Times: B3C5S3
Customers: BP10-BU20-CU10-MP40-
    OR20
Union Shop: N
Year Established: 1901
Terms: Net 30

Statement: Science Press, a Division of
The Mack Printing Group, is a full-service
commercial printer, offering Xyvision composition through black and white and four-
color prep work, and printing, mostly on
new Miller and Heidelberg presses.

GC
Scott SD Printing Co Inc
Marketing Director
145 Hudson St
New York NY 10013
212-226-7100; Fax: 212-226-7977

GC
Screenprint Inc
Marketing Director
PO Box 1332
Wilmington MA 01887
617-935-6395; Fax: 617-658-2307

GC
Seiple Lithograph Co
Marketing Director
2213 Cleveland Ave NW
Canton OH 44709
216-456-3433

BK
Service Printing Company
Jerry Edelstein, President
2725 Miller St
San Leandro CA 94577
510-352-7890

Min: 1000 Max: 25000 Opt: 5000
Sizes: 245X
Bindings: HCX-CBX-PBX-SSX-SBX
Items: BCDFLN
Equipment: SO
Turnaround Times: B2
Year Established: 1925
Terms: Net 30

Statement: They specialize in producing
training and technical manuals for the
computer/technical industry. Their routine
delivery is 5 to 7 working days from receipt
of camera-ready copy.

CS
Service Web Offset Corporation
Norine Satterlee, Marketing Director
2500 S Dearborn St
Chicago IL 60616
312-567-7000; Fax: 312-567-9121
800-621-1567
Send RFQ's to: Norine Satterlee

Min: 25000 Max: 500000 Opt: 250000
Sizes: 2356
Bindings: SSX
Items: CDKP
Services: 4C-RP
Equipment: 2WO
Annual Sales: 4
Number of Employees: 140
Turnaround Times: S2
Customers: BU76-OT21
Union Shop: Y
Year Established: 1943
Terms: Net 30

Statement: Service Web has specialized in
full-color printing for over 50 years and
has a friendly, professional staff you can
count on. Our offices, computerized high-
speed presses, and complete bindery
facilities are at one location for maximum
efficiency.

SERVICES				BINDINGS		SD	Side Stitching
4C	4-Color Printing	RA	Rachwal System	HC	Hardcover	SB	Spiral Binding
CS	Color Separations	AF	Acid-Free Paper	CB	Comb Binding	WB	Wire-O Binding
DA	Design / Artwork	RP	Recycled Paper	LB	Loose-Leaf Binding	I	In-House
FF	Fulfillment/Mailing	TY	Typesetting	OB	Otabind	O	Out of House
LM	List Maintenance	TD	Typeset w/ Disk	PB	Perfect Binding	X	Unknown
OC	OptiCopy System	WA	Warehousing	SS	Saddle Stitching		

Sexton Printing **MS**

Marketing Director
250 E Lothenbach Ave
Saint Paul MN 55118
612-457-9255; Fax: 612-457-7040
Send RFQ's to: Mike Poquette

Min: 500 Max: 50000 Opt: 10000
Sizes: 123456
Bindings: HCO-CBO-LBO-OBO-PBO-
    SDO-SBO-WBO-SSI
Items: ABCDEFIKLNOPS1
Services: 4C-DA-LM-OC-RP-TY-TD-FF
Equipment: 11SO(11x15 to 28x40)-
    LP(10x15)
Annual Sales: 3
Number of Employees: 90
Turnaround Times: B2C3S2
Customers: BU10-CU10-NP27-SP3-BP10-
    MP26-OR10-SC4
Union Shop: N
Year Established: 1950
Terms: Net 30 established customers,
    COD, 50% down on new

Statement: Sexton specializes in printing
newsletters and magazines for clubs, associations, and businesses.

Shea Communications **MG**

Jim Mitchell, Marketing Director
Oklahoma Graphics
5400 NW 5th St
Oklahoma City OK 73127
405-945-6200; Fax: 405-945-6277
Send RFQ's to: John McKinnon

Min: 100000 Max: 12000000 Opt: 5000000
Sizes: 123456
Bindings: SSX
Items: DIPQRWX
Services: 4C-CS-FF-WA
Annual Sales: 6
Number of Employees: 500
Turnaround Times: S1-2
Customers: BU95-MP5
Union Shop: N
Terms: Net 30 with approval

Shepard Poorman Communications **GC**

Russ Jehs, Marketing Director
7301 N Woodland Dr
Indianapolis IN 46278
317-293-1500; Fax: 317-293-2722
Send RFQ's to: Terri Price

Min: 5000 Max: 0 Opt: 2000000
Bindings: HCO-PBO-SBO-SSI-SDI-WBI
Items: ABCDEFHIKLMNOPQ3
Services: 4C-CS-FF-OC-RP-TY-TD-WA-
    AF
Equipment: 6SO(40)-WO(38x23)-OP
Annual Sales: 6
Number of Employees: 500
Turnaround Times: B2C3S1
Customers: BU70-CU1-MP15-BP10-OR4
Union Shop: N
Year Established: 1978
Terms: Net 30

Statement: Shepard Poorman Communications is a full service commercial printer,
specializing in corporate literature,
catalogs, publications, directories/guides
and calendars.

Sales Offices:

Chicago
2300 N Barrington Rd
Hoffman Estate IL 60195
708-490-5384; Fax: 708-882-9775

Corporate Woods
700 Forest Edge Dr
Vernon Hills IL 60061
708-913-9800; 708-913-9891

Columbus OH
150 E Broad St
Columbus OH 43215
614-469-4666; Fax: 614-469-3566

Atlanta GA
Tower Pl
3340 Peachtree Rd #1800
Atlanta Ga 30326

The Sheridan Press **MS**

Roger Hansen
450 Fame Ave
Hanover PA 17331
717-632-3535; Fax: 717-633-8900
800-352-2210
Send RFQ's to: Pat Gentry

PRINTED ITEMS					
A	Annual Reports	G	Galley Copies	N	Workbooks	U	Greeting Cards
B	Books	H	Journals	O	Yearbooks	V	Labels / Stickers
C	Booklets	I	Magazines	P	Brochures	W	Maps
D	Catalogs	J	Mass-Market Books	Q	Calendars	X	Newspapers
E	Cookbooks	K	Newsletters	R	Comic Books	Y	Postcards
F	Directories	L	Software Manuals	S	Direct Mail Letters	Z	Posters
		M	Textbooks	T	Envelopes	1	Stationery

Min: 500 Max: 20000 Opt: 4000
Sizes: 345
Bindings: PBI-SSI-SDI-HCO-CBO-SBO-
    WBO
Items: HI
Services: 4C-FF-LM-OC-AF-TY-TD-WA
Equipment: 9SO(to 41)
Annual Sales: 5
Number of Employees: 340
Turnaround Times: B2C4S2
Customers: BP80-CU10-MP5-NP5
Union Shop: N
Year Established: 1915
Terms: Net 30

Statement: The Sheridan Press is a full service printer with a strong emphasis on
quality and performance to customer
schedules.

Sheridan Printing Company MS
Marketing Director
1425 Third Ave
Phillipsburg NJ 08865
908-454-0700; Fax: 908-454-2554

Min: 1000 Max: 50000 Opt: 0
Sizes: 123456
Bindings: CBI-PBI-SDI-CBO-LBO-OBO-
    SBO-WBO
Items: BCDFHIKLMNOP
Services: 4C-FF-AF-RP-TY-TD
Equipment: 4SO(23x35,25x38,25x37.25)
Annual Sales: 3
Number of Employees: 75
Turnaround Times: B2C5S2
Customers: BP10-BU30-MP10-NP50
Union Shop: Y
Year Established: 1956
Terms: Net 30

Sidney Printing Works Inc GC
Marketing Director
2611 Colerain Ave
Cincinnati OH 45214
513-542-4000; Fax: 513-542-4741

Sinclair Printing Co GC
Marketing Director
4005 Whiteside
Los Angeles CA 90063
213-264-4000; Fax: 213-264-3716

Min: 5000 Max: 150000 Opt: 50000
Sizes: 24
Bindings: HCO-CBO-LBO-OBO-PBI-SSI-
    SDI-SBO-WBO
Items: BCDEFNP
Services: 4C-OC-TD-WA
Equipment: 5SO(40)-2WO(35)-OP(Heat-
    set web 26)
Annual Sales: 4
Number of Employees: 175
Turnaround Times: B3C3S3
Customers: BP3-BU90-CU3-MP3
Union Shop: N
Year Established: 1929
Terms: Net 30

Statement: A blend of full service, tradition, and high-tech.

Sioux Printing Inc GC
Marketing Director
1212 S Cliff Ave
Sioux Falls SD 57105
605-336-1370; Fax: 605-338-7691

SLC Graphics MG
James Killian, Marketing Director
50 Rock St
Hughestown Borough
Pittston PA 18640
717-655-9681

Min: 100000 Max: 400000 Opt: 100000
Sizes: 23456X
Bindings: PBI-SSI
Items: I
Services: 4C-CS-FF-WA
Equipment: 3WO(22x36,38x54)
Annual Sales: 4
Number of Employees: 200
Turnaround Times: S4
Customers: MP100
Union Shop: Y
Year Established: 1970
Terms: Net 30

SERVICES				BINDINGS		SD	Side Stitching
4C	4-Color Printing	RA	Rachwal System	HC	Hardcover	SB	Spiral Binding
CS	Color Separations	AF	Acid-Free Paper	CB	Comb Binding	WB	Wire-O Binding
DA	Design / Artwork	RP	Recycled Paper	LB	Loose-Leaf Binding	I	In-House
FF	Fulfillment/Mailing	TY	Typesetting	OB	Otabind	O	Out of House
LM	List Maintenance	TD	Typeset w/ Disk	PB	Perfect Binding	X	Unknown
OC	OptiCopy System	WA	Warehousing	SS	Saddle Stitching		

Sleepeck Printing Co GC
Marketing Director
815 25th Ave
Bellwood IL 60104
708-544-8900; Fax: 708-544-8928

Smith Lithographic Arts Inc GC
Marketing Director
424 W 6th St
Tustin CA 92680
714-838-6511; Fax: 714-731-8815

Smith Print Inc GC
Marketing Director
Longwater Dr
Norwell MA 02061
617-871-0640; Fax: 617-878-2040

R E Smith Printing GC
Andrew Morrow, Sales Manager
900 Jefferson St
PO Box 6300
Fall River MA 02724
508-679-2131; Fax: 508-677-4466

Min: 1000 Max: 100000 Opt: 20000
Sizes: 35
Bindings: CBO-LBO-PBO-SSI-SDI-SBO-
   WBO
Items: ADFIKLNPQYZ
Services: 4C-TY
Equipment: 11LP(to 27x41)-9SO(to
   28x40)-2WO
Annual Sales: 3
Number of Employees: 125
Turnaround Times: B3C3S3
Customers: BU75-CU5-MP5-NP3-SC3-
   SP2-OT5
Union Shop: N
Year Established: 1964
Terms: Net 30

Statement: R E Smith is a division of Close
Manufacturing.

Smith-Edwards-Dunlap GC
Harold Coopersmith, VP
2867 E Allegheny Ave
Philadelphia PA 19134
215-425-8800

Min: 0 Max: 0 Opt: 0
Bindings: SSI
Items: ACDHIKNPSZ
Services: 4C-FF-TY
Equipment: LP-SO-WO
Annual Sales: 5
Number of Employees: 410

Snohomish Publishing BK
Marketing Director
114 Ave C
Snohomish WA 98290
206-568-1242

Min: 500 Max: 50000 Opt: 3000
Sizes: 25
Bindings: HCO-PBI-SSI
Items: ABCDEFIKLNPQX
Services: 4C-TY-TD
Equipment: 2SO(18x23,23x29)-2WO
   (22x36)
Annual Sales: 3
Number of Employees: 30
Turnaround Times: B4C3S3
Customers: BP5-SP35-OT60
Union Shop: N
Year Established: 1965
Terms: 1/2 down, 1/4 delivery, net 30

Statement: We provide the author with the
guidance and facilities to produce their
self-published book.

Southam Paragon Graphics GC
Andre' Beaudet, Group President
140 Industrial Blvd
Candiac PQ Canada J5R 1R2
514-659-9121; Fax: 514-659-7512

**Southeastern Printing Company
Inc** GC
R F "Tony" Anthony, Marketing Director
3601 SE Dixie Hwy
Stuart FL 34997
407-287-2141; Fax: 407-288-3988
Parent Company: Mastec Inc

PRINTED ITEMS	G	Galley Copies	N	Workbooks	U	Greeting Cards
A Annual Reports	H	Journals	O	Yearbooks	V	Labels / Stickers
B Books	I	Magazines	P	Brochures	W	Maps
C Booklets	J	Mass-Market Books	Q	Calendars	X	Newspapers
D Catalogs	K	Newsletters	R	Comic Books	Y	Postcards
E Cookbooks	L	Software Manuals	S	Direct Mail Letters	Z	Posters
F Directories	M	Textbooks	T	Envelopes	1	Stationery

Min: 5000 Max: 100000 Opt: 50000-75000
Sizes: 23456
Bindings: HCX-CBX-PBX-SSX-SDX-
WBX
Items: ABCDEHIPQUZ
Services: 4C-CS-DA-FF-OC-RP-TY-TD
Equipment: SO-OP
Annual Sales: 4
Number of Employees: 100+
Turnaround Times: B3C6S2
Customers: BU85-BP5-MP5-CU5
Union Shop: N
Year Established: 1970
Terms: Net 30

Statement: To acquire and maintain customers by delivering products and services that exceed the requirements of those customers through an ongoing committment of superior consistency.

MS
Southern California Graphics
Marketing Director
8432 Stellar Dr
Culver City CA 90230
213-559-3600

MG
Sowers Printing Company
Frances S Russomanno, Marketing
Director
220 N Tenth St
PO Box 479
Lebanon PA 17042-0479
717-272-6667; Fax: 717-274-2928
800-233-7028
Send RFQ's to: Estimating Dept

Min: 5000 Max: 150000 Opt: 50000
Sizes: 123456X
Bindings: HCO-CBO-LBO-OBO-PBO-
SSI-SDO-SBO-WBO
Items: ABCDFHIKLMNP
Services: 4C-DA-FF-WA
Equipment: 3SO(28x40)-WO(23⁹⁄₁₆x38)
Annual Sales: 3
Number of Employees: 75
Turnaround Times: B3C4S2
Customers: BP10-BU10-CU15-MP5-NP50-
OR5-OT5
Union Shop: Y
Year Established: 1895
Terms: Net 30
Statement: Dedicated salespeople and CSR's work with you from concept to

delivery. Specialties are publications, journals, catalogs, and direct mail. Art/design, electronic and conventional prepress, sheetfed and heatset web presses.

GC
Spartan Printing
Marketing Director
320 S Dickey St
Sparta IL 62286
618-443-2154; Fax: 618-443-4449

Parent Company: World Color Press

GC
Speakes-Hines & Thomas Inc
Marketing Director
3366 Remy
PO Box 11120
Lansing MI 48906
517-321-0740; Fax: 517-321-9340
800-292-2630
Send RFQ's to: Mark Buchko

Min: 20000 Max: 5000000 Opt: 200000
Sizes: 12345
Bindings: LBI-SSI-HCO-CBO-OBO-PBO-
SDO-SBO-WBO
Items: CDFINPX
Services: 4C-CS-OC-RP-TY-TD
Equipment: 2SO(25x38,35x15)-2WO(38)
Annual Sales: 5
Number of Employees: 100
Turnaround Times: B4C6S3
Union Shop: Y
Year Established: 1884

SP
Specialty Printers of America
Marketing Director
700 E Parker St
Scranton PA 18509
717-342-2203; Fax: 717-342-4349

CA
Spencer Press
Mark R Turner
90 Spencer Dr
Wells ME 04090
207-646-9926; Fax: 207-646-5021
800-343-5690
Send RFQ's to: Mark R Turner

Min: 50000 Max: 5000000 Opt: 0

SERVICES			BINDINGS		SD	Side Stitching
4C	4-Color Printing	RA Rachwal System	HC	Hardcover	SB	Spiral Binding
CS	Color Separations	AF Acid-Free Paper	CB	Comb Binding	WB	Wire-O Binding
DA	Design / Artwork	RP Recycled Paper	LB	Loose-Leaf Binding	I	In-House
FF	Fulfillment/Mailing	TY Typesetting	OB	Otabind	O	Out of House
LM	List Maintenance	TD Typeset w/ Disk	PB	Perfect Binding	X	Unknown
OC	OptiCopy System	WA Warehousing	SS	Saddle Stitching		

Sizes: 125X
Bindings: SSX
Items: ACDELPQWX
Services: 4C-CS-LM-RP
Equipment: 6WO(Various)
Annual Sales: 6
Number of Employees: 480
Customers: 50% retail advertising, 50%
    direct mail
Union Shop: N
Year Established: 1940
Terms: Net 30 prior to credit approval

Statement: Spencer Press is one of the
largest privately owned printing companies
in the nation, dedicated to very personalized service to their customers. Spencer is
committed to being technologically aggressive with state-of-the-art equipment from
prepress through mailing and distribution
of the finished product. Various sizes and
formats can be produced on their 6 web offset presses.

St Croix Press Inc   MG
Dean O Lindquist, Marketing Director
1185 S Knowles Ave
New Richmond WI 54017
800-826-6622; Fax: 715-243-7555
Send RFQ's to: Dean O Lindquist

Min: 1000 Max: 100000 Opt: 10000-40000
Sizes: 25X
Bindings: PBX-SSX
Items: DFHIPQ
Services: 4C-CS-OC-RP-TY-TD
Equipment: 3SO(25x40)-3WO(17.5x22)
Annual Sales: 4
Number of Employees: 200
Turnaround Times: B3S2
Customers: MP80
Union Shop: Y
Year Established: 1960
Terms: 2% 10, net 30

Statement: Distinctive color lithography
and printing.

St Joseph Printing Ltd   GC
Frank Gagliano, Vice President
50 MacIntosh Blvd
Concord ON Canada L4K 4P3
416-660-3111

Min: 0 Max: 0 Opt: 0

Bindings: SSI
Items: DIQ

Standard Printing Service   BK
C Bernady, Executive Director
162 N State St
Chicago IL 60601
312-346-0499

Min: 0 Max: 0 Opt: 0
Bindings: HCI
Items: BFM
Services: TY
Equipment: SO-WO

Standard Publishing   CA
Roger Flom, Marketing Director
8121 Hamilton Ave
Cincinnati OH 45231
513-931-4050; Fax: 513-931-0904
800-543-1301
Send RFQ's to: Commercial Sales
Parent Company: Standex International

Min: 8000 Max: 750000 Opt: 175000
Sizes: 12345X
Bindings: PBX-SSX-SDX
Items: DHIJKLMNPQRSZ
Services: 4C-FF-TY-DA-LM-AF-RP-TD-
    WA
Equipment: 3SO(28,38,55)-3WO
    (38x22.75,38x23$^9/_{16}$)
Annual Sales: 4
Number of Employees: 450
Turnaround Times: B2S1
Customers: BU75-NP10-OR5-OT10
Union Shop: Y
Year Established: 1869
Terms: Net 30

John Stark Printing Co Inc   GC
C A Kelley, Marketing Director
12969 Manchester Rd
St Louis MO 63131
314-966-6800; Fax: 314-966-4296
800-452-4POP
Send RFQ's to: Patrick J Ryan
Parent Company: A Cliff Kelley Company

Min: 0 Max: 0 Opt: 0
Sizes: 23456
Bindings: HCO-CBO-LBO-SSO-SDO-
    SBO-WBO

PRINTED ITEMS							
A	Annual Reports	G	Galley Copies	N	Workbooks	U	Greeting Cards
B	Books	H	Journals	O	Yearbooks	V	Labels / Stickers
C	Booklets	I	Magazines	P	Brochures	W	Maps
D	Catalogs	J	Mass-Market Books	Q	Calendars	X	Newspapers
E	Cookbooks	K	Newsletters	R	Comic Books	Y	Postcards
F	Directories	L	Software Manuals	S	Direct Mail Letters	Z	Posters
		M	Textbooks	T	Envelopes	1	Stationery

Items: ACDFKLPSVZ
Services: 4C-CS-FF-RP
Equipment: SO-WO-OP(43x60,28x41,
    25x38,17.5x22)
Annual Sales: 4
Number of Employees: 85
Turnaround Times: B2S2
Customers: BU100
Union Shop: Y
Year Established: 1960
Terms: Net 30

Statement: One call does it all. We offer
the latest "state-of-the-art" equipment,
with presses ranging from 38" to 60" in
four and six colors.

Sales Offices:

Dallas Area
Rick Jones
2201 N Collins #190
Arlington TX 76011

Kansas City Area
Andy Fox
11317 Acuff Ln
Lenexa KS 66215

New York Area
Dan Grieco
379 Meadowbrook Ave.
Ridgewood NJ 07450

Allan O'Keefe
1614 Bronson Rd
Fairfield CT 06430

Southern California Area
David Coats
2401 North Baker St
Santa Ana CA 92706

GC
State Printing Co
Marketing Director
1210 Key Rd
Columbia SC 29201
803-799-9550; Fax: 803-252-2852

GC
Stein Printing Co
Marketing Director
2161 Monroe Dr NE
Atlanta GA 30324
404-875-0421; Fax: 404-872-8814

CA
Stephenson Lithograph Inc
Sandy Stephenson, VP Sales Marketing
5731 General Washington Dr
Alexandria VA 22312
703-642-9000; 800-336-4637

Min: 2000 Max: 2000000 Opt: 0
Sizes: 2345X
Bindings: PBI-SSI
Items: ABCDFHIKNPSWXZ
Services: 4C-CS-FF
Equipment: 7SO(40)-2WO
Annual Sales: 4
Number of Employees: 180

CA
Stevens Graphics
Debbie Storey, Marketing Director
713 R D Abernathy Blvd
Atlanta GA 30310
404-753-1121; Fax: 404-752-0514
Send RFQ's to: John Wehrspann
Parent Company: BellSouth Corporation

Min: 75000 Max: 0 Opt: 0
Sizes: 5
Bindings: PBX-SSX
Items: DF
Services: 4C-OC-RA
Equipment: 2LP(68)-SO(38)-9WO(71,
    75,68)
Annual Sales: 6
Number of Employees: 750
Turnaround Times: B2S2
Customers: BP100
Union Shop: Y
Year Established: 1887
Terms: Net 15

Statement: Stevens Graphics is a directory
and business to business catalog printer
specializing in perfect bound books printed
on lightweight uncoated groundwood
papers from 1-color to 4-color.

GC
Stevens Graphics Inc
Marketing Director
100 W Oxmoor Rd
PO Box 908
Birmingham AL 35201
205-942-0511

Min: 0 Max: 0 Opt: 0
Bindings: HCX-PBX-SSX

SERVICES				BINDINGS		SD	Side Stitching
4C	4-Color Printing	RA	Rachwal System	HC	Hardcover	SB	Spiral Binding
CS	Color Separations	AF	Acid-Free Paper	CB	Comb Binding	WB	Wire-O Binding
DA	Design / Artwork	RP	Recycled Paper	LB	Loose-Leaf Binding	I	In-House
FF	Fulfillment/Mailing	TY	Typesetting	OB	Otabind	O	Out of House
LM	List Maintenance	TD	Typeset w/ Disk	PB	Perfect Binding	X	Unknown
OC	OptiCopy System	WA	Warehousing	SS	Saddle Stitching		

Items: BCDFNR
Equipment: LP-SO

Statement: (Office in Atlanta, GA, See listing). Print Bibles, catalogs, directories, and coloring books.

Stewart Publishing & Printing Co   **BK**
David Stewart, Marketing Director
Rt 1 Box 646
Marble Hill MO 63764
314-238-4273; Fax: 314-238-2010
Send RFQ's to: David Stewart
Parent Company: Napsac Reproductions

Min: 50 Max: 5000 Opt: 1000
Sizes: 123456
Bindings: HCX-CBX-LBX-PBX-SSX-WBX
Items: BCEFHJKPSZ12
Services: DA-TY-TD
Equipment: SO(12x18)-OP(11x17 Risograph Digital Duplicator 6300)
Annual Sales: 1
Number of Employees: 4
Turnaround Times: B4C6S1
Customers: BU15-CU5-NP10-SP60-OT10
Union Shop: N
Year Established: 1978
Terms: 50% down, balance on delivery

Statement: To help self-publishers to produce a quality product at reasonable cost and to successfully market their books.

The Stinehour Press   **BK**
Stephen Stinehour, Marketing Director
PO Box 159
Lunenburg VT 05906-0159
800-331-7753; Fax: 802-328-3960
Send RFQ's to: Judy Friend

Min: 100 Max: 50000 Opt: 1000-10000
Sizes: 123456
Bindings: PBI-SSI-HCO-CBO-LBO-OBO-SDO-SBO-WBO
Items: ABCDEFHMQUYZ
Services: 4C-CS-DA-FF-TY-TD-WA-LM-AF-RP
Equipment: 4SO(28x40)-2LP(22x28)
Annual Sales: 3
Number of Employees: 65
Turnaround Times: B4C5S4
Customers: BP40-CU10-OR40-SP10
Union Shop: N

Year Established: 1950
Terms: Net 15

Statement: Committment to high standards of service and quality in the printing of books and publications of lasting value.

Strathmore Printing   **CA**
Howard Stevens, Marketing Director
2000 Gary Ln
Geneva IL 60134
708-232-9677; Fax: 708-232-0198
Send RFQ's to: Chang Park

Min: 10000 Max: 5000000 Opt: 100000-500000
Sizes: 123
Bindings: PBX-SSX-SDX
Items: DFIKL
Services: 4C-DA-FF-LM-OC-RP-TY-TD-WA
Equipment: 3WO(half webs)
Annual Sales: 4
Number of Employees: 105
Turnaround Times: B1C1S1
Customers: BP5-BU75-CU5-MP5-NP2-OR2-SC1-OT5
Union Shop: N
Year Established: 1914
Terms: Net 30

Straus Printing Company   **GC**
Michael Elliott, Marketing Director
1028 E Washington Ave
Madison WI 53703
608-251-3222; Fax: 608-251-1083
Send RFQ's to: Michael Elliott

Min: 1000 Max: 10000000 Opt: 0
Sizes: 123456X
Bindings: HCO-CBO-LBI-OBO-PBI-SSI-SDI-SBI-WBI
Items: ACDEFIKLMNOPQUVWXYZ123
Services: 4C-CS-DA-FF-RP-WA
Equipment: 3SO(2-28x40,20x26)-WO(17x26)
Annual Sales: 3
Number of Employees: 68
Turnaround Times: B2S2
Customers: BU80-CU2-MP2-OR2-BP2-SP2-OT8-NP2
Union Shop: N
Year Established: 1920

PRINTED ITEMS					
A	Annual Reports	G	Galley Copies	N	Workbooks	U	Greeting Cards
B	Books	H	Journals	O	Yearbooks	V	Labels / Stickers
C	Booklets	I	Magazines	P	Brochures	W	Maps
D	Catalogs	J	Mass-Market Books	Q	Calendars	X	Newspapers
E	Cookbooks	K	Newsletters	R	Comic Books	Y	Postcards
F	Directories	L	Software Manuals	S	Direct Mail Letters	Z	Posters
		M	Textbooks	T	Envelopes	1	Stationery

Statement: To satisfy our customer's needs by being a superior commercial printer with a dedicated team of employees who take pride in their work. We build partnerships by providing outstanding customer service offering innovative solutions and by utilizing advanced technologies.

GC

Strine Printing Co Inc
Marketing Director
I83 Industrial Park
PO Box 149
York PA 17405
717-767-6602; Fax: 717-764-3459

BK

The Studley Press
Thomas Reardon, President
PO Box 214
151 E Housatonic St
Dalton MA 01226
413-684-0441

Min: 1000 Max: 20000 Opt: 10000
Sizes: 235
Bindings: HCI-CBO-PBI-SSO-SBO
Items: ABCHIP
Services: 4C-DA-TY-TD
Equipment: SO
Terms: Net 30

GC

Stylecraft Printing Co
Marketing Director
8472 Rhonda Dr
Canton MI 48187
313-525-0001; Fax: 313-455-9461

GC

Sullivan Brothers Printers Inc
Marketing Director
117 Marginal St
Lowell MA 01851
508-458-6333; Fax: 508-458-1420

MG

Sullivan Graphics Inc
Marketing Director
100 Winners Cir
Brentwood TN 37027
615-377-0377; 800-621-7746

Min: 200000 Max: 2000000 Opt: 0
Sizes: 25X
Items: DIX
Services: 4C-CS-DA-FF-TY
Equipment: 24LP-20WO
Annual Sales: 6
Number of Employees: 2500
Year Established: 1926

Statement: Sullivan prints magazines, tabloids, and catalogs in satellite plants in Dunkirk, New York; Lufkin, Texas; Marengo, Iowa; Pittsburgh, California; York, Pennsylvania; Stevensville, Ontario; and Sylacauga, Alabama.

CA

Sun Graphics
Larry Reed, Marketing VP
1818 Broadway
Parsons KS 67357
316-421-6200; Fax: 316-421-2089
800-835-0588
Send RFQ's to: Larry Reed

Min: 0 Max: 0 Opt: 0
Bindings: SSI
Items: DP
Services: 4C-CS-TY

Statement: Primarily color separators.

GC

Sun Graphix
Robert Shimko, Exec VP
Division of Geiger Bros
Mountain Hope Ave
PO Box 1850
Lewiston ME 04241
800-284-5580; Fax: 207-777-7084

Min: 0 Max: 0 Opt: 0
Sizes: 25X
Bindings: LBI-SSI-SBI-CBI
Items: BQ
Equipment: LP-50WO
Annual Sales: 3
Number of Employees: 100

BK

The Sun Hill Press
Darrell Hyder, Marketing Director
23 High St
North Brookfield MA 01535
508-867-7274; Fax: Same

Min: 50 Max: 5000 Opt: 2000

SERVICES				SD	Side Stitching
4C	4-Color Printing	RA	Rachwal System	SB	Spiral Binding
CS	Color Separations	AF	Acid-Free Paper	WB	Wire-O Binding
DA	Design / Artwork	RP	Recycled Paper	I	In-House
FF	Fulfillment/Mailing	TY	Typesetting	O	Out of House
LM	List Maintenance	TD	Typeset w/ Disk	X	Unknown
OC	OptiCopy System	WA	Warehousing		

BINDINGS			
HC	Hardcover		
CB	Comb Binding		
LB	Loose-Leaf Binding		
OB	Otabind		
PB	Perfect Binding		
SS	Saddle Stitching		

Bindings: HCO-PBO-SSI
Items: BCPTU1
Services: DA-TY
Equipment: 2LP(10x15,21x28)
Annual Sales: 1
Number of Employees: 1
Turnaround Times: B6C6S6
Union Shop: N
Year Established: 1972
Terms: Net 30

Statement: We produce ephemeral and book printing by letterpress exclusively and either designed and set by us, or from plates from supplied art, in fine press tradition.

Sun Litho Inc **GC**
Marketing Director
7950 Haskell Ave
Van Nuys CA 91406
818-787-4100; Fax: 818-787-0164

Sun Lithographing & Printing Co **GC**
Marketing Director
2105 W 2300 South
Salt Lake City UT 84119
801-972-6120; Fax: 801-972-0351

Sun Printing **GC**
Marketing Director
2260 Stumbo Rd
Mansfield OH 44906
800-735-6313

Sundance Press **MS**
Marketing Director
817 East 18th St (85719)
PO Box 26605
Tucson AZ 85726-6605
602-622-5233; 800-528-4827

Min: 1000 Max: 10000 Opt: 10000
Sizes: 5
Bindings: SSI
Items: I
Equipment: SO

Sung in Printing America Inc **GC**
Marketing Director
901 Mariner's Island Blvd #525
San Mateo CA 94404
415-578-0206; Fax: 415-578-0805

Sweet Printing Co **GC**
Marketing Director
97 Commerce St
PO Box 679
Glastonbury CT 06033
203-633-5276; Fax: 203-633-3876

Swenson Offset Service **GC**
Marketing Director
204 G St
Antioch CA 94509
510-778-2442; Fax: 510-778-7281

John S Swift Company Inc **GC**
Bryan M Swift
1248 Research Blvd
PO Box 28252
St Louis MO 63132
314-991-4300; Fax: 314-991-3080
Send RFQ's to: Bryan M Swift
Min: 100 Max: 15000 Opt: 2000
Sizes: 123456X
Bindings: CBI-PBI-SSI-SDI-SBI-WBI-
    HCO-LBO-OBO
Items: ABCDEFHKLMNOPWYZ1
Services: 4C-RA-AF-RP-TY-TD-WA
Equipment: 25SO(11x17,42x60)-3WO(23)
Annual Sales: 4
Number of Employees: 200
Turnaround Times: B1C4S1
Customers: BP5-BU50-CU5-SC5-OT35
Union Shop: N
Year Established: 1912
Terms: Net 30

Statement: Our goal is to produce the highest quality printing with the best service, while being competitive. We specialize in book, catalog and commercial lithography. All our manufacturing plants have typesetting and desktop capabilities. We are able to generate camera ready copy from the customer's magnetic media.

PRINTED ITEMS						
A	Annual Reports	G	Galley Copies	N	Workbooks	U	Greeting Cards
B	Books	H	Journals	O	Yearbooks	V	Labels / Stickers
C	Booklets	I	Magazines	P	Brochures	W	Maps
D	Catalogs	J	Mass-Market Books	Q	Calendars	X	Newspapers
E	Cookbooks	K	Newsletters	R	Comic Books	Y	Postcards
F	Directories	L	Software Manuals	S	Direct Mail Letters	Z	Posters
		M	Textbooks	T	Envelopes	1	Stationery

John S Swift, Chicago   GC

Marketing Director
17 N Loomis St
PO Box 7261
Chicago IL 60607
312-666-7020

Min: 100 Max: 15000 Opt: 5000
Sizes: 23456X
Bindings: CBI-PBI-SSI-SBI
Items: ABCDKPQSTWYZ1
Services: 4C-TY
Equipment: SO-WO

John S Swift, Cincinnati   GC

Marketing Director
2524 Spring Grove
Cincinnati OH 45214
513-721-4147
Min: 100 Max: 15000 Opt: 5000
Sizes: 23456X
Bindings: CBI-PBI-SSI-SBI
Items: ABCDKPQSTWYZ1
Services: 4C-TY
Equipment: SO-WO

John S Swift, Teterboro   GC

Marketing Director
U S Route 46
Teterboro NJ 07608
201-288-2050

Min: 100 Max: 15000 Opt: 5000
Sizes: 23456X
Bindings: CBI-PBI-SSI-SBI
Items: ABCDKPQSTWYZ1
Services: 4C-TY
Equipment: SO-WO

Syracuse Lithographing Co   GC

Marketing Director
163 Solar St
Syracuse NY 13204
315-422-1600; Fax: 315-476-7806

T

TAN Books & Publishers Inc   BK

Marketing Director
2119 N Central Ave
Rockford IL 61101
815-987-1834; Fax: 815-987-1833
Send RFQ's to: Marlene King

Min: 1000 Max: 80000 Opt: 20000
Sizes: 235
Bindings: CBI-PBI-SSI-HCO-SBO-WBO
Items: ABCDEFKMNPS
Services: 4C-TY-WA
Equipment: SO
Number of Employees: 59
Turnaround Times: B4C5S3
Customers: BP75-BU15-CU10
Union Shop: N
Year Established: 1967
Terms: 2% 20, net 30

Statement: We provide uncompromising
quality and excellent service at a fair price.

Tapco Inc   MG

Marketing Director
Fort Dix-Pemberton Rd
Pemberton NJ 08068
609-894-2282
Min: 10000 Max: 1000000 Opt: 0
Sizes: 25X
Bindings: SSI
Items: CDIKNPX
Services: 4C-TY
Equipment: SO-WO

Statement: Their specialties are tabloids,
mini-tabs, and digest-sized books.

Tapemark   SP
Jeff Wigen, Marketing Director
223 E Marie Ave
West St Paul MN 55118
612-455-1611; Fax: 612-455-8703
800-328-0135
Send RFQ's to: Jeff Wigen

Items: V
Services: 4C-DA-TY-TD
Equipment: OP
Annual Sales: 4

SERVICES				BINDINGS		SD	Side Stitching
4C	4-Color Printing	RA	Rachwal System	HC	Hardcover	SB	Spiral Binding
CS	Color Separations	AF	Acid-Free Paper	CB	Comb Binding	WB	Wire-O Binding
DA	Design / Artwork	RP	Recycled Paper	LB	Loose-Leaf Binding	I	In-House
FF	Fulfillment/Mailing	TY	Typesetting	OB	Otabind	O	Out of House
LM	List Maintenance	TD	Typeset w/ Disk	PB	Perfect Binding	X	Unknown
OC	OptiCopy System	WA	Warehousing	SS	Saddle Stitching		

Number of Employees: 150
Customers: BU95-OT5
Union Shop: N
Terms: Net 30

BK

Taylor Publishing Company
Michael Zahn, Marketing Director
1550 W Mockingbird Ln
Dallas TX 75235
214-819-8501; Fax: 214-630-1852
800-947-0507
Send RFQ's to: Jack Smith
Parent Company: Insilco

Min: 300 Max: 25000 Opt: 15000
Sizes: 356
Bindings: HCX-PBX-SSX
Items: ABCDEF
Services: 4C-CS-FF-TY-TD-RP-WA
Equipment: 20SO
Annual Sales: 6
Number of Employees: 1500
Turnaround Times: B5C5S5
Customers: BP10-BU5-CU20-OR5-SC60
Union Shop: Y
Year Established: 1939
Terms: Net 30

Statement: Taylor Publishing Company is a
full service printer and publisher. Taylor is
divided into three divisions. Since 1939,
Taylor has produced high quality, limited
edition titles in our Fine Books Division,
mass market titles in our Trade Books
Division, and college and school yearbooks
in our Yearbook Division. Please contact
our Fine Books Division for quotes and
specifications for all your printing needs.

BK

TEACH Services
Timothy Hullquist, Marketing Director
Rte 1 Box 182
Brushton NY 12916
518-358-2125; Fax: 518-358-3028
Send RFQ's to: Wayne Reid

Min: 1000 Max: 250000 Opt: 5000
Sizes: 12345
Bindings: HCX-CBX-LBX-OBX-PBX-
    SSX-SBX-WBX
Items: BCDEFJKLMNP
Services: 4C-CS-DA-FF-LM-RP-TY-TD-
    WA
Equipment: SO-WO
Annual Sales: 2

Number of Employees: 13
Turnaround Times: B4C5S4
Customers: BP60-NP30-SP10
Union Shop: N
Year Established: 1984
Terms: 1/3 down, 1/3 proof, 1/3 delivery

Statement: To provide quality books at affordable prices.

GC

Technical Communication
    Services
Marketing Director
110 West 12th Ave
North Kansas City MO 64116
816-842-9770; Fax: 816-842-0628
Send RFQ's to: David Teeters

Min: 200 Max: 5,000 (PM) 10,000 (SS)
    Opt: 1500
Sizes: 2345
Bindings: CBX-PBX-SSX-SDX-WBX
Items: BCDEFHKLMS
Services: FF-AF-RP-TD-WA
Equipment: 13SO(38)
Annual Sales: 3
Number of Employees: 60
Turnaround Times: B2C4S2
Customers: BP60-NP40
Union Shop: N
Year Established: 1945
Terms: Net 30

Statement: Quality for our client is always
number one — with prices to match.

GC

Textile Printing Co
Marketing Director
PO Box 9296
Chattanooga TN 37412
615-894-1110; Fax: 615-899-4665

BK

Thomson-Shore
Jim Holefka, Marketing Director
7300 W Joy Rd
Dexter MI 48130-0305
313-426-3939; Fax: 313-426-6216 or 6219
Send RFQ's to: Ned Thomson

Min: 25 Max: 7500 Opt: 1500
Sizes: 12345
Bindings: HCX-CBX-PBX-SSX
Items: BCDEFHJKL

PRINTED ITEMS					
A	Annual Reports	G	Galley Copies	N	Workbooks	U	Greeting Cards
B	Books	H	Journals	O	Yearbooks	V	Labels / Stickers
C	Booklets	I	Magazines	P	Brochures	W	Maps
D	Catalogs	J	Mass-Market Books	Q	Calendars	X	Newspapers
E	Cookbooks	K	Newsletters	R	Comic Books	Y	Postcards
F	Directories	L	Software Manuals	S	Direct Mail Letters	Z	Posters
		M	Textbooks	T	Envelopes	1	Stationery

Services: 4C-RA-AF-RP
Equipment: 9SO(28x35 to 38x50)
Annual Sales: 4
Number of Employees: 280
Turnaround Times: B3C4S3
Customers: BP35-CU40-NP10-SP15
Union Shop: N
Year Established: 1972
Terms: Net 30

Statement: We do not have sales people. We rely on quality and service to sell our product.

MS
Times Litho Inc
David W Pearson
1829 Pacific Ave
PO Box 7
Forest Grove OR 97116
503-359-0300; Fax: 503-357-3754
Send RFQ's to: David W Pearson

Min: 5000 Max: 500000 Opt: 50000
Sizes: 25
Bindings: HCO-CBO-LBO-OBO-PBO-
    SDO-SBO-WBO-SSI
Items: CDFHIKPRS
Services: 4C-TY-WA
Equipment: 3WO(17x23,23x36,23x38)
Annual Sales: 4
Number of Employees: 110
Turnaround Times: B3C4S1
Customers: BU40-CU5-MP55
Union Shop: N
Year Established: 1950
Terms: Net 30

Statement: We are a communications resource company using knowledge, information, and printing technology to solve our customer's communications challenges.

MS
Times Printing
Marketing Director
453 Fifth St
PO Box 325
Random Lake WI 53075
414-994-4396

Min: 500 Max: 1000000 Opt: 100000
Sizes: 123456X
Bindings: HCO-CBI-LBI-PBI-SSI-SDI-
    SBO-WBO
Items: ABCDEFGHIKLNPQRSTWXYZ1
Services: 4C-DA-FF-AF-TY-TD-WA

Equipment: 6SO(10x15,28x41)-2WO(to
    22x37)
Annual Sales: 3
Number of Employees: 125
Turnaround Times: B3C2S2
Customers: BU40-CU1-MP50-NP1-OR1-
    SC1-SP6
Union Shop: N
Year Established: 1954
Terms: Net 30

GC
Todd Publications Inc
Nancy Todd-Catherman, Marketing Director
205 E Third
PO Box 269
Smithville TX 78957
512-237-3546; Fax: 512-237-5358
Send RFQ's to: Sharon Haney, Sales Manager

Min: 2500 Max: 2000000 Opt: 100000
Sizes: 3456
Bindings: HCO-PBO-SSO
Items: BCDFIJKLMNPXZ
Services: 4C-CS-FF-OC-WA
Equipment: 3SO(23x25,43x60)-
    2WO(23x36)
Annual Sales: 2
Number of Employees: 25
Turnaround Times: B3C2S2
Customers: CU25-MP50-NP5-State20
Union Shop: N
Year Established: 1975
Terms: COD

Statement: We strive to provide quality products for our customers within a reasonable time frame. Our emphasis is on our service and educating publishers to the printing industries.

BK
Starr Toof Printing
S Toof Brown Jr
670 S Cooper St
PO Box 14607
Memphis TN 38104
901-278-2200; Fax: 901-274-6191

Min: 0 Max: 0 Opt: 0
Bindings: CBI
Items: E
Year Established: 1864

SERVICES			BINDINGS		SD	Side Stitching
4C	4-Color Printing	RA Rachwal System	HC	Hardcover	SB	Spiral Binding
CS	Color Separations	AF Acid-Free Paper	CB	Comb Binding	WB	Wire-O Binding
DA	Design / Artwork	RP Recycled Paper	LB	Loose-Leaf Binding	I	In-House
FF	Fulfillment/Mailing	TY Typesetting	OB	Otabind	O	Out of House
LM	List Maintenance	TD Typeset w/ Disk	PB	Perfect Binding	X	Unknown
OC	OptiCopy System	WA Warehousing	SS	Saddle Stitching		

Statement: Toof prints and distributes many community cookbooks and fundraising cookbooks.

Town House Press BR
Marketing Director
552 Weatherfield
Pittsboro NC 27312
919-542-6242; Fax: 919-542-3922

Min: 100 Max: 5000 Opt: 2000
Sizes: 123456X
Bindings: HCO-CBO
Items: BCEFHMN
Services: 4C-CS-DA-AF-TY-TD
Annual Sales: 2
Number of Employees: 2
Turnaround Times: B4
Customers: BP40-NP35-OR10-SP15
Union Shop: N
Year Established: 1970
Terms: Net 30 (qualified clients)

Statement: We broker book printing services for small publishing houses, nonprofit publishers, and self-publishing ventures.

Tracor Publications GC
Doug McBride, Marketing Manager
6500 Tracor Ln
Austin TX 78725
512-929-2222

Min: 50 Max: 50000 Opt: 10000
Sizes: 123456X
Bindings: CBI-PBI-SSI
Items: ABCDEFHIKNO
Services: 4C-DA-TY
Equipment: SO

Statement: Tracor is a commercial printer specializing in magazine production and fulfillment as well as the printing of annual reports, sales literature, and technical manuals.

Trade Litho Inc CS
Marketing Director
5301 NW 37th Ave
Miami FL 33142
305-633-9779; Fax: 305-633-2848
800-367-5871

Min: 2500 Max: 200000 Opt: 10000
Items: DP
Services: 4C-CS

Trend Offset Printing Service Inc GC
Robert Pittman, Marketing Director
3791 Catalina St
Los Alamitos CA 90720
714-826-2360; Fax: 310-598-2716
Send RFQ's to: Robert Pittman

Min: 10000 Max: 10000000 Opt: 150000
Sizes: 12345
Bindings: PBX-SSX
Items: BCDFHIKLNRSX
Services: 4C-CS-DA-LM-OC-RP-TY-TD-WA
Equipment: 11WO(35,17.5)
Annual Sales: 6
Number of Employees: 400
Turnaround Times: B2S1
Customers: BU20-CU10-MP40-SC10-SP20
Union Shop: N
Year Established: 1960
Terms: Net 30

Statement: Trend maintains an excellent service staff. We keep the pricing within budgets and are a one-stop shop.

Tri-Graphic Printing BK
Doug Doane, VP Sales
485 Industrial Ave
Ottawa ON Canada K1G 0Z1
613-731-7441; Fax: 613-731-3741
800-267-9750
Send RFQ's to: Doug Doane

Min: 2000 Max: 1000000 Opt: 50000
Sizes: 123456
Bindings: HCI-LBI-PBI-SSI-SDI-SBO-WBO
Items: BDEFHILMNPQ
Services: 4C-OC-AF-RP-TY-TD
Equipment: 7SO(11x15,28x40)-WO (40x46)
Annual Sales: 4
Number of Employees: 115
Turnaround Times: B3C3S3
Customers: BP30-BU30-CU20-MP10-OT10
Union Shop: N
Year Established: 1968
Terms: Net 30

PRINTED ITEMS					
A	Annual Reports	G	Galley Copies	N	Workbooks	U	Greeting Cards
B	Books	H	Journals	O	Yearbooks	V	Labels / Stickers
C	Booklets	I	Magazines	P	Brochures	W	Maps
D	Catalogs	J	Mass-Market Books	Q	Calendars	X	Newspapers
E	Cookbooks	K	Newsletters	R	Comic Books	Y	Postcards
F	Directories	L	Software Manuals	S	Direct Mail Letters	Z	Posters
		M	Textbooks	T	Envelopes	1	Stationery

Statement: We service publishers of trade books, college course calendars, and wholesale trade catalogs. We also have facilities for inserting, labelling, and mailing.

GC

Triangle Booklet Company
Vicky Goodin, Marketing Director
325 Hill Ave
Nashville TN 37210
800-843-9529; Fax: 800-845-4767
Send RFQ's to: Dan Woods
Parent Company: Triangle Wholesale Printing and Evvelope Company

Min: 250 Max: 25000 Opt: 5000
Sizes: 123456X
Bindings: LBO-PBI-SSI-SDI-CBI-SBI-WBI
Items: ABCDEFKNPSTY1
Services: TY
Equipment: SO-WO-LP-OP
Annual Sales: 2
Number of Employees: 30
Turnaround Times: B3S2
Customers: BP5-OT95
Union Shop: N
Year Established: 1979
Terms: Net 30, no discount

Statement: We are a full service commercial wholesale printer, which operates through a network of national distributors in more than 40 states. Triangle Wholesale Printing was established in 1936 and currently specializes in 1 and 2 color flat sheet printing including edge-glued carbonless forms, books, booklets, pamphlets, envelopes, letterheads, index tabs, advertising and memo pads.

GC

Triangle Printers Inc
Marketing Director
3737 Chase Ave
Skokie IL 60076
312-465-3400; Fax: 708-674-1230

GC

Trojan Lithograph Inc
Marketing Director
22828 68th Ave S
Kent WA 98032
206-251-3900; Fax: 206-251-3929

GC

TSO General Corporation
Donald R Skahan, President
4402 11th St
Long Island City NY 11101
718-937-0680

Min: 500 Max: 50000 Opt: 25000
Sizes: 235
Bindings: CBI-LBI-PBI-SSI-SBI-WBI
Items: ACDFIKLNPSY
Equipment: 3SO(28x40)
Annual Sales: 4
Number of Employees: 30
Turnaround Times: B4C4
Customers: BP20-BU20-CU20-MP20-NP20
Union Shop: N
Year Established: 1969
Terms: Net 30

Statement: TSO is a cost-competitive medium to high quality sheetfed printer with major emphasis on quick turnaround service.

GC

Tweddle Litho Co
James L Foster, Marketing Director
24700 Maplehurst
Clinton Township MI 48036-1336
810-307-3700; Fax: 810-307-3708
Send RFQ's to: James L Foster

Min: 100 Max: 2000000 Opt: 50000
Sizes: 12345
Bindings: HCO-CBO-LBO-OBO-PBO-SSO-SDO-SBO-WBO
Items: BCDEFJKLMN
Services: 4C-DA-FF-LM-RP-TY-TD-WA
Equipment: SO(28x40)-2WO(36,38)
Annual Sales: 5
Number of Employees: 95
Turnaround Times: B4C5S3
Customers: BU94-CU3-SC3
Union Shop: Y
Year Established: 1954
Terms: Net 30

Statement: Our goal is to reduce customer costs for graphic arts expenses and produce quality products to match customer delivery expectations.

SERVICES			BINDINGS		SD	Side Stitching
4C	4-Color Printing	RA Rachwal System	HC	Hardcover	SB	Spiral Binding
CS	Color Separations	AF Acid-Free Paper	CB	Comb Binding	WB	Wire-O Binding
DA	Design / Artwork	RP Recycled Paper	LB	Loose-Leaf Binding	I	In-House
FF	Fulfillment/Mailing	TY Typesetting	OB	Otabind	O	Out of House
LM	List Maintenance	TD Typeset w/ Disk	PB	Perfect Binding	X	Unknown
OC	OptiCopy System	WA Warehousing	SS	Saddle Stitching		

U

Uniform Printing & Supply Inc `GC`
Marketing Director
210 S Progress Dr E
Kendallville IN 46755
219-347-3000; Fax: 219-347-3499
800-382-2424

Unique Printers & Lithographers `GC`
Marketing Director
5500 W 31st St
Cicero IL 60650
312-242-1433; Fax: 708-656-2176

United Lithographic Services Inc `MG`
Ken Garner, Marketing Director
2818 Fairfax Dr(22042)
PO Box 191
Falls Church VA 22046
703-560-5700
800-368-6100

Min: 0 Max: 0 Opt: 0
Sizes: 5
Bindings: SSI
Items: I
Services: 4C-TY-TD
Annual Sales: 3
Number of Employees: 150

United Lithograph `GC`
Marketing Director
48 Third Ave
Somerville MA 02143
617-776-6400

Min: 1000 Max: 10000 Opt: 5000
Items: BCK

United Printing Inc `GC`
Marketing Director
55 Colorado Ave
Warwick RI 02888
401-739-5600; Fax: 401-732-1537

Universal Graphics `GC`
Marketing Director
14461 E Eleven Mile Rd
Warren MI 48093
810-779-8660; Fax: 810-779-9665
Parent Company: World Color Press

Universal Litho Inc `GC`
Marketing Director
3500 Sprinkle Rd
Kalamazoo MI 49002
616-345-1175; Fax: 616-345-6999

Universal Press `CA`
Marketing Director
10 Park Ln
Providence RI 02907
401-944-2700

Min: 5000 Max: 2000000 Opt: 0
Sizes: 2345X
Bindings: PBI-SSI
Items: ACDIPUXZ
Services: 4C-CS-FF
Equipment: 5SO(26x40)-2WO(23x38)
Annual Sales: 4
Number of Employees: 175

Statement: Capturing the art of color is the specialty of our house: from separations right through to final delivery.

Universal Printers Ltd `BK`
Herb Krushel, Marketing Director
706 Bradford St
PO Box 804
Winnipeg MB Canada R3C 2N8
204-775-8486; Fax: 204-783-0090
Send RFQ's to: Herb Krushel

Min: 3000 Max: 1000000 Opt: 250000
Sizes: 123456X
Bindings: PBX
Items: BCDEFJ
Equipment: 2OP(36.5x60)
Annual Sales: 3
Number of Employees: 50
Turnaround Times: B4
Customers: BP82-BU6-CU2-NP5-OR4
Union Shop: Y
Year Established: 1954
Terms: Net 30

PRINTED ITEMS			
A Annual Reports	G Galley Copies	N Workbooks	U Greeting Cards
B Books	H Journals	O Yearbooks	V Labels / Stickers
C Booklets	I Magazines	P Brochures	W Maps
D Catalogs	J Mass-Market Books	Q Calendars	X Newspapers
E Cookbooks	K Newsletters	R Comic Books	Y Postcards
F Directories	L Software Manuals	S Direct Mail Letters	Z Posters
	M Textbooks	T Envelopes	1 Stationery

Statement: Our company specializes in producing mass-market paperback books. That is our only product.

Universal Printing Company **MS**
Wayne Parker, Marketing Director
3100 NW 74th Ave
Miami FL 33122
305-592-5387; Fax: 305-592-5104

Min: 20000 Max: 250000 Opt: 75000
Sizes: 25
Bindings: PBX-SSX
Items: CDEFHIJKNX
Services: 4C-OC-RP-WA
Equipment: 2WO(22x38)-2OP(non-heat-set 21x35)
Annual Sales: 3
Number of Employees: 125
Turnaround Times: B2S2
Customers: BU5-MP65-OR15-SC5-OT10
Union Shop: N
Year Established: 1961
Terms: Net 30

Universal Printing Company **GC**
Thomas Emerson, President
1701 Macklind Ave
Saint Louis MO 63110
314-771-6900

Min: 0 Max: 0 Opt: 0
Bindings: PBI-SSI
Items: ABCDFHIKNPS
Services: 4C-CS
Equipment: LP-SO-WO
Annual Sales: 5
Number of Employees: 400

US Press Inc **CS**
Michael R Jetter, VP Marketing
PO Box 640
Valdosta GA 31603-0640
912-244-5634; Fax: 912-247-4405
800-227-7377
Send RFQ's to: Michael R Jetter
Min: 500 Max: 100000 Opt: 5000-25000
Sizes: 25
Bindings: SSX
Items: ACDKPYZ
Services: 4C-CS-RP-TY
Equipment: 5SO(various)

Annual Sales: 3
Number of Employees: 40
Turnaround Times: S3
Customers: BU85-OT15
Union Shop: N
Year Established: 1986
Terms: 50% with order, balance COD, Free freight with 100% pre-payment

Statement: US Press Inc specializes in short-run, low cost, high quality process color printing nationwide. From color laminated postcards to catalog sell sheets, folded brochures to posters, and catalogs to packet folders, we can help you to meet your color printing needs start to finish.

V

Vail Ballou Press **BK**
William Long, Marketing Director
Pine Camp Dr
PO Box 1005
Binghamton NY 13902-1005
607-723-7981; Fax: 607-771-4634
Send RFQ's to: Rick Weidner
Parent Company: Maple Press Co

Min: 500 Max: 250000 Opt: 0
Sizes: 235
Bindings: HCI-CBI-PBI-SBI
Items: BCDE
Services: FF-TY-TD
Equipment: SO-WO
Annual Sales: 6
Number of Employees: 1200
Year Established: 1903
Terms: To be established

Valco Graphics **MS**
Robert Valentine, Sales Manager
480 Andover Park East
Seattle WA 98188
206-575-3500

Min: 5000 Max: 500000 Opt: 50000
Sizes: 23456
Bindings: HCO-CBO-LBO-PBI-SSO-SDO-SBO-WBO
Items: DEFFHIKLMNOPRSWX
Services: 4C-TY-TD-WA
Equipment: 2SO(to 25x38)-2WO(22x36)
Annual Sales: 3

SERVICES					BINDINGS			SD	Side Stitching
4C	4-Color Printing	RA	Rachwal System		HC	Hardcover		SB	Spiral Binding
CS	Color Separations	AF	Acid-Free Paper		CB	Comb Binding		WB	Wire-O Binding
DA	Design / Artwork	RP	Recycled Paper		LB	Loose-Leaf Binding		I	In-House
FF	Fulfillment/Mailing	TY	Typesetting		OB	Otabind		O	Out of House
LM	List Maintenance	TD	Typeset w/ Disk		PB	Perfect Binding		X	Unknown
OC	OptiCopy System	WA	Warehousing		SS	Saddle Stitching			

Number of Employees: 30
Turnaround Times: B4C2S2
Customers: CU10-MP50-SC5-SP10
Union Shop: Y
Year Established: 1927
Terms: Net 30

Statement: Valco specializes in quick turn-around, medium-length runs, pleasing quality publications and tabloids.

BK
Van Volumes
Russell L Tate, President
311 Park St
PO Box 449
Palmer MA 01069
413-283-8556; Fax: 413-283-7884
Send RFQ's to: Russell L Tate

Min: 100 Max: 3000 Opt: 1000
Sizes: 12345
Bindings: CBO-PBI-SSI-SDI-SBO
Items: ABCEFGHLNZ1
Equipment: 3SO(14x20)
Annual Sales: 1
Number of Employees: 9
Turnaround Times: B3
Customers: BP80-CU1-SP19
Union Shop: N
Year Established: 1984
Terms: 50% deposit, balance net 30

Statement: We print and bind books at reasonable rates for a good product. We provide film lamination and 2-cover colors.

GC
Vaughan Printing
Marketing Director
411 Cowan St
Nashville TN 37207
615-256-2244

GC
The Veitch Printing Corp
Marketing Director
1740 Hempstead Rd
Lancaster PA 17601
717-291-1100; Fax: 717-291-1722

BK
Versa Press Inc
Steve Kennell, VP Sales
RR 1 Spring Bay Rd
PO Box 2460
East Peoria IL 61611-2460
309-822-8272; Fax: 309-822-8141
800-447-7829
Send RFQ's to: Steve Kennell

Min: 1000 Max: 20000 Opt: 5000
Sizes: 12345
Bindings: HCO-CBO-OBO-SDO-SBO-WBO-PBI-SSI
Items: BMN
Services: CS-AF-RP
Equipment: 4SO(40)
Annual Sales: 3
Number of Employees: 80
Turnaround Times: B3
Customers: BP80-CU20
Union Shop: N
Year Established: 1937
Terms: Net 30 with approved credit

Statement: Our goal is to provide the publishing trade with quality paperback bound books at a competitive price.

BK
Vicks Lithograph & Printing
Marketing Director
Commercial Dr
PO Box 270
Yorkville NY 13495
315-736-9344

Min: 1000 Max: 35000 Opt: 10000
Sizes: 456
Bindings: CBI-PBI-SSI-SDI-SBI
Items: BHLMN
Services: 4C-AF
Equipment: 9SO(to 64)-2WO(22x36,24x38)
Annual Sales: 3
Number of Employees: 75
Turnaround Times: B4
Customers: BP40-BU40-CU10-NP10
Union Shop: Y
Year Established: 1957
Terms: Net 30

BK
Victor Graphics
Marketing Director
200 N Bentalou
PO Box 4446
Baltimore MD 21223
410-233-8300; Fax: 410-233-8304

PRINTED ITEMS			
A Annual Reports	G Galley Copies	N Workbooks	U Greeting Cards
B Books	H Journals	O Yearbooks	V Labels / Stickers
C Booklets	I Magazines	P Brochures	W Maps
D Catalogs	J Mass-Market Books	Q Calendars	X Newspapers
E Cookbooks	K Newsletters	R Comic Books	Y Postcards
F Directories	L Software Manuals	S Direct Mail Letters	Z Posters
	M Textbooks	T Envelopes	1 Stationery

Send RFQ's to: Ross Agee

Min: 1000 Max: 30000 Opt: 8000
Sizes: 123456
Bindings: HCX-CBX-LBX-OBX-PBX-
SSX-SDX-SBX-WBX
Items: BDFHLN
Services: FF-OC-AF-TY-TD
Equipment: 2SO-3WO
Number of Employees: 85
Turnaround Times: B3C5S3
Customers: BP50-CU10-NP40
Union Shop: N
Year Established: 1983
Terms: Net 30

Statement: Paperback and casebound
books produced on very short schedules at
very competitive prices.

Viking Press **BK**
Chip Furmann, Marketing Director
7000 Washington Ave S
Eden Prairie MN 55344
612-941-8780; Fax: 612-941-2154
800-765-7327

Min: 5000 Max: 100000 Opt: 35000
Sizes: 2356
Bindings: OBX-PBX-SSX
Items: BCDFQW
Services: AF-RP-TD
Equipment: 5SO(19x25,28x40)-4WO
(19,38,45)
Annual Sales: 6
Number of Employees: 265
Turnaround Times: B5S4
Customers: BP70-BU30
Union Shop: N
Year Established: 1963
Terms: Net 30 with approved credit
Statement: Our speciality is one and two
color softcover books. We go directly from
disk to imposed film. We will be ISO cer-
tified in 1994.

Vogue Printers **GC**
Pete Deperte, Sales Manager
2421 Green Bay Rd
North Chicago IL 60064
708-689-4044

Min: 10 Max: 100000 Opt: 0
Sizes: 2345X
Bindings: CBI-PBI-SSI-SBX

Items: ABCDEFKNPSTYZ1
Services: 4C-DA-TY-TD
Equipment: SO
Terms: Net 30

W

Wagners Printers **GC**
Deborah Wilson, Marketing Director
1515 E Kimberly Rd
Davenport IA 52807
319-386-1533; Fax: 319-386-6963
Send RFQ's to: Chris Black

Min: 5000 Max: 1500000 Opt: 0
Sizes: 123456
Bindings: SSX-SBX
Items: ACDFKPQ
Services: 4C-CS-OC-RP-WA
Equipment: 4SO(Max 40)-2WO(Max 26.5)
Annual Sales: 4
Number of Employees: 180
Turnaround Times: S2
Union Shop: N
Terms: Net 30 subject to credit approval

Waldman Graphics **CA**
Marketing Director
9100 Pennsauken Highway
Pennsauken NJ 08110
609-662-9111; Fax: 609-665-1789
800-543-0955

Min: 1000 Max: 100000 Opt: 25000
Sizes: 123456X
Bindings: SSI-CBO-HCO-LBO-OBO-
PBO-SSO-SBO-WBO
Items: ACDFHIKLPQZ2
Services: 4C-CS-OC-TY-TD
Equipment: 4SO(28x40)-OP(Silk-screen)
Annual Sales: 4
Number of Employees: 260
Turnaround Times: B3C4S3
Customers: BP20-BU40-CU5-MP20-OR15
Union Shop: N
Year Established: 1972

Statement: Waldman Graphics maintains
an overriding commitment to quality while
offering service, experienced craftsmanship, and the very latest technological in-
novations for producing all manner of publications and publishing promotional

SERVICES				BINDINGS		SD	Side Stitching
4C	4-Color Printing	RA	Rachwal System	HC	Hardcover	SB	Spiral Binding
CS	Color Separations	AF	Acid-Free Paper	CB	Comb Binding	WB	Wire-O Binding
DA	Design / Artwork	RP	Recycled Paper	LB	Loose-Leaf Binding	I	In-House
FF	Fulfillment/Mailing	TY	Typesetting	OB	Otabind	O	Out of House
LM	List Maintenance	TD	Typeset w/ Disk	PB	Perfect Binding	X	Unknown
OC	OptiCopy System	WA	Warehousing	SS	Saddle Stitching		

materials. As the region's most award-winning prepress and printing company, we specialize in producing the finest multicolor sheetfed printing, UV coating, and state-of-the-art electronic prepress capabilities, including desktop publishing.

Waldon Press Inc **GC**
Frank Wise
216 West 18th St
New York NY 10011
212-691-9220; Fax: 212-675-9239
Send RFQ's to: Frank Wise

Min: 3000 Max: 500000 Opt: 25000-200000
Sizes: 123456
Bindings: HCO-CBO-LBO-OBO-PBO-
    SSO-SDO-SBO-WBO
Items: ABCDFHIKLMNOPTYZ1
Services: 4C-CS-DA-FF-RP-TY-TD-WA
Equipment: 4SO(23x29,28x40)-7WO
    (22.75x36)
Annual Sales: 3
Number of Employees: 36
Turnaround Times: B2C4S1
Customers: BP5-BU65-CU3-NP15
Union Shop: Y
Year Established: 1949
Terms: Net 30

Statement: Our goal is to provide our clients with the fastest possible service while maintaining reasonable prices and excellent quality.

Wallace Press **BK**
Tom Franke, Sales Manager
4600 W Roosevelt Rd
Hillside IL 60162
708-449-8611; Fax: 708-449-5999
Send RFQ's to: Tom Franke

Min: 5000 Max: 2000000 Opt: 500000
Sizes: 25X
Bindings: PBI-SSI
Items: ABCDFKLNPQ
Services: 4C-CS-DA-FF-LM-OC-RP-TY-
    TD
Equipment: 2SO(28x41)-4WO(22,24,38)
Annual Sales: 5
Number of Employees: 270
Turnaround Times: B3C3
Customers: BP10-BU70-NP10-SP10
Union Shop: Y
Year Established: 1908

Terms: Net 30

Statement: A division of Wallace Computer Services, we offer not just products, but solutions.

Walsworth Publishing Co Inc **BK**
Saliie Buck, Marketing Director
306 North Kansas Ave
Marceline MO 64658
816-376-3543; Fax: 816-258-7798
800-369-2147(Customer Service)
800-369-2646(Administrative Offices)
Send RFQ's to: Estimating Department:
    Janet Lange, Manager

Min: 500 Max: 10000 Opt: 5000
Sizes: 123456
Bindings: HCI-CBI-PBI-SSI-SBI-WBI
Items: BEFMO
Services: 4C-CS-DA-FF-OC-AF-RP-TY-
    TD-WA
Equipment: 22SO(25x38,28x40)
Annual Sales: 1
Number of Employees: 1400
Turnaround Times: B4C5S4
Customers: BP40-CU60
Union Shop: N
Year Established: 1941
Terms: Net 30

Statement: Produce high quality, short run books, meeting customer needs and schedule and develop long term working relationships.

Walter's Publishing **BK**
Wayne J Dankert, General Manager
215 Fifth Ave SE
Waseca MN 56093
507-835-3691; 800-447-3274

Min: 200 Max: 3000 Opt: 500
Sizes: X
Bindings: CBI
Items: E
Services: TY
Customers: NP-OR-SC-(fundraisers)
Terms: Net 90

Statement: Walters produces personalized standard-format cookbooks for fundraising organizations. Send for their free cookbook fundraising kit.

PRINTED ITEMS							
A	Annual Reports	G	Galley Copies	N	Workbooks	U	Greeting Cards
B	Books	H	Journals	O	Yearbooks	V	Labels / Stickers
C	Booklets	I	Magazines	P	Brochures	W	Maps
D	Catalogs	J	Mass-Market Books	Q	Calendars	X	Newspapers
E	Cookbooks	K	Newsletters	R	Comic Books	Y	Postcards
F	Directories	L	Software Manuals	S	Direct Mail Letters	Z	Posters
		M	Textbooks	T	Envelopes	1	Stationery

Looking For a Book Manufacturer Who Can Provide...

- A Full Range of Typesetting Capabilities
- Quality Scanning of Black and 4/Color Originals
- A Full Range of House Cover and Text Stocks
- Quality One, Two and 4/Color Printing
- Complete In-House Bindery
- A Variety of Trim Sizes
- Excellent Customer Service

You Just Found One!

Publishing Company

306 N. Kansas Ave.
Marceline, MO 64658
800-369-2646
816-258-7798 (Fax)

Watermark Press   **GC**

Marketing Director
11419-10 Cronridge Dr
Owings Mills MD 21117
410-356-2000

Watkins Printing Company   **CA**

Marketing Director
1401 East 17th Ave
Columbus OH 43211
614-297-8270; Fax: 614-291-1961

Min: 20000 Max: 2000000 Opt: 0
Sizes: 2345
Bindings: CBI-SSI-SDI
Items: ACDI
Services: 4C-FF-TY
Equipment: SO-WO

Watt/Peterson   **BK**

Bruce Warner, Marketing Director
15020 27th Ave N
Plymouth MN 55447
612-553-1617; Fax: 612-553-0956
800-328-3328
Send RFQ's to: Bruce A Warner

Min: 0 Max: 0 Opt: 0
Sizes: 123456
Bindings: HCO-CBO-LBO-OBO-PBO-
  SSI-SDO-SBO-WBO
Items: ABCDFIKNPQUZ
Services: 4C-CS-WA
Equipment: 4SO(28x40)
Annual Sales: 4
Number of Employees: 140
Customers: BP2-BU77-CU1-OR20
Union Shop: N
Year Established: 1963

Statement: Watt/Peterson is a commercial
color sheetfed printer, experienced in the
printing of annual reports, brochures,
hardbound books, posters, greeting cards,
and fine art prints. We utilize a totally digital prepress system and in-house m/c
blending facility.

Waverly Press   **MS**

Patricia Chalfant, Marketing Manager
940 Elkridge Land Rd
Linthicum MD 21090
410-850-0500; 800-638-5198

Parent Company: Cadmus Communica-
  tions Corp
Min: 1000 Max: 125000 Opt: 0
Sizes: 345
Bindings: HCO-CBO-LBO-PBI-SSI-SDO-
  SBO-WBO
Items: HI
Services: 4C-DA-FF-LM-AF-TY-TD-WA
Equipment: SO-WO
Annual Sales: 5
Number of Employees: 610
Union Shop: N
Year Established: 1889
Terms: Net 30

Statement: Waverly is a full service peri-
odical printer which continually strives to
live up to our corporate goal: without
blemish.

Web Inserts   **GC**

Marketing Director
4213 Ridgegate Dr
Duluth GA 30136
404-662-5391; Fax: 404-416-8341
Parent Company: World Color Press

Web Specialties   **CA**

Vernon Carson, VP Marketing
401 S Milwaukee Ave
Wheeling IL 60090
708-459-0800

Min: 100000 Max: 2000000 Opt: 1500000
Sizes: X
Bindings: SSI
Items: CDEPSTVY
Services: 4C-CS-DA
Equipment: 22WO(various)
Annual Sales: 4
Number of Employees: 45
Turnaround Times: C2
Union Shop: N
Year Established: 1983
Terms: Net 30

Statement: We are manufacturer's reps for
specialty printers of mini-catalogs and
other unique direct response formats.

PRINTED ITEMS	G Galley Copies	N Workbooks	U Greeting Cards
A Annual Reports	H Journals	O Yearbooks	V Labels / Stickers
B Books	I Magazines	P Brochures	W Maps
C Booklets	J Mass-Market Books	Q Calendars	X Newspapers
D Catalogs	K Newsletters	R Comic Books	Y Postcards
E Cookbooks	L Software Manuals	S Direct Mail Letters	Z Posters
F Directories	M Textbooks	T Envelopes	1 Stationery

BK

Webcom Ltd
Michael Collinge, Marketing Manager
3480 Pharmacy Ave
Scarborough ON Canada M1W 3G3
416-496-1000; Fax: 416-496-1537

Send RFQ's to: Chris Wilkins, Sales
    Manager

Min: 1000 Max: 250000 Opt: 5000
Sizes: 123456
Bindings: HCO-LBO-PBI-SBI-SSO-SBO-
    WBO
Items: BDEFHJLMN
Services: 4C-FF-OC-AF-RP-TY-TD-WA
Equipment: 2SO(29)-3WO(35,38)
Annual Sales: 4
Number of Employees: 175
Turnaround Times: B4
Customers: BP50-BU35-CU10-SP5
Union Shop: N
Year Established: 1961
Terms: Net 25

Statement: One of the few North
American printers capable of Otabinding a
book, this company manufactures books,
catalogs, and directories for the North
American market.

GC

Webcraft Technologies
Marketing Director
Route 1 & Adams Station
North Brunswick NJ 08902
908-297-5100; 800-283-4044

Min: 25000 Max: 2000000 Opt: 0
Sizes: 23456X
Items: DPST
Services: 4C-CS
Equipment: 25WO
Annual Sales: 6
Number of Employees: 1900
Year Established: 1969

Statement: Webcraft is the world's largest
printer of direct mail forms.

BK

Webcrafters
W Jerome Frautschi, VP Sales
PO Box 7608 (53707)
2211 Fordem Ave
Madison WI 53704
608-244-3561; Fax: 608-244-5120

800-356-8200
Min: 10000 Max: 2000000 Opt: 0
Sizes: 2345X
Bindings: HCI-PBI-SSI
Items: BCDFILMN
Services: 4C
Equipment: SO-WO
Annual Sales: 6
Number of Employees: 620
Turnaround Times: B4
Year Established: 1893

GC

F A Weber & Sons
Robert A Weber, President
175 S Fifth Ave
PO Box 449
Park Falls WI 54552
715-762-3707

Min: 500 Max: 100000 Opt: 60000
Sizes: 123456
Bindings: CBX-LBX-PBX-SSX
Items: ABCDEFIKOPQSTUVWYZ1
Services: 4C-CS-DA-AF-TY-WA
Equipment: 4SO(10x14,18x23,25x38)
Annual Sales: 2
Number of Employees: 30
Turnaround Times: B5C5S4
Customers: BU50-MP20-NP20-SP10
Union Shop: N
Year Established: 1965
Terms: Net 30

Statement: A general commercial printer
who prints book jackets and covers.

BK

Weber Printing Company
Duane A Weber, Marketing Director
2200 N Cotner Blvd
Lincoln NE 68505-2999
402-466-1951; Fax: 402-466-2243
Send RFQ's to: Duane A Weber

Min: 100 Max: 25000 Opt: 1000-5000
Sizes: 123456
Bindings: HCO-PBI-SSI-SDI
Items: BDEFJKLNPQSTYZ12
Services: 4C-AF-TY-TD-WA-DA-RP
Equipment: SO
Annual Sales: 1
Number of Employees: 10
Turnaround Times: B3S3
Customers: BU50-NP50
Union Shop: N

SERVICES				BINDINGS			SD	Side Stitching
4C	4-Color Printing	RA	Rachwal System	HC	Hardcover		SB	Spiral Binding
CS	Color Separations	AF	Acid-Free Paper	CB	Comb Binding		WB	Wire-O Binding
DA	Design / Artwork	RP	Recycled Paper	LB	Loose-Leaf Binding		I	In-House
FF	Fulfillment/Mailing	TY	Typesetting	OB	Otabind		O	Out of House
LM	List Maintenance	TD	Typeset w/ Disk	PB	Perfect Binding		X	Unknown
OC	OptiCopy System	WA	Warehousing	SS	Saddle Stitching			

Year Established: 1947
Terms: 30% down, balance on completion

Statement: Our goal is to satisfy and do business with you.

The Wessel Company Inc **GC**
Clay Jacobs, Marketing Director
1201 Kirk St
Elk Grove Village IL 60007
708-595-7011; Fax: 708-595-1435
Send RFQ's to: Clay Jacobs

Min: 100000 Max: 0 Opt: 1000000
Items: PQSTVWY23
Services: 4C-WA
Equipment: 4WO
Annual Sales: 5
Number of Employees: 110
Customers: BU100
Union Shop: N
Terms: Net 30

Westerman Print Co **GC**
Marketing Director
2116 Colerain Ave
Cincinnati OH 45214
513-721-6492; Fax: 513-721-6293

Western Publishing Co Inc **BK**
Marketing Director
Book Printing Division
1220 Mound Ave
Racine WI 53404
414-633-2431; Fax: 414-631-1436
800-453-1222

Min: 10000 Max: 2000000 Opt: 0
Sizes: 2345X
Bindings: HCI-PBI-SSI
Items: BCDLNRZ
Services: 4C-CS-DA-FF-TY-TD-WA
Equipment: CB-LP-SO(to 55x78)-7WO(to 35x50)
Annual Sales: 6
Number of Employees: 2500
Turnaround Times: B4
Year Established: 1907

Statement: This publisher also prints books for other publishers.

Western Web Printing **BK**
Jerry Schmuck, Marketing Director
4005 S Western Ave
Sioux Falls SD 57105
605-339-2383; Fax: 605-339-1523
Send RFQ's to: Jerry Schmuck
Parent Company: Sioux Falls Shopping News

Min: 25000 Max: 5000000 Opt: 1000000
Sizes: 1245
Bindings: SSI-PBO
Items: BCDFKX
Services: 4C-DA-FF-LM-RP-TY-TD-WA
Equipment: SO(19x25)-3WO(22.75x35)
Annual Sales: 3
Number of Employees: 100
Turnaround Times: B3S2
Customers: NP1-SP99
Union Shop: N
Year Established: 1973
Terms: Net 30

Westview Press **BK**
Tari Warwick, Marketing Director
5500 Central Ave
Boulder CO 80301
303-444-3541; Fax: 303-449-3356
Send RFQ's to: Jolene Lelwica

Min: 100 Max: 5000 Opt: 1000
Sizes: 12345X
Bindings: HCO-CBO-LBO-OBO-PBI-SSI-SDO-SBO-WBO
Items: ABDEFGMN
Services: FF-AF-RP-TY-TD-WA
Equipment: 3SO(13x18 maximum)
Annual Sales: 2
Number of Employees: 13
Turnaround Times: B3C5S3
Customers: BP95-SP5
Union Shop: N
Year Established: 1982
Terms: 50% up front — 50% completion new customers, then net 30

White Arts Inc **GC**
William McClanahan, Marketing Director
1203 E St Clair St
Indianapolis IN 46202
317-638-3564; Fax: 317-638-6793
800-748-0323

PRINTED ITEMS						
A	Annual Reports	G	Galley Copies	N	Workbooks	U	Greeting Cards
B	Books	H	Journals	O	Yearbooks	V	Labels / Stickers
C	Booklets	I	Magazines	P	Brochures	W	Maps
D	Catalogs	J	Mass-Market Books	Q	Calendars	X	Newspapers
E	Cookbooks	K	Newsletters	R	Comic Books	Y	Postcards
F	Directories	L	Software Manuals	S	Direct Mail Letters	Z	Posters
		M	Textbooks	T	Envelopes	1	Stationery

Send RFQ's to: William McClanahan

Min: 1000 Max: 100000 Opt: 10000
Sizes: 23456
Bindings: SSX
Items: ACDKLNPQSZ3
Services: 4C-AF-RP-TD
Equipment: 5SO(28x40)-2LP(13x20)
Annual Sales: 4
Number of Employees: 100
Turnaround Times: B3C4S2
Customers: BP5-BU70-CU8-MP5-NP2-
 OR10
Union Shop: N
Year Established: 1945
Terms: 1st order cash; net 30 after D & B
 report approved

Statement: At White Arts Inc, our determination to continually improve our products, process, and services creates a superior value for our customers. The products we produce can be purchased from any number of sources. That's why our continuing investment to expand our state-of-the-art facilities that always meet or exceed OSHA and environmental regulations, and training our people in the latest technology is a long-term commitment to our customers — and is the cornerstone of our mission. A supplier you can count on now and in the future.

White Printing Inc   GC
Marketing Director
2101 Central Ave
PO Box 36369
Charlotte NC 28205
704-333-1123; Fax: 704-335-0553

Whitehall Printing Company   BK
Mike Hirsch, President
4244 Corporate Square
Naples FL 33942
800-321-9290; Fax: 800-643-6439
Send RFQ's to: Mike Hirsch

Min: 1000 Max: 10000 Opt: 0
Sizes: 2345X
Bindings: PBI-OEI(openbak Eurobind)
Items: BDEFHLMN
Services: 4C-CS-OC-AF-RP-TD
Equipment: SO-WO
Turnaround Times: B4

Union Shop: N
Year Established: 1959

Statement: A quality book manufactured by caring people yielding a small but fair profit.

Wickersham Printing Co Inc   BK
John W Yurchak, Marketing Director
2959 Old Tree Dr
Lancaster PA 17603
717-299-5731; Fax: 717-393-7469
800-437-7171
Send RFQ's to: Jason P Yurchak

Min: 1 Max: 50000 Opt: 50-5000
Sizes: 123456
Bindings: HCO-CBI-LBI-OBO-PBI-SSI-
 SBO-WBO
Items: BCDEFGHKLN
Services: FF-RA-AF-RP
Equipment: 2SO(42x60)-2WO(21,22.75)-
 4OP
Annual Sales: 3
Number of Employees: 89
Turnaround Times: B3C4S3
Customers: BP80-BU10-CU5-OR5
Union Shop: Y
Year Established: 1859
Terms: 2% 10, net 30

Statement: Our Mission: Continuing to provide reliable products with responsible, timely, and flexible services at a competitive price. Our commitment is to be helpful, courteous, and professional to our customers, to each other, and to our suppliers. In carrying out our mission, we place high value on customer satisfaction, improving technology, empowered employees, profitable operations, and building community.

Wicklander Printing Corp   GC
Marketing Director
1550 S State St
Chicago IL 60605
312-225-1300; Fax: 312-225-7266

Williams Printing Co   GC
Marketing Director
1240 Spring St NW
Atlanta GA 30309
404-875-6611; Fax: 404-872-4025

SERVICES				BINDINGS			
4C	4-Color Printing	RA	Rachwal System	HC	Hardcover	SD	Side Stitching
CS	Color Separations	AF	Acid-Free Paper	CB	Comb Binding	SB	Spiral Binding
DA	Design / Artwork	RP	Recycled Paper	LB	Loose-Leaf Binding	WB	Wire-O Binding
FF	Fulfillment/Mailing	TY	Typesetting	OB	Otabind	I	In-House
LM	List Maintenance	TD	Typeset w/ Disk	PB	Perfect Binding	O	Out of House
OC	OptiCopy System	WA	Warehousing	SS	Saddle Stitching	X	Unknown

GC

Williamson Printing Corp
Marketing Director
6700 Denton Dr
Dallas TX 75235
214-352-1122; Fax: 214-352-5698

BK

The Wimmer Companies
Glen Wimmer, President
PO Box 18408
4210 BF Goodrich Blvd
Memphis TN 38118
901-362-8900
Min: 3000 Max: 50000 Opt: 0
Sizes: 23
Bindings: HCO-CBI-LBO-PBO-SSO-
SDO-SBO-WBO
Items: BCEIPW
Services: 4C-FF-WA
Equipment: 2WO(25x38)
Annual Sales: 3
Number of Employees: 100
Turnaround Times: B6
Customers: BU20-NP40-OR20-SP10-OT10
Union Shop: N
Year Established: 1950

Statement: We serve cookbook fundraisers
(design, consultation, manufacturing, fulfillment, storage, and distribution).

BR

Windsor Associates
Marketing Director
4655 Cass St #314
PO Box 90282
San Diego CA 92109
619-270-1000

Min: 500 Max: 20000 Opt: 0
Sizes: 12356
Bindings: HCX-CBX-PBX
Items: BCDEFLMO
Services: 4C-CS-DA-TY
Equipment: SO-WO
Turnaround Times: B5C4
Customers: BU10-CU10-SC10-SP50-OT20
Union Shop: N
Year Established: 1978
Terms: Open

Statement: They offer complete self-
publishing services: editing, typesetting,
art, design, printing, binding, distribution,
and copyright.

GC

Winslow Printing Co
Marketing Director
1225 N 7th St
Minneapolis MN 55411
612-522-3868; Fax: 612-522-0025

GC

Winston Printing Co Inc
Marketing Director
8095 N Point Blvd
Winston-Salem NC 27106
910-759-0051; Fax: 910-759-0304

BK

Winter Sun Inc
Mark Bruno, Marketing Director
11036 N 23 Ave
Phoenix AZ 85029
602-870-3987; Fax: 602-866-9633
Send RFQ's to: Mark Bruno

Min: 0 Max: 0 Opt: 0
Sizes: 2356
Bindings: CBX-SSX-SDX
Items: BCDKNSTV1
Services: DA-FF-TY-WA
Equipment: SO(12x18)
Annual Sales: 1
Number of Employees: 5
Turnaround Times: S2
Customers: BP100
Union Shop: N
Year Established: 1970
Terms: COD

GC

Winthrop Printing Co Inc
Marketing Director
235 Old Colony Ave
Boston MA 02127
617-268-9660; Fax: 617-268-6735

MG

Wisconsin Color Press Inc
Thomas R Bayer, Marketing Director
5400 W Good Hope Rd
Milwaukee WI 53223
414-353-5400; Fax: 414-353-0189
Send RFQ's to: Richard Bortolotti or Ed
Hammett
Parent Company: Midcontinent Printing
Inc

PRINTED ITEMS						
A	Annual Reports	G	Galley Copies	N	Workbooks	U	Greeting Cards
B	Books	H	Journals	O	Yearbooks	V	Labels / Stickers
C	Booklets	I	Magazines	P	Brochures	W	Maps
D	Catalogs	J	Mass-Market Books	Q	Calendars	X	Newspapers
E	Cookbooks	K	Newsletters	R	Comic Books	Y	Postcards
F	Directories	L	Software Manuals	S	Direct Mail Letters	Z	Posters
		M	Textbooks	T	Envelopes	1	Stationery

Min: 0 Max: 0 Opt: 0
Sizes: 2X
Bindings: PBX-SSX
Items: CDI
Services: 4C-CS-OC-RP-TY-TD-WA
Equipment: 4WO(23x36)
Annual Sales: 5
Number of Employees: 450
Turnaround Times: B3S3
Customers: BU15-MP85
Union Shop: Y
Year Established: 1926
Terms: Net 30

Statement: The heart of every printing company is the pressroom. Quality of print cannot be understated and Wisconsin Color Press has the latest technology available. Our state-of-the-art equipment is complimented by talented press operators, each having over a decade of experience. Continuous training is provided, both on and off the job to ensure consistency and continuous improvement.

Sales Offices:

Winconsin Color Press Inc
2340 Des Plaines Ave #419
Des Plaines IL 60018
708-390-7990; Fax: 708-390-6087

Winconsin Color Press Inc
c/o The Empire State Building #6404
New York NY 10018
212-967-5400; Fax: 212-967-5411

Wisconsin Color Press Inc
3113 Delta River Dr
Lansing MI 48906
517-323-9324; Fax: Same

Wisconsin Color Press Inc
10002 Wild Dunes Dr
Austin TX 78747
512-280-5100; 512-280-5101

Wisconsin Color Press Inc
1714 Kestwick Cir
Birmingham Al 35226
205-823-3827; Fax: 205-988-4359

GC
Wolfer Printing
Michael Wellman, Marketing Director
6670 Flotilla St
Commerce CA 90040-1816
213-721-5411; Fax: 213-720-1245
Send RFQ's to: Michael Wellman

Min: 25000 Max: 0 Opt: 0
Sizes: 123456
Bindings: HCO-CBO-OBO-LBO-PBO-
    SSO-SDO-SBO-WBO
Items: ACDFKLPQX
Services: 4C-CS-OC-RP
Equipment: 2WO(22.75x38)
Annual Sales: 5
Number of Employees: 90
Turnaround Times: B2S2
Customers: BU85-MP15
Union Shop: Y
Year Established: 1913
Terms: Net 30

Statement: Wolfer Printing Company is a six-color heatset printer striving to satisfy the marketplace with excellent service, quality, and value for the dollar.

GC
Wood & Jones Printers
Hanna Wood, Marketing Director
1260 Lincoln Ave #800
Pasadena CA 91105
818-797-5700; Fax: 818-797-5868
Send RFQ's to: J J Gish

Min: 250 Max: 10000 Opt: 5000
Sizes: 2345
Bindings: SSX
Items: ABCDFIKLPSTUVY12
Services: 4C-CS-DA-AF-RP-TY-TD-WA
Equipment: 3LP(10x15)-5SO(24x36)
Annual Sales: 2
Number of Employees: 16
Turnaround Times: S2
Customers: BP5-BU50-SC20-OT25
Union Shop: N
Year Established: 1907
Terms: Net 30 on approved credit

Statement: Wood & Jones was founded in 1907 by Bert Wood, a pressman from England, and Fred Jones, a typesetter from Canada. They combined their skills to open a quality printing business in the prosperous community of Pasadena. Various members of Bert's family have continued to run Wood & Jones in its present location for over 80 years. New technology has been added as it was developed; lithography, phototypesetting, photo duplicating, computerized typesetting and bookkeeping systems, and most recently, desktop publishing. Throughout its history, the hallmark of Wood & Jones

SERVICES			BINDINGS		SD	Side Stitching
4C	4-Color Printing	RA Rachwal System	HC	Hardcover	SB	Spiral Binding
CS	Color Separations	AF Acid-Free Paper	CB	Comb Binding	WB	Wire-O Binding
DA	Design / Artwork	RP Recycled Paper	LB	Loose-Leaf Binding	I	In-House
FF	Fulfillment/Mailing	TY Typesetting	OB	Otabind	O	Out of House
LM	List Maintenance	TD Typeset w/ Disk	PB	Perfect Binding	X	Unknown
OC	OptiCopy System	WA Warehousing	SS	Saddle Stitching		

has always been to produce fine quality printing with prompt personal service for our customers. At the same time, it has always been important to create a friendly and cooperative work environment. We are committed to upholding our prestigious reputation. Our goal is to increase our business share through teamwork, open communication and the kind of enthusiasm that Bert and Fred brought with them to Pasadena when they started Wood & Jones.

The Wood Press   GC
Robert E Wood
515 East 41st St
Paterson NJ 07509
201-684-4472

Min: 5000 Max: 2000000 Opt: 0
Sizes: 2345X
Items: PSVZ
Services: 4C-CS
Equipment: SO-WO

Statement: They specialize in printing multicolor direct mail pieces, roll labels, and pharmaceutical packaging.

World Color Press Inc   GC
Ann Marie Bushell, Marketing Director
101 Park Ave
New York NY 10178
212-986-2440; Fax: 212-455-9266
Send RFQ's to: Ed Fishel
Parent Company: Printing Holdings, LP

Min: 100000 Max: 4000000 Opt: 30000000
Sizes: 123456
Bindings: HCO-CBI-LBI-OBO-PBI-SSI-
  SDI-SBO-WBO
Items: ACDFIQRS
Services: 4C-CS-FF-LM-OC-RA-TY-TD
Equipment: SO-WO-OP
Annual Sales: 6
Number of Employees: 6500
Turnaround Times: B2S2
Customers: MP40-OT60
Union Shop: Yes/No
Year Established: 1903

Statement: We will provide our customers with a level of service that exceeds their expectations. We will produce work of such high quality that it will challenge the standards of the industry. All World Color Press employees will have the opportunity to grow personally and financially. We will endeavor to produce a consistent return on investment for our shareholders. We will contribute to the welfare of our communities. We will continue to invest in the finest technology and talent in anticipation of the needs of our customers. We will conduct business in a fair and honorable way. When we make mistakes, we will act quickly to right them, learning from the experiences so they do not occur twice. We envision printing as a critical way in which we take part in the exchange of ideas between people. It is never a mere process. Our goal is to be the world's best printer. Our standards reflect that goal.

Sales Offices:

Boston
PO Box 1157
33 Mayflower Ave
Duxbury MA 02331-1157
617-934-9227; Fax: 617-934-9228

Chicago
Oak Brook Executive Plaza
1211 West 22nd St #905
Oak Brook IL 60521
708-990-0711; Fax: 708-990-0723

Dallas
5024 Lakeshore Blvd
PO Box 560186
The Colony TX 75056-0186
214-625-2626; Fax: 214-625-1180

Florida
1200 North Fereal Highway
Boca Raton FL 33432
407-393-1712; Fax: 407-362-4375

Los Angeles
1925 Century Park East #650
Los Angeles CA 90067
310-553-1466; Fax: 310-553-4123

Minnesota
6458 City West Parkway #100
Eden Prairie MN 55344
612-943-3997; Fax: 612-941-0327

New York
101 Park Ave, 19th Fl
New York NY 10178
212-986-2440; Fax: 212-455-9266

PRINTED ITEMS					
A	Annual Reports	G	Galley Copies	N	Workbooks	U	Greeting Cards
B	Books	H	Journals	O	Yearbooks	V	Labels / Stickers
C	Booklets	I	Magazines	P	Brochures	W	Maps
D	Catalogs	J	Mass-Market Books	Q	Calendars	X	Newspapers
E	Cookbooks	K	Newsletters	R	Comic Books	Y	Postcards
F	Directories	L	Software Manuals	S	Direct Mail Letters	Z	Posters
		M	Textbooks	T	Envelopes	1	Stationery

San Jose
2001 Gateway Pl #580
San Jose CA 95110
408-453-3366; Fax: 408-453-3371

Seattle
1818 Westlake Ave N #422
Seattle WA 98109
206-284-9343; Fax: 206-284-9546

Washington DC
1330 Braddock Pl #310
Alexandria VA 22314
703-549-2394; Fax: 703-549-2423

BK

Worzalla Publishing Company
John Prais, Marketing Director
3535 Jefferson St
Stevens Point WI 54481
715-344-9600; Fax: 715-344-2578
Send RFQ's to: John Butkus

Min: 2000 Max: 100000 Opt: 10000-20000
Sizes: 123456
Bindings: HCI-CBI-OBO-PBI-SSI-SDI
Items: BDEFLMN
Services: 4C-FF-OC-RA-TD-WA
Equipment: 7SO(40,63)-WO(23⁹/₁₆x36)
Annual Sales: 5
Number of Employees: 350
Turnaround Times: B3C4S3
Customers: BP95-BU5
Union Shop: N
Year Established: 1892
Terms: Net 30

Statement: Our mission is to consistently manufacture a high quality printed and bound product at reasonable cost, while providing fast and flexible service to our customers so as to attract and retain their business.

GC

Henry Wurst Inc
Daniel S Davis, Vice President
1331 Saline St
North Kansas City MO 64116
816-842-3113

Min: 25000 Max: 2000000 Opt: 500000
Sizes: 25
Bindings: SSI
Items: ADFIPQS
Services: 4C-CS-FF
Equipment: 7WO(23x38)

Annual Sales: 6
Number of Employees: 400
Turnaround Times: C2S2
Customers: BP1-BU89-MP10
Union Shop: N
Year Established: 1952
Terms: Net 20

Y

GC

Y/Z Printing Co
Terry A Zerphey
90 South Wilson Ave
Elizabethtown PA 17022
717-367-4151; Fax: 717-367-0515
Send RFQ's to: Duane Albright

Min: 100 Max: 25000 Opt: 10000
Sizes: 123456
Bindings: HCO-PBO-SBO-WBO-CBI-LBI-SSI-SDI
Items: ABCDEFGHKMNPSTUVYZ123
Services: 4C-DA-AF-RP-TY-TD-WA
Equipment: 2LP(to 14x20)-6SO(to 20x29)
Annual Sales: 2
Number of Employees: 25
Turnaround Times: B3C4S3
Customers: BU60-CU20-NP10-SP10
Union Shop: N
Year Established: 1980
Terms: 1% 10, net 30

Statement: A full service commercial printer offering: desktop publishing, illustrative artists, prepress, offset printing, bindery and direct mailing, all at a fair price with service and quality a priority.

GC

Ye Olde Genealogie Shoppe
Ray Gooldy, Owner
9605 Vandergriff Rd
PO Box 39128
Indianapolis IN 46239
317-862-3330
Send RFQ's to: Ray Gooldy

Min: 25 Max: 1000 Opt: 300
Sizes: 35X
Bindings: HCX-CBX-LBX-SDX-SBX
Items: BCDIKNPSTW
Services: SO
Customers: OT(genealogists)
Year Established: 1980

SERVICES			BINDINGS		SD	Side Stitching
4C	4-Color Printing	RA Rachwal System	HC	Hardcover	SB	Spiral Binding
CS	Color Separations	AF Acid-Free Paper	CB	Comb Binding	WB	Wire-O Binding
DA	Design / Artwork	RP Recycled Paper	LB	Loose-Leaf Binding	I	In-House
FF	Fulfillment/Mailing	TY Typesetting	OB	Otabind	O	Out of House
LM	List Maintenance	TD Typeset w/ Disk	PB	Perfect Binding	X	Unknown
OC	OptiCopy System	WA Warehousing	SS	Saddle Stitching		

Terms: Quotation by job

Statement: They specialize in printing family genealogies.

Yorktown Printing & Pennysaver **GC**
Marketing Director
1520 Front St
Yorktown Heights NY 10598
914-962-3871; Fax: 914-962-5123

Youngstown Lithographing Inc **GC**
Marketing Director
380 Victoria Rd
Youngstown OH 44515
216-793-2471; Fax: 216-793-8471

PRINTED ITEMS			
A Annual Reports	G Galley Copies	N Workbooks	U Greeting Cards
B Books	H Journals	O Yearbooks	V Labels / Stickers
C Booklets	I Magazines	P Brochures	W Maps
D Catalogs	J Mass-Market Books	Q Calendars	X Newspapers
E Cookbooks	K Newsletters	R Comic Books	Y Postcards
F Directories	L Software Manuals	S Direct Mail Letters	Z Posters
	M Textbooks	T Envelopes	1 Stationery

Chapter 6

Low-Cost Short-Run Color Printers

The following printers specialize in printing low-cost short runs of full-color catalog sheets, posters, postcards, and other such items. Many of these printers do gang-runs (that is, they print a number of 4-color jobs at the same time). Nonetheless, most do a good job for a low cost.

CX

3PI Prepress
102 Elden St #15 Rm E
Herndon VA 22070-4809
703-435-6001; Fax: 703-742-3990

Equipment: SO-WO(Gravure, Flexo)
Annual Sales: 5
Number of Employees: 150

Statement: 5 plants

CX

Action Nicholson Color
6519 Eastland Rd
Cleveland OH 44142
216-234-5370; Fax: 216-234-7850

Equipment: SO-WO(Gravure, Flexo)
Annual Sales: 3
Number of Employees: 100

Statement: 2 plants

CX

Accu-Color
11786 Westline Ind Dr
St Louis MO 63146
314-993-5669; Fax: 314-993-6357

Equipment: SO-WO(Flexo)
Annual Sales: 4
Number of Employees: 125

Statement: 2 plants

CX

Allied Color
63 Soundview Rd
Guilford CT 06437
203-453-6501; Fax: 203-453-3872

Equipment: SO-WO
Annual Sales: 3
Number of Employees: 51

SERVICES				BINDINGS			
4C	4-Color Printing	RA	Rachwal System	HC	Hardcover	SD	Side Stitching
CS	Color Separations	AF	Acid-Free Paper	CB	Comb Binding	SB	Spiral Binding
DA	Design / Artwork	RP	Recycled Paper	GB	Glue Binding	WB	Wire-O Binding
FF	Fulfillment/Mailing	TY	Typesetting	LB	Loose-Leaf Binding	I	In-House
LM	List Maintenance	TD	Typeset w/ Disk	PB	Perfect Binding	O	Out of House
OC	OptiCopy System	WA	Warehousing	SS	Saddle Stitching		

CX

American Color
402 N 24th St
Phoenix AZ 85008
602-275-4347; Fax: 602-244-1467

Equipment: SO-WO(Gravure, Flexo)
Annual Sales: 6
Number of Employees: 850

Statement: 15 plants

CX

Apollo Graphics Ltd
Roy Innella
1085 Industries Blvd
Southampton PA 18966
215-953-0500; Fax: 215-953-1144
800-522-9006

Min: 100 Max: 1000000 Opt: 500000
Sizes: 123456
Bindings: HCO-CBI-LBI-OBO-PBO-SSI-
   SDI-SBO-WBO
Items: ABCDEFGHKLNOPQWYZ
Services: 4C-CS-DA-FF-LM-AF-RP-TY-
   TD-WA
Equipment: 2SO(28x41)
Annual Sales: 2
Number of Employees: 35
Turnaround Times: B2
Customers: BU100
Union Shop: N
Year Established: 1985
Terms: Net 30

Statement: To provide extremely high
quality process color printing and color
separations at a reasonable cost to the customer.

CX

Applied Graphics Technologies
50 W 23rd St
New York NY 10010
212-627-4111; Fax: 212-627-8602

Equipment: SO-WO(Gravure, Flexo)
Annual Sales: 6
Number of Employees: 1700

Statement: 14 plants

CX

Balzer/Shopes
175 S Hill Dr
Brisbane CA 94005
415-468-6550; Fax: 415-468-7139

Equipment: SO-WO
Annual Sales: 3
Number of Employees: 106

CX

Black Dot Graphics
6115 Official Rd
Crystal Lake IL 60014
815-459-8520; Fax: 815-459-7259

Equipment: SO-WO(Gravure)

Statement: 10 plants

CX

Blanks Color Imaging
2343 N Beckley
Dallas TX 75208
214-741-3905; Fax: 214-741-6105

Equipment: SO-WO(Gravure, Flexo)
Annual Sales: 4
Number of Employees: 225

CX

Capitol Engraving
808 Lea Ave
Nashville TN 37203
615-244-6603; Fax: 615-255-9477

Equipment: SO-WO(Gravure)
Annual Sales: 3
Number of Employees: 90

Statement: 2 plants

CX

Carey Color
775 S Progress Dr
Medina OH 44256
216-725-5637; Fax: 216-725-2546

Equipment: SO-WO(Gravure, Flexo)
Annual Sales: 4
Number of Employees: 120

Statement: 4 plants

PRINTED ITEMS					
A	Annual Reports	G	Galley Copies	N	Workbooks
B	Books	H	Journals	O	Yearbooks
C	Booklets	I	Magazines	P	Brochures
D	Catalogs	J	Mass-Market Books	Q	Calendars
E	Cookbooks	K	Newsletters	R	Comic Books
F	Directories	L	Software Manuals	S	Direct Mail Letters
		M	Textbooks	T	Envelopes

U	Greeting Cards
V	Labels / Stickers
W	Maps
X	Newspapers
Y	Postcards
Z	Posters
1	Stationery

CX

Catalog King
1 Entin Rd
Clifton NJ 07014
201-472-1221; Fax: 201-472-5270

Min: 5000 Max: 1000000 Opt: 50000
Sizes: 5
Bindings: SSI
Items: DPS
Services: 4C-CS-DA-TY

CX

CGI
5 Old Concord Rd
Burlington MA 01803
617-229-2345; Fax: 617-272-2469

Equipment: SO-WO(Gravure, Flexo)
Annual Sales: 3
Number of Employees: 100

CX

Characters
2501 Central Pkwy #A5
Houston TX 77092
713-683-6666; Fax: 713-681-6501

Equipment: SO-WO(Gravure, Flexo)
Annual Sales: 3
Number of Employees: 110

Statement: 2 plants

CX

Chicago Color Express
1111 Pasquinelli Dr
Westmont IL 60559
708-323-2232; Fax: 708-323-2833
800-872-0720

Min: 250 Max: 10000 Opt: 0
Sizes: 25
Bindings: DPZ
Items: 4C-CS
Equipment: SO

CX

Chicago Color Express
Tom Palmer, Marketing Director
512 N Franklin St
Chicago IL 60610
312-222-0662; Fax: 312-222-0824

Send RFQ's to: Tom Palmer
Min: 250 Max: 20000 Opt: 50000
Sizes: 123456
Bindings: HCO-CBO-LBO-OBO-PBO-
    SSO-SDO-SBO-WBO
Items: CDKPSUVYZ12
Services: 4C-RP
Equipment: SO(14X20)
Annual Sales: 2
Number of Employees: 6
Customers: BU100
Union Shop: N
Terms: Net 30 or COD

Statement: Four color printing can be a complicated, expensive and time consuming process, but not at Chicago Color Express!

CX

Chroma-Graphics
8300 NE Underground Dr
Kansas City MO 64161
816-454-7891; Fax: 816-453-5083

Equipment: SO-WO(Gravure, Flexo)
Annual Sales: 3
Number of Employees: 51

CX

Classic Litho Arts
6140 Roosevelt Rd
Oak Park IL 60304
708-386-7779; Fax: 708-386-3808

Equipment: SO-WO(Gravure, Flexo)
Annual Sales: 3
Number of Employees: 60

Statement: 2 plants

CX

Color Express
5885 Rickenbacker Rd
Commerce CA 90040
213-724-1588; Fax: 213-724-1689
800-272-0507
Send RFQ's to: Roger Vingino

Min: 250 Max: 10000 Opt: 0
Sizes: 25
Bindings:
Items: DPZ
Services: 4C-CS

SERVICES			BINDINGS			
4C	4-Color Printing	RA Rachwal System	HC	Hardcover	SD	Side Stitching
CS	Color Separations	AF Acid-Free Paper	CB	Comb Binding	SB	Spiral Binding
DA	Design / Artwork	RP Recycled Paper	GB	Glue Binding	WB	Wire-O Binding
FF	Fulfillment/Mailing	TY Typesetting	LB	Loose-Leaf Binding	I	In-House
LM	List Maintenance	TD Typeset w/ Disk	PB	Perfect Binding	O	Out of House
OC	OptiCopy System	WA Warehousing	SS	Saddle Stitching		

Equipment: SO

Statement: We do short run four-color work with a 2-3 turnaround, very competitively priced.

Color Express **CX**
100 N Sixth St Ste #119
Minneapolis MN 55403
612-333-3932; Fax: 612-333-3432
800-833-0323

Min: 250 Max: 10000 Opt: 0
Sizes: 25
Items: DPZ
Services: 4C-CS
Equipment: SO

Color Express **CX**
1 Baltimore Place #G10
Atlanta GA 30308
404-872-6618; Fax: 404-872-6637
800-872-0720

Min: 250 Max: 10000 Opt: 0
Sizes: 25
Items: DPZ
Services: 4C-CS
Equipment: SO

Color Express **CX**
2 Jenner St #120
Irvine CA 92718
714-727-9257; Fax: 714-727-3499

Min: 250 Max: 10000 Opt: 0
Sizes: 25
Items: PDZ
Services: 4C-CS
Equipment: SO

Color Service **CX**
595 Monterey Pass Rd
Monterey Park CA 91754
818-282-4137; Fax: 818-282-8125

Equipment: SO-WO
Annual Sales: 3
Number of Employees: 77

Colorbrite **CX**
1011 Plymouth Ave
Minneapolis MN 55411
612-522-6711; Fax: 612-522-9263

Equipment: SO-WO(Gravure, Flexo)
Annual Sales: 4
Number of Employees: 100

Colorhouse **CX**
13010 Cty Rd #6
Minneapolis MN 55441
612-553-0100; Fax: 612-550-3600

Equipment: SO-WO(Gravure, Flexo)
Annual Sales: 4
Number of Employees: 100

Colorlith Corporation **CX**
R Celleme, Marketing Director
777 Hartford Ave
Johnston RI 02919
401-521-6000; Fax: 401-751-9436
800-556-7171
Send RFQ's to: J Moura

Min: 500 Max: 0 Opt: 1000000
Sizes: 123456
Bindings: SSX
Items: ABCDEPQSUVW1234
Services: 4C-CS-AF-RP-TY-TD-WA
Equipment: 5SO(19x25,25x38)-WO
　(17.75x26)
Annual Sales: 2
Number of Employees: 40
Turnaround Times: S3
Customers: BU95-OT5
Union Shop: N
Year Established: 1958
Terms: 1% 10, net 30

Statement: We are full service, long and short-run. Photography to bindery.

Colotone Group **CX**
260 Branford Rd
Branford CT 06471
203-483-0608; Fax: 203-481-3063

Equipment: SO-WO(Gravure)

PRINTED ITEMS						
A	Annual Reports	G	Galley Copies	N	Workbooks	U Greeting Cards
B	Books	H	Journals	O	Yearbooks	V Labels / Stickers
C	Booklets	I	Magazines	P	Brochures	W Maps
D	Catalogs	J	Mass-Market Books	Q	Calendars	X Newspapers
E	Cookbooks	K	Newsletters	R	Comic Books	Y Postcards
F	Directories	L	Software Manuals	S	Direct Mail Letters	Z Posters
		M	Textbooks	T	Envelopes	1 Stationery

Annual Sales: 6
Number of Employees: 215

Statement: 4 plants

Communicolor   **CX**
Karen Reinhard, Marketing Director
PO Box 400
Newark OH 43058-0400
614-928-6110; Fax: 614-928-1061
800-848-7040
Send RFQ's to: Bill McWilliams

Min: 25000 Max: 0 Opt: 500000-5000000
Services: 4C-CS-DA-TY(Dupont Cros-
   field)
Equipment: 22WO(17,21,22,28
Annual Sales: 6
Number of Employees: 610
Customers: MP45OT55
Union Shop: N

Statement: Masterful use of personaliza-
tion technology (7 methods to choose
from) coupled with innovative formats and
high-quality printing (up to 12 colors) on
coated or uncoated stocks, sets Communicolor apart as a professional in
production of responsive direct mail packages. Achieve the competitive difference,
call Communicolor for your direct mail
needs today.

Contemporary Color Graph   **CX**
200 Executive Dr #F
Brentwood NY 11717
516-242-7915

Dayal Graphics   **CX**
111 Stephen St
Lemont IL 60439
708-257-2511; Fax: 708-257-2946

Items: DP
Services: 4C-CS

Designer Color Systems   **CX**
11971 Borman Dr
St Louis MO 63146
314-432-8800

Equipment: SO-WO(Gravure)
Annual Sales: 3
Number of Employees: 100

Direct Press/Modern Litho   **CX**
Bob Dodson, Marketing Director
386 Oakwood Road
PO Box 8104
Huntington Station NY 11746
516-271-7000; Fax: 516-271-7008
Send RFQ's to: Bob Dodson

Min: 2500 Max: 0 Opt: 0
Sizes: 123456X
Bindings: HCO-CBO-OBO-LBO-PBO-
   SSI-SDO-SBO-WBO
Items: DPYZ
Services: 4C-CS-RP-TY-TD
Equipment: 3SO(28x40)-WO(17.75x23)
Annual Sales: 4
Number of Employees: 250
Turnaround Times: B6S6
Customers: BU98-CU1-NP1
Union Shop: Y
Year Established: 1917
Terms: 1/2 deposit, balance COD

Statement: A full service printer of color
catalog, brochures, direct mail pieces,
postcards. Offer a complete package from
photography to printed sheets. Sales offices and photo studios throughout the
country.

Eastern Rainbow   **CX**
Industrial Park
Derry NH 03038
603-432-2547; Fax: 603-432-1993

Equipment: SO-WO(Flexo)
Annual Sales: 3
Number of Employees: 100

Enteron Group   **CX**
815 S Jefferson St
Chicago IL 60607
800-497-2695; Fax: 312-922-2086

Equipment: SO-WO(Gravure)
Annual Sales: 6
Number of Employees: 350

Statement: 4 plants

SERVICES				BINDINGS			
4C	4-Color Printing	RA	Rachwal System	HC	Hardcover	SD	Side Stitching
CS	Color Separations	AF	Acid-Free Paper	CB	Comb Binding	SB	Spiral Binding
DA	Design / Artwork	RP	Recycled Paper	GB	Glue Binding	WB	Wire-O Binding
FF	Fulfillment/Mailing	TY	Typesetting	LB	Loose-Leaf Binding	I	In-House
LM	List Maintenance	TD	Typeset w/ Disk	PB	Perfect Binding	O	Out of House
OC	OptiCopy System	WA	Warehousing	SS	Saddle Stitching		

Flower City Printing Inc **CX**

Mark Ashworth, Marketing Director
4800 Dewey Ave
Rochester NY 14612
716-663-9000; Fax: 716-663-4908
Send RFQ's to: Mark Ashworth

Min: 1,000 Max: 1000000 Opt: 100000
Sizes: 123456
Bindings: SSI-SDI-HCO-CBI-LBI-OBO-
    PBO-SBO-WB0
Items: ABCDEFKLMNOPQUYZS24
Services: 4C-CS-DA-FF-RP-TD-WA-TY-
    OC
Equipment: 5SO(28x40,39x55,44x64)
Annual Sales: 4
Number of Employees: 125
Turnaround Times: B3C4S3
Customers: BP5BU75SP10
Union Shop: N
Year Established: 1971
Terms: Net 30

Statement: Flower City is in the communications business. We produce the
highest possible quality color printing and
packaging materials, in support of our
clients' business and communication needs.
We recognize the client as the driving
force. Our vision of Flower City Printing
includes: hard work; innovative ideas;
partnering with clients and suppliers; exceeding expectations; proactive and
responsive communication; a safe, fun, and
profitable workplace to insure that we will
continue to make a difference in our
clients' and employee's lives.

Flying Color Graphics **CX**

1001 W North St
Pontiac IL 61764
815-842-2811; Fax: 815-844-1044

Equipment: SO-WO(Gravure, Flexo)
Annual Sales: 4
Number of Employees: 130

Statement: 4 plants

Gamma One **CX**

12 Corporate Dr
North Haven CT 06473
203-234-0440; Fax: 203-234-2015

Equipment: SO-WO(Gravure)
Annual Sales: 4
Number of Employees: 135

Graphic Art Service **CX**

550 Commerce Park Dr
Marietta GA 30060
404-422-5500; Fax: 404-422-5428

Equipment: SO-WO(Gravure, Flexo)
Annual Sales: 4
Number of Employees: 190

Statement: 4 plants

Graphic Production Center **CX**

5865 Jimmy Carter Blvd
Norcross GA 30071
404-903-0472; Fax: 404-903-0474

Min: 0 Max: 0 Opt: 5000
Items: DP
Services: 4C-CS
Equipment: SO

Graphic Technology **CX**

1724 Whitehead Rd
Baltimore MD 21207
410-298-6100; Fax: 410-298-2545

Equipment: SO-WO(Flexo)
Annual Sales: 5
Number of Employees: 280

Statement: 4 plants

Graphics Express **CX**

99 Bedford St
Boston MA 02111
800-653-3323; Fax: 617-426-4114

Equipment: SO-WO(Flexo)
Annual Sales: 3
Number of Employees: 105

Statement: 3 plants

PRINTED ITEMS							
A	Annual Reports	G	Galley Copies	N	Workbooks	U	Greeting Cards
B	Books	H	Journals	O	Yearbooks	V	Labels / Stickers
C	Booklets	I	Magazines	P	Brochures	W	Maps
D	Catalogs	J	Mass-Market Books	Q	Calendars	X	Newspapers
E	Cookbooks	K	Newsletters	R	Comic Books	Y	Postcards
F	Directories	L	Software Manuals	S	Direct Mail Letters	Z	Posters
		M	Textbooks	T	Envelopes	1	Stationery

CX

Hi-Tech Color House
Albert Stevens, President
5901 N Cicero Ave
Chicago IL 60646
312-588-8200
800-621-4004
Send RFQ's to: Albert Stevens

Min: 1000 Max: 2000000 Opt: 7500
Sizes: 25
Bindings: SSI
Items: DPZ
Services: 4C-CS
Equipment: 2SO

CX

Horan Engraving
44 W 28th St
New York NY 10001
212-689-8585; Fax: 212-889-2969

Equipment: WO
Annual Sales: 4
Number of Employees: 128

CX

Imaging Systems
600 Haddon Ave
Collingswood NJ 08108
609-854-5252; Fax: 609-854-4840

Equipment: SO-WO(Gravure, Flexo)
Annual Sales: 3
Number of Employees: 65

CX

Instant Web
7951 Powers Blvd
Chanhassen MN 55317
612-474-0961; Fax: 612-474-6467

Min: 50000 Max: 2000000 Opt: 0
Sizes: 5X
Items: PST
Services: 4C-FF
Equipment: 20WO
Annual Sales: 5
Number of Employees: 200

CX

Intaglio Vivi-Color
4910 River Rd
Schiller Park IL 60176
708-671-3131; Fax: 708-671-5579

Equipment: SO-WO(Gravure, Flexo)
Annual Sales: 5
Number of Employees: 450

Statement: 13 Plants

CX

Kolor View Press
Shelly Pichler, Marketing Director
112 W Olive
Aurora MO 65605
417-678-2135; Fax: 417-678-3626
800-225-6567
Send RFQ's to: Customer Service Department
Parent Company: MWM Dexter Inc

Min: 500 Max: 1000000 Opt: 10000
Sizes: 123456
Bindings: SSX-SBX-WBX-SDX
Items: ABCDEFKPQUWZY1
Services: 4C-CS-TY-DA-FF-LM-OC-RP-
   TD-WA-AF
Equipment: LP(11x17)-3SO(40)
Annual Sales: 4
Number of Employees: 150
Turnaround Times: S4
Customers: BU90-NP10
Union Shop: N
Year Established: 1922
Terms: 50% down, balance COD

Statement: Provide quality full-color printing, competitive prices and guaranteed turnaround. 5,10,15 day Kolor Kwik Service or your order is FREE! Economical five week service is also available.

CX

Kreber Graphics
670 Harmon Ave
Columbus OH 43223
614-228-3501; Fax: 614-228-3523

Equipment: SO-WO(Gravure, Flexo)
Annual Sales: 4
Number of Employees: 150

Statement: 2 plants

SERVICES				BINDINGS			
4C	4-Color Printing	RA	Rachwal System	HC	Hardcover	SD	Side Stitching
CS	Color Separations	AF	Acid-Free Paper	CB	Comb Binding	SB	Spiral Binding
DA	Design / Artwork	RP	Recycled Paper	GB	Glue Binding	WB	Wire-O Binding
FF	Fulfillment/Mailing	TY	Typesetting	LB	Loose-Leaf Binding	I	In-House
LM	List Maintenance	TD	Typeset w/ Disk	PB	Perfect Binding	O	Out of House
OC	OptiCopy System	WA	Warehousing	SS	Saddle Stitching		

Kwik International Color   **CX**
229 W 28th St
New York NY 10001
212-643-0200; Fax: 212-643-0201

Equipment: SO-WO(Gravure, Flexo)
Annual Sales: 6
Number of Employees: 120

Lanman Companies   **CX**
120 Q St NE
Washington DC 20002
202-269-5400; Fax: 202-269-0206

Equipment: SO-WO(Gravure)
Annual Sales: 6

Statement: 3 plants

Magna Graphic   **CX**
2528 Palumbo Dr
Lexington KY 40509
606-268-1211; Fax: 606-268-9546

Equipment: SO-WO(Gravure)
Annual Sales: 4
Number of Employees: 140

Statement: 2 plants

McGrew Color Graphics   **CX**
Kathy Neuman-St Clair, Marketing
    Director
1615 Grand Ave
Kansas City MO 64108
800-877-7700; Fax: 816-221-3154
Send RFQ's to: Gordon Crape
Parent Company: Pet Inc

Min: 1000 Max: 0 Opt: 0
Sizes: 123456
Bindings: SSX-SDX-SBX-WBX
Items: ADFKPQUYZ
Services: 4C-CS-DA-OC-TY-TD
Annual Sales: 4
Number of Employees: 160
Turnaround Times: S3
Customers: BU100
Union Shop: Y
Year Established: 1956
Terms: Net 30

Statement: McGrew is committed to excellence in product quality and customer service, while providing a working environment that encourages and rewards diligence, innovation and professional growth.

MCP   **CX**
2320 N 11th St
Milwaukee WI 53206
414-374-5660; Fax: 414-374-4037

Equipment: SO-WO(Gravure, Flexo)
Annual Sales: 4
Number of Employees: 147

Statement: 2 plants

Mitchell Graphics   **CX**
2230 E Mitchell
Petosky MI 49770
616-347-5650; Fax: 616-347-9255
800-841-6793
Send RFQ's to: Customer Service

Min: 1500 Max: 50000 Opt: 5000
Sizes: 5
Bindings: SSI
Items: DPZ
Services: 4C-TY
Equipment: 3SO
Annual Sales: 2
Number of Employees: 50
Union Shop: N
Terms: Variety of options

Statement: We specialize in producing quality full-color direct mail products at an affordable price.

Multi-Ad Services   **CX**
1720 Detweiller Dr
Peoria IL 61615
309-692-1530; Fax: 309-692-8518

Equipment: SO-WO
Annual Sales: 5
Number of Employees: 450

PRINTED ITEMS							
A	Annual Reports	G	Galley Copies	N	Workbooks	U	Greeting Cards
B	Books	H	Journals	O	Yearbooks	V	Labels / Stickers
C	Booklets	I	Magazines	P	Brochures	W	Maps
D	Catalogs	J	Mass-Market Books	Q	Calendars	X	Newspapers
E	Cookbooks	K	Newsletters	R	Comic Books	Y	Postcards
F	Directories	L	Software Manuals	S	Direct Mail Letters	Z	Posters
		M	Textbooks	T	Envelopes	1	Stationery

CX

MultiPrint Co Inc
Monica Pratl, Marketing Director
5555 W Howard St
Skokie IL 60077
800-858-9999; Fax: 708-677-7544
Send RFQ's to: Sales Department
Parent Company: Great Lakes Graphics
    Inc

Min: 500 Max: 50000 Opt: 250000
Sizes: 123456
Bindings: PBO-SBO-WBO-SDO-SSI-
    HCO-CBO-LBO-OBO
Items: ABCDEFKPQSTUVYZ124
Services: 4C-CS-DA-TY-TD-WA-RA
Equipment: 4SO(29x40)-OP
Annual Sales: 4
Number of Employees: 150
Turnaround Times: B4C5S2
Union Shop: N
Year Established: 1932
Terms: 50% deposit, 1%-10, net 30

Statement: We also offer a "pricelist" for
fullcolor printing of brochures, catalogs,
sales sheets, posters, and postcards as well
as a "free idea sampler" kit.

CX

NEC
1504 Elm Hill Pike
Nashville TN 37210
615-367-9110; Fax: 615-360-7952

Equipment: SO-WO(Gravure, Flexo)
Annual Sales: 4
Number of Employees: 175

CX

Ohio Valley Litho Color
Econocolor Catalog
7405 Industrial Road
Florence KY 41042
606-525-7405; Fax: 606-525-7654
800-877-7405

Min: 5000 Max: 250000 Opt: 20000
Items: DPZ
Services: 4C-CS

CX

One Color Communications
PO Box 8277
Emeryville CA 94662
510-652-9005; Fax: 510-547-4129

Equipment: WO(Gravure, Flexo)
Annual Sales: 4
Number of Employees: 90

CX

Peerless Engraving
823 Main St #T
Little Rock AR 72201
800-880-7671; Fax: 501-375-5633

Equipment: SO-WO
Annual Sales: 2
Number of Employees: 110

CX

Penn Colour Graphics
99 Buck Rd
Huntington Valley PA 19006
215-364-4000; Fax: 215-364-2596

Items: D
Services: 4C-CS-DA
Equipment: SO

CX

Phototype Color Graphics
7890 Airport Hwy
Merchantville NJ 08109
609-663-4100; Fax: 609-663-6631

Equipment: SO-WO(Gravure, Flexo)
Annual Sales: 4
Number of Employees: 112

CX

Pinecliffe Printers
1815 N Harrison
Shawnee OK 74801
405-275-7351; Fax: 405-275-4977
Sizes: 5
Services: 4C-CS
Equipment: SO

SERVICES			BINDINGS			
4C	4-Color Printing	RA Rachwal System	HC	Hardcover	SD	Side Stitching
CS	Color Separations	AF Acid-Free Paper	CB	Comb Binding	SB	Spiral Binding
DA	Design / Artwork	RP Recycled Paper	GB	Glue Binding	WB	Wire-O Binding
FF	Fulfillment/Mailing	TY Typesetting	LB	Loose-Leaf Binding	I	In-House
LM	List Maintenance	TD Typeset w/ Disk	PB	Perfect Binding	O	Out of House
OC	OptiCopy System	WA Warehousing	SS	Saddle Stitching		

Potomac Graphic Industries CX
508 W 26th St
New York NY 10001
212-924-4880; Fax: 212-627-0866

Equipment: SO-WO(Gravure)
Annual Sales: 5
Number of Employees: 280

Statement: 2 plants

Prepsat CX
PO Box 289
Franklin KY 42135
502-586-5773; Fax: 502-586-4558

Equipment: SO-WO(Gravure, Flexo)
Annual Sales: 4
Number of Employees: 210

Statement: 4 plants

The Press CX
18780 West 78th St
Chanhassen MN 55317
612-937-9764; Fax: 612-937-5633
800-336-2680

Min: 2000 Max: 2000000 Opt: 0
Sizes: 25
Bindings: SSI
Items: ACDNPSUWYZ
Services: 4C-CS-FF
Equipment: 3SO(19x26)-7WO(20x23,38)

Quality House of Graphic CX
4747 Van Dam St
Long Island City NY 11101
718-784-7400; Fax: 718-937-5179

Equipment: SO-WO(Gravure)
Annual Sales: 4
Number of Employees: 200

Rapidocolor Corporation CX
101 Brandywine Parkway
PO Box 2540
West Chester PA 19380
800-872-7436; Fax: 215-344-0506

Send RFQ's to: Leslie Bullitt
Min: 50 Max: 25000 Opt: 0
Items: PVYZ
Services: 4C
Equipment: SO

Service Web Offset Corporation CX
Norine Satterlee, Marketing Director
2500 S Dearborn St
Chicago IL 60616
312-567-7000; Fax: 312-567-9121
800-621-1567
Send RFQ's to: Norine Satterlee

Min: 25000 Max: 500000 Opt: 250000
Sizes: 2356
Bindings: SSX
Items: CDKP
Services: 4C-RP
Equipment: 2WO
Annual Sales: 4
Number of Employees: 140
Turnaround Times: S2
Customers: BU76-OT21
Union Shop: Y
Year Established: 1943
Terms: Net 30

Statement: Service Web has specialized in full-color printing for over 50 years and has a friendly, professional staff you can count on. Our offices, computerized high-speed presses and complete bindery facilities are at one location for maximum efficiency.

Stevenson Photo Color CX
5325 Ridge Ave
Cincinnati OH 45213
513-351-5100; Fax: 513-351-8100

Equipment: SO-WO(Gravure, Flexo)
Annual Sales: 4
Number of Employees: 125

Trade Litho Inc
Faith Levine, Marketing Director
5301 NW 37th Ave
Miami FL 33142
305-633-9779; Fax: 305-633-2848
800-367-5871

PRINTED ITEMS					
A	Annual Reports	G	Galley Copies	N	Workbooks	U	Greeting Cards
B	Books	H	Journals	O	Yearbooks	V	Labels / Stickers
C	Booklets	I	Magazines	P	Brochures	W	Maps
D	Catalogs	J	Mass-Market Books	Q	Calendars	X	Newspapers
E	Cookbooks	K	Newsletters	R	Comic Books	Y	Postcards
F	Directories	L	Software Manuals	S	Direct Mail Letters	Z	Posters
		M	Textbooks	T	Envelopes	1	Stationery

Send RFQ's to: Eduardo Tefel

Min: 2500 Max: 0 Opt: 0
Sizes: 123456X
Bindings: HCO-CBO-LBO-OBO-PBI-SSI-
   SDI-SBO-WBO
Items: ACDEFHIKLNOPQSUWZ
Services: 4C-CS-DA-AF-RP-TY-TD-WA-
   FF
Equipment: 4SO(25x38,28x40)
Annual Sales: 3
Number of Employees: 80
Turnaround Times: B2
Customers: BU60-CU5-MP30-OT5
Union Shop: N
Year Established: 1973
Terms: Net 30

Statement: Provide quality printing,
delivered in a timely manner, competitively
priced.

CX
TSI Graphics
1035 Hanley Ind Ct
St Louis MO 63144
314-968-6800; Fax: 314-968-2739

Equipment: SO-WO(Gravure, Flexo)
Annual Sales: 4
Number of Employees: 337

Statement: 4 plants

CX
Tukaiz Innovative
2917 Latoria Ln
Franklin Park IL 60131
708-455-1588; Fax: 708-455-0577

Equipment: SO-WO(Gravure, Flexo)
Annual Sales: 4
Number of Employees: 150

CX
Ultra-Color Corporation
1814 Washington Ave
Saint Louis MO 63103
314-241-0300; Fax: 314-241-6032

Min: 1000 Max: 50000 Opt: 10000
Sizes: 25
Bindings: SSI
Items: DP
Services: 4C-CS-DA-TY

CX
US Press
Michael R Jetter, VP Marketing
1628A James P Rodgers Dr
PO Box 640
Valdosta GA 31603-0640
912-244-5634; Fax: 912-247-4405
800-227-7377
Send RFQ's to: Michael R Jetter

Min: 500 Max: 10000 Opt: 5000-25000
Sizes: 25
Bindings: SSI
Items: ACDKPYZ
Services: 4C-CS-RP-TY
Equipment: 5SO(various)
Annual Sales: 3
Number of Employees: 40
Turnaround Times: S3
Customers: BU85-OT15
Union Shop: N
Year Established: 1986
Terms: 50% with order, balance COD, free
freight with 100% pre-payment

Statement: US Press Inc specializes in
short-run, low cost, high quality process
color printing nationwide. From color
laminated postcards to catalog sell sheets,
folded brochures to posters, and catalogs
to packet folders, we can help you to meet
your color printing needs start to finish.

CX
Wace USA
2 N Riverside Plz
Chicago IL 60606
312-876-0533; Fax: 312-876-0120

Equipment: SO,WO(Gravure, Flexo)
Annual Sales: 6
Number of Employees: 1800

Statement: 22 Plants

CX
Warson Graphics Inc
1265 N Warson Rd #A
St Louis MO 63132

SERVICES				BINDINGS			
4C	4-Color Printing	RA	Rachwal System	HC	Hardcover	SD	Side Stitching
CS	Color Separations	AF	Acid-Free Paper	CB	Comb Binding	SB	Spiral Binding
DA	Design / Artwork	RP	Recycled Paper	GB	Glue Binding	WB	Wire-O Binding
FF	Fulfillment/Mailing	TY	Typesetting	LB	Loose-Leaf Binding	I	In-House
LM	List Maintenance	TD	Typeset w/ Disk	PB	Perfect Binding	O	Out of House
OC	OptiCopy System	WA	Warehousing	SS	Saddle Stitching		

Book Publishing Resource Guide

by Marie Kiefer

Over 10,000 KEY CONTACTS for marketing your books, including:

- ▼ 453 Book Wholesalers
- ▼ 174 Distributors
- ▼ 154 Publishers who distribute books
- ▼ 194 Sales representatives
- ▼ 176 Book clubs
- ▼ 381 Chainstore buyers
- ▼ 880 Mail order catalogs that carry books
- ▼ 77 Book remainder dealers
- ▼ 230 Mail list brokers and managers
- ▼ 154 Card decks
- ▼ 24 Foreign sales agents and distributors
- ▼ 47 Book fairs and conventions
- ▼ 25 Cooperative marketing programs
- ▼ 25 Fulfillment services
- ▼ 102 Book marketing and publicity services
- ▼ 144 Publicity/marketing directories
- ▼ 77 Bibliographies and publisher listings
- ▼ 1146 Newspaper editors and book reviewers
- ▼ 2921 Magazine editors and book reviewers
- ▼ 709 Syndicated columnists

Softcover: $25.00
Shipping: $4.50

A ll the information you need to target your promotions to these key book marketing contacts: Company name, person to contact, address, phone number and their subject interests.

It also indicates, for instance, which wholesalers carry audio or video tapes as well as books, and which newspapers buy first or second serial rights.

Pinpoint the BEST MARKETS and MEDIA for your books — quickly and inexpensively!
No other directory offers so much for so little!

ADLIB
PUBLICATIONS

Mail check or money order:
51 West Adams, Fairfield, IA 52556
(515) 472-6617   Fax: (515) 472-3186

Use your VISA, MasterCard or American Express Card:
1-800-669-0773

- **Customs** – Your shipment will have to clear customs. In most cases, you will find it easier and more cost-effective to work with a shipping broker rather than try to clear the shipment yourself. Nonetheless, this is just another stumbling block you don't have to deal with when working with printers here.

The primary benefit of printing overseas is that the job can cost half as much as the same job printed in the United States. Yet, when making a decision to print overseas, you should not overlook the hidden costs involved. Generally speaking, printing overseas pays off only if you have the experience to work closely with the printers or their local representatives *and* intend to develop a long-term relationship with them. The savings for a one-shot deal rarely justifies the added expenditure of your time and efforts in organizing and monitoring the job.

Guidelines for Working with Foreign Printers

For those of you who do plan to use overseas printers, the following guidelines give you some pointers on how best to work with them.

- Clearly define all of your specifications in writing. Make sure the printer understands exactly what you expect. When sending requests for quotation, be sure to include samples of the paper, binding, and print quality you expect.
- Examine bids carefully for any differences from your specifications.
- Check to see that the standard operating systems of the printing trade in that country match your own assumptions. Write to the local printers association for their standard terms and conditions.
- Get references and check them out thoroughly.
- Have them send you samples of their work. Indeed, ask them to make a dummy of your proposed job so you can see exactly how they see your project. Inspect those samples very carefully.
- Monitor all stages of the manufacturing process. Be sure to get updated regularly on the progress of your job.
- Note that each country has its busy season when printers are less likely to offer you the best prices and delivery. The busy season for Japanese book printers is August through December (because their school year starts in April), while the busy season for most other Asian printers is March through September (when they are printing books for the school and holiday seasons).

- You'll probably want to install a fax machine or telex machine to facilitate quick and inexpensive communication between you and the printer.

- Set a production schedule that allows plenty of leeway for delays. Allow at least four extra weeks in your marketing plan.

- Make sure the shipment of books is insured.

- Arrange to have a shipping broker clear the books or other printed items through customs. Note that while books and magazines are duty-free, postcards and greeting cards are charged a duty. Check with customs to see if any duty applies to whatever you are having printed.

- When possible, work with their sales representatives in this country. Let them deal with the language barrier and time differences.

- Whenever possible, work toward establishing a lasting relationship with the overseas printer. As one Japanese printer noted in *Publishers Weekly*, "If U.S. publishers were willing to think in the long term, and to guarantee work for the slow season as well as the peak time, we would certainly look forward to having that sort of customer." Asians especially appreciate such customer loyalty and reciprocate with better service.

Overseas Printers: The Listings

On the following eight pages, we list more than 180 foreign printers of books, catalogs, magazines, and other bound publications. Most of the information in this section is taken from material we gathered from foreign trade offices and printers who exhibited at the American Booksellers Convention during the past few years. Because this information is compiled from less up-to-date sources (rather than surveys), some of the addresses may no longer be correct.

Since we did not survey any of these printers, we list only their names, addresses, and phone numbers — and any special details that made the printers stand out in our minds. Note, however, that the PrintBase data files do contain more details about each of these overseas printers. For more information about the PrintBase data files, write to **Ad-Lib Publications, 51 1/2 West Adams, P O Box 1102, Fairfield IA 52556**, or call toll-free **800-669-0773**.

Overseas Printers: A Listing

Australia

Mandarin Offset
Joy Willis
261 Pacific Hwy
N Sydney 2060 Australia
2-959-4347; Fax: 2-925-0911

Tien Wah Press Australia Ltd
7/130 Pacific Hwy
St Leonards NSW 2065
Sydney ND 00000 Australia
02-436-0255; Fax: 02-438-5381

China

Wenwu Jardine Printing Company
No 21 Bei St
Xihuangchenggen, Hicheng District
Beijing China
66-2595

Columbia

Carvajal SA
Calle 229 North No 6A-40
PO Box 46
Cali Columbia
675-011; Fax: 661-5881

Statement: Specialty: pop-up books. They
have done pop-ups for Hallmark Cards
and Intervisual Communications.

Ediciones Laser Saenz Hurtado
Raul E Urquijo Puerto, Manager
Calle 11 No. 22-01
PO Box 34905
Bogota Columbia
247-2640; Fax: 201-6780

Ediciones Lerner
Jack Grimberg, Manager
Calle 8-B, No 68A-41
PO Box 8304
Bogota Columbia
571-262-8200; Fax: 571-262-4459

Specialty: pop-up books and other hand-intensive books.

Editorial Presencia Ltda
Carlos Ossa, Manager
Calle 23 No 24-20
Bogota Columbia
404-977-0067; Fax: 404-977-6989

Educar
Calle 44 No 15-28
PO Box 45616
Bogota Columbia
287-1055; Fax: 288-5788

Impresora Feriva Ltda
Calle 18 No 3-33
PO Box 4342
Cali Columbia
283-1595; Fax: 283-5788

Intergraficas Ltda
Carrera 34A No 10-47
Bogota Columbia
247-0445

Op Graficas Ltda
Calle 11 No 22-51
Bogota Columbia
201-211

Rei Nades Ltda
Avenida 40A No 13-09
PO Box 21834
Bogota Columbia
287-5334

Roto/Offset - Cano Isaza Y Cia
Avenida Carrera 68, #23-71
PO Box 3341
Bogota Columbia
571-290-5555; Fax: 572-262-2323

Saenz Y Cia Ltda
Calle 55 No 13-19
PO Box 024000
Bogota Columbia
217-9788; Fax: 249-8911

Susaeta Ediciones
Calle 50 No 46A-6
PO Box 1742
Medellin Columbia
288-4422

Tecimpre SA
Calle 15A No. 69-84
PO Box 82514
Bogota Columbia
292-4900; Fax: 292-5558

Tempora Impresores SA
Avenida El Dorado No 79-34
PO Box 98916
Bogota Columbia
295-0511; Fax: 295-4713

Costa Rica

Trejos International
Alvaro Trejos, President
PO Box 10096
San Jose 1000 Costa Rica
506-24-2411

France

Industrial Coatings Group Inc
Walter Strohmberg, European S&S
91 Rue du Faubourg St Honore
75008 Paris France

Hong Kong

Angel Wise Co
Unit 7, 9th Floor, Fonda Ind Bldg
37-39 Au Pui Wan St
Fotan Shatin NT Hong Kong
6935375; Fax: 852-6935366

Asher & Co (Hong Kong) Ltd
3rd Fl S China Morning Post
28 Tong Chong St
Quarry Bay Hong Kong
5-618145; Fax: 852-5-659420

Asia Magazines Ltd
Morning Post Bldg
Tong Chong St
Hong Kong
5-652222; Fax: 852-5-695441

Asian Sources Publications Ltd
9th Fl First Comm Bldg
33-35 Leighton Rd
Causeway Bay Hong Kong
8310222; Fax: 852-8730488

Asiaweek Ltd
13th Fl S Somerset House
28 Tong Chong St
Quarry Bay Hong Kong
5-630232; Fax: 852-5-657730

Billion Publishing Ltd
Unit B, 7th Fl Chung Sun Bldg
88 Hing Fat St
Causeway Bay Hong Kong
849110; Fax: 852-845285

Bookbuilders
Ian Green, Managing Director
Tonic Industrial Ctr 10F B13
19 Lam Hing St
Kowloon Bay Hong Kong
3-7968123; Fax: 3-7968267

C & C Joint Printing Company
Daniel Tam, Sales Manager
75 Pau Chung St
Tokwawan
Kowloon Hong Kong
3-7135175

Cameron Printing Co Ltd
Flat A1-A4, 12th Fl
256-264 Texaco Rd
Tsuen Wan NT Hong Kong
4992121; Fax: 852-4136198

Caritas Printing Trading Centre
3rd Fl Blk D, Caritas House
2 Caine Rd
Hong Kong
5-261148; Fax: 852-8450438

Caryl Brill & Associates Ltd
Albion Plaza #1105, Granville Rd
Tsimshatsui
Kowloon Hong Kong
7221160; Fax: 852-7225556

Chiapton Co Ltd
Tung Ying Bldg #506
100 Nathan Rd
Kowloon Hong Kong
3-683257; Fax: 852-7231255

Climax Paper Converters Ltd
Climax Bldg, 14 Dai Fu St
Tai Po Ind Estate
Tai Po NT Hong Kong
6582388; Fax: 852-6506057

Colorcraft Ltd
Daniel K C Chung, Director
502-3 Citicorp Ctr
18 Whitfield Road
Causeway Bay Hong Kong
5-786301; Fax: 5-8072005

The Commercial Press (HK) Ltd
Kiu Ying Bldg
2D Finnie St
Quarry Bay Hong Kong
5-651371; Fax: 852-5-651113

The Company Marque Int'l Ltd
Wang Kee Bldg 7th Fl
34-37 Connaught Rd
Central Hong Kong
84631100; Fax: 852-8100478

Computers Forsm (HK) Ltd
3rd Fl C & D, Mercantile Whse
16-24 Ta Chuen Ping St
Kwai Chung NT Hong Kong
4277201; Fax: 852-4201580

Cosmos Printing Press Ltd
Flat B, 3rd Fl. Aik San Fty Bldg
14 Westland Rd
Quarry Bay Hong Kong
5-620203; Fax: 852-5-650203

Creative Printing
William Shue, Managing Director
5/F, Wah Ha Factory Bldg #B
8 Shipyard Ln
Quarry Bay Hong Kong
5-632187

Dah Hua Printing Press
Bernard Chan, Sales Director
9/F Chai Wan Industrial Bldg
26 Lee Chung St
Chaiwan Hong Kong
5-560221

Dai Nippon Printing Company
2-5/F, Tsuen Wan Industrial Ctr
220-248, Texaco Rd
Tsuen Wan NT Hong Kong
0-499-1031

Dow Jones Printing (Asia)
GPO Box 9825
2/F AIA Bldg, 1 Stubbs Rd
Hong Kong
5-8910794

Earl & Associates Ltd
13/F Jubilee Commercial Bldg
42-46 Gloucester Rd
Hong Kong
5-286275

Effendi (Asia) Ltd
Unit 2&3, 21st Fl Fullagar Bldg
234 Aberdeen Main Rd
Aberdeen Hong Kong
8731110; Fax: 852-8141942

Everbest Printing Company Ltd
Block C5 10th Fl KoFai Industry Bldg
7 Ko Fai Rd
Yau Tong Kowloon Hong Kong
852-727-4433

Geoghegan Publishing Co Ltd
Rm 13A, Haywood Mansion
17 Paterson St
Causeway Bay Hong Kong
8955827; Fax: 852-5-776215

Gold Winston Printing Co Ltd
Alan Lai, Sr Sales Manager
7/F, Shing Dao Industrial Bldg
232, Aberdeen Main Rd
Aberdeen Hong Kong
852-814-1355; Fax: 852-814-7033

Statement: A member of the Roman Group.

Golden Cup Printing Company
5 Seapower Industrial Centre
177 Hoi Bun Rd
Kwun Tong Kowloon Hong Kong
852-3-434255; Fax: 852-3-415426

Great Wall Graphics LTD
Philip Rosenberg, Director
22 Stanley St 16th Fl
Central Hong Kong
852-5-257021

Harmonic Hall
Ian Yiu, President
Join-in Hang-Sing Ctr, #320
71-75 Container Port Rd
Kwai Chung NT Hong Kong
0-480-4697

US Office:
Harmonic Hall
1513 Sixth St
Santa Monica CA 90401

Hip Shing Offset Printing
95 How Ming St 1/F
Kwun Tong
Kowloon Hong Kong
3-434254

Hong Kong Educational Publish
Kiu Ying Bldg
2D Finnie St
Quarry Bay Hong Kong
Fax: 852-5-651113

Hung Hing OffSet Printing Co Lt
Matthew Yum, Director
Hung Hing Printing Ctr
17-19 Dai Hei St Tai Po Ind
Est NT Hong Kong Hong Kong
852-650-8682; Fax: 852-650-2070

Interasia Publications Ltd
13th Fl 200 Lockhart Rd
Wanchai Hong Kong
5-749317; Fax: 852-5-26846

International Jewelry Box Co
Superluck Ind Ctr 4th Fl
57 Sha Tsui Rd
Tsuen Wan NT Hong Kong
4145112; Fax: 852-4165629

J P Printing Press
3/F East Sun Industrial Bldg
16 Shing Yip St
Kwun Tong Kowloon Hong Kong
3-440255

Jardine Printing
Alex Wong, Marketing Manager
5/F, 260 King's Rd
North Point
Hong Kong Hong Kong
0-411-3371; Fax: 0-413-5181

Joint Publishing (HK) Co Ltd
Chung Sheung Bldg Ground Fl
9 Queen Victoria St
Central Hong Kong
5-230105; Fax: 852-8104201

Kai Hang Keyboard & Printing Co
Marvel Bldg 7th Fl Blk B
17-23 Kwai Fung Crescent
Kwai Chung NT Hong Kong
4299373; Fax: 852-4209502

Koon Wah Printing PTE Ltd
Tom Earl, Director
Division of Outback Traders Ltd
B01, G/F 45 Tai Hong St
Lei King Wan Hong Kong
852-5-812051

Kwong Fat Offset Printing Co Lt
Alun Stern, Sales Director
10th Fl Bl ABD & 14th Fl Bl B
Wah Ha Factory Bldg 8 Shipyard L
Quarry Bay Hong Kong
5622144-7; Fax: 852-565-7736

Leefung-Asco Printers Ltd
Yang's Ind Bldg 1st Fl
830 Lai Chi Kok Rd
Kowloon Hong Kong
7430310; Fax: 852-7851010

Liang Yu Printing Factory LTD
Eric Hui, General Manager
Hip Shing Industrial Bldg
9 Sai Wan Ho St
Shaukiwan Hong Kong
5-677563; Fax: 5-8858099

Libra Press
56 Wong Chuk Hang Road #5D
Hong Kong Hong Kong
5-528147

Longman Group (Far East) Ltd
Cornwall House, 18th Fl
Tong Chong St
Quarry Bay Hong Kong
8118168; Fax: 852-5-657440

Mandarin Offset
David Martin
6th Floor
22A Westlands Rd
Quarry Bay Hong Kong
852-5-636251; Fax: 852-5-657417

New York Office:
Mandarin Offset
David Pryor, Sales Director
475 Park Ave S
New York NY 10016
212-481-7170; Fax: 212-683-0882

MC Packaging (KH) Ltd
Tai Po Ind. Estate
1 Dai Fu St
Tai Po NT Hong Kong
6578008; Fax: 852-6507038

New Island Printing Co Ltd
Somerset House, 3-5th Fl
28 Tong Chong St
Quarry Bay Hong Kong
6578008; Fax: 852-6507038

Ods Uarco Business Forms Ltd
Tai Po Ind Estate
12 Dai Fu St
Tai Po NT Hong Kong
6531033; Fax: 852-6515928

Off Duty Publications Ltd
Park Comm Centre 14th Fl
8 Shelter St
Causeway Bay Hong Kong
5-777215; Fax: 852-8901761

Oxford University Press
Warwick House, 18th Fl
Tong Chong St
Quarry Bay Hong Kong
5-651351; Fax: 852-5-658491

Pacific Offset Printing
Timmy Ng, General Manager
Block A, 6/F Wah Ha Factory Bldg
8 Shipyard Ln
Quarry Bay Hong Kong
852-5-616146

Paper Communication Intl
Alam Ng
4A Dragon Industrial Bldg
93 King Lam St, Cheung Sha Wan
Kowloon Hong Kong
852-786-4191; Fax: 852-786-4498

Permanent Typesetting & Print-
ing Co Ltd
Stanley W C So, Director
Flat E, 15th Fl, Ho Lee Commercial Bldg
38-44 D'Aguliar St
Central Hong Kong
8525246173-4; Fax: 8452419, 5229046

Pointex Company Ltd
10/F, Wyndham St
Hong Kong
5-215392

Professional Printers (HK) Ltd
65 Wyndham St 1st Fl
Central Hong Kong
852-5-210195; Fax: 852-58453681

Review Publishing Co
7th Fl Centre Point
181 Goucester Rd
Wanchai Hong Kong
8328406; Fax: 852-8345571

Roman Financial Press Ltd
26/F International Bldg
141 Des Voeux Rd Central
Hong Kong
850-6000; Fax: 850-6050

Statement: Financial Printing

Sheck Way Tong Printing Press
653 King's Rd 1/F
North Point Hong Kong
5-614959

Sing Cheong Printing Company
651 King's Rd 1/F
North Point Hong Kong
852-5-626317; Fax: 852-5-659467

South China Printing Company
James Binnie, Director
Aik San Factory Bldg 14th Fl
14 Westlands Rd, Quarry Bay
North Point Hong Kong
5-628261; Fax: 5-811-0221

South Sea International Press
Jeeves Norombaba, General Mgr
20/F Eastern Ctr Bldg #3
1065 King's Rd
Quarry Bay Hong Kong

Tai Cheong Calendar & Printing
151-153 Kweilin St Ground Fl
Shamshuipo
Kowloon Hong Kong
5-651371; Fax: 852-5-651113

Timers Ringier (HK) Ltd
Tai Po Ind Estate
11-13 Dai Kwai St
Tai Po NT Hong Kong
6542666; Fax: 852-6501993

Times-Ringier Ltd
Rm B, 9th Fl, Cindic Tower
128 Gloucester Rd
Wanchai Hong Kong
5-834-1033; Fax: 5-834-0790

Toppan Moore Data Products Ltd
Hong Kong Computer Ctr 12th Fl
54-62 Lockhart Rd
Wanchai Hong Kong
5-201522; Fax: 852-8650349

Toppan Printing Company (HK) Ltd
Walter Lee, General Manager
Toppan Bldg
22 Westlands Rd
Quarry Bay Hong Kong
5-610101; Fax: 852-8110598

Travel & Trade Publishing Asia
Capital Centre, 16th Fl
5-19 Jardine's Bazaar
Causeway Bay Hong Kong
8903067; Fax: 852-8952378

Travel Publishing Asia Ltd
Roy Howard, Managing Director
1801 World Trade Ctre
Causeway Bay Hong Kong
5-8903067;

Union Paper Box & Printing Press
Union Ind Bldg 27 Ko Fai Rd
Yau Tong
Kowloon Hong Kong
7750225; Fax: 852-7727683

Valdar Paper Products Ltd
3/F How Ming Industrial Bldg
99 How Ming St Kwun Tong
Kowloon Hong Kong
389-4351; Fax: 341-9531

Valiant Printing Ltd
David Pearce, Mktg & Sales Director
2/F, 3/F, 5/F Liven House
61-63 King Yip St
Kwun Tong Kowloon Hong Kong
341-6214; Fax: 341-5879

Veribest (Far East) Ltd
Luxe Bldg Rm 1501
594-596 Nathan Rd
Kowloon Hong Kong
3-322493; Fax: 852-7700184

Wan Li Book Co Ltd
18th Fl
499 King's Rd
North Point Hong Kong
5-647511; Fax: 852-5-655539

Wing King Tong Company
Alex Yan, General Manager
188-202 Texaco Rd 3rd Fl
Leader Industrial Ctr
Tsuen Wan NT Hong Kong
4112287; Fax: 852-4110330

Wing Yiu Printing Co
Melbourne Industrial Bldg 16
Westlands Rd 6/F Blk A
Quarry Bay
852-561-0283; Fax: 852-565-8233

Winson Printing Company
2/F Aiksan Factory Bldg
14 Westlands Rd
Quarry Bay Hong Kong
5-635257

Wishing Printing Company
8 Shipyard Lane, Block C, 1/F
Quarry Bay Hong Kong
5-614131

Yee King Silk-Screen Printing
Hoi Luen Ind Ctr, 2nd Fl
55 Hoi Yuen Rd
Kwun Tong Kowloon Hong Kong
3-410389; Fax: 852-7-978923

Yee Tin Tong Printing Press
Jim Viney, General Manager
Morning Post Bldg 4F
Tong Chong St
Quarry Bay Hong Kong
852-5-652222; Fax: 852-5-659833

Hungary

Kultura Budapest
Agnes Tompa, Export Sales Mgr
Printing Department
1389 Budapest 62, PO Box 149
Budapest Hungary
361-180-3194; Fax: 361-180-3306

Statement: Hungarian Foreign Trading
Company represents printers of all sorts.

Indonesia

P T Victory Offset Prima
Jalan Raya Pegangsaan Dua #17
Jacarta Utara 14250
Indonesia
489-6084; Fax: 489-6032

Ireland

Datapage International Ltd
John O'Neil, President
1DA Centre
E Wall Rd
Dublin Ireland
Fax: 3531787060

ICPC Ltd
Seamus McCague, Managing Director
Greencastle Parade
Dublin 17 Ireland
011-3531474; Fax: 011-3537474

Master Photo Limited
4 Heytesbury Ln
Ballsbridge
Dublin 4 Ireland
3531-680357

Scanway Graphics International
Vincent Gillen, Managing Director
Unit 4
Clonshaugh Industrial Estate
Dublin 17 Ireland
3531478833; Fax: 3531478242

Italy

American Pizzi Offset Corp
141 East 44th St
New York NY 10017
212-986-1658; Fax: 212-286-1877

Statement: A subsidiary of Arti Grafiche
Amilcare Pizzi of Milan, Italy.

Italgraphics International Inc
Marketing Director
1100 E Broadway
Glendle CA 91205
818-246-5067

Mondadori-AME Publishing Ltd
740 Broadway 6th Fl
New York NY 10003
212-505-7900; Fax: 212-420-9721

New Interlitho USA
Elaine Freedman, President
276 5th Ave
New York NY 10001
212-686-4088 Fax: 212-686-4087

New Interlitho SPA
Via Curiel 19
20090 Trezzano Sui Naviglio
Milano Italy
02-445-6741 Fax: 02-445-0653

Japan

Dai Nippon Printing Company
1-12 Ichigaya Kagacho
Shinjuku-ku
Tokyo Japan
03-266-2111

New York Office:
DNP America
2 Park Ave, #1405
New York NY 10016
212-686-1919 Fax: 212-686-3250

California Office:
DNP America
50 California St #777
San Francisco CA 94111
415-788-1618; Fax: 415-495-4481

Nissha Printing Co LTD
3 Mibu Hanaicho
NaKaKyo-Ku Kyoto Japan
075-811-811

Korea

Amko Color Graphics
D Kwak, President
156-8 Chodong, Chung Ku
Seoul, Korea
272-7975; Fax: 011-122-2731

New York Office:
Amko Color Graphics
301 Maidson Ave 5th Fl
New York NY 10017
212-687-0400; Fax: 212-687-0467

Sung In Printing
60-4 Garibong-Dong, Guro-Gu
PO Box 38
Seoul Korea
02-864-6121; Fax: 02-866-6156

US Representative:
Overseas Printing Corporation
Hal Belmont, President
2806 Van Ness Ave
San Francisco CA 94109-1426
415-441-7725; Fax: 415-441-7618

Malaysia

Web Printers Sdn Bhd
Ashraf Ali, Marketing Manager
PO Box 40
46700 Petaling Jaya
Delangor Malaysia
03-756-3577

Mexico

Impresora Donneco Internacional
Reynosa Mexico

Statement: A division of RR Donnelley &
Sons.

Jinno Internationale
Yoh Jinno, President
Nebraska 170
Col Napoles 03810 Mexico

US Office:
Jinno International
Yoh Jinno, President
3 Christine Dr
Chestnut Ridge NY 10977
914-735-4666; Fax: 914-735-2906

Litoart
Rafael Sevilla, Manager
Ferrocarril de Cuernavaca 683
DF 11520 Mexico
525-250-3482

Netherlands

Mandarin Offset
Patricia Ahlstrom
Bovendorpsstraat 5
NL-7038
CH Zeddam Netherlands
3108345-1691; Fax: 3108345-1754

Singapore

Chroma Graphics PTE Ltd
Thomas KP Chan, Mkt/Sales Director
Blk 12 Larong Bakar Batu
#06-07/09
Singapore 1334 Singapore
7423706

COS Printers
Alfred Ang, Managing Director
5 Kian Teck Way
Singapore 2262 Singapore
265-9022; Fax: 265-9074

CAMGraphics PTE Ltd
HK Foong, Director
52 Kallang Bahru, #03-02
Singapore 1233 Singapore

Chin Chang Press
Poh But Chay, Manager
113 Eunos Ave 3
#04-08 Gordon Industrial Bldg
Singapore 1440 Singapore
746-6025; Fax: 748-9108

Chong Moh Offset Printing Ptd
Havelock Flatted Factory 1/F
554 Havelock Rd #101
Singapore 0316 Singapore
862-2701; Fax: 862-4335

Colourwork Press
Cho Jock Min, Managing Director
21 Mandai Estate
Singapore 0512 Singapore
368-4048; Fax: 368-9107

Continental Press
Danny Ng, Proprietor
1160 Depot Rd #03-01/07
Telok Blangah JTC Industrial Estate
Singapore 0410 Singapore
65-278-5816; Fax: 65-272-2398

Craft Print
Dora Chan, Managing Director
18 Pasir Panjang Rd #06-13/17
Singapore 0511 Singapore
65-278-5816; Fax: 65-272-2398

Des Meyer Press
2021 Bukit Batok #23
Industrial Park A #04-210
Singapore Singapore
65-5660629; Fax: 65-5661952

Dominie Press
T H Oh, Managing Director
Blk 1200 Depot Close #01-21/27
Telok Blangah Industrial Estate
Singapore 0410 Singapore
65-273-0755; Fax: 65-273-0060

Specialties: Jigsaw puzzles, pop-up cards, and items involving handwork.

Eurasia Press Pte Ltd
Allan Fong, Marketing Director
10,12 & 14 Kampong Ampat
Singapore 1336 Singapore
280-5522; Fax: 280-0593

FEP International
Richard Toh, General Manager
348 Jalan Boon Lay Jurong
Singapore 2261 Singapore
265-0311; Fax: 265-5103

Fong & Sons Printers
Tony Fong, Marketing Director
1090 Lower Delta Rd #0110-011
Tiong Bahru JTC Flatted Factory
Singapore 0316 Singapore
272-0154; Fax: 273-3855

General Printing Services
Lee Joachim, Managing Director
2 Soon Wing Rd #03-09
Soon Wing Industrial Bldg
Singapore 1334 Singapore
747-2657; Fax: 746-4681

Hiap Seng Press
Eddie Yam, Managing Director
77 Lorong 19 Geylang #01-00
Singapore 1438 Singapore
747-5391; Fax: 743-3251

Ho Printing Company
11-15 Harper Rd
Singapore 1336 Singapore
65-280-9322; Fax: 65-289-6065

Hofer Press
Chan Ying Lock, Director
3 Gul Crescent
Singapore 2262 Singapore
65-861-2755; Fax: 65-861-6438

Specialty: Guide books and maps.

Hoong Fatt Press
George S W Wong, Manager
9 Harper Rd
Singapore 1336 Singapore
65-288-2243; Fax: 65-280-5943

Huntsmen Offset Printing
Heung Yam Yuen, General Manager
12 Fan Yoong Rd
Jurong Town
Singapore 2262 Singapore
265-0600; Fax: 265-8575

Specialties: Bibles, diaries, and directories.

Hup Khoon Press
Ng Eng Kiat, Managing Director
Block 173 Stirling Rd #01-104
Singapore 0314 Singapore
65-479-8350; Fax: 65-473-1118

International Press Company
Kim-Hui Tan, Export Mktg Manager
32 Kallang Pl
Singapore 1233 Singapore
65-298-3800; Fax: 65-297-1668

Jin Jin Printing Industry
See Cheng Tee, Managing Director
Block 1013
Eunos Ave 5 #01-02
Singapore 1440 Singapore
65-745-5166; Fax: 65-747-1232

Khai Wah Litho Pte LTD
Jeannie Lim, Managing Director
16 Kallang Pl #07-02
Singapore 1233 Singapore
65-296-8644; Fax: 65-297-1540

Kim Hup Lee Printing
Lim Geok Khoon, Director
22 Lim Teck Boo Rd
Kim Hup Lee Building
Singapore 1953 Singapore
65-283-3306; Fax: 65-288-9222

Kin Keong Printing Co Pte LTD
H L Wong, Executive Director
Block 3015A
Ubi Road 1 #05-04/07
Singapore 1440 Singapore
65-744-0915; Fax: 65-743-0692

Kok Wah Press Pte Ltd
3 Gul Crescent
Singapore 2262 Singapore
861-7966

Kyodo-Shing Loong Printing
Yoshiki Ishii, General Manager
112 Neythal Rd
Singapore 2262 Singapore
265-2955; Fax: 264-4939

Lenco Printing
Poon Yeow Wah, Managing Director
37 Kallang Pudding Rd #04-14
Tong Lee Bldg, Block B
Singapore 1334 Singapore
747-5573; Fax: 743-3928

Statement: Also boxes and other packaging.

Lolitho Pte
Peter Loh, General Manager
5055 Ang Mo Kio Industrial Park
2 #01-1141
Singapore 2056 Singapore
273-8390; Fax: 456-9297

MCD Pte
Philip Chew, Director
18 Pasir Panjang Rd #04-01
PSA Multi Storey Complex
Singapore 0511 Singapore
273-6888; Fax: 273-8328

Monocrafts Pte
Alan Sim, Managing Director
34 Kallang Pl
Singapore 1233 Singapore
298-1088; Fax: 298-2624

Specialty: Diaries.

Nissin Printcraft
Wong Yat Bor, Managing Director
No 3 Kallang Way 3
Singapore 1334 Singapore
747-2992

Specialties: Diaries and calendars.

PacPress Industries Ptd LTD
Sai Yong Mong, Director
25 Tannery Ln #01-25/27
Singapore 1334 Singapore
746-3939; Fax: 748-9242

Saik Wah Press Pte LTD
Anthony Tham, Marketing Director
50 Kallang Bahru #01-21
Singapore 1233 Singapore
65-292-8759; Fax: 65-296-0638

Sin Sin Lithographers
Michael Tsu, Director
Block 115-A
Commonwealth Dr #02-30
Singapore 0314 Singapore
474-5692; Fax: 473-3817

Sing Chew Press
Tan Kay Tiak, Managing Director
38 Genting Ln #02-01/04
Metal House Bldg
Singapore 1334 Singapore
746-6611; Fax: 746-5209

Singapore National Printers
Henry Pang, Marketing Manager
303 Upper Serangoon Rd
Singapore 1334 Singapore
282-0611; Fax: 288-7246

Stamford Press Pte LTD
S Kesava, Marketing Manager
48 Lorong 21 Geylang #04-03
Singapore 1438 Singapore
65-748-5111; Fax: 65-746-7310

Sydney Press Indusprint
Ng Ghin, Managing Director
Block 3011 Bedok N Ave 4
#04-2012 Bedok Industrial Park
Singapore 1648 Singapore
445-6006; Fax: 449-7634

Tien Mah Litho Printing
David Chew, Marketing Director
2 Jalan Jentera
Jurong Town
Singapore 2261 Singapore
264-0988; Fax: 265-0549

Tien Wah Press
84 Wooster #505
New York NY 10012
212-274-8090; Fax: 212-274-0771

Statement: A division Dai Nippon

Specialties: Thin casebound books, pop-ups and special handwork.

Times Offset
Ricky Ang, General Manager
Koon Wah Printing
18 Tuas Ave 5
Singapore 2263 Singapore
862-3333; Fax: 862-1313

Times Printers Pte Ltd
Ricky Ang, General Manager
Commercial Printing Group
2 Jurong Port Rd
Singapore 2261 Singapore
65-265-8855; Fax: 65-268-5979

Times Publishing Group
Colin Hodkinson, Gen Mkt Mgr
2 Jurong Port Rd
Singapore 2261 Singapore
268-5979

Toppan Printing Company
Ee Swee Kiat, General Manager
38 Liu Fang Rd
Jurong Town
Singapore 2262 Singapore
265-1811; Fax: 265-8298

Viva Lithographers
Michael Oh, Managing Director
22 Pasir Panjang Rd #06-22/23
PSA Multi-Storey Complex
Singapore 0511 Singapore
272-1880; Fax: 273-5425

Welpac Printing & Packaging
Joe Yip, Sales Manager
36 Lok Yang Way
Singapore 2262 Singapore
261-5555; Fax: 265-7961

Spain

Novograph SA
Ctra de Irun, km 12,450
Madrid Spain 341-734

US Office:
Novograph SA
30 W 26th St
New York NY 10010
212-675-0364

Taiwan

Morion Company
Simon Huang, General Manager
PO Box 43-396
Taipei 10657 Taiwan
02-704-1715; Fax: 02-704-8197

Taipei Yung Chang Printing
49-15 Chuanyuan Rd
Peitou (112)
Taipei Taiwan

Thailand

Thai Watana Panich Press
Thira T Suwan, Managing Director
891 Rama 1 Rd
Bangkok 10500 Thailand
662-215-0060; Fax: 662-215-2360

United Kingdom

Anthony Rowe Ltd
Bumper's Way, Bristol Road
Chippenham WI
SN14 6QA England
0249-659705; Fax: 0249-443103

BAS Printers Ltd
Paul Gumn, Sales Director
Over Wallop
Stockbridge
Hants SO2O 8JD England
264-781711; Fax: 264-781116

Bourne Offset Ltd
Tim Staples, Managing Director
2 The Ridgeway
Iver Buckinghamshire
SL0 9HR England
753-652004

Cradley Print
Chris Jordan, Managing Director
PO Box 34
Chester Road Cradley Heath Warl
W Midlands B64 6AB England

Mandarin Offset
Mike Weller
32 Hanway St
London W1P 9DD England
071-631-3281; Fax: 071-631-1489

Pensord Press
David Seabrook, Director
The Old Tram Rd
Southend-on-Sea
Essex SS2 6UN England
495-223721

The Phoenix Press
John Oliver, Managing Director
Unit 6 Birch Rd
Eastbourne
E Sussex BN23 6PE England
323-411090; Fax: 323-411060

Thamesmouth Printing Group
Ian Cuthbert, Managing Director
17-21 Hovefields Ave
Burnt Mills Ind Est, Basildon
Essex SS13 1EB England

Tien Wah Press Ltd
16 the Ivories
6-8 Northampton St
London N-1 2NY England
01-354-3323; Fax: 01-359-8777

All Languages Graphics Inc
Arthur Raskin, President
Div of Liguistic Systems
116 Bishop Allen Dr
Cambridge MA 02139
617-864-3900;

Yugoslavia

Gorenjski Tisk Printing Company
Boris Kozely, Export Manager
Mdse Pijadeja St No 1
64000 Kranj Yugoslavia
064-23-341; Fax: 064-24-475

Print Brokers

Chanticleer Company
568 Broadway #1005A
New York NY 10012
212-941-1522

Codra Enterprises
1083 Bedmar St
Carson CA 90746
310-223-1626; Fax: 310-223-1612

Faron Melrose Inc
Penny Melrose, President
19925 Stevens Creek Boulevard
Cupertino CA 95014
408-773-8022; Fax: 408-773-8023

Four Colour Imports
George Dick, President
2843 Brownsboro Rd #102
Louisville KY 40206
502-896-9644; Fax: 502-896-9594

Statement: US representative of Cayfosa
Industria Grafica, D W Friesen, Everbest
Printing and Coloursplendor Graphics.

Imago Sales (USA)
Joseph Braff, President
310 Madison Avenue #2103
New York NY 10017
212-370-4411; Fax: 212-370-4542

Interprint
Ken Coburn, President
2447 Petaluma Boulevard N
Petaluma CA 94952
707-765-6116; Fax: 707-765-6018

Trans-Pacific Publishers
712 Sartori Dr
Petaluma CA 94956
707-762-8839

Chapter 8

Other Publishing Services

This chapter features other publishing suppliers and services that you might want to use in designing, producing, and marketing your books. For a more complete list of such services, see the **Book Publishing Resource Guide**. To order, call Ad-Lib Publications toll-free: **800-669-0773**.

Art

Dunn & Associates
PO Box 870
Hayward WI 54843
715-634-4857

Service: Advertising design

Dynamic Graphics Inc
6000 N Forest Park Dr
PO Box 1901
Peoria IL 61656-1901
309-688-8800; Fax: 309-688-3075
800-255-8800

Service: Clip Art Service

Wheeler Arts
66 Lake Park
Champaign IL 61821-7101
217-359-6816

Service: Clip art

Audio/Video

Abbey Tape Duplicators
9525 Vassar Ave
Chatsworth CA 91311
818-882-5210; 800-346-3827

Service: Audio/video duplication

Allied Film & Video
7375 Woodward Ave
Detroit MI 48202
313-871-2222; Fax: 313-871-5907

Service: Video duplication

Statement: Video duplication and warehousing.

Carpel Video
429 E Patrick St
Frederick MD 21701
301-694-3500; 800-238-4300

Service: Video duplication

Cassette Productions Unlimited
5796 Martin Rd
Baldwin Park CA 91706-6299
818-969-6881; 800-345-0145

Service: Audio/video duplication

Custom Screen Prints
7225 Woodland Dr
PO Box 68881
Indianapolis IN 46268-0881
317-297-1888; 800-582-2000

Service: Video duplication

Statement: Video duplication and distribution.

Ditto Video
4950 Cohasset Rd #10
Chico CA 95926
916-343-4886

Service: Video duplication

Dove Enterprises
4520 Hudson Dr
Cuyahoga Falls OH 44224
216-928-9160; 800-233-3683

Service: Audio/video duplication

The Dub Centre
10304 S Dolfield Rd
Owings Mills MD 21117
301-363-4810; 800-382-0080

Service: Audio/video duplication

Duplication Depot
215B Central Ave
Farmingdale NY 11735
800-727-0608

Service: Video duplication

Evatone
4801 Ulmerton Rd (34622)
PO Box 7020
Clearwater FL 34618-7020
813-572-7000; 800-382-8663

Service: Audio/video duplication, compact disc

Karol Media
Gene Dodge, Vice President
350 N Pennsylvania Ave
PO Box 7600
Wilkes-Barre PA 18702
717-822-8899; Fax: 717-822-8226
800-526-4773

Service: Video duplication

Statement: Duplication and marketing of videos, films, etc.

Mother Dubbers
13626 Gamma Rd
Dallas TX 75234
214-980-4840

Service: Audio/video duplication

Reel Time Duplicators
2093 Faulkner Rd NE
Atlanta GA 30324
404-321-4830; 800-323-2061

Service: Video duplication

Sifford Video Services
2815 Evans St
PO Box 7917
Hollywood FL 33081
305-920-5054; 800-727-2679

Service: Video duplication

Sifford Video Services
121 Lyle Ln
PO Box 101510
Nashville TN 37210-1510
615-248-1010; 800-251-1009

Service: Video duplication

STS Film Video Productions
935 Bullion St
Salt Lake City UT 84123
800-654-4870

Service: Video duplication

Statement: Video and film production.

Binding/Bindery

Ad Industries of California
Joel Fishman
12160 Sherman Way
North Hollywood CA 91605
818-765-4200
Service: Bindery
Statement: Casemaking, collating, diecutting, plastic comb, shrink wrapping.

American Sample Co
822 San Julian St
Los Angeles CA 90014
213-627-1441; Fax: 213-688-7544

Service: Bookbinding and related work

Area Trade Bindery
157 W Providencia Ave
Burbank CA 91502
818-849-5881; Fax: 818-849-3733
800-225-1343

Service: Binding

Statement: Also 818-849-1163 and 818-846-5581.

Bayless Bindery Inc
501 SW 7th St
Renton WA 98055
206-226-6395

Service: Bookbinding and related work

Bee Bindery Inc
15 S Throop St
Chicago IL 60607
312-666-6210; Fax: 312-666-5804

Service: Bookbinding and related work

Bindagraphics Inc
Ralph Possion
2701 Wilmarco Ave
Baltimore MD 21223
401-362-7200; Fax: 410-362-7233
800-326-0300

Service: Bookbinding and related work

Bound to Stay Bound Books Inc
W Morton Rd
Jacksonville IL 62650
217-245-5191; Fax: 217-245-0424

Service: Bookbinding and related work

California Sample Service Corp
1611 S Hope St
PO Box 15264
Los Angeles CA 90015
213-748-6333; Fax: 213-748-3940

Service: Bookbinding and related work

Continental Bindery Corp
700 Fargo Ave
Elk Grove Village IL 60007
708-439-6811; Fax: 708-439-6847

Service: Bookbinding and related work

John H Dekker & Sons
Howard Dekker, President
2941 Clydon Ave SW
Grand Rapids MI 49509
616-538-5160; Fax: 616-538-0720

Service: Book binding

Demco Inc
500 E North St
De Forest WI 53532
608-846-3716; Fax: 608-846-3701

Service: Bookbinding and related work

Dunn & Company
Carlso Llanso
75 Green St
PO Box 968
Clinton MA 01510
508-368-8505; Fax: 508-368-7867

Service: Book binding

Statement: Dunn and Company converts books from case to paper and paper to case. We cancel and tip pages. We also do all sorts of repair work.

Durham Exchange Club Industries
1717 E Lawson St
Durham NC 27703
919-596-1341

Service: Bookbinding and related work

Econo Clad Books
2101 NW Topeka Blvd
Topeka KS 66608
913-233-4252; Fax: 913-233-3129

Service: Bookbinding and related work

John D Ellis Bindery Inc
141 Manufacturing St
Dallas TX 75207
214-748-0736; Fax: 214-748-5047

Service: Bookbinding and related work

AV Emmott & Sons Bookbinders Inc
5700 Mitchelldale St
Houston TX 77092
713-956-0211; Fax: 713-956-7545

Service: Bookbinding and related work

The Form House Inc
Robert B Volkert
7200 S Leamington Ave
Chicago IL 60638
708-594-7300

Service: Binding service and material

Gazette Press Inc
16 School St
Yonkers NY 10701
914-963-8300; Fax: 914-476-1052

Service: Bookbinding and related work

General Binding Corporation
One GBC Plaza
Northbrook IL 60062
708-272-3700; Fax: 708-272-7087

Service: Binding

General Bookbinding Co
8844 Mayfield Rd
Chesterland OH 44026
216-729-9411; Fax: 216-729-9415

Service: Bookbinding and related work

General Products
4045 N Rockwell St
Chicago IL 60618
312-463-2424; Fax: 312-463-3028
800-888-1934

Service: Bookbinding and related work

Graphic Converting
4500 W Division St
Chicago IL 60651
312-235-5100

Service: Bindery

Statement: Collating, folding, diecutting, drilling, shrink wrapping.

Haddon Craftsmen Inc
1001 Wyoming Ave
Scranton PA 18509
717-348-9211; Fax: 717-348-9266

Service: Bookbinding and related work

Heckman Bindery Inc
1010 N Sycamore St
North Manchester IN 46962
219-982-2107; Fax: 219-982-1104
800-334-3628

Service: Bookbinding and related work

Hertzberg-New Method
617 E Vandalia Rd
Jacksonville IL 62650
217-243-5451

Service: Bookbinding and related work

Information Conservation Inc
6204 Corporate Park Dr
Browns Summit NC 27214
910-375-1202

Service: Bookbinding and related work.

Inter-City Manufacturing Co Inc
Bill Mitchell
7401 Alabama St
St Louis MO 63111

314-351-3100

Service: Bindery

Statement: Casemaking, collating, plastic comb, round cornering, scoring

Robert F Lewis Inc
64 Rt 4 W
Woodstock VT 05091
802-457-1205; Fax: 802-457-1227

Service: Bookbinding and related work

Library Binding Service
2134 E Grand Ave
Des Moines IA 50317
515-262-3191; Fax: 515-262-3839

Service: Bookbinding and related work

Lo Gatto Book Binding Inc
Medo Lo Gatto
390 Paterson Ave
East Rutherford NJ 07073
201-438-4344; Fax: 201-438-1775

Service: Bookbinding and related work

Statement: Short run edition, cloth and leather binding. Gold stamping restoration.

Looseleaf Inc
418 Harding Industrial Dr
Nashville TN 37211
615-834-7600; Fax: 615-331-7833

Service: Bookbinding and related work.

Macke Brothers Inc
10355 Spartan Dr
Cincinnati OH 45215
513-771-7500; Fax: 513-771-3830

Service: Bookbinding and related work

Mark Binding Corporation
1900 Delaware Ave
Des Moines IA 50317
515-263-1800; Fax: 515-263-1202

Service: Binding

Mountain States Bindery Co
1818 W 2300 S
Salt Lake City UT 84119
801-972-2300; Fax: 801-972-2883

Service: Bookbinding and related work

Muscle Bound Bindery
Gerald L Hanson
701 Plymouth Ave N
Minneapolis MN 55411
612-522-4406; Fax: 612-522-0927

Service: Book binding

Statement: Bookbindery offering otabind, perfectbind, sidesewn flatback casebinding, folding, drilling, cutting, laminating, and shrinkwrap.

National Library Bindery Co
100 Hembree Park Dr
Roswell GA 30076
404-442-5490; Fax: 404-442-0183

Service: Bookbinding and related work

Nicholstone Companies
Tony Riviezzo, Sr Vice President
418 Harding Industrial Dr
Nashville TN 37211
615-834-7600; Fax: 615-331-7833

Service: Book binding

NSC International/EBSCO Industries
Bud West
Rt 5 #171
Hot Springs AR 71913
501-623-8861

Service: Binding

Statement: We offer the fastest service on plain 19 ring, odd size, and custom imprinted, as fast as, 48 hours turnaround on imprinted combs.

Oxford Bookbinding Co
3101 Red Lion Rd
Philadelphia PA 19114
215-632-0400; Fax: 215-632-0166

Service: Bookbinding and related work

Pease Bindery
Sandy Harms
111 Oak Creek Dr
Lincoln NE 68528
402-476-1303; Fax: 402-476-2978

Service: Book binding

Statement: Pease Bindery is a full service bindery that specializes in Smyth sewn case bound book manufacturing. Our commitment to our customers is to provide high quality, cost efficient book manufacturing. Pease is also dedicated to providing superior customer service because we realize our customers are the lifeblood of our company.

Printers Bindery Inc
345 Hudson St
New York NY 10014
212-924-4200

Service: Bookbinding and related work

Printing Plus
1710A NW 86th St
Des Moines IA 50325
515-223-1711

Service: Bookbinding

The Riverside Group
Frank Liedtke
50 Ajax Rd
Rochester NY 14624
716-328-1128

Service: Bindery

Statement: Binding, casemaking, plastic comb, saddlestitching, wire-o binding.

Roger's Binding & Mailing
Kenn Hughes, Sr Estimator
1905 Riverview Dr
San Bernardino CA 92408
909-799-5400

Service: Bookbinding

Schuler Sales & Service
25 Whitney Way
Mahwah NJ 07430
201-385-6330; Fax: 201-848-0368

Service: Bookbinding and related work

Souther Binders
205 Industrial Blvd
Dalton GA 07511
706-277-2227; Fax: 706-277-1839
800-241-4698

Service: Bookbinding and related work

Specialties Bindery
Mark Lee
4815 Lawrence St
Hyattsville MD 20781
301-699-8800

Service: Bindery

Statement: Banding, drilling, plastic comb, spiral binding, wire-o binding.

Spiral Binding of Illinois Inc
4200 W Wrightwood Ave
Chicago IL 60639-2095
312-772-4442; Fax: 312-772-1585
800-782-8798

Service: Bindery service

Spirial Binding Co Inc
1 Maltese Dr
Totowa NJ 07512
201-256-0666; Fax: 201-256-5981

Service: Bookbinding and related work

Star Binding & Trimming
1109 Grand Ave
North Bergen NJ 07047
201-864-2200

Service: Bookbinding and related work

Stauffer Edition Binding (SEBCO)
Bob Stauffer, President
1000 Monterey Pass Rd
PO Box 74
Monterey Park CA 91754
213-263-9434; Fax: 213-263-3218

Service: Book binding

Statement: We specialize in hard cover bookbinding only. Established in 1974. Automated binding line and cover making.

Superior Tube Co
PO Box 159
Collegeville PA 19426
610-489-5200

Service: Bookbinding and related work

Wert Bookbinding Inc
RR 2 Box 2000
Grantville PA 17028
717-469-0626; Fax: 717-469-0629
800-344-9378

Service: Bookbinding and related work

Bar Code

Accession Inc
Richard Oaksford
PO Box 2299
Lynnwood WA 98036-2299
206-672-8897; 800-531-6029

Service: Bar codes

Bar Code Graphics Inc
Andrew Verb
343 W Erie St
Chicago IL 60610
312-664-0700; Fax: 312-664-4939
800-662-0701

Service: Labels, eletronic bar code

Statement: Bar code film masters, labels, EP symbols, and verifiers.

Film Masters Inc
Deborah Maniaci, Account Executive
1457 W 26th St
Cleveland OH 44113
216-621-FILM(3456); Fax: 216-621-7908
800-621-AUPC(2872)

Service: Bar codes

Fotel/GGX
John Nachtrib, President
6 Grace Ave
Great Neck NY 11021
516-487-6370; Fax: 516-487-6449

Service: Bar codes, labels

Images by Design
Fred or Scott Bauries
2277 Science Pkwy
Okemos MI 48864
517-349-4635; Fax: 517-349-7608

Service: Bar code masters

Statement: UPC and Bookland EAN bar code masters. Same day turnaround.

International Artwork
1111 W El Camino Real #109-316
Sunnyvale CA 94087
408-625-5093; Fax: 408-625-0855

Service: Bar codes

Precision Photography
Bruce Sitka
1150 N Tustin Ave
Anaheim CA 92807
714-632-9000; Fax: 714-630-6581
800-872-9977

Service: Bar codes

Publication Identification and Processing Systems
436 E 87th St
New York NY 10128-6502
212-996-6000; Fax: 212-410-7477

Service: Bar codes

ScanLine Graphics Inc
10044 S Pioneer Blvd
Santa Fe Springs CA 90670
310-801-4500; 800-932-7801

Service: Label, electronic bar code

Symbology Inc
Carla Sandess, Marketing Director
PO Box 14849
Minneapolis MN 55414-0849
612-331-6200; Fax: 612-331-3500

Service: Bar codes

Bookbinding Supplier

The Davey Company Inc
Alfred Broks, VP Sales
164 Laidlaw Ave
Jersey City NJ 07306
201-653-0606; Fax: 201-653-0872

Service: Bookbinding supplier

Industrial Coatings Group
Charlotte Collier, Marketing
220 Broad St
Kingsport TN 37660
615-247-2131; Fax: 615-247-2134
800-251-7520

Service: Bookbinding supplier

Business Cards

Heritage Business Cards
Division of Utley Brothers Inc
567 Robbins Dr
Troy MI 48083
313-585-1700; Fax: 313-585-1048

Service: business cards

Mercer Color Corp
Drawer 113
Coldwater OH 45828
800-248-6665; Fax: 419-678-3144

Service: Business cards

Professional Lithography
630 New Ludlow Rd
South Hadley MA 01075
413-532-9473

Service: Business cards

Zip Business Cards
PO Box 935
Norwalk OH 44857
419-668-0930; Fax: 419-668-0864

Service: Business cards

Casemaking

DVC Industries
1440 Fifth Ave
Bay Shore NY 11706
516-968-8500

Service: Casemaking

Kleer-Vu Plastics
Kleer-Vu Dr
PO Box 449
Brownsville TN 38012
901-772-2500; Fax: 901-772-4632
800-365-5827

Service: Casemaking

Card Decks

Collated Products Corporation
220 Eastview Dr
Cleveland OH 44131
216-741-1288

Service: Card deck printer

Scoville Press
14505 27th Ave N
Minneapolis MN 55447
612-553-1400; Fax: 612-553-0042

Service: Card pack printers

Solar Press
1120 Frontenac Rd
Naperville IL 60563
708-983-1400; 800-323-2751

Service: Card pack printer

Color Separations

American Color Printing Inc
889 Chalmers St
Detroit MI 48215
313-331-1500

Four Colour Imports Ltd
2843 Brownsboro Rd
Louisville KY 40206
502-896-9644; Fax: 502-896-9594

Service: Color separation

Statement: Four Colour Imports is the direct representative of Everlast Printing

Co in Hong Kong and Friesen Printers in Canada.

Kieffer-Nolde
160 E Illinois St
Chicago IL 60611
312-337-5500; Fax: 312-337-3724
800-621-8314

Service: Color separations, desk top publishing

Separacolor International Inc
27278 Eastvale Rd
Palos Verdes Peninsula CA 90274
310-541-6610

Service: Color separations

TSI Graphics
Richard Whitsitt
70 Jackson Dr
Cranford NJ 07016
908-272-2520; Fax: 908-272-2043

Service: Color separations

Unigraphic Color Corp
180 W Main Street
Plymouth PA 18651
717-779-9543; Fax: 717-779-9549

Service: 4-color, color separations

Unitron Color Graphics of NY INC
Aaron Fertig
4710 32nd Pl
Long Island City NY 11101
718-784-9292

Service: Color separation

Design/Graphics

Abacus Graphics
4751 Morning Canyon Rd
Oceanside CA 92056
619-724-7750; Fax: 619-724-8788

Service: Design

Alexander Teshin Associates
2146 Leghorn St
Mountain View CA 94043
415-961-0400; Fax: 415-961-0464

Service: Design graphics

American National Can Company Graphic Arts Center
2400 Maywood Dr
Bellwood IL 60104
708-544-4414; Fax: 708-547-4472

Service: Graphics

Antler & Baldwin Design Group
Sallie Baldwin, President
7 East 47th St
New York NY 10017
212-751-2031; Fax: 212-355-0327

Service: Books designers

Archetype Graphicdesign/ Publishing
111 Ingalls St
Santa Cruz CA 95060
408-425-8131; Fax: 408-423-2159

Service: Design

Associated Graphics
13 Henry Johnson Blvd
Albany NY 12210
518-465-1497; 800-836-6310

Service: Graphics

Bessas and Ackerman
Jo Ellen Ackerman
16 Marion Ave
Cliffside Park NJ 07010
201-886-2049; Fax: 201-886-8443

Service: Design

Statement: Graphics, design.

Color Graphic Press Inc
1120 46th Rd
Long Island City NY 11101
718-392-2727; 800-621-6680

Service: Art design services

Statement: Specializing in color combo runs.

Comp-Type Inc
Cynthia Frank
155 Cypress St
Fort Bragg CA 95437
707-964-9520; Fax: 707-964-7531

Service: Design services

Statement: Complete editorial and book production services. Call or write.

Graphic Productions Inc

Karen Hanke
975 E 22nd St #D
Wheaton IL 60187-4386
708-462-1446; Fax: 708-462-1748

Service: Cover design, page layout

Statement: We are a graphic design firm specializing in book production. Our designers and illustrators can work within your budget to create the professional cover design and page layout you desire. We can work from your disk or typeset from your manuscript. The finished product camera-ready for the printer.

Graphics International

Don Izod, President
20475 Bunker Hill Dr
Cleveland OH 44126
216-333-9988

Service: Design graphics

H & M Graphics

21 Gregory Rd
Wallingford CT 06492
203-269-4464
Service: Design

Statement: Graphics, logo designs

Robert Howard Graphic Design

Robert Howard
631 Manfield Dr
Fort Collins CO 80525
303-225-0083

Service: Design

Statement: Design of book covers, photography, illustrations, ads, and more.

Impressions Division of Edwards Brothers

William Kasdorf, President
2016 Winnebago St
Madison WI 53704
608-244-6218; Fax: 608-244-7050

Service: Design, typesetting

Larry Milam Illustration

Larry Milam
Solo Pro Studio
3530 SE Hawthorne Blvd
Portland OR 97214
503-236-9121

Service: Design

Statement: Illustrations, design

Leon Bolognese & Associates

135 Connecticut Ave
Freeport NY 11520
516-379-3405

Service: Book designers

Orange Ball Corporation

Brenda Mitchell-Powell
20 Vani Ct
Westport CT 06880
203-221-1224; Fax: 203-221-1087

Service: Design

Statement: Editorial services, graphic design, camera-ready productions.

Print by Design

3617 Thousand Oaks Blvd #222
Thousand Oaks CA 91362
805-495-8938

Service: Design

Statement: Graphics, design

Ted Goff Illustration

847 Vermont #102
Oakland CA 94610
415-272-9264

Service: Design

Statement: Graphics illustrations

Versa Type Inc

249 E Ocean Blvd #504
Long Beach CA 90802
310-432-4086; Fax: 310-437-0754

Service: Art design services

Patrick J Welsh

PO Box 463
Williamstown NJ 08094
609-728-0264

Service: Design graphics

Desktop Publishing

International Computaprint Corp

William R Haskitt
12712 Dupont Cir
Tampa FL 33626
813-855-4635; 800-872-2828

Service: Desktop/color publishing

Diskette Duplication

Cycle Software Services
B Norton, President
6552 Edenvale Blvd
Eden Prairie MN 55346
612-949-2606; Fax: 612-949-2604

Service: Diskette duplication

Displays

Ad-Lib Publications
Jeanette Beasley
PO Box 1102
Fairfield IA 52556
515-472-6617; Fax: 515-472-3186
800-669-0773

Serice: Display

Statement: Stocks 6x9 and 8½x11 plain white countertop displays

City Diecutting Inc
17 Cotters Lane
East Brunswick NJ 08816
908-390-9599; Fax: 908-390-8654

Service: Prepack counter displays

Editing

Cross Pind Editing Group
333 Hook Rd
Katonah NY 10536
914-232-1258; Fax: 914-232-1258

Service: Editing

Sternwheeler Press
Sharon Carrigan
PO Box 381
Maryhurst OR 97036
503-636-7580

Service: Editing

Envelopes

Shipman Printing Industries
Mike Fiore
2424 Niagara Falls Blvd
Niagara Falls NY 14302
716-731-3281; Fax: 716-731-9620
800-462-2114

Service: Envelopes

Statement: Shipman produces high quality litho printed envelopes, web cut sheets and commercial printing to the trade only for resale with the union label if requested.

Folding Machines

Dick Moll & Sons
415 Constance Dr
Warminster PA 18974
215-443-7517; Fax: 215-443-8310

Service: Folding machines

Fullfilment 800#

Upper Access, Inc
Steve & Lisa Carlson
PO Box 457
Hinesburg VT 05461
802-482-2988; Fax: 802-482-3125
800-356-9315

Service: 800 # FF for retail sales

Statement: No front money, volume discount, 24 hours, 7 days a week, US and Canada, all major credit cards.

Fulfillment Service

APR Fulfillment
5350 Dickman Rd
Battle Creek MI 49016
616-968-2221; Fax: 616-968-0340
800-562-9733

Service: Fulfillment

Associations Book Distributors
Maureen P Schwab
503 Thomson Park Dr
Mars PA 16046
412-772-0070; Fax: 412-772-5281

Service: Fulfillment

Statement: Book fulfillment for associations.

Central Distribution Services
Wayne Nemecek, Manager
J J Keller & Associates Inc
7273 US Hwy 45
Neenah WI 54956
414-727-7511; 800-558-5011

Service: Fulfillment

Statement: Book fulfillment.

Express Fulfillment Services
2515 E 43rd St
Chattanooga TN 37407
615-867-9081; Fax: 615-867-5526

Service: Full service fulfillment

WH Freeman & Company
4419 W 1980 S
Salt Lake City UT 84104-4702
801-973-4660

Service: Fulfillment

Gage Distribution Company
164 Commander Blvd
Agincourt ON M1S 3C7 Canada
416-293-8141

Service: Fulfillment

Statement: Book fulfillment center.

Georgetown Book Warehouse
Lois M Fraser, Director
34 Armstrong Ave
Georgetown ON L7G 4R9 Canada
905-873-2750; Fax: 905-873-6170

Service: Book distribution.

Statement: On site customs clearance, book order fulfillment services, reforwarding.

Integrated Distribution Services
Harold Levine, President
195 McGregor St
Manchester NH 03102
603-623-0305; Fax: 603-669-7945

Service: Fulfillment

Statement: Complete book fulfillment services.

Intrepid Productions
1331 Red Cedar Cir
Fort Collins CO 80524
303-493-3793

Service: Fulfillment

IPP Shipping & Warehousing
Floyd Campbell, Vice President
211 E Harrison St
PO Box 1517
Danville IL 61834-1517
217-442-1190; Fax: 217-442-9141

Service: Fulfillment

J V West
Ben Rose, President
MTL Inc
PO Box 11950
Reno NV 89510-9959
702-359-9811

Service: Fulfillment

Statement: Fulfills for 70 book publishers.

Mercedes Distribution Center
62 Imlay St
Brooklyn NY 11231-1298
718-522-7111; Fax: 718-852-5341

Service: Fulfillment

Statement: Book fulfillment cente.r

Midpoint National Inc
2215 Harrison St
Kansas City MO 64108
816-842-8420

Service: Fulfillment

National Fulfillment Services
Holmes Corporate Center
100 Pine Ave
Holmes PA 19043-1484
610-532-4700; Fax: 610-596-3232
800-345-8112

Service: Fulfillment

Statement: Book and magazine fulfillment.

Palmer Publications
Charles E Spanbauer, President
PO Box 296
Amherst WI 54406
715-824-3214

Service: Fulfillment

Statement: Single-copy mail order shipping for books.

Progressive Distribution Ctrs Inc
6490 Lynch Rd
Detroit MI 48234
313-571-8600

Service: Fulfillment

Statement: Book fulfillment services.

Publishers Storage & Shipping
E B Quick
46 Development Rd
Fitchburg MA 01420
508-345-2121; Fax: 508-348-1233

Service: Fulfillment

Statement: Book storage and order fulfillment. Presently serving about 200.

Southwest School Book Depository
Jeffrey A Falke, President
1815 Monetary Ln
Carrollton TX 75006
214-245-8588

Service: Fulfillment

Statement: Serves 40 publishers.

Whitehurst & Clark Inc
100 Newfield Ave
Edison NJ 08837
908-225-2727; 908-225-1562

Service: Fulfillment

Statement: Warehousing, packing, and shipping.

Glossary Service

Lexik House
David Barnhart
75 Main St
Cold Spring NY 10516
914-265-2822

Service: Glossary service

Indexing

Alberta M Morrison
Freelance Book Indexing
498 Sixth St
Freedom PA 15042
412-774-4929

Service: Indexing

William J Richardson Associates
13072 Camino de Valle
Poway CA 92064
619-451-1715; Fax: 619-535-0426

Service: Indexing

Labels

Z-Label Systems Inc
110 E 16th Ave
North Kansas City MO 64116
816-474-4455; Fax: 816-474-3894
800-225-8992

Service: Labels

Statement: Stock continuous labels, laser printer labels, continuous tags

Looseleaf Binders

20th Century Plastics
Tom Hurley, Vice President
3628 Crenshaw Blvd
Los Angeles CA 90016
213-731-0900

Service: Looseleaf binders

Statement: Looseleaf binders, folders, plastic inserts

Ago Plastics Inc
Jeffrey Lamb,
5900 Decatur St
Glendale NY 11385
718-366-9700; Fax: 718-386-7415

Service: Looseleaf binders

Statement: Looseleaf binding, others.

American Thermoplastic Company
Customer Service
106 Gamma Dr
Pittsburgh PA 15238
412-261-6657; Fax: 412-642-7464
800-245-6600

Service: Looseleaf binders

Statement: Specialists in the manufacture of custom-imprinted ring binders and related loose-leaf products.

Binder Products
2814 Clearwater St
Los Angeles CA 90039
213-661-2171

Service: Looseleaf binders

Statement: Looseleaf binders, software packaging.

Dilley Manufacturing Company
211 E 3rd St
Des Moines IA 50309
515-288-7289; 800-247-5087

Service: Looseleaf binders

Statement: Looseleaf binders, folders, software packaging.

Eckhart & Company
2206 Production Dr
Indianapolis IN 46241
800-443-3791

Service: Looseleaf binders

Statement: Looseleaf binding, spiral, wire-o, comb-binding.

General Loose Leaf Bindery
3811 Hawthorne Ct
Waukegan IL 60087
708-244-9700; 800-621-0493

Service: Looseleaf binders

Gentile Brothers Folder Factory
116-A High St
PO Box 429
Edinburg VA 22824
703-984-8852; 800-368-5270

Service: Looseleaf binders

Statement: Looseleaf binders, folders, and more.

Hexon Corporation
Tom Krebaum
700 E Whitcomb
Madison Heights MI 48071
810-585-7585

Service: Looseleaf binders

Statement: Looseleaf binding, other bindings.

Malo & Weste Corporation
Marc Aron, Vice President
809 3rd Ave
Asbury Park NJ 07712

Service: Looseleaf binders

Statement: Looseleaf binders, folders, audio/video packaging.

Moore American Graphics
8904 S Harlem Ave
Bridgeview IL 60455
708-599-2200; Fax: 708-599-3109
800-323-2700

Service: Looseleaf binders

Statement: Looseleaf binders, folders, software packaging, displays.

National Loose Leaf
15505 Cornet Ave
Santa Fe Springs CA 90670
310-926-4511; Fax: 310-926-0222
800-421-6184

Service: Looseleaf binders

NCS
Rt 5 #171
Hot Springs AR 71913
501-623-8861; Fax: 501-525-1527
800-331-5295

Service: Looseleaf binders

Statement: We offer one color screened binders with a 48 hour turnaround up to 500, upon receipt of artwork.

Peterson Electronics Die Company
Vinyl Case Division
199 Liberty Ave
Mineola NY 11501
516-747-3833

Service: Looseleaf binders

Quality Looseleaf Manufacturing
2701 Wilmarco Ave
Baltimore MD 21223-3339
410-362-3700; Fax: 301-242-4533

Service: Looseleaf binders

Silvanus Products
40 Merchant St
PO Box 427
St Genevieve MO 63670
314-883-3521; Fax: 314-883-3160
800-822-2788

Service: Looseleaf binders

Simon Products Company
201 Mittel Dr
Wood Dale IL 60191-1196

Service: Looseleaf binders

Statement: Looseleaf binders, folders.

Universal Bookbindery Inc
1200 N Colorado
San Antonio TX 78207
210-734-9502

Service: Looseleaf binders

Statement: Looseleaf binding, wire-o, comb-binding, slip cases.

Vinyl Industrial Products
1700 Dobbs Rd
Saint Augustine FL 32086-5223
904-824-0824
800-342-8324; 800-874-0855

Service: Looseleaf binders

Statement: Looseleaf binders with clear overlay

Laminating

Graphic Laminating Inc
6185 Cochran Rd
Solon OH 44139
216-498-3400; Fax: 216-498-3410
800-345-5300

Service: Laminating

Lawyer

BZ / Rights & Permissions
Barbara Zimmerman
125 W 72nd St 5th Fl
New York NY 10023
212-580-0615

Service: Lawyer

Statement: Clears rights for literary works, film, TV, music, photographs.

Jamison & Kent
Charles A Kent
1428 DeLa Vina St
Santa Barbara CA 93101
805-965-4561

Service: Lawyer

Statement: Publishing lawyer.

Peter H Karlen
1205 Prospect
La Jolla CA 92037
619-454-9696

Service: Lawyer

Statement: Publishing lawyer.

Sheila J Levine
210 W 101st St
New York NY 10025
212-866-5353;

Service: Lawyer

Statement: Publishing-related legal services.

Roy Ward
601 N Humphreys St
Flagstaff AZ 86002
602-774-2773

Service: Lawyer

Statement: Publishing lawyer.

Marketing Consultant/ Promotions

About Books
Marilyn Ross
425 Cedar St
Buena Vista CO 81211-1500
719-395-2459; Fax: 719-395-8374

Service: Marketing, editing, design

Statement: This publishing consulting firm-headed by the bestselling authors of **The Complete Guide to Self-Publishing,** Tom and Marilyn Ross, offers "turnkey" service. They help start-up publishing companies evaluate book projects, do editing, and offer design/typesetting services. These experts also develop successful National Book Marketing Plans and powerful promotional packages.

Cypress
Cynthia Frank
155 Cypress St
Fort Bragg CA 95437
707-964-9520; Fax: 707-964-7531

Service: Production, promotion

Statement: Complete editorial, design, production, marketing and promotion services to independent publishers. Editorial services include manuscript evaluation, editing, rewriting, copymarking and proofing. Production services include book, cover and page design and make-up to camera-ready. We broker printing and offer marketing and promotional services for selected titles.

Final Draft Commercial Writing
Costa Mesa CA 92701
714-546-1201; Fax: 714-953-1494

Service: Marketing service

Statement: Copywriting for ads, news releases, catalogs, blurbs, etc.

The Huenefeld Company
John Huenefeld
41 North Rd #201
Bedford MA 01730
617-275-1070

Service: Marketing consultant

Kaufmann & Associates
Deborah Kaufmann
3422 E Howell
Seattle WA 98122
206-324-4258

Service: Marketing consultant

Statement: Marketing and promotion services.

Laing Communications
Norman P Bolotin
16250 NE 80th St
Redmond WA 98052
206-869-6313; Fax: 206-869-6318

Service: Marketing consultant

Media Management Services
Ed Meell, President
10 N Main St #301
MorrisvillePA 19067
215-493-1211
800-637-8509; 800-523-5948

Service: Marketing service

Northwest Publishers Consortium
Heather Kibbey
15800 SW Boones Ferry Rd #A-14
Lake Oswego OR 97035
503-697-7964; Fax: 503-635-9656

Service: Marketing service

Statement: Promotion a la carte for independent publishers, including exhibiting.

Planned Television Arts
Rick Frishman
25 W 43rd St
New York NY 10036
212-921-5111

Service: Marketing services

Statement: Book and author publicity, especially radio and TV tours.

Practical Book Marketing Service
28 Brookwood Rd
S Orange NJ 07079
201-763-1283

Service: Marketing service

Statement: Book marketing ad agency.

PrePRESS SOLUTIONS
Product & Training Information
11 Mt Pleasant Ave
East Hanover NJ 07936
201-887-8000; Fax: 201-884-6210
800-631-8134

Service: Scanners, programs, etc

Statement: PrePRESS SOLUTIONS is an innovative direct response distribution company dedicated to providing the global prepress community a'la carte access to the broadest array of productivity boosting products and services, including installation, training maintenance, and help line capability, at the lowest possible price.

Publishers Distribution Service
Jerrold Jenkins, Marketing Director
6893 Sullivan Rd
Grawn MI 49637
616-276-5196; Fax: 616-276-5197

Service: Marketing services

Radio-TV Interview Report
Bradley Communications
101 W Baltimore Ave
PO Box 1206
Lansdowne PA 19050-1206
215-259-1070

Service: Marketing service

Sensible Solutions
Judith Appelbaum
271 Madison Ave #1007
New York NY 10016
212-687-1761; Fax: 212-986-3218

Service: Marketing service

Statement: Headed by Judith Appelbaum, author of **How to Get Happily Published.**

Vocational Marketing Services
Michael S Walsh
18510 Carpenter St
17600 S Williams St #6
Thornton IL 60476-1077
708-877-2814

Service: Marketing service

Statement: Co-op mailings, show displays and distribution to vocation/education.

Media Insurance

Media/Professional Insurance
2300 Main St
Kansas City MO 64108
816-471-6118

Service: Media insurance

Miscellaneous

Dixie Graphics Co
PO Box 1290
Nashville TN 37202
615-832-7000

Service: Print trade manufacturing

Newsletter

Hessey Inc
424 Lafayette St
PO Box 100226
Nashville TN 37224-9904
615-244-7180; Fax: 615-255-6911
800-822-7180

Service: Specializes in newsletters.

Lewis Color Lithographers
30 Joe Kennedy Blvd
Statesboro GA 30458
912-681-6824; Fax: 912-681-8817
800-346-0371

Service: Newsletters/booklets

Paper

Alling and Cory
Richard Brown, VP Sales
2717 Jackson Ave
Long Island City NY 11101
718-784-6200; Fax: 718-392-8726
800-221-0650

Service: Paper supplier

Appleton Papers Inc
825 E Wisconsin Ave
Appleton WI 54912
414-734-9841

Service: Carbonless paper

Baldwin Paper Company
161 Ave of the Americas
New York NY 10013
212-255-1600; Fax: 212-463-7095
800-221-3213

Service: Paper supplier

Bulkley Dunton
George J Doehner
1115 Broadway 6th Fl
New York NY 10010-2890
212-337-5600; 800-443-0539

Service: Paper merchant

Cincinnati Cordage & Paper Co
Paul B Stueve
800 E Ross Ave
Cincinnati OH 45217
513-242-3600; Fax: 513-242-3871

Service: Paper merchant

Statement: Paper Distribution — Representing many mills. Manufacturing paper for the publishing industry.

Finch, Pruyn & Company
One Glen St
Glens Falls NY 12801
518-793-2541; Fax: 518-793-7364
800-833-9981

Service: Paper supplier

Fraser Paper Ltd
Paul Beaudoin
9 W Broad St
PO Box 10055
Stamford CT 06904
203-359-2544; Fax: 203-965-7197

Service: Paper supplier

Georgia-Pacific Corp
Communication Papers Division
133 Peachtree St NE
Atlanta GA 30303
414-521-4000; 800-727-3738

Service: Paper

P H Glatfelter Company
R H Boyer, Marketing VP
228 S Main St
Spring Grove PA 17362
717-225-4711

Service: Paper supplier

Hammermill Papers

Steve Pacheco
International Paper
6400 Poplar Ave
Memphis TN 38197-7000
901-763-7830; Fax: 901-763-6673
800-242-2148

Service: Paper

Statement: Hammermill Papers manufactures several grades of high quality printing papers suitable for the book publishing market.

Holliston Mills - Pajco

Hwy 11 W
Church Hill TN 37642
615-357-6141; Fax: 615-357-8840

Service: Cover materials, etc

Statement: Holliston Mills and Pajco Products manufacture a complete line of papers.

Hopper Papers

8400 E Prentice Ave
Englewood CO 80111
303-771-1420

Service: Recycled printing paper

International Paper

Kirke Vernon
1290 Ave of the Americas 9th FL
New York NY 10104
212-459-7300

Service: Paper manufacturers

Statement: Coated.

James River Corporation

John Munkenbeck, Manager
Book Publishing Papers
41 E 42nd St
New York NY 10017
212-818-0650

Service: Paper supplier

Lindenmeyr Central

Robert G McBride
Three Manhattanville Rd
Purchase NY 10577-2110
914-696-9300

Service: Paper merchant

Lindenmeyr Paper Company

100 Park Ave
New York NY 10017
212-551-3900

Service: Paper supplier

The Madden Corp

9 Rockfeller Plaza
New York NY 10020
212-246-9373

Service: Paper manufacturer

Madison Paper Industries

Main St
Madison ME 04950
207-696-3307

Service: Paper manufacturer

Mail-Well Envelope

23 Inverness Way E
Englewood CO 80112
303-790-8023; Fax: 303-799-7494
800-688-9355

Service: Paper and paper products

McKeon Paper Inc

Michael Moloney
2008 Renaissance Blvd #200
King of Prussia PA 19406
215-270-0987; Fax: 215-270-0211

Service: Paper merchant

Mead Paper

Fine Paper Division
Courthouse Plaza NE
Dayton OH 45463
513-495-6323

Service: Carbonless paper

Penntech Papers

William E Curtis, VP Sales
181 Harbor Dr
Stamford CT 06902
203-356-1850; Fax: 203-324-3782
800-243-9455

Service: Manufacturer of fine paper

Statement: Penntech Papers, a division of Williamette Industries Inc. is a manufacturer of commercial, business and book publishing papers, offering a variety of end uses such as letterhead, direct mail pieces trade and text books, brochures, annual reports and more! Contact Penntech Papers today for samples or more information about the diversity of Penntech Papers.

Repap Sales Corportation

99 Park Ave
New York NY 10016
212-687-7111; Fax: 212-697-0496
800-553-3331

Service: Paper

Statement: Repap is the fourth largest coated paper manufacturer in North America. Repap manufactures virgin and recycled, groundwood and frees heet products to the highest quality standards. Repap has sales offices located throughout the US and Canada.

ROL Paper Inc
Barry McKeon
628 E B St
King of Prussia PA 19406
215-270-9208

Service: Paper merchant

Roosevelt Paper Company
7601 State Rd
Philadelphia PA 19136
215-331-5000; Fax: 215-331-8388
800-523-3470

Service: Paper and paper products

Statement: Specialists in magazine-web and sheet.

Rourke Eno
261 Weston St
Hartford CT 06120
203-522-8211; Fax: 203-527-9953

Service: Paper merchants

Stora Papyrus Newton Falls Inc
Thomas M Hanley
PO Box 253
Newton Falls NY 13666
315-848-3321; Fax: 315-848-2081
800-448-8900

Service: Paper supplier

Statement: Manufacturers of coated papers in gloss and matte finishes.

S D Warren Company
Henry Mollenhauer, Manager
Publishing Department
225 Franklin Street
Boston MA 02110
617-423-7300; Fax: ext. 138

Service: paper supplier

Websource
Donald J Heller
161 Ave of the Americas
New York NY 10013
212-255-1600; Fax: 212-463-7095

Service: Paper merchant

Westvaco Corp
Westvaco Building
299 Park Ave
New York NY 10171
212-688-5000; Fax: 212-751-0130

Service: Paper manufacturer

Weyerhaeuser Paper Company
PO Box 829
Valley Forge PA 19482
215-251-9220

Service: Paper manufacturer

Zellerbach Publishing Papers
Al Grainick
Lexington Ave
New York NY 10170
212-297-6133; 800-448-0068

Service: Paper merchant

Proofreading

Private Eyes Proofreading
3020 Hozoni Rd
Prescott AZ 86301
602-445-6723; Fax: 602-445-6723

Service: Proofreading
Statement: Proofreading and copyediting

Norma L Sheldon
7421 E Holly St
Scottsdale AZ 85257
602-945-6573

Service: Proofreading

Photos

ABC Pictures
1867 E Florida
Springfield MO 65803
417-869-9433

Service: Photos

Statement: Stock photography.

Apco Apeda Rik Shaw
525 W 52nd St
New York NY 10019
212-586-5755

Service: Photos

Statement: B&W photos, color, transparencies.

Midwest Photo Company
W H Butler, Sales Manager
4900 G St
PO Box 686
Omaha NE 68101-0686
402-734-7200; Fax: 402-734-4319
800-228-7208

Service: Photos

Statement: B&W and color photo reprints, slides, enlargements, etc.

Photolabels USA
C E Nevin, VP Marketing
333 Kimberly Dr
Carol Stream IL 60188-1842
708-690-0132

Service: Photos

Publishers' Photographic Services
Pine Ridge Dr
Ellicott City MD 21042
410-750-6225

Service: Photos

Prepress

Advanced Polymer Technologies
3581 Big Ridge Rd
Spencerport NY 14559
716-352-9700; Fax: 716-352-9676

Service: Prepress

Automated Graphic Systems
Mark Edgar
4590 Graphics Dr
White Plains MD 20695
301-274-4441; Fax: 301-843-6339
800-678-8760

Service: Prepress, electronic

Battle Creek Litho Inc
549 Major Ave
Battle Creek MI 49015
616-964-0177; Fax: 616-964-3210

Service: Prepress

Beaumont Graphic Ltd
7800 Bonhomme Ave
St Louis MO 63105
314-454-9755; Fax: 314-534-1565

Service: Prepress

Capitol Engraving Co
808 Lea Ave
Nashville TN 37203

615-244-6603; Fax: 615-255-9477

Service: Prepress

Cardinal Communications Group Inc
545 W 45th St
New York NY 10036
212-489-1717; Fax: 212-581-9690

Service: Prepress

Celo Valley Books
Diana Donovan
346 Seven Mile Ridge Rd
Brunsville NC 28714
704-675-5918

Service: Prepress design and output

Chromagen
1100 University Ave #206
Rochester NY 14607
716-256-6218

Service: Prepress services

Citiplate Inc
275 Warner Ave
Roslyn Heights NY 11577
516-484-2000; Fax: 516-484-9775

Service: Prepress

Clarinda Color
1780 W 7th St
St Paul MN 55116
612-699-2771; Fax: 612-699-9008
800-522-5018

Service: Prepress, film

CNW Inc
4710 Madison Rd
Cincinnati OH 45227
513-321-2775; Fax: 513-321-2618
800-327-5900

Service: Color separations, engraving

Statement: Another Fax: 513-321-2013.

Collins Miller & Hutchings
Division Wace USA
225 W Superior St 1st & 2nd Fl
Chicago IL 60610
312-943-0400; Fax: 312-943-6186

Service: Prepress

Color Response Inc
3101 Stafford Dr
Charlotte NC 28208
704-392-1153; Fax: 704-392-1018

Service: Prepress

Colorbrite Inc
1001 Plymouth N
Minneapolis MN 55411
612-522-6711

Service: Prepress

Colour Graphics Corp
3355 Republic
Minneapolis MN 55426
612-929-0357

Service: Prepress

Colour Image Corp
2343 Miramar Ave
Long Beach CA 90815
310-498-3731; Fax: 310-498-6187

Service: Prepress

Container Graphics Corp
200 Mackenan Dr
Cary NC 27511
919-481-4200; Fax: 919-469-4897

Service: Prepress

R R Donnelley & Sons Co
Jerry D Butler or Gary S Davis
Corporate Headquarters
77 W Wacker Dr
Chicago IL 60601
312-326-8000

Service: Prepress and composition

Edwards Brothers Inc
John Pugsley
2500 S State St
PO Box 1007
Ann Arbor MI 48106-1007
313-769-1000

Service: Prepress and composition

Excelsior Processing & Engraving
1466 Curran Hwy
North Adams MA 01247
413-664-4321; Fax: 413-663-4676

Service: Prepress

Graphic Art Systems Corp
2150 N Lincoln St
Burbank CA 91504
818-848-5571; Fax: 818-841-0834

Service: Prepress

Imperial Metal & Chemical Co
717 Main St
Holyoke MA 01040
413-533-7181

Service: Prepress

Information International Inc
5933 W Slauson Ave
Culver City CA 90230
310-390-8611; Fax: 310-391-7724

Service: Prepress

Jahn & Ollier Engraving Co
817 W Washington Blvd
Chicago IL 60607
312-666-7080; Fax: 312-666-2652

Service: Prepress

LaserMaster
6900 Shady Oak Rd
Eden Prairie MN 55344
612-944-6069; Fax: 612-944-6932
800-950-6868

Service: Prepress

Lasky Company
Ronald H Barnhard
7 E Willow St
Millburn NJ 07041
201-376-9200; Fax: 201-376-3832

Service: Prepress and composition

Lehigh Press-Colortronics
361 Bonnie Ln
Elk Grove Village IL 60007
708-364-8000; Fax: 708-364-8170

Service: Prepress

Liberty Engraving Co
1112 S Wabash Ave
Chicago IL 60605
312-786-0600

Service: Prepress

Lincoln Graphics Inc
1110 S Cornell Ave
Cherry Hill NJ 08002
609-662-3433; Fax: 609-665-9415

Service: Prepress

LSI-KALA
2325 N Burdick St
Kalamazoo MI 49007
616-381-3820; Fax: 616-382-9070

Service: Prepress

MacKay Gravura Systems Inc
PO Box 37160
Louisville KY 40233
502-637-8731; Fax: 502-636-2647

Service: Prepress

Matthews International Corp
PO Box 4999
Pittsburgh PA 15206
412-665-2500; 412-665-2550

Service: Prepress

Charles Mc Henry Printing Co
PO Box 68
Greensburg PA 15601
412-834-7600; Fax: 412-836-7759

Service: Prepress

Media Graphics Corp
427 S LaSalle St
Chicago IL 60605
312-922-6800; Fax: 312-922-6633

Service: Prepress

Munder Color Co
2771 Galilee Ave
Zion IL 60099
708-872-5462; Fax: 708-872-0265

Service: Prepress

National Correct Color Service
640 Scribner Ave NW
Grand Rapids MI 49504
616-459-9548; Fax: 616-459-2891
800-253-4670

Service: Prepress

NEC Inc
1504 Elm Hill Pike
Nashville TN 37210
615-367-9110; Fax: 615-360-7952

Service: Prepress

New England Book Components Inc
Jerry G Creteau
125 Industrial Park Rd
Hingham MA 02043
617-749-8500; Fax: 617-479-2106

Service: Prepress

Northwestern Colorgraphics Inc
1457 Earl St
Menasha WI 54952
414-722-3375; Fax: 414-722-6925

Service: Prepress

Paragon Publishing Systems
10 Corporate Dr
Bedford NH 03110
693-471-0077; Fax: 603-668-9383
800-431-1403

Service: Prepress system

Photo Mechanical Services Inc
333 W 78th St
Minneapolis MN 55420
612-881-3200; Fax: 612-881-5076

Service: Prepress

Printing Developments Inc
2010 Indiana St
Racine WI 53405
414-554-9425; Fax: 414-554-7828

Service: Prepress

Progress Graphics Inc
418 Summit Ave
Jersey City NJ 07306
201-653-0717; Fax: 201-653-8209

Service: Prepress

Publishing Professionals
Scot Patterson
681 W 28th Ave (97405)
PO Box 1587
Eugene OR 97440
503-683-3244; Fax: 503-683-0871

Service: Prepress

Statement: We are a full-service book production company, specializing in PC-based page design and composition. On full-service projects we can provide copyediting, proofreading, cover and jacket design, illustration, developmental permissions, and indexing Contact Scot Patterson for a bid package checklist.

Quad\Graphics
Duplainville Rd
Pewaukee WI 53072-4195
414-246-9200; Fax: 800-888-9010

Service: Prepress and composition

Quality House of Graphics
4747 Van Dam St
Long Island City NY 11101
718-784-7400; Fax: 718-937-5179

Service: Platemaking

Revere Graphic Service
815 S Jefferson St
Chicago IL 60607
312-922-8816; Fax: 312-922-2086

Service: Prepress

Rochester Empire Graphics Corp
150 Research Blvd
Rochester NY 14623
716-272-1100; Fax: 716-272-8846

Service: Prepress, color separation

Schawkgraphics Inc
1600 E Sherwin Ave
Des Plaines IL 60018
708-296-6000; Fax: 708-296-9466
800-621-1909

Service: Prepress

Signature Graphics
Jeffrey C Keyser
116 W Cooper St
Mechanicsburg PA 17055
717-766-1197; Fax: 717-766-1197

Service: Electronic prepress

Statement: A full-service electronic prepress company supporting sc itex APR. workflow; offer systems consulting/configuration; authorized reseller for: Adobe Systems, IBM, Linotype-Heil and others. Jobs range from business cards to 4/C catalogs and magazines.

Stevenson Photo Color Co Inc
5325 Ridge Ave
Cincinnati OH 45213
513-351-5100; Fax: 513-351-8100

Service: Prepress

Tapsco Inc
Herbert B Landau
PO Box 131
Akron PA 17501-3702
717-859-2006; 800-548-3795

Service: Prepress, typesetting

Statement: Subsidiary of Lancaster Press Inc

Trece Mark Inc
806 Race Rd W
Baltimore MD 21221
410-687-5252

Service: Prepress

Ultra Scan Inc
2 Industrial Park
Hudson NH 03051
603-880-1465; Fax: 603-880-0051

Service: Prepress

Wace USA Graphic Arts
2 N Riverside Plaza
Chicago IL 60606
312-876-0533; Fax: 312-876-0120

Service: Prepress

Waldman Graphics
Richard Greene
9100 Pennsauken Hwy
Pennsauken NJ 08110
609-662-9111; Fax: 609-665-1789
800-543-0955

Service: Prepress and composition

Statement: Printers, typographers, prepress specialists.

Western Lithotech
3433 Tree Ct Industrial Blvd
St Louis MO 63122
314-225-5031; Fax: 314-825-4681

Service: Prepress

Print Consultant

Good Print Consultant
Howard L Good
112 Surrey Ln
Westfield NJ 07090
908-232-3418; Fax: 908-232-3418

Service: Print production consultant

Statement: Effective supervision of printing projects.

Special

O'Neil Data Systems Inc
Scott Pruner
12655 Beatrice St
Los Angeles CA 90066
310-448-6400; Fax: 310-577-7483

Service: Computerized book manufacturer

Rietmulder Associates
Jim Rietmulder
988 Siddonsburg Rd
Lewisberry PA 17339
717-938-6000; Fax: 717-938-0190
Service: Black and white scans to disk

Statement: Reflective art scanned on Autokon 1000 to Mac or PC floppies, syquest, opticals for publishers of printed or electronic material.

Software

Ad Express Inc
William D Rilling
400 Technecenter Dr
Milford OH 45150
513-248-8110; Fax: 513-248-8102

Service: Electronic ad delivery service

Statement: Software for your PC or Mac that allows you to deliver your camera-ready ads to newspapers across the US in minutes.

Auto-Graphics Inc
John L Varnau, VP Sales & Mktg
3201 Temle Ave
Pomona CA 91768
909-595-7204 ext 307; Fax: 909-595-3506
800-776-6939

Service: CD-ROM Electronic publishing
and database services

Statement: Software developer of the SGML Smart Editor for editorial solutions operating on DOS platforms, Novell networks and Unix-based Client/Server systems founded in 1950. Specializing in CD-ROM electronic publishing and database services for publishers of directories, catalogs, dictionaries, manuals and other complex reference works. Full composition services for computerized typesetting on Xyvision and Macintosh systems.

The Edge
Connie Lowery
913 Lyttleton St
PO Box 1195
Camden SC 29020
803-432-7674; Fax: 803-425-5064

Service: Computer software

Statement: Selection of the month club system—PC-based system for handling.

The Mail Order Accountant
Goldsmith & Associates
48 Shattuck Square #86
Berkeley CA 94704
415-540-8396

Service: Computer software
Statement: Software for mail order fulfillment.

The Mail Order Wizard
The Haven Corporation
802 Madison St
Evanston IL 60202
312-869-3434; 800-782-8278

Service: Computer software

Statement: Software for fulfillment of mail orders

Metagroup Consultants
Marek Karon
12103 S Brookhurst St #E-410
Garden Grove CA 92642-3065
714-638-8663

Service: Computer software

Statement: Mail Business Manager, Version 3.5 is a comprehensive PC computer.

PIIGS Software
Lisa Carlson
Upper Access
PO Box 457
Hinesburg VT 05461
800-356-9315

Service: Computer software

Statement: Software for book publishers: order entry, fulfillment, inventory.

RH Communications
PO Box 26225
Colorado Springs CO 80936
719-592-9204; Fax: 719-592-0960
Send RFQ's to: Ron Hillestad

Service: Computer software

Statement: Publishers Business System computer software programs are designed for the publisher's business management needs. PBS provides invoicing and bookkeeping support, including information management which makes product and processing more efficient.

Upper Access, Inc
Steve & Lisa Carlson
PO Box 457
Hinesburg VT 05461
802-482-2988; Fax: 802-482-3125
800-356-9315

Service: Software service

Statement: Publishers Invoice and Information Generating System: PIIGS. Industry specific tool. Royalties, inventory, consignment, commisions, maillist, etc.

Typesetting

A-1 Composition Co
208 S Jefferson St
Chicago IL 60661
312-236-8733; Fax: 312-236-1807
Service: Typesetting

A-R Editions Inc
801 Deming Way
Madison WI 53717
608-836-9000

Service: Typesetting

Statement: Typesetting books.

Advanced Typographics Inc
461 Ferry St SE
Salem OR 97301
503-364-2100

Service: Typesetting

AlphaGraphics Corporation
3760 N Commerce Dr
Tucson AZ 85705
602-293-9200; Fax: 602-888-9641

Service: Typesetting

Statement: Typesetting, prepress, data conversion, mailing list maintenace.

Arrow Typographers Inc
214 Liberty St
Neward NJ 07102
201-622-0111; Fax: 201-623-3521

Service: Typesetting

B & J Typesetting
6720 Emerald St
Boise ID 83704
208-376-1771; Fax: 208-378-0563

Service: Typesetting

Black Dot Graphics Inc
6115 Official Rd
Crystal Lake IL 60014
815-459-8520; Fax: 815-459-8062

Service: Typesetting

Books International
5555 Oakbrook Pky #340
Norcross GA 30093
404-242-6223; Fax: 404-242-6209

Service: Typesetting

Statement: Typesetting books.

Bryd Data Imaging Group
5408 Port Royal Rd
Springfield VA 22151
703-321-8610; Fax: 703-321-7634

Service: Typesetting

C C S Graphics Associate
11755 Slauson Ave #B
Santa Fe Springs CA 90670
310-695-1555

Service: Typesetting, design

Carter Printing Co
PO Box 6901
Richmond VA 23230
804-359-9206; Fax: 804-359-4907

Service: Typesetting

Central Graphics Inc
725 13th St
San Diego CA 92101
619-234-6633; Fax: 619-234-6817

Service: Typesetting

Chambers & Sons Inc
PO Box 719
Baltic CT 06330
203-822-8213; Fax: 203-822-9758

Service: Typesetting

Chronicle Type & Design
1255 23rd St NW #785
Washington DC 20037
202-466-1090; Fax: 202-296-2691
800-448-2771

Service: Typesetting and design

Clarinda Co
220 N 1st St
Clarinda IA 51632
712-542-5131; Fax: 712-542-5059
800-831-5821

Service: Typesetting

Coghill Composition Company
Jim Coghill
1627 Elmdale Ave
Richmond VA 23224
804-231-0224; 800-368-3594

Service: Typesetting

Statement: Typesetting books

Col D'var Graphics
Nanette Workman
1626 N Prospect Ave #908
Milwaukee WI 53202

Service: Typesetting

Com Com Inc
834 N 12th St
Allentown PA 18102
610-437-9656; Fax: 610-437-1387

Service: Typesetting

Composing Room Inc
841 Chestnut St #C-40
Philadelphia PA 19107
215-829-9611; Fax: 215-922-1685

Service: Typesetting

Dahl & Curry Inc
1320 Yale Pl
Minneapolis MN 55403
612-339-0615; Fax: 612-339-0993

Service: Typesetting

Davidson Group
111 N Jefferson St
Chicago IL 60661
312-559-8973; Fax: 312-559-9029

Service: Typesetting

Design Typographers Inc
1101 Taft Ave
Berkley IL 60163
312-944-0010; Fax: 312-944-4007

Service: Typesetting

Digicon Graphics Inc
40 Gardenville Pkwy
West Seneca NY 14224
716-668-1619; Fax: 716-668-5838

Service: Typesetting

Dynatype
501 E Harvard St
Glendale CA 91205
818-243-1114; Fax: 818-243-0734

Service: Typesetting

Electra Typography
340 Brannan St #202
San Francisco CA 94107
415-495-3730; Fax: 415-495-3587

Service: Typesetting

Electronic Publishing Center
LeRoy St
New York NY 10014
212-727-2655

Service: Typesetting, prepress

Excelsior Printing Co
60 Roberts Dr
North Adams MA 01247
413-663-3771; Fax: 413-663-3465

Service: Typesetting

Fry Communications Inc
Chris Yates
800 W Church Rd
Mechanicsburg PA 17055-3198
717-766-0211
800-334-1429

Service: Typesetting

Statement: Typesetting books and magazines.

Graffolio
Sue Knopf
1528 Mississippi St
La Crosse WI 54601
608-784-8064; Fax: 608-784-8064

Service: Book design and typesetting

Statement: We design and typeset books — from manuscript (or Mac or DOS disk) to camera-ready copy. Also: one and two-color brochures, forms, catalogs, manuals, graphics, cartoons, scanning and copyediting. Ask for our brochure, samples and flyer, "Save Time, Money and Your Sanity on Your Next Book Project."

Graphic Sciences Corporation
830 First Ave NE
Cedar Rapids IA 52402
319-366-3583; Fax: 319-366-0735

Service: Typesetting

Graphics Unlimited
3000 2nd St N
Minneapolis MN 55411
612-588-7571; Fax: 612-588-8783

Service: Typesetting

Heather Marks Designs
Cliff Sanderlin
10522 Robbers Roost Rd
Edmonds WA 98020
206-546-8983

Service: Typesetting

Heritage Publishing Company
Derek Wood
2402 Wildwood Ave
Sherwood AR 72116
501-835-5000; Fax: 501-835-5834
800-643-8822

Service: Typesetting

Statement: Typesetting and composition services. Over 30 years experience.

Interchange Inc
PO Box 16244
Saint Louis Park MN 55416
612-929-6669

Service: Typesetting

Intergraphics
106A S Columbus St
Alexandria VA 22314
703-683-9414

Service: Typesetting

International Computaprint Corp
12712 Dupont Cir
Tampa FL 33626
813-855-4635; Fax: 813-855-2309

Service: Typesetting

K-W Publishing Services
Mike Kelly
11532 Alkaid Dr
San Diego CA 92126-1370
619-566-6489

Service: Typesetting

L Grafix Inc
Rick Sanders
108 NW 9th #201
Portland OR 97209
503-248-9713; Fax: 503-274-7828

Service: Typesetting, book formating, print and bind

Statement: Since 1980 L Grafix has been a consistent supplier to book, family history, manual and catalog publishers and self publishers.

Maryland Composition Co Inc
Barry Hart, Marketing VP
6711 Dover Rd
Baymeadow Industrial Park
Glen Burnie MD 21060
410-760-7900; Fax: 410-760-5295

Service: Typesetting

MonoLith
54 Granby St
Bloomfield CT 06002
203-242-3006; Fax: 203-242-8441

Service: Typesetting

NAMECO-Worcester Stamp
75 Webster St
Worcester MA 01603
508-756-7138; Fax: 508-756-4307

Service: Typesetting

Napp Systems Inc
360 S Pacific St
San Marcos CA 92069
619-744-4387; Fax: 609-489-1853
800-854-2860

Service: Typesetting

PDR Royal
50 W 23rd St 5th Fl
New York NY 10010
212-477-3300; Fax: 212-647-0209

Service: Typesetting

Pearson Typographers Corp
1101 N Taft Ave
Berkeley IL 60163
708-449-5200; Fax: 708-449-5221

Service: Typesetting

Pointer's International Corp
8224 White Settlement Rd
Ft Worth TX 76108
817-246-4931; Fax: 817-246-0301

Service: Typesetting

Recorder Sunset Press
99 S Van Ness Ave
San Francisco CA 94103
415-554-1000; Fax: 415-861-2278

Service: Typesetting

Regency Thermographers Inc
55 W 27th St
New York NY 10001
212-447-1700; Fax: 212-779-3426

Service: Typesetting

Regency Typographic Service Inc
2867 E Alegherry Ave
Philadelphia PA 19134
215-425-8810; Fax: 215-634-0780

Service: Typesetting

Sans Serif Inc
Sharon Curtis
2378 E Stadium Blvd
Ann Arbor MI 48104
313-971-1050; Fax: 313-971-7534

Service: Typesetting

Science Typographers Inc
5 Industrial Blvd
Medford NY 11763
516-924-4747

Service: Typesetting

Shadow Canyon Graphics
2276 S Pebble Beach Ct
PO Box 2765
Evergreen CO 80439
303-670-0401; Fax: 303-670-0406

Service: Typesetting, proofreading

Skilset/Alpha Graphix Inc
5290 W Washington
Los Angeles CA 90016
213-937-5757

Service: Typesetting

Southern California Print Corp
Mike Flanigan
1915 Midwick Dr #B
Altadena CA 91001
818-398-3501; Fax: 818-398-3565

Service: Camera ready paper and film

Statement: We offer Linotronic camera-ready paper and film. Special offer for Ad-Lib readers. Mention code AD040794 and pay $1 per page!

Southwestern Typographics
2820 Taylor St
Dallas TX 75226
214-748-0661

Service: Typesetting

Stanton Publications Services
Don Leeper
2402 University Ave #701
St Paul MN 55114
612-642-9241; Fax: 612-378-2141

Service: Typesetting

Statement: Typesetting books, design.

Superior/Premier Graphics Inc
571 W Polk St
Chicago IL 60607
312-294-2670; Fax: 312-922-2086

Service: Typesetting

Synergistic Data Systems Inc
Dale Schroeder
3975 E Foothill Blvd #329
Pasadena CA 91107
818-351-8622; Fax: 818-351-7717

Service: Typesetting

Statement: High-resolution (1000 DPI) laser typesetting

Tegra Varityper
11 Mt Pleasant Ave
East Hanover NJ 07936
201-887-8000; Fax: 201-887-0731

Service: Typesetting

TPH Graphics
1177 W Baltimore St
Detroit MI 48202
313-875-1950; Fax: 313-875-6046
Service: Typesetting

TSI Graphics Inc
1300 S Raney St
Effingham IL 62401
217-347-7733; Fax: 217-342-9611

Service: Typesetting

Type House & Duragraph Inc
3030 N 2nd St
Minneapolis MN 55411
612-588-7511; Fax: 612-588-8783

Service: Typesetting

Typetronix Inc
2607 Hanson St
Ft Myers Fl 33901
813-337-2227; Fax: 813-337-2225

Service: Typesetting

Statement: Typesetting books.

Typographic House
63 Melcher St
Boston MA 02210
617-482-1719; Fax: 617-357-7324

Service: Typesetting

Typography Plus
1601 Prudential Dr
Dallas TX 75235
214-630-2800; Fax: 214-630-0713

Service: Typesetting

Villager Communications Inc
Charles Thomes
757 S Snelling Avenue
Saint Paul MN 55116-2296
612-699-1462; Fax: 612-699-6501

Service: Typesetting, cover design

Statement: Typesetting, cover design, typographic design. Customers include Hazelden, Red Leaf Press, Thomas Register.

Willens-Michigan Corp
1959 E Jefferson Ave
Detroit MI 48207
313-567-8900; Fax: 313-567-0072

Service: Typesetting

York Graphic Services
Jamie Barr, Marketing
3600 W Market St
York PA 17404
717-792-3551; Fax: 717-792-0532

Service: Typesetting, prepress, etc

Wholesalers

Quality Books
Michael L Huston, Manager, Vendor
    Relations
1003 W Pines Rd
Oregon IL 61061-9680
800-323-4241; Fax: 815-732-4499

Service: Library marketing and distribution

Statement: Direct-selling distributor of small press books and independently produced special interest videos to libraries throughout the United States and Canada. Quality books also provides Publisher's Cataloging in publication for small publishers.

Unique Books
Richard Capps
4230 Grove Ave
Gurnee IL 60031
708-623-9171; Fax: 708-623-7238

Service: Library distributor

Statement: One of the largest small press and special interest video distributors to the library market.

Glossary

The definitions for this glossary have been adapted for the most part from Braun-Brumfield's *Book Manufacturing Glossary*. Write to them for a copy of the complete booklet as well as their paper samples book and paper bulk chart.

A

Across the Grain — The direction 90 degrees, or at a right angle, to the paper grain.

Acid-Free Paper — Paper which is free from acid or other ingredients likely to have a destructive effect. Libraries prefer books with acid free paper, because these books last longer.

Actual Value Shipment — A truck shipment insured for the actual value of the commodity rather than the amount specified by ICC regulations. In the case of books, the standard insurance is $1.65 each, rather than the actual value, which may be more or less. An actual value shipment is at a higher freight rate.

Advance Copies — Finished books sent to a customer, usually by air mail, prior to bulk shipment of the balance of the order.

Aluminum Plate — A thin sheet of aluminum used in lithography for some press plates; used for both surface-type and deep-etch offset plates.

Alterations — In composition, changes made in the copy after it has been set in type.

Antique — A natural or cream-white color of paper.

Antique Finish — A term describing the surface, usually on book and cover papers, that has a natural rough finish.

Appendix — An addition to the back matter of a book listing material related to the subject but not necessarily essential to its completeness.

Art Work — A general term used to describe photographs, drawings, paintings, hand lettering, and the like, prepared to illustrate printed matter.

Author's Alterations (A.A.'s) — Changes from original copy or author's corrections.

B

BMI (Book Manufacturing Institute) — A graphic arts industry organization consisting of book manufacturers and related companies.

Back Lining — (1) A paper or fabric strip used to reinforce the spine of a casebound book after rounding and backing. It provides the means for a firm connection between book and case. (2) The paper stiffening used in the backbone of a case, between the binder's boards.

Back Matter — Material printed at the end of a book; usually includes appendix, addenda, glossary, index, and bibliography.

Back Up — To print the reverse side of a sheet already printed on one side. Printing is said to back up when the printing areas on both sides are exactly opposite one another.

Backbone — The back of a bound book connecting the two covers; also called the Spine.

Back Cover — Back outside surface of a casebound or softcover book.

Back Margin — The distance between live matter of a left hand page and live matter of a right hand page — always measured in even PICAS, for example: 8, 10, or 12 picas. This measurement is critical in producing a good design and continuity to your printed product. It is also used as a guide in stripping and becomes the bible for finishing and bindery.

Band — (1) A strip of paper, printed or unprinted, which wraps around loose sheets (in lieu of binding with a cover) or assembled pieces. (2) The operation of putting a paper band around loose sheets or assembled pieces. (3) Metal straps wrapped around skids of cartons to secure the contents to the skid for shipment.

Basis Weight — A term used to distinguish the various weights of paper. For example, a basis weight of 60 pounds means a ream (500 sheets) of such paper in size 25" x 38" weighs 60 pounds. The 25" x 38" size is standard for book papers. Other kinds of paper (bond, bristol, cover, newsprint, etc.) determine their basis weight on different base sizes, and thus weights are not comparable with book paper weights.

Benday — A method of laying a screen (dots, line, and other textures) on artwork or plates to obtain various tones and shadings.

Bill of Lading (B/L) — The document that originates a shipment; it contains all the necessary information for the carrier to handle the shipment in transit, such as special instructions for protection from the elements as well as delivery information. A Bill of Lading is: a) The contract of carriage, b) Documentary evidence of title, and c) Receipt for the goods.

Bind—To join pages of a book together with thread, wire, adhesive or other means; to enclose them in a cover when so specified.

Bind Margin—The gutter or inner margin from the binding to the beginning of the printed area.

Binder's Boards—A stiff, high-grade composition board used in book binding inside the cloth of the case; more dense than chipboard.

Binding Edge—Edge of sheet or page that is nearest the saddle of the book. Right-hand pages are bound at the left side, left-hand pages at the right side. Ample space between the line matter and the binding edge should be allowed.

Black and White—Originals and reproductions displayed in monochrome (single color) as distinguished from polychrome or multicolor.

Blank—An unprinted page.

Bleed—Any part of the printed area (usually a photograph or some non-illustrative artwork; never headlines or copy) that extends beyond the trim edge of the page.

Blind Embossing—A design which is stamped without gold leaf or ink, giving a bas-relief effect.

Blind Stamp—A design which is impressed (stamped) without foil or ink, giving a bas-relief effect.

Blowup—An enlargement of the original size.

Blues or **Blueprint**—See: Silverprint.

Book Cloth—Cotton gray goods woven as for any other fabric and finished in one of three ways; starch-filled, pyroxlin impregnated, or plastic coated. Cloth comes in different weights and weaves. The quality of the cloth is determined by the number of threads per inch and the tensile strength of the threads.

Booklet—Any pamphlet sewn, wired, or bound with adhesive, containing a few pages and generally not produced for permanence.

Book Paper—A class or group of papers having common physical characteristics that, in general, are most suitable for book manufacture. Book paper is made to close tolerances on caliper (pages per inch).

Broadside—A broadside may be printed on both sides or on portions of a single fold. Product size is usually 22" x 16 ½" folded to 8 ¼" x 11" — heads not trimmed.

Buckram—A book cloth which can be identified by heavy, coarse threads; available in a number of grades.

Bulk—(1) The stacked thickness of paper usually expressed as pages per inch (ppi). (2) The thickness of a book exclusive of cover.

Bulking Dummy—Unprinted sheets folded in the signature size and signature number of a given job to determine the actual bulk; used to establish dimensions for cover art preparation or by the binder to determine case size for a casebound book.

C

C1S (Coated One Side) — Cover or text paper which has been coated on one side only; usually used for covers and dust jackets.

C2S (Coated Two Sides) — Cover or text paper which has been coated on both sides.

Calendered Paper — Paper which has been glazed during manufacture by passing it through a stack of polished metal rollers called calenders.

Caliper — The thickness of a single sheet of paper; usually expressed in thousandths of an inch.

Camera Ready Copy — Customer furnished material suitable for photographing and reproduction. See: Copy.

Case Bound — A term denoting a book bound with a stiff or hard cover.

Casebinding — A method of binding in which the cover is made separately and consists of rigid or flexible boards covered with cloth, paper or other material, in such a manner that the covering material surrounds the outside and edges of the board. Covers always project beyond the edges of the text pages.

Check Copy — (1) A folded and gathered but unbound copy of a book sent to a customer for approval before binding. (2) The gathered, trimmed copy which is inspected and approved prior to any binding operations; used as a guide in bindery for assembling in the proper sequence, including inserts, furnished items, etc.

Chipboard — A less expensive and less durable single ply substitute for binder's boards; also used as a backing for padded forms. It is more likely to absorb moisture or to warp than binder's boards.

Close Register — Used to describe low trap allowance requiring more press printing position accuracy; also known as Tight Register.

Clothbound — See: Casebound.

Coated Paper — Paper having a surface coating which produces a smooth gloss finish. When producing four-color process printing, high quality halftones or heavy ink coverages, coated papers will produce the best results. Magazines, books, booklets, trade journals, and catalogs use these papers for high-quality printing.

Cold Type — Text composition prepared for photomechanical reproduction with a typewriter, by hand-lettering, or by photocomposition. Hot type is set using melted lead to form the type. All typesetting not using melted lead is known as Cold Type.

Collate — In binding, collating is the gathering (assembling) of sections (signatures in proper sequence) for binding.

Color Key — Color proofing material from 3-M (available as positive or negative) of a light-sensitive polyester-base film, supplied in any one of a number of colors.

Color Process Work – A reproduction of color made by means of photographic separations. The printing is done using cyan, magenta, yellow, and black inks, each requiring its own negative. This is the means of full color reproduction in printing; also called Four Color.

Color Separation – (1) In photography, the division of colors of a continuous-tone full color original, each of which will be reproduced by a separate printing plate carrying a color. (2) In lithographic platemaking, manual separation of colors by handwork performed directly on the printing surface. (3) For some kinds of color reproduction, a paste-up artist can prepare separate overlays for each color.

Comb – See: Plastic Comb Binding.

Composition – The assembling of characters into words, lines, and paragraphs of texts or body matter for reproduction by printing.

Confirming Proof – A proof confirming to the customer that the page, as shown by the proof, is the way the page will print. No approval will be required or expected.

Contact Print – A photographic print made from either a negative or a positive, exposed in contact with sensitized paper or film, instead of by projection.

Continuous Tone – A photographic image without a halftone dot screen containing gradient tones from black to white.

Copy – The photograph, paste-up, art, or other material furnished for reproduction. A better term is Original, since it is from this material that reproduction originates. Most commonly, the term Photocopy is used. See also Camera Ready Copy.

Corner Marks – Open parts of squares placed on original copy as a positioning guide.

Covers – **Cover 1**: Outside front cover. **Cover 2**: Inside front cover. **Cover 3**: Inside back cover. **Cover 4**: Outside back cover.

Crop Marks – Marks along the margins of an illustration indicating the part of the illustration to be reproduced.

Cropping – The process of defining the reproduction image area of line and continuous-tone art by drawing crop marks or of continuous-tone art by producing windows in negatives by means of dropout masks.

Cross Grain – A fold at right angles to the binding edge of a book, or at right angles to the direction of the grain in the paper or board; also called Against the Grain.

Customer Furnished (CF) – Any material supplied by the customer, for example, paper, printed covers, or separations.

Cutoff – The paper dimension fixed by the size of the press cylinder, which limits the cutoff at right angles to the travel of the web.

D

Data Conversion—Taking the data produced on one system (e.g., a word processor) and transferring it to another system (e.g., a phototypesetter) by changing the codes and the format via an interface, but leaving the text alone, thereby saving proofreading and keyboarding time. Data may be delivered for conversion over telephone lines or by sending the media (disks) through the mail.

Deboss —General term covering the process of lowering the surface with the use of a male die.

Density—(1) The specific gravity of paper or weight per unit volume. (2) A measure of the degree of blackness. (3) The blackness and weight of type set in phototypesetting.

Die Cutting—The use of sharp steel rules to cut special shapes like labels, boxes and containers, from printed or unprinted material. Die-cutting can be done on flat-bed or rotary presses.

Digest—A size of publication that is usually 5¼ x 8¼ inches.

Dividers—Tabbed sheets of index or other heavy stock used to identify and separate specific sections of a book; used in looseleaf and bound books (tipped in on bound books).

Dot Gain—The slight enlargement of a halftone dot during exposure, development, or on the printing press.

Dot Spread—In printing, a defect in which dots print larger than they should, causing darker tones or colors.

Drill(ing)—Punching of holes in folded sections, trimmed or untrimmed, or in finished books, which will permit their insertion over rings or posts in a binder.

Drop-Out—A halftone negative exposed to eliminate the extreme highlight dots so the white background of the artwork will not produce a printing dot on the negative. This technique can also be used to eliminate printing of shadow dots.

Dull Coated Paper—Coated paper that has a dull finish. Such paper may have a dull finish on one side for text and a glossy finish on the other side for fine color work or halftones. The dull finish eliminates glare for reading solid copy.

Dummy—(1) A preliminary drawing or layout showing the positioning of illustrations and text as they are to appear in the final reproduction. (2) A set of blank pages made up in advance to show the size, shape, form, and general style of a piece of printing.

Duotone—A term for a two-color halftone reproduction from a one-color original, requiring two halftone negatives for opposite ends of the gray scale, one emphasizing highlights and the other emphasizing shadows. One plate usually is printed in dark ink; the other in a lighter one.

Dust Jacket—The printed or unprinted wrapper, usually paper, placed around a case bound book.

Dylux (Silverprint)—A pre-press proof of negatives made by photographic techniques. A DuPont product.

E

Edition Binding—See: Casebinding.

Embossed Finish—Paper with a raised or depressed surface resembling wood, cloth, leather, or other pattern.

Embossing—(1) Impressing an image in relief to achieve a raised or depressed surface, either over printing or on blank paper for decorative purposes. (2) The swelling of the image on an offset blanket, due to its absorbing solvents from the ink. (3) A finish on paper or cloth.

Emulsion Side—The side of a photographic film to which the emulsion is applied, and on which the image is developed; the side on which scratching or scribing can be done.

Endsheets—Four pages each at the beginning and end of a casebound book; one leaf of each being solidly pasted against the inside board of the case. Stock is stronger and heavier than text stock; may be white or colored stock, printed or unprinted. Other common terms frequently used are Endpapers, Endleaves or Lining Paper.

Estimate—A price provided to a customer based on the specifications outlined on the estimate form, it is normally sent prior to entry of an order, and prices may change if the order specifications are not the same as the estimate specifications.

Exposure—A step in photographic processes during which light produces an image on the light-sensitive coating on film or plates; in photography, called Shot; in platemaking, called Burn.

F

F & G—A term used to refer to a folded and gathered, but unbound, copy of a book; sometimes called a Check Copy.

F.O.B. (Free on Board)—The f.o.b. point is usually the inland point of departure, or the port of shipment. The buyer pays all shipping charges beyond the f.o.b. point.

Filling In—A condition in offset lithography where ink fills the area between the halftone dots or plugs up the type; also known as Plugging or Filling Up.

Film Lamination—See: Laminate.

Finish — The general surface properties of paper determined by various manufacturing techniques.

Finishing — Any post-press operations such as folding, binding, etc.

Flat — A stripped flat, in photographic platemaking for offset lithography, is an assembled (stripped) composite of negatives or positives from which a printing plate is made.

Foil — Tissue-thin material faced with metal or pigment used in book stamping with a Stamping Die.

Foil emboss — Embossing the image with the addition of foil.

Foldout — An insert which is wider than the page width of a publication. In some cases, one or more vertical folds are required so that it will occupy the same area as the page. Foldouts are used to accommodate large illustrations, charts, and the like.

Folio — (1) A page number. (2) Sometimes used to refer to a sheet that has been folded once.

Foreword — A statement forming part of the front matter of a book, often written by an expert other than the author to give the book greater promotability and authority.

Four Color Process Printing — The overprinting of specially made plates of four separate colors of transparent ink (magenta, cyan, yellow, and black). The result is a finished printed page which matches the color proof by combining to reproduce all colors.

Four-Sided Trim (Trim 4) — After the job is printed and folded, a trim will be taken off of all four sides to remove any reference or registration marks and give a clean edge to the pile of sheets.

Front Flap — The inside fold on the front of a dust jacket.

Front Matter — The pages preceding the text of a book.

Frontispiece — An illustration facing the title page of a book; also called Frontis.

Fulfillment — A system of storage and mailing upon customer request of title and quantity specified.

Full Color Printing — See: Four Color Process Printing.

G

Galley — Typeset material before it has been arranged into page form.

Galley Copies — Bound galleys (generally printed before proofing and final typeset corrections) which are used in prepublication review mailings.

Galley Proof — Proof of typeset material in galley form prior to page makeup.

Gatefold – A four page insert having foldouts on either side of the center spread.

Gathering – Collecting, by hand or machine, the signatures of a book in the sequence in which they are to be bound; also called Collating.

Grain – (1) In paper, the machine direction in papermaking along which the majority of fibers are aligned. This governs some paper properties such as increased size change with relative humidity across the grain and better folding qualities along the grain.

Gray Scale – A strip of standard gray tones ranging from white to black, placed at the side of original copy during photography, or beside the negative or positive during plate exposure to measure the tonal range obtained.

Gripper Margin – The margin on the forward or leading edge of paper held by grippers of the printing press which feeds and controls paper as it is printed. Because this area cannot be used for a printed image, allowance must be made for gripper edges.

Gutter – (1) In multi-column composition, the space between columns on a page. (2) Short for Gutter Margin.

Gutter Bleed – Occurs when a page bleeds or prints to the binding edge.

Gutter Margin – In binding, the blank space where two pages meet; the inside margin at the binding edge; also called Back Margin or Bind Margin.

H

Hairline Register – The joining or butting of two or more colors with no color overlapping. Also called Close Register or Tight Register.

Halftone – The reproduction of a continuous tone original, such as a photograph, in which detail and tone values are represented by a series of evenly spaced dots of varying size and shape. The dot areas vary in direct proportion to the intensity of the tones they represent. A halftone screen is placed in front of the negative during photography.

Halftone Screen – A screen placed in front of the negative material in a process camera to break up a continuous tone image into dot formation.

Hand Tip – To attach a leaf, foldout, etc. to a signature or bound book by hand operations of gluing and placement of the item.

Handwork – Any operation which can only be accomplished by hand. This includes hand operations in Composition or Plate Prep, as well as those performed in Bindery.

Hard Copy – Any output from a machine, or as a result of machine processing, which is readable copy on paper or film. Examples are computer printouts and phototypesetting output on film or paper.

Hardbound — Another term for Casebound.

Headband — A small band of silk or cotton glued at the top or bottom (or both) of a casebound book to fill the gap normally formed between the spine of the book and the cover. The only purpose is decorative.

Hickey — An imperfection in offset press work due to a number of causes such as dirt on the press, hardened specks of ink, or any dry hard particle working into the ink or onto the plate or offset blanket. It is characterized by a solid center area surrounded by a white ring.

High Bulk Paper — A paper specifically manufactured to retain a thickness not found in other papers of the same basis weight.

House Sheet — Paper stocked by the printer.

I

Illustrations — The drawings, photographs, etc., used to supplement the text of printed matter.

Imposition — The laying out of pages in press form so that they will be in the correct sequence after the printed sheet is folded.

Index — A list at the end of a book showing individual terms from the text contents in alphabetical order and listing the pages on which each entry appears.

Insert — (1) In stripping, a section of film carrying printing detail that is spliced into a larger piece of film. (2) In printing, a page, etc., that is printed separately and then placed into, or bound with, the main publication. (3) In typesetting, copy to be added.

Inside Delivery — Delivery made inside a door or garage on the ground level. It will not necessarily include breaking of skid bands and unloading cartons. There is an additional charge for Inside Delivery.

J

Jacket — See: Dust Jacket.

Jacketing — The application of dust jackets on finished casebound books.

Joint — The flexible hinge where the cover of a casebound book meets the spine, permitting the cover to open without breaking the spine of the book or breaking apart signatures; also called Hinge.

K

Keyline — In artwork, an outline drawing of finished art to indicate the exact shape, position, and size for such elements as halftones and type.

Knock Out — See: Reverse

Kraft – Paper or board made from unbleached woodpulp by the sulfate process; it is brown in color. Kraft paper is often used to wrap books and journals for mailings of individual copies.

L

Laid Paper – Paper which, when held up to the light, shows fine parallel lines (wire-marks) and crosslines (chain-marks).

Laminate – Bonding plastic film by adhesives or heat and pressure to a sheet of paper; especially used to protect the covers of paperback books and improve their appearance.

Layout – (1) The drawing or sketch of a proposed printed piece; the working diagram for a printer to follow. (2) Brief for Layout Sheet. (3) Another term for Plate Prep.

Layout Sheet – The imposition form; indicates the sequence and positioning of negatives on the flat which corresponds to printed pages on the press sheet. Pages are not sequential on the Layout Sheet – it is a folding imposition. Once the sheet is folded, pages will be in consecutive order.

Leaf – (1) Each separate piece of paper in a book with a page on each side. (2) A pigmented stamping material used to decorate cases.

Library Binding – A book bound in accordance with the standards of the American Library Association, having strong endpapers, muslin reinforced end signatures, sewing with four-cord thread, canton flannel backlining, and covers of Library or Caxton Buckram cloth with round corners.

Lightweight Paper – Paper in the 17 to 35 lb. weight range.

Line Copy – Any copy suitable for reproduction without using a halftone screen; copy composed of lines or dots as distinguished from copy composed of continuous tones. Lines or dots may be small and close together so as to simulate tones, but are still regarded as line copy if they can be faithfully reproduced without a halftone screen.

Line Drawing – A drawing containing no grays or middle tones. In general, any drawing that can be reproduced without the use of halftone techniques.

Linen Finish – Book Cloth which has a two-tone effect due to the white threads which show through the color.

Lint – Small fuzzy particles in paper.

Lithography – A generic term for any printing process in which the image area and nonimage area exist on the same plane (plate) parts of which have been chemically treated to repel ink.

Long Grain – Paper made with the machine direction of fibers in the longest dimension of the sheet.

Long Run — A print run in excess of 20,000 copies.

Loose Register — Color or other copy that fits loosely where positioning (register) is not critical.

Low Bulk Paper — A paper with a smooth surface; a thin sheet.

M

M — The abbreviation for a quantity of 1000.

Machine Finish (m.f.) — A term applied to paper which has been made smooth and somewhat glossy by passing through several rolls of the calendering machine.

Magenta — Process red, a purplish red, one of the four process colors.

Makeready — (1) On an offset press, all work done before running a job such as adjusting the feeder and side guide, putting a plate on press and ink in the fountain, etc. (2) Any machine adjustments made for size, bulk, etc., prior to performing the required operation. (3) The process of setting up a press to emboss or foil stamp.

Manuscript (mss) — A written or typewritten work which the typesetter follows as a guide in setting copy.

Matte Finish — Dull paper finish without gloss or luster.

Mechanical — Camera-ready copy (usually for a cover or dust jacket) showing exact placement of every element and carrying actual or simulated type and artwork.

Mechanical Binding — Individual leaves fastened by means of an independent binding device such as Plastic Comb, Wire-O, or Spiral.

Moire — The undesirable wave-like or checkered effect that results when a halftone is photographed through a screen. The second screen must be angled at 15 degrees away from that of the halftone to avoid this effect.

Multi-level die — Generally, a brass machine — cut die with distinct levels such as 2 level, 3 level, etc.

N

Natural — A paper color such as cream, white, or ivory.

Natural Finish — A book cloth characterized by a soft, slightly fuzzy appearance due to to the finishing process.

Negative — Photographic image on film in which black values in the original subject are transparent and white values are opaque; light grays are dark and dark grays are light.

Neutral pH Paper — See: Acid-free Paper.

Newsprint — Paper made mostly from ground wood pulp and small amounts of chemical pulp; used for newspaper printing.

Nick — A small tear on the head of a saddlestitched book that occurs during the trim operation.

Notch Bind — An adhesive binding similar to Perfect Binding. Pieces of text stock (notches) are removed on the binding edge during folding to allow greater adhesive penetration without trimming the spine; primarily used for adhesive binding of coated paper.

O

Odd Sizes — Any nonstandard paper or book size.

Offset — In printing, the term refers to the transfer of the image from the plate to a rubber blanket to stock as in photo-offset lithography.

Offset Lithography — Lithography produced on an offset lithographic press. A right-reading plate is used and an intermediate rubber-covered offset cylinder transfers the image from the plate cylinder to the paper, cloth, metal or other material being printed.

Offset Paper — Paper which is strong enough to resist the tacky inks and considerable moisture encountered in offset printing.

Opacity — That property of paper which minimizes the show-through of printing from the back side or the sheet under it.

Opaque — (1) An area or material which completely blocks out unwanted light; a filter may be opaque to only certain colors. (2) A red or black liquid used to block out or cover unwanted clear or grey areas on a negative. (3) White opaque used to cover unwanted black images in an original copy (on white paper). (4) To paint out areas on a negative which are not to print. (5) In paper, the property which makes it less transparent.

Opti-Copy — A computerized camera that uses a Slo-Syn numerical control to position images on film in proper position for printing on a flat-size piece of film; the exposed film contains the matter to be printed on one side of the press sheet.

Original — The artwork, mechanical, or other material furnished for printing reproduction; usually refers to photographs or drawings for halftone reproduction. More commonly called Photocopy.

Out of Register — (1) Descriptive of pages on both sides of the sheet which do not back up accurately. (2) Two or more colors are not in the proper position when printed; register does not match.

Overlay — In artwork, a clear acetate sheet or tissue with color separated as it is to be photographed.

Overrun — Additional copies above the number ordered to be printed.

P

PMS (Pantone Matching System) — An ink color system widely used in the graphic arts. There are approximately 500 basic colors for both coated and uncoated paper. The color number and formula for each color are shown beneath the color swath in the ink book.

PMT (Photomechanical Transfer Prints) — Camera-generated positive prints used for pasteup and for making paper contacts without the need for a negative.

Page Makeup — The hand or electronic assembly of the elements that comprise a page.

Page Proof — Proof of type in page form.

Pagination — The numbering of pages of a book.

Paperbound — A paper covered book; also called Paperback or Soft Cover.

Pasteup — The assembling of type elements, illustrations, etc., into final page form ready for photographing. See also: Mechanical and Page Makeup.

Perfect Binding — A binding method, usually for paper-covered books, with adhesive (glue) the only binding medium; also known as Adhesive Binding.

Perfecting Press — A printing press that prints both sides of the paper in one pass through the press.

Perforate — To make slits in the paper during folding, at the fold, to prevent wrinkles and allow air to escape. Books with perfect bind are perforated on the spine fold to aid in binding.

Phototypesetting — The process of setting type via a photographic process directly onto film or paper film.

Pica — Printer's unit of measurement used in printing, principally for measuring lines. One pica is equal to 12 points and is approximately 1/6 of an inch. The pica is used to measure the width and length of pages, columns, slugs, and so on.

Pin Register — The use of accurately positioned holes and special pins on copy, film, plates, and presses to insure proper register or fit of colors.

Pinhole — A small, unwanted transparent area in the developed emulsion of a negative or black area on a positive; usually due to dust or other defects on the copy, copyboard glass, or the film.

Plastic Comb Binding — A type of mechanical binding using a piece of rigid vinyl plastic sheeting diecut in the shape of a comb or rake and rolled to make a cylinder of any thickness. The book is punched with slots along the binding edge through which this comb is inserted.

Plastic Shrink Wrap — A method of packaging in plastic film. The material to be wrapped is inserted into a folded roll of polyethylene film which is heat-sealed around it. The package then goes through a heat tunnel where the film shrinks tightly around the package.

Plate — Brief for printing plate; a thin sheet of metal that carries the printing image whose surface is treated so that only that image is ink receptive.

Plate Prep — Those operations after camera and before plate making; includes opaquing of negatives, strip-ins of halftones, stripping of negative flats, and any other operations needed to make negative flats ready for platemaking.

Point — A unit of measurement based on the pica used to measure type sizes. A point measures .013 of an inch. There are approximately 72 points to an inch.

Positive — Photographic image usually made from a negative in which the dark and light values are the same as the original. A positive on paper is called a print; one on a transparent base, such as film, is called a positive transparency.

Prep — In the prep department, work is assembled and made ready for the printing process. Litho camera, stripping, proofing, and platemaking are done in the prep department of a printing company.

Prepress — All manufacturing operations prior to press.

Press Proof — Actual press sheets to show image, tone value, and color. A few sheets are run and approval received from the customer prior to printing the job.

Printer's Error (PE) — Usually refers to a typesetting or other compositor error as distinguished from an author's alteration. Correction costs are not chargeable to the customer.

Process Color Separation Negative — A color picture is actually printed with the four process colors. To obtain the negatives for each of these colors requires a complicated darkroom process. Basically, filters are used to block out all but the desired color for each color negative and for each a different angle screen is employed. Usually, special color correction masks must be used to improve colors for each negative.

Process Color — (1) Yellow (2) Cyan Blue (3) Magenta Red (4) Black. When these colors are used in various strengths and combinations, they make it possible to produce thousands of colors with a minimum of photography, platemaking, and presswork.

Progressive Proofs — Proofs of plates for color printing showing each color separately and also the combined colors in the order they are to print (second color on first; third on second and first; fourth on third, second, and first): used to indicate color quality and as a guide for printing; also called Progs.

Proportional Wheel — A circular scale used to determine proportional reductions and enlargements of copy. Linear scales may be used for the same purpose.

R

Recto — A right-hand page of a book, usually odd-numbered.

Recycled Paper — Paper made from old paper pulp; used paper is cooked in chemicals and reduced back to pulp after it is de-inked. Recycled paper might also include pulp made from mill waste.

Register — (1) Exact correspondence in the position of pages or other printed matter on both sides of a sheet or in relation to other matter already ruled or printed on the same side of the sheet. (2) In photo-reproduction and color printing, the correct relative position of two or more colors so that no color is out of its proper position.

Register Marks — (1) Small crosses, guides, or patterns placed on originals before reproduction; used for positioning the negatives for stripping or for color register. (2) Similar marks added to a negative flat to print along the margins of a press sheet; used as a guide for correct alignment, backing, and color register in printing.

Reprints — Articles for which negatives are relaid so the articles are produced in booklet form.

Reproduction Proof (Repro) — Carefully printed proofs from type forms; used as camera-ready copy for reproduction.

Reverse — Type appearing in white on a black or color background or in a dark area of a photograph.

Ring Binder — A looseleaf mechanism comprised of a metal housing to which heavy wire rings are attached. These rings open in the center. Such binders are used in looseleaf binding.

Rounding and Backing — In binding, the process of rounding gives books a convex spine and a concave fore-edge. The process of backing makes the spine wider than the rest by the thickness of the covers, thus providing a shoulder against which the boards of the front and back covers fit (i.e., the crease or joint).

Rub-Off — (1) Ink on printed sheets, after sufficient drying, which smears or comes off on the fingers when handled. (2) Ink which comes off the cover during shipment and transfers to other covers or to the shipping carton or mailer; also called Scuffing.

Run — The total number of copies ordered.

Running Head — A headline or title repeated at the top of each page for the quick reference of the reader.

S

Saddlestitch — A binding method which inserts sections into sections, then fastens them with wires (stitches) through the middle fold of the sheets. The limiting factor in this type of binding is bulk (thickness) Also called: Saddlewire.

Screened Print – Halftone illustration made on photographic paper. It can be mounted with line copy on a page layout so that the entire page can be photographed at the same time as line copy for reproduction.

Screened Print – A print made from continuous tone copy which was screened during exposure.

Screentone – A halftone film having a uniform dot size over its area and rated by its approximate printing dot size value, such as 20%; also called Screen Tint.

Sculptured Die – A hand-cut die, usually of brass, which embosses many levels through the use of curves and varying depths. An example would be the embossing of a front view of a man.

Self-Cover – A cover of the same paper as the inside text pages.

Sheet-Fed Press – A printing press which takes paper previously cut into sheets as opposed to paper in a continuous roll (web press).

Short Grain Paper – Paper made with the direction of the majority of fibers in the shortest sheet dimension.

Show Through – The ability to see the printing from the back of a sheet because the ink is too oily, too opaque, or the paper too transparent.

Shrink Wrap – See: Plastic Shrink Wrap.

Sidestitching – Method of mechanical binding in which a booklet or a signature is stitched at the sides. "At the sides" means that the booklet or signature is stitched in the closed position. The pages therefore cannot be opened to their full width.

Sidesewing – An entire book is sewn together along the binding edge without any sewing of individual sections as is done with Smyth Sewing. A sidesewn book will not lie open flat.

Signature – A sheet of paper printed on both sides and folded to make up part of a publication. For example, a sheet of paper with 2 printed pages on each side is folded once to form a 4-page signature. One with 4 pages on each side is folded twice to form an 8-page signature, and so on up to a 64-page signature with 32 pages on each side of the sheet.

Silkscreen Printing – A piece of silk is stretched on a frame and blocked out in the nonprinting areas. A rubber squeegee pushes ink or paint through the porous areas of the design onto the material to be printed.

Silverprint – A paper print made from a single negative or a flat, used primarily as a proof, to check content and/or positioning; also called Brownline, Brownprint, Bluelines, Blues, or Van Dykes.

Single-Color Press – A printing press capable of printing only one color at a time.

Sizing—The treatment of paper which gives it resistance to the penetration of liquids (water) or vapors.

Skid—(1) A platform support made of wood, on which sheets of paper are delivered, and on which printed sheets or folded sections are stacked. Also used to ship materials, usually in cartons, which have been strapped (banded) to the skid.

Slip Case—A decorated slide box in which the finished book or books are inserted so that the spine(s) remains visible.

Smearing—A press condition in which the impression is slurred and unclear because too much ink was used or sheets were handled or rubbed before the ink was dry.

Smyth Sewing—A method of fastening side-by-side signatures so that each is linked with thread to its neighbor as well as saddle sewn through its own center fold. Smyth-sewn books open flat. The stitching is on the back of the fold.

Softcover—Another term for paperback or paperbound books.

Spec (Specification) Sheet—A form which is the primary source document for estimate and order specifications. Pricing is based on these written specifications and orders are entered based on the data on this form. Also known as a Request for Quotation or RFQ.

Special Handling—Intended for preferential handling in dispatch and transportation of third and fourth-class mail. A special fee is assessed for each piece in addition to the regular postage.

Spine—The back of a book connecting the two covers.

Spiral Binding—A type of mechanical binding which uses a continuous wire of corkscrew or spring coil form run through round holes punched in the binding edge; the wire can be exposed, semi-concealed, or concealed.

Split Bind—Refers to an order with two or more bind types such as perfect and case.

Spot Varnish—Press varnish applied to a portion of the sheet as opposed to an overall application of the varnish.

Stamping—Pressing a design onto a book cover using metal foil colored foil or ink, applied with metal dies.

Stamping Die—Die used for flat foil stamping. Image of the die is raised with "dead area" relieved usually .060".

Stapling—Binding a book or loose sheets with one or more wire staples.

Stripping—(1) The act of positioning or inserting copy elements in negative or positive film to make a complete negative; the positioning of photographic negatives or positives on a lithographic flat or form imposition.

T

Tare Weight — The weight of packing material (cartons, skids, pallets, etc.).

Text — The body matter of a page or book as distinguished from the headings.

Text Paper — A general term applied to high quality antique or laid papers made in white and colors; used for books, booklets, etc.

Three-M(3M) — See: Color Key.

Tip-Ins — Any separate page (or pages) pasted in a book such as foldouts, frontispiece, etc.; it may be pasted by hand or machine.

Tip-Ons — Endsheets or other material attached to the outside of folded sections by machine application of a thin strip of adhesive.

Tissue Overlay — A thin translucent paper placed over artwork for protection and correction. Register marks are used on the overlay to ensure proper register for corrections. Any light, inexpensive onionskin that does not have oily characteristics may serve as an overlay.

Tolerance — The acceptable amount of variance from stated specifications.

Transparency — (1) A monochrome or full-color photographic positive or picture on a transparent support; the image intended for viewing and reproduction by transmitted light. (2) The quality that allows images to be seen through a sheet.

Trim — To cut away the folded or uneven edges to form a smooth, even edge and permit all pages to open.

Trim Margin — The margin of the open side away from the bind. Also called the Outside Margin.

Trim Size — The finished size of a book after binding and trimming.

Turnaround — Time from the acceptance or beginning of a job until the completed job is delivered.

Typo — Short for typographical error.

U

Underinked — Not enough ink used resulting in light printing.

Underrun — A shortage in the number of copies completed; a quanity less than the amount ordered.

Undertrimmed — Trimmed to a size smaller than the specified trim size.

Upright — In bookbinding, a book bound on its long dimension as opposed to oblong binding which is on the short dimension.

V

Vandyke — See: Silverprint.

Varnish — A thin, protective coating applied to a printed sheet for protection or appearance; generally cheaper than lamination, but with less gloss and providing less protection.

Vellum Finish — (1) In paper, a toothy finish which is relatively absorbent for fast ink penetration. (2) A smooth finish, solid color book cloth.

Velox — Print of a photograph or other continuous-tone copy which has been prescreened before paste-up.

Verso — A left-hand page of a book, usually even numbered.

Vignette — An illustration in which the background fades gradually away until it blends into the unprinted paper.

W

Washup — The process of cleaning the rollers, form or plate, and sometimes the fountain of a printing press.

Web Offset Printing — Lithographic printing from rolled stock. The paper manufacturer refers to the roll as a web. The printer uses the term "web" not only for the roll but for the paper itself as it feeds through the press.

Widow — A short single line at the top of a page or column, usually the last line of a paragraph. Widows are to be avoided in good typesetting. Also a single word or syllable standing alone as the last line of a paragraph. An orphan, on the other hand, is a single line at the bottom of a page or column from a paragraph that runs to the next page or column.

Wire-O-Binding — A type of mechanical binding which uses a series of double wire loops formed from single continuous wire, run through longitudinal slots along the bind edge, which must be crimped closed after insertion of trimmed text sheets.

With the Grain — A term applied to folding paper parallel to the grain of the paper.

Wrap — (1) To place jackets on finished books. (2) A folded section, such as four pages, into which another folded section is hand inserted; the outside section wraps around the inside section; (3) To package in kraft paper or plastic film (shrink wrapping).

Wrap-Around Mailer — A single piece corrugated pad which is wrapped or rolled around a book, then stapled closed at both ends; also called: Book-Wrap Mailer.

Wrinkles — Creases in paper occurring during printing or folding.

Indexes

An Introduction to the Indexes

Because of the amount of details contained in each complete listing, we have provided indexes to highlight those printers who are capable of providing the type of service or support that any user might require. The following pages of indexes are divided into six sections, as follows:

- **Index of Book Printers, Etc.** — Indicates which printers can print books, magazines, catalogs, in long or short runs.

- **Index of Other Printed Items** — Indicates which printers can print annual reports, calendars, comics, computer documentation, cookbooks, directories, galley copies, journals, maps, mass-market paperbacks, postcards, posters, textbooks, workbooks, and yearbooks.

- **Index of Binding Capabilities** — Indicates which printers can do case binding, comb binding, loose-leaf binding, perfect binding, spiral binding, and wire-o binding.

- **Index of Printing Equipment** — Indicates which printers use Cameron belt, letterpress, gravure, sheetfed offset, and web offset presses.

- **Index of Printing Services** — Indicates which printers can print full-color publications, do color separations, design and lay out material, provide fulfillment, maintain lists, typeset copy, or warehouse books. Also indicates which printers stock acid-free and/or recycled papers. And, it also indicates which printers use union labor.

- **Index of Printers by State** — Lists printers by their location so you can find printers who are located near you who might be able to provide the services you need.

Index: Book Printers, Etc.

This set of indexes lists those printers who can print books, catalogs, and magazines. The set of indexes that follows, lists printers of other items and bound publications.

Book Printers

aBCD, Seattle WA
Academy Books, Rutland VT
Accurate Web Inc, Central Islip NY
Action Printing, Fond Du Lac WI
Ad Infinitum Press, Mount Vernon NY
Adams Press, Chicago IL
Advanced Duplicating & Printing,
    Minneapolis MN
Adviser Graphics, Red Deer AB
Alcom Printing Group Inc, Bethlehem PA
Alonzo Printing, Hayward CA
AM Graphics & Printing, San Marcos CA
American Litho, Hayward CA
American Printers & Litho, Niles IL
Americomp, Brattleboro VT
Anundsen Publishing Company, Decorah
    IA
Apollo Graphics Ltd, Southampton PA
Arcata Graphics/Fairfield, Fairfield PA
The Argus Press Inc, Niles IL
Associated Printers, Grafton ND
Austin Printing Co Inc, Akron OH
Automated Graphic Systems, White
    Plains MD
Baker Johnson Inc, Dexter MI
Bang Printing Company, Brainerd MN
Banta Company, Menasha WI
Banta-Harrisonburg, Harrisonburg VA
Barton Press Inc, West Orange NJ
Bawden Printing Inc, Eldridge IA
Bay Port Press, Chula Vista CA
Bay State Press, Framingham MA
Beacon Press Inc, Richmond VA
Beacon Press, Seattle WA
Ben-Wal Printing, Pomona CA
Berryville Graphics, Berryville VA
Bertelsmann Printing and Manufactur-
    ing, New York NY
Best Gagne Book Manufacturers, Toron-
    to ON
Blake Printery, San Luis Obispo CA
Blue Dolphin Press Inc, Grass Valley CA
Bolger Publications/Creative Printing,
    Minneapolis MN
The Book Press Inc, Brattleboro VT

Book-Mart Press Inc, North Bergen NJ
BookCrafters, Chelsea MI
BookCrafters Inc, Fredericksburg VA
Booklet Publishing Company Inc, Elk
    Grove Village IL
BookMasters Inc, Ashland OH
Boyd Printing Co Inc, Albany NY
Braceland Brothers Inc, Philadelphia PA
Braun-Brumfield Inc, Ann Arbor MI
Brennan Printing, Deep River IA
Brunswick Publishing Corp, Lawren-
    ceville VA
R L Bryan Company, Columbia SC
BSC Litho, Harrisburg PA
Buse Printing & Advertising, Phoenix AZ
C & M Press, Denver CO
Cal Central Press, Sacramento CA
Caldwell Printers, Arcadia CA
Camelot Publishing Co, Ormond Beach
    FL
Canterbury Press, Rome NY
Capital City Press, Montpelier VT
Capital Printing Company Inc, Austin TX
Carpenter Lithographic Co, Springfield
    OH
Carter Rice, Boston MA
Clarkwood Corporation, Totowa NJ
Coach House Press, Toronto ON
The College Press, Collegedale TN
Colonial Graphics, Paterson NJ
Color World Printers, Bozeman MT
ColorCorp Inc, Littleton CO
Colorlith Corporation, Johnston RI
Colortone Press, Landover MD
Colotone Riverside Inc, Dallas TX
Columbus Bookbinders & Printers,
    Columbus OH
Combined Communications
Comfort Printing, St Louis MO
Community Press, Provo UT
Coneco Litho Graphics, Glen Falls NY
Consolidated Printers Inc, Berkeley CA
Cookbook Publishers Inc, Lenexa KS
Cookbooks by Mom's Press, Kearney NE
Cooley Printers & Office Supply, Monroe
    LA
Corley Printing Company, Earth City MO
Country Press, Mohawk NY

Book Printers

Country Press Inc, Lakeville MA
Courier Stoughton Inc, Stoughton MA
Courier-Kendallville Corp, Kendallville IN
Courier-Westford Inc, Westford MA
Crane Duplicating Service, West Barnstable MA
Crusader Printing Co, East Saint Louis IL
Cushing-Malloy, Ann Arbor MI
Daamen Printing Company, West Rutland VT
Data Reproductions Corp, Rochester Hills MI
Dataco A WBC Company, Albuquerque NM
Dellas Graphics, Syracuse NY
Delmar Printing & Publishing, Charlotte NC
Delta Lithograph Company, Valencia CA
Des Plaines Publishing Co, Des Plaines IL
Desaulniers Printing Company, Milan IL
Deven Lithographers, Brooklyn NY
Dharma Press, Oakland CA
Diamond Graphics Tech Data, Milwaukee WI
Dickinson Press Inc, Grand Rapids MI
Directories America, Chicago IL
Diversified Printing, Brea CA
Dollco Printing, Ottawa ON
R R Donnelley & Sons Co, Crawfordsville IN
RR Donnelley & Sons Co, Harrisonburg, Harrisonburg VA
R R Donnelley & Sons Co, Willard, Willard OH
Dragon Press, Delta Junction AK
Eastwood Printing Co, Denver CO
EBSCO Media, Birmingham AL
Edison Lithographing, North Bergen NJ
Edwards Brothers Inc, Ann Arbor MI
Edwards Brothers Inc, Lillington NC
Eerdmans Printing Co, Grand Rapids MI
Electronic Printing, Brentwood NY
Eureka Printing Co Inc, Eureka CA
Eusey Press Inc, Leominster MA
Eva-Tone Inc, Clearwater FL
Evangel Press, Nappanee IN
The Everton Publishers, Logan UT
Faculty Press Inc, Brooklyn NY
Federated Lithographers, Providence RI
Fleming Printing Co, St Louis MO
Flower City Printing Inc, Rochester NY
The Forms Man Inc, Deer Park NY
Friesen Printers, Altona MB
Futura Printing Inc, Boynton Beach FL
Gateway Press Inc, Louisville KY
Geiger Brothers, Lewiston ME
George Lithograph, San Francisco CA

Geryon Press Limited, Tunnel NY
Gilliland Printing Inc, Arkansas City KS
Golden Belt Printing Inc, Great Bend KS
Golden Horn Press, Berkeley CA
Goodway Graphics of Virginia, Springfield VA
Gorham Printing, Rochester WA
Gowe Printing Company, Medina OH
Graphic Printing, New Carlisle OH
Graphic Ways Inc, Buena Park CA
Graphics Ltd, Indianapolis IN
Great Impressions Printing & Graphics, Dallas TX
Great Lakes Lithograph Company, Cleveland OH
Gregath Publishing Co, Wyandotte OK
Griffin Printing & Lithograph, Glendale CA
Griffin Printing and Lithograph Company, Sacramento CA
GRT Book Printing, Oakland CA
GTE Directories Corp, Des Plaines IL
GTE Directories Printing Corp, Mt Prospect IL
Gulf Printing, Houston TX
Guynes Printing Company, Albuquerque NM
Haddon Craftsmen, Scranton PA
Harlo Printing Co, Detroit MI
Harmony Printing, Liberty MO
Hart Graphics, Austin TX
Hawkes Publishing, Salt Lake City UT
Henington Publishing Company, Wolfe City TX
Hennegan Company, Cincinnati OH
Heritage Printers Inc, Charlotte NC
D B Hess Company, Woodstock IL
Hignell Printing, Winnepeg MB
A B Hirschfeld Press, Denver CO
Hoechstetter Printing, Pittsburgh PA
Holladay-Tyler Printing, Glenn Dale MD
Rae Horowitz Book Manufacturers, Fairfield NJ
Independent Printing Company, New York NY
Independent Publishing Company, Sarasota FL
Interform Corporation, Bridgeville PA
Interstate Printing Company, Omaha NE
IPC Publishing Services, Saint Joseph MI
Jersey Printing Company, Bayonne NJ
JK Creative Printers, Quincy IL
The Job Shop, Woods Hole MA
The Johnson & Hardin Company, Cincinnati OH
Jostens Printing & Publishing, State College PA
Jostens Printing & Publishing, Winston-Salem NC
Jostens Printing & Publishing, Minneapolis MN

Book Printers

Jostens Printing & Publishing, Clarksville TN
Jostens Printing & Publishing, Topeka KS
Jostens Printing & Publishing, Visalia CA
Julin Printing, Monticello IA
Kimberly Press, Santa Barbara CA
Kirby Lithographic Co Inc, Arlington VA
KNI Incorporated, Anaheim CA
Kolor View Press, Aurora MO
The Company J Krehbiel Company, Cincinnati OH
La Crosse Graphics, LaCrosse WI
Lancaster Press, Lancaster PA
Les Editions Marquis, Montmagny PQ
Lighthouse Press Inc, Manchester NH
Litho Productions, Madison WI
Litho Specialties, St Paul MN
Lithocolor Press, Westchester IL
Lithograph Printing Company, Memphis TN
Lithoid Printing Corporation, East Brunswick NJ
The Little River Press Inc, Miami FL
John D Lucas Printing Company, Baltimore MD
Mackintosh Typography Inc, Santa Barbara CA
The Mad Printers, Mattituck NY
Malloy Lithographing, Ann Arbor MI
Maple-Vail Book Manufacturing Group, York PA
Maquoketa Web Printing, Maquoketa IA
Mark IV Press, Hauppauge NY
Marrakech Express, Tarpon Springs FL
Maverick Publications Inc, Bend OR
The Mazer Corporation, Dayton OH
MBP Lithographics, Hillsboro KS
McArdle Printing Co, Upper Marlboro MD
McClain Printing Company, Parsons WV
McDowell Publications, Utica KY
McNaughton & Gunn Inc, Saline MI
Meaker the Printer, Phoenix AZ
Media Publications, Santa Clara CA
Mercury Printing Company, Memphis TN
Merrill Corporation, Los Angeles CA
Mitchell Press, Vancouver BC
MMI Press, Harrisville NH
Monument Printers & Lithographer Inc, Verplanck NY
Moran Printing Company, Orlando FL
Moran Printing Inc, Baton Rouge LA
Morgan Press, Dobbs Ferry NY
Morgan Printing, Austin TX
Morningrise Printing, Costa Mesa CA
MultiPrint Co Inc, Skokie IL
Murphy's Printing Company, Campbell CA

National Publishing Company, Philadelphia PA
National Reproductions Corp, Madison Heights MI
Naturegraph Publishers, Happy Camp CA
Louis Neibauer Co Inc, Warminster PA
Newsfoto Publishing Company, San Angelo TX
Neyenesch Printers, San Diego CA
Nimrod Press Inc, Boston MA
Northlight Studio Press, Barre VT
Nystrom Publishing Company, Osseo MN
Oaks Printing Company, Bethlehem PA
Odyssey Press Inc, Dover NH
Offset Paperback Manufacturers, Dallas PA
OMNIPRESS, Madison WI
Optic Graphics Inc, Glen Burnie MD
Original Copy Centers, Cleveland OH
Outstanding Graphics, Kenosha WI
The Ovid Bell Press Inc, Fulton MO
The Oxford Group, Norway ME
Pantagraph Printing, Bloomington IL
Paraclete Press, Brewster MA
Patterson Printing Company, Benton Harbor MI
Paust Inc, Richmond IN
Pearl-Pressman-Liberty Communications Group, Philadelphia PA
Peconic Companies, Mattituck NY
Pentagram Press, Minneapolis MN
PFP Printing Corporation, Gaithersburg MD
Phillips Brothers Printers Inc, Springfield IL
Pine Hill Press Inc, Freeman SD
Pioneer Press, Wilmette IL
Plus Communications, Saint Louis MO
Port City Press Inc, Baltimore MD
Practical Graphics, New York NY
The Press of Ohio, Kent OH
Princeton Academic Press, Lawrenceville NJ
Prinit Press, Dublin IN
The Printing Company, Indianapolis IN
Printing Corporation of America, Pompano Beach FL
Printright International, Palmdale CA
Printwest, Regina SK
Progress Printing, Lynchburg VA
ProLitho, Provo UT
Publishers Express Press, Ladysmith WI
Publishers Press, Salt Lake City UT
Quebecor Book Group, New York NY
Quebecor Printing (USA) Corp, Boston MA
Quebecor Printing Inc, Richmond Hill ON
Quebecor/Semline, Braintree MA
Quinn-Woodbine Inc, Woodbine NJ
RBW Graphics, Owen Sound ON

Repro-Tech, Fairfield CT
Rich Printing Company, Nashville TN
Ringier America, Itasca IL
Ringier America/Brookfield, Brookfield
    WI
John Roberts Company, Minneapolis MN
Ronalds Printing, Vancouver BC
Ronalds Printing, Montreal PQ
Rose Printing Company, Tallahassee FL
Roxbury Publishing Company, Los An-
    geles CA
Royle Printing, Sun Prairie WI
Sanders Printing Company, Garretson
    SD
The Saratoga Printing Co, Saratoga
    Springs NY
Schiff Printing, Pittsburgh PA
Science Press, Ephrata PA
Service Printing Company, San Leandro
    CA
Sexton Printing, Saint Paul MN
Shepard Poorman Communications, In-
    dianapolis IN
Sheridan Printing Company, Phil-
    lipsburg NJ
Sinclair Printing Co, Los Angeles CA
Snohomish Publishing, Snohomish WA
Southeastern Printing Co, Stuart FL
Sowers Printing Company, Lebanon PA
Standard Printing Service, Chicago IL
Stephenson Lithograph Inc, Alexandria
    VA
Stevens Graphics, Birmingham AL
Stewart Publishing & Printing Co,
    Marble Hill MO
The Stinehour Press, Lunenburg VT
The Studley Press, Dalton MA
Sun Graphix, Lewiston ME
The Sun Hill Press, North Brookfield MA
John S Swift Company Inc, St Louis MO
John S Swift/Chicago, Chicago IL
John S Swift/Cincinnati, Cincinnati OH
John S Swift/Teterboro, Teterboro NJ
TAN Books & Publishers Inc, Rockford
    IL
Taylor Publishing Company, Dallas TX
TEACH Services, Brushton NY
Technical Communication Services,
    North Kansas City MO
Thomson-Shore, Dexter MI
Times Printing, Random Lake WI
Todd Publications Inc, Smithville TX
Town House Press, Pittsboro NC
Tracor Publications, Austin TX
Trend Offset Printing Service Inc, Los
    Alamitos CA
Tri-Graphic Printing, Ottawa ON
Triangle Booklet Company, Nashville TN
Tweddle Litho Co, Clinton Township MI
United Lithograph, Somerville MA
Universal Printers Ltd, Winnipeg MB

Universal Printing Company, St Louis
    MO
Vail Ballou Press, Binghamton NY
Van Volumes, Palmer MA
Versa Press Inc, East Peoria IL
Vicks Lithograph & Print, Yorkville NY
Victor Graphics, Baltimore MD
Viking Press, Eden Prairie MN
Vogue Printers, North Chicago IL
Waldon Press Inc, New York NY
Wallace Press, Hillside IL
Walsworth Publishing Co Inc, Marceline
    MO
Watt/Peterson, Plymouth MN
Webcom Ltd, Scarborough ON
Webcrafters, Madison WI
F A Weber & Sons, Park Falls WI
Weber Printing Company, Lincoln NE
Western Publishing Co Inc, Racine WI
Western Web Printing, Sioux Falls SD
Westview Press, Boulder CO
Whitehall Printing Company, Naples FL
Wickersham Printing Co Inc, Lancaster
    PA
The Wimmer Companies, Memphis TN
Windsor Associates, San Diego CA
Winter Sun Inc, Phoenix AZ
Wood & Jones Printers, Pasadena CA
Worzalla Publishing Company, Stevens
    Point WI
Y/Z Printing Co, Elizabethtown PA
Ye Olde Genealogie Shoppe, In-
    dianapolis IN

Catalog Printers

The following printers all can print
catalogs.

A-1 Business Service Inc, St Paul MN
aBCD, Seattle WA
Academy Books, Rutland VT
Accurate Web Inc, Central Islip NY
Action Printing, Fond Du Lac WI
Adams Press, Chicago IL
Adviser Graphics, Red Deer AB
Alan Lithograph Inc, Inglewood CA
Alcom Printing Group Inc, Bethlehem PA
Alden Press, Elk Grove Village IL
Alden Press/Wm Feathers Printer, Ober-
    lin OH
All-Star Printing, Lansing MI
Alonzo Printing, Hayward CA
AM Graphics & Printing, San Marcos CA
American Litho, Hayward CA
American Printers & Litho, Niles IL
American Signature Memphis Division,
    Olive Branch MS
American Signature Graphics, Dallas TX

Catalog Printers

American Signature, Lincoln NE
American Signature/Foote, Atlanta GA
Americomp, Brattleboro VT
Amidon & Associates Inc, St Paul MN
Apollo Graphics Ltd, Southampton PA
Arandell Corporation, Menomonee Falls WI
The Argus Press Inc, Niles IL
Arizona Lithographers Inc, Tucson AZ
Artcraft Press, Waterloo WI
Associated Printers, Grafton ND
Austin Printing Co Inc, Akron OH
Automated Graphic Systems, White Plains MD
Baker Johnson Inc, Dexter MI
Bang Printing Company, Brainerd MN
Banta Company, Menasha WI
Banta Publications Group, Hinsdale IL
Banta-Harrisonburg, Harrisonburg VA
Barton Press Inc, West Orange NJ
Bawden Printing Inc, Eldridge IA
Bay Port Press, Chula Vista CA
Bay State Press, Framingham MA
Beacon Press Inc, Richmond VA
Beacon Press, Seattle WA
Ben-Wal Printing, Pomona CA
Blake Printery, San Luis Obispo CA
Blue Dolphin Press Inc, Grass Valley CA
Bolger Publications/Creative Printing, Minneapolis MN
The Book Press Inc, Brattleboro VT
Book-Mart Press Inc, North Bergen NJ
BookCrafters, Chelsea MI
BookCrafters/Fredericksburg, Fredericksburg VA
Booklet Publishing Company Inc, Elk Grove Village IL
Braceland Brothers Inc, Philadelphia PA
Bradley Printing Company, Des Plaines IL
Braun-Brumfield Inc, Ann Arbor MI
Brookshore Lithographers, Elk Grove Village IL
Brown Printing, Waseca MN
Brown Printing/Franklin, Franklin KY
Brown Printing/East Greenville Division, East Greenville PA
R L Bryan Company, Columbia SC
BSC Litho, Harrisburg PA
Buse Printing & Advertising, Phoenix AZ
The William Byrd Press, Richmond VA
C & M Press, Denver CO
Cal Central Press, Sacramento CA
Caldwell Printers, Arcadia CA
California Offset Printers, Glendale CA
Calsonic Miura Graphics, Irvine CA
Camelot Publishing Co, Ormond Beach FL
Canterbury Press, Rome NY
Capital City Press, Montpelier VT

Capital Printing Company Inc, Austin TX
Carlson Color Graphics, Ocala FL
Carpenter Lithographic Company, Springfield OH
Catalog King, Clifton NJ
Champagne Offset Co Inc, Natick MA
Chicago Color Express, Chicago IL
Chicago Press Corp, Chicago IL
Citizen Printing, Beaver Dam WI
The College Press, Collegedale TN
Colonial Graphics, Paterson NJ
Color Express, Commerce CA
Color Express, Minneapolis MN
Color Express, Atlanta GA
Color Express, Irvine CA
Color World Printers, Bozeman MT
ColorCorp Inc, Littleton CO
ColorGraphics Printing Corp, Tulsa OK
Colorlith Corporation, Johnston RI
Colortone Press, Landover MD
Colotone Riverside Inc, Dallas TX
Combined Communication Services, Mendota IL
Comfort Printing, St Louis MO
Communicolor, Newark OH
Concord Litho Company Inc, Concord NH
Coneco Litho Graphics, Glen Falls NY
Consolidated Printers Inc, Berkeley CA
Continental Web, Itasca IL
Cookbooks by Mom's Press, Kearney NE
Cooley Printers & Office Supply, Monroe LA
Corley Printing Company, Earth City MO
Country Press, Mohawk NY
The Craftsman Press Inc, Seattle WA
Crane Duplicating Service, West Barnstable MA
Crusader Printing Co, East Saint Louis IL
Cushing-Malloy, Ann Arbor MI
Daamen Printing Company, West Rutland VT
Danbury Printing & Litho Inc, Danbury CT
Danner Press Corporation, Canton OH
Dartmouth Printing, Hanover NH
Data Reproductions Corp, Rochester Hills MI
Dataco A WBC Company, Albuquerque NM
Dayal Graphics, Lemont IL
Dellas Graphics, Syracuse NY
Delmar Printing & Publishing, Charlotte NC
Delta Lithograph Company, Valencia CA
Democrat Printing & Litho, Little Rock AR
Des Plaines Publishing Company, Des Plaines IL
Desaulniers Printing Company, Milan IL
Deven Lithographers, Brooklyn NY

Catlog Printers

Dharma Press, Oakland CA
Diamond Graphics Tech Data,
    Milwaukee WI
Dickinson Press Inc, Grand Rapids MI
The Dingley Press, Lisbon ME
Direct Graphics Inc, Sidney OH
Direct Press/Modern Litho, Huntington
    Station NY
Directories America, Chicago IL
Diversified Printing, Brea CA
Dollco Printing, Ottawa ON
Donihe Graphics Inc, Kingsport TN
R R Donnelley & Sons Co, Craw-
    fordsville IN
Dragon Press, Delta Junction AK
Dynagraphics, Carlsbad CA
E & D Web Inc, Cicero IL
Eagle Web Press, Salem OR
Eastwood Printing Co, Denver CO
EBSCO Media, Birmingham AL
Edison Lithographing, North Bergen NJ
Editors Press Inc, Hyattsville MD
Edwards Brothers Inc, Ann Arbor MI
Edwards Brothers Inc, Lillington NC
Eerdmans Printing Co, Grand Rapids MI
Electronic Printing, Brentwood NY
EP Graphics, Berne IN
EU Services, Rockville MD
Eureka Printing Co Inc, Eureka CA
Eusey Press Inc, Leominster MA
Eva-Tone Inc, Clearwater FL
Evangel Press, Nappanee IN
Faculty Press Inc, Brooklyn NY
Federated Lithographers, Providence RI
Fetter Printing, Louisville KY
First Impressions Printer & Lithog-
    raphers Inc, Elk Grove IL
Fisher Printers Inc, Cedar Rapids IA
Fisher Printing Co, Galion OH
Fleming Printing Co, St Louis MO
Flower City Printing Inc, Rochester NY
Focus Direct, San Antonio TX
The Forms Man Inc, Deer Park NY
Fort Orange Press Inc, Albany NY
Frye & Smith, Costa Mesa CA
Futura Printing Inc, Boynton Beach FL
Gateway Press Inc, Louisville KY
Gaylord Printing, Detroit MI
George Lithograph, San Francisco CA
Geryon Press Limited, Tunnel NY
Gilliland Printing Inc, Arkansas City KS
Glundal Color Service, East Syracuse NY
Golden Belt Printing Inc, Great Bend KS
Goodway Graphics of Virginia,
    Springfield VA
Gorham Printing, Rochester WA
Gowe Printing Company, Medina OH
Graftek Press, Woodstock IL
Graphic Arts Center, Portland OR
Graphic Arts Publishing, Boise ID

Graphic Litho Corporation, Lawrence MA
Graphic Printing, New Carlisle OH
Graphic Production Ctr, Norcross GA
Graphic Ways Inc, Buena Park CA
Graphics Ltd, Indianapolis IN
Gray Printing, Fostoria OH
Great Impressions Printing & Graphics,
    Dallas TX
Great Lakes Lithograph Company,
    Cleveland OH
Great Northern/Design Printing, Skokie
    IL
Gregath Publishing Co, Wyandotte OK
Griffin Printing & Lithograph, Glendale
    CA
Griffin Printing and Lithography Com-
    pany, Sacramento CA
GRIT Commercial Printing Services, Wil-
    liamsport PA
Gulf Printing, Houston TX
Guynes Printing Company, Albuquerque
    NM
Harlo Printing Co, Detroit MI
Harmony Printing, Liberty MO
Hart Graphics, Austin TX
Hart Press, Long Prairie MN
Heartland Press Inc, Spencer IA
Hennegan Company, Cincinnati OH
D B Hess Company, Woodstock IL
Hi-Tech Color House, Chicago IL
Hignell Printing, Winnepeg MB
Hinz Lithographing Company, Mt
    Prospect IL
A B Hirschfeld Press, Denver CO
Hoechstetter Printing, Pittsburgh PA
Hollady-Tyler Printing Corp, Glenn Dale
    MD
Imperial Litho/Graphics Inc, Phoenix AZ
Inland Printing Company, Syosset NY
Intelligencer Printing, Lancaster PA
Interform Corporation, Bridgeville PA
Interstate Printing Company, Omaha NE
IPC Publishing Services, Saint Joseph MI
Japs-Olson Company, Minneapolis MN
Jersey Printing Company, Bayonne NJ
JK Creative Printers, Quincy IL
The Job Shop, Woods Hole MA
Johns Byrne Co, Niles IL
The Johnson & Hardin Company, Cin-
    cinnati OH
Johnson Graphics, East Dubuque IL
Jostens Printing & Publishing, State Col-
    lege PA
Jostens Printing & Publishing, Winston-
    Salem NC
Jostens Printing & Publishing, Min-
    neapolis MN
Jostens Printing & Publishing,
    Clarksville TN
Jostens Printing & Publishing, Topeka
    KS
Jostens Printing & Publishing, Visalia CA

Catalog Printers

Julin Printing, Monticello IA
K-B Offset Printing, State College PA
Kaufman Press Printing, Syracuse NY
Keys Printing Co Inc, Greenville SC
KNI Incorporated, Anaheim CA
Kolor View Press, Aurora MO
Kordet Graphics Inc, Oceanside NY
The Company J Krehbiel Company, Cincinnati OH
La Crosse Graphics, LaCrosse WI
Lancaster Press, Lancaster PA
Lasky Company, Milburn NJ
Lehigh Press Inc, Cherry Hill NJ
Les Editions Marquis, Montmagny PQ
Lighthouse Press Inc, Manchester NH
Litho Productions Inc, Madison WI
Litho Specialties, St Paul MN
Lithocolor Press, Westchester IL
Lithograph Printing Company, Memphis TN
Lithoid Printing Corporation, East Brunswick NJ
The Little River Press Inc, Miami FL
Long Island Web Printing, Jericho NY
John D Lucas Printing Company, Baltimore MD
Mack Printing Group Inc, Easton PA
Mackintosh Typography Inc, Santa Barbara CA
Mail-O-Graph, Kewanee IL
Malloy Lithographing, Ann Arbor MI
Maquoketa Web Printing, Maquoketa IA
Marathon Communications Group, Wausau WI
Mark IV Press, Hauppauge NY
Marrakech Express, Tarpon Springs FL
Maverick Publications Inc, Bend OR
MBP Lithographics, Hillsboro KS
McArdle Printing Co, Upper Marlboro MD
McClain Printing Company, Parsons WV
The McFarland Press, Harrisburg PA
McGill Jensen, St Paul MN
McGrew Color Graphics, Kansas City MO
McKay Printing Services, Michigan City IN
McNaughton & Gunn Inc, Saline MI
McQuiddy Printing Company, Nashville TN
Meaker the Printer, Phoenix AZ
Media Printing, Miami FL
Media Publications, Santa Clara CA
Meehan-Tooker & Company, East Rutherford NJ
Mercury Printing Company, Memphis TN
Merrill Corporation, Los Angeles CA
Metroweb, Erlanger KY
Mitchell Graphics, Petosky MI

Mitchell Press, Vancouver BC
MMI Press, Harrisville NH
Moebius Printing, Milwaukee WI
Monument Printers & Lithographer Inc, Verplanck NY
Moran Printing Company, Orlando FL
Moran Printing Inc, Baton Rouge LA
Morgan Press, Dobbs Ferry NY
Morgan Printing, Austin TX
Morningrise Printing, Costa Mesa CA
Motheral Printing, Fort Worth TX
William H Muller Printing Co, Santa Clara CA
MultiPrint Co Inc, Skokie IL
Murphy's Printing Co, Campbell CA
National Graphics Corp, Columbus OH
National Lithographers Inc, Miami FL
National Reproductions Corp Madison Heights MI
Nationwide Printing, Burlington KY
Louis Neibauer Co Inc, Warminster PA
Neyenesch Printers, San Diego CA
The Nielsen Lithographing Co, Cincinnati OH
Nimrod Press Inc, Boston MA
Noll Printing Company, Huntington IN
Northeast Graphics, New York NY
Northlight Studio Press, Barre VT
Northprint International Inc, Grand Rapids MN
Northwest Web, Eugene OR
Nystrom Publishing Company, Osseo MN
Oaks Printing Company, Bethlehem PA
Ohio Valley Litho Color, Florence KY
Omaha Printing Company, Omaha NE
Original Copy Centers, Cleveland OH
Outstanding Graphics, Kenosha WI
The Ovid Bell Press Inc, Fulton MO
The Oxford Group, Norway ME
Padgett Printing Corporation, Dallas TX
Page Litho Inc, Detroit MI
Panorama Press Inc, Clifton NJ
Paraclete Press, Brewster MA
Park Press, South Holland IL
Parker Graphics, Fuquay-Varina NC
Patterson Printing Company, Benton Harbor MI
Paust Inc, Richmond IN
Pearl-Pressman-Liberty Communication Group, Philadelphia PA
Peconic Companies, Mattituck NY
Penn Colour Graphics, Huntington Valley PA
PennWell Printing, Tulsa OK
Penton Press Publishing, Berea OH
The Perlmuter Printing Co, Cleveland OH
Perry Printing Corp, Waterloo WI
Perry/Baraboo, Baraboo WI
Petty Co Inc, Effingham IL
PFP Printing Corporation, Gaithersburg MD

Catalog Printers

Phillips Brothers Printers Inc, Springfield IL

Pine Hill Press Inc, Freeman SD

Pioneer Press, Wilmette IL

Plain Talk Printing Company, Des Moines IA

Plus Communications, Saint Louis MO

Port City Press Inc, Baltimore MD

Port Publications, Port Washington WI

Practical Graphics, New York NY

The Press of Ohio, Kent OH

The Press, Chanhassen MN

Prinit Press, Dublin IN

Print Northwest, Tacoma WA

The Printer Inc, Osseo MN

The Printing Company, Indianapolis IN

Printing Corporation of America, Pompano Beach FL

Printright International, Palmdale CA

Printwest, Regina SK

Progress Printing, Lynchburg VA

ProLitho, Provo UT

Promotional Printing Corporation, Houston TX

Publishers Express Press, Ladysmith WI

Publishers Press, Salt Lake City UT

Quad/Graphics Inc, Saratoga Springs NY

Quad/Graphics, Lomira, Lomira WI

Quad/Graphics, Pewaukee, Pewaukee WI

Quad/Graphics, Sussex, Sussex WI

Quebecor Book Group, New York NY

Quebecor Printing (USA) Corp, Boston MA

Quebecor Printing Inc, Saint Paul MN

Quebecor Printing Inc, Richmond Hill ON

Quebecor/Dallas, Dallas TX

Quebecor/Semline, Braintree MA

Quinn-Woodbine Inc, Woodbine NJ

RBW Graphics, Owen Sound ON

Repro-Tech, Fairfield CT

Rich Printing Company Inc, Nashville TN

Ringier America, Itasca IL

Ringier America Inc, Corinth MS

Ringier America/Brookfield, Brookfield WI

Ringier America/Jonesboro, Jonesboro AR

Ringier America/Phoenix, Phoenix AZ

John Roberts Company, Minneapolis MN

Ronalds Printing, Vancouver BC

Ronalds Printing, Montreal PQ

Rose Printing Company, Tallahassee FL

S Rosenthal & Company Inc, Cincinnati OH

Royle Printing, Sun Prairie WI

RPP Enterprises Inc, Libertyville IL

Sanders Printing Company, Garretson SD

The Saratoga Printing Co, Saratoga Springs NY

Schelin Printing Co, St Louis MO

Schiff Printing, Pittsburgh PA

Science Press, Ephrata PA

Service Printing Company, San Leandro CA

Service Web Offset Corporation, Chicago IL

Sexton Printing, Saint Paul MN

Shea Communications, Oklahoma City OK

Shepard Poorman Communications, Indianapolis IN

Sheridan Printing Company, Phillipsburg NJ

Sinclair Printing Co, Los Angeles CA

R E Smith Printing, Fall River MA

Smith-Edwards-Dunlap, Philadelphia PA

Snohomish Publishing, Snohomish WA

Southeastern Printing Company Inc, Stuart FL

Sowers Printing Company, Lebanon PA

Speakes-Hines & Thomas Inc, Lansing MI

Spencer Press, Wells ME

St Croix Press Inc, New Richmond WI

St Joseph Printing Ltd, Concord ON

Standard Publishing, Cincinnati OH

John Stark Printing Co Inc, St Louis MO

Stephenson Lithograph Inc, Alexandria VA

Stevens Graphics, Atlanta GA

Stevens Graphics Inc, Birmingham AL

Stinehour Press, Lunenburg VT

Strathmore Printing, Geneva IL

Straus Printing Company, Madison WI

Sullivan Graphics Inc, Brentwood TN

Sun Graphics, Parsons KS

John S Swift Company Inc, St Louis MO

John S Swift/Chicago, Chicago IL

John S Swift/Cincinnati, Cincinnati OH

John S Swift/Teterboro, Teterboro NJ

TAN Books & Publishers Inc, Rockford IL

Tapco Inc, Pemberton NJ

Taylor Publishing Company, Dallas TX

TEACH Services, Brushton NY

Technical Communication Services, North Kansas City MO

Thomson-Shore, Dexter MI

Times Litho Inc, Forest Grove OR

Times Printing, Random Lake WI

Todd Publications Inc, Smithville TX

Tracor Publications, Austin TX

Trade Litho Inc, Miami FL

Trend Offset Printing Service Inc, Los Alamitos CA

Tri-Graphic Printing, Ottawa ON

Triangle Booklet Company, Nashville TN

TSO General Corporation, Long Island City NY

Catalog Printers

Tweddle Litho Co, Clinton Township MI
Ultra-Color Corporation, Saint Louis MO
Universal Press, Providence RI
Universal Printers, Winnipeg MB
Universal Printing Company, Miami FL
Universal Printing Company, Saint
    Louis MO
US Press Inc, Valdosta GA
Vail Ballou Press Binghamton NY
Valco Graphics, Seattle WA
Victor Graphics, Baltimore MD
Viking Press, Eden Prairie MN
Vogue Printers, North Chicago IL
Wagners Printers, Davenport IA
Waldman Graphics, Pennsauken NJ
Waldon Press Inc, New York NY
Wallace Press, Hillside IL
Watkins Printing Company, Columbus
    OH
Watt/Peterson, Plymouth MN
Web Specialties, Wheeling IL
Webcom Ltd, Scarborough ON
Webcraft Technologies, North Brunswick
    NJ
Webcrafters, Madison WI
F A Weber & Sons, Park Falls WI
Weber Printing Company, Lincoln NE
Western Publishing Co Inc, Racine WI
Western Web Printing, Sioux Falls SD
Westview Press, Boulder CO
White Arts Inc, Indianapolis IN
Whitehall Printing Company, Naples FL
Wickersham Printing Co Inc, Lancaster
    PA
Windsor Associates, San Diego CA
Winter Sun Inc, Phoenix AZ
Wisconsin Color Press Inc, Milwaukee WI
Wolfer Printing, Commerce CA
Wood & Jones Printers, Pasadena CA
World Color Press Inc, New York NY
Worzalla Publishing Company, Stevens
    Point WI
Henry Wurst Inc, North Kansas City MO
Y/Z Printing Co, Elizabethtown PA
Ye Olde Genealogie Shoppe, In-
    dianapolis IN

Magazines

The following printers all can print
magazines.

Academy Books, Rutland VT
Adviser Graphics, Red Deer AB
Alan Lithograph Inc, Inglewood CA
All-Star Printing, Lansing MI
Alonzo Printing, Hayward CA
AM Graphics & Printing, San Marcos CA
American Signature Memphis Division,
    Olive Branch MS
American Signature Graphics, Dallas TX
American Signature/Foote, Lincoln NE
American Signature/Foote, Atlanta GA
American Web Inc, Denver CO
Americomp, Brattleboro VT
The Argus Press Inc, Niles IL
Arizona Lithographers, Tucson AZ
Artcraft Press, Waterloo WI
Associated Printers, Grafton ND
Austin Printing Co Inc, Akron OH
Automated Graphic Systems, White
    Plains MD
Banta Publications Group, Hinsdale IL
Banta-Harrisonburg, Harrisonburg VA
Bay Port Press, Chula Vista CA
Bay State Press, Framingham MA
Beacon Press Inc, Richmond VA
Beacon Press, Seattle WA
Blake Printery, San Luis Obispo CA
Bolger Publications/Creative Printing,
    Minneapolis MN
Brown Printing, Waseca MN
Brown Printing/Franklin, Franklin KY
Brown Printing/East Greenville Division,
    East Greenville PA
R L Bryan Company, Columbia SC
BSC Litho, Harrisburg PA
Buse Printing & Advertising, Phoenix AZ
The William Byrd Press, Richmond VA
Cal Central Press, Sacramento CA
Caldwell Printers, Arcadia CA
California Offset Printers, Glendale CA
Calsonic Miura Graphics, Irvine CA
Canterbury Press, Rome NY
Capital Printing Company Inc, Austin TX
Carlson Color Graphics, Ocala FL
Citizen Printing, Beaver Dam WI
The College Press, Collegedale TN
Color World Printers, Bozeman MT
Colortone Press, Landover MD
Colotone Riverside Inc, Dallas TX
Combined Communication Services,
    Mendota IL
Comfort Printing, St Louis MO
Community Press, Provo UT
Continental Web, Itasca IL
Country Press, Mohawk NY

Magazine Printers

The Craftsman Press Inc, Seattle WA
Crusader Printing Co, East Saint Louis IL
Lew A Cummings Printing Company Inc, Manchester NH
Danner Press Corporation, Canton OH
Dartmouth Printing, Hanover NH
Dataco A WBC Company, Albuquerque NM
Delmar Printing & Publishing, Charlotte NC
Democrat Printing & Litho, Little Rock AR
Des Plaines Publishing Company, Des Plaines IL
Desaulniers Printing Company, Milan IL
Deven Lithographers, Brooklyn NY
Dharma Press, Oakland CA
Diamond Graphics Tech Data, Milwaukee WI
Directories America, Chicago IL
Dollco Printing, Ottawa ON
R R Donnelley & Sons Co, Crawfordsville IN
Dragon Press, Delta Junction AK
Dynagraphics, Carlsbad CA
Eastwood Printing Co, Denver CO
EBSCO Media, Birmingham AL
Edison Lithographing, North Bergen NJ
Editors Press Inc, Hyattsville MD
EP Graphics, Berne IN
Faculty Press Inc, Brooklyn NY
Fetter Printing, Louisville KY
First Impressions Printer & Lithographers Inc, Elk Grove IL
Fisher Printers Inc, Cedar Rapids IA
Fleming Printing Co, St Louis MO
Frye & Smith, Costa Mesa CA
Futura Printing Inc, Boynton Beach FL
Gateway Press Inc, Louisville KY
Glundal Color Service, East Syracuse NY
Gowe Printing Company, Medina OH
Graftek Press, Woodstock IL
Graphic Litho Corporation, Lawrence MA
Graphic Ways Inc, Buena Park CA
Graphics Ltd, Indianapolis IN
Gray Printing, Fostoria OH
Great Lakes Lithograph Company, Cleveland OH
Greenfield Printing & Publishing, Greenfield OH
Gregath Publishing Company, Wyandotte OK
GRIT Commercial Printing Services, Williamsport PA
Gulf Printing, Houston TX
Guynes Printing Company, Albuquerque NM
Harmony Printing, Liberty MO
Hart Graphics, Simpsonville SC

Hart Press, Long Prairie MN
Hawkes Publishing, Salt Lake City UT
Heartland Press Inc, Spencer IA
Hennegan Company, Cincinnati OH
Hinz Lithographing Company, Mt Prospect IL
A B Hirschfeld Press, Denver CO
Hoechstetter Printing, Pittsburgh PA
Holladay-Tyler Printing Corp, Glenn Dale MD
Independent Publishing Company, Sarasota FL
Inland Printing Company, Syosset NY
Insert Color Press, Ronkonkoma NY
Intelligencer Printing, Lancaster PA
Interform Corporation, Bridgeville PA
IPC Publishing Services, Saint Joseph MI
Jersey Printing Company, Bayonne NJ
JK Creative Printers, Quincy IL
Johns Byrne Co, Niles IL
The Johnson & Hardin Company, Cincinnati OH
Johnson Graphics, East Dubuque IL
Jostens Printing & Publishing, State College PA
Jostens Printing & Publishing, Winston-Salem NC
Jostens Printing & Publishing, Minneapolis MN
Jostens Printing & Publishing, Clarksville TN
Jostens Printing & Publishing, Topeka KS
Jostens Printing & Publishing, Visalia CA
Judd's Incorporated, Strasburg VA
Judd's Incorporated, Washington DC
Julin Printing, Monticello IA
K-B Offset Printing, State College PA
Kaufman Press Printing, Syracuse NY
Kimberly Press, Santa Barbara CA
The C J Krehbiel Company, Cincinnati OH
La Crosse Graphics, LaCrosse WI
Lancaster Press, Lancaster PA
The Lane Press, Burlington VT
Lasky Company, Milburn NJ
Lehigh Press Inc, Cherry Hill NJ
Lighthouse Press Inc, Manchester NH
Lithocolor Press, Westchester IL
Lithoid Printing Corporation, East Brunswick NJ
The Little River Press Inc, Miami FL
John D Lucas Printing Company, Baltimore MD
Mack Printing Group Inc, Easton PA
Mackintosh Typography Inc, Santa Barbara CA
Maquoketa Web Printing, Maquoketa IA
Marathon Communications Group, Wausau WI
Marrakech Express, Tarpon Springs FL
Mars Graphic Services, Westville NJ

Magazine Printers

MBP Lithographics, Hillsboro KS
McArdle Printing Co, Upper Marlboro MD
McClain Printing Company, Parsons WV
The McFarland Press, Harrisburg PA
McGill Jensen, St Paul MN
McKay Printing Services, Michigan City IN
Meaker the Printer, Phoenix AZ
Mercury Printing Company, Memphis TN
Metroweb, Erlanger KY
Mitchell Press, Vancouver BC
Moebius Printing, Milwaukee WI
Moran Printing Company, Orlando FL
Motheral Printing, Fort Worth TX
William H Muller Printing Co, Santa Clara CA
National Graphics Corp, Columbus OH
National Lithographers Inc, Miami FL
Louis Neibauer Co Inc, Warminster PA
Neyenesch Printers, San Diego CA
Nies/Artcraft Printing Company, Saint Louis MO
Nimrod Press Inc, Boston MA
Noll Printing Company, Huntington IN
Northeast Graphics, New York NY
Northprint International Inc, Grand Rapids MN
Northwest Web, Eugene OR
Nystrom Publishing Company, Osseo MN
Oaks Printing Company, Bethlehem PA
Omaha Printing Company, Omaha NE
Outstanding Graphics, Kenosha WI
The Ovid Bell Press Inc, Fulton MO
The Oxford Group, Norway ME
Padgett Printing Corporation, Dallas TX
Paraclete Press, Brewster MA
Park Press, South Holland IL
Paust Inc, Richmond IN
Pendell Printing, Midland MI
PennWell Printing, Tulsa OK
Penton Press Publishing, Berea OH
The Perlmuter Printing Co, Cleveland OH
Perry Printing Corp, Waterloo WI
Perry/Baraboo, Baraboo WI
PFP Printing Corporation, Gaithersburg MD
Pioneer Press, Wilmette IL
Plain Talk Printing Co, Des Moines IA
Port Publications, Port Washington WI
Practical Graphics, New York NY
The Press of Ohio, Kent OH
Print Northwest, Tacoma WA
The Printer Inc, Osseo MN
The Printing Company, Indianapolis IN
Printing Corporation of America, Pompano Beach FL
Printwest, Regina SK

Progress Printing, Lynchburg VA
Promotional Printing Corporation, Houston TX
Publishers Choice Reprints, The Woodlands TX
Publishers Press, Salt Lake City UT
Publishers Printing Co, Shepherdsville KY
Quad Graphics - Thomaston, Thomaston GA
Quad/Graphics Inc, Saratoga Springs NY
Quad/Graphics, Lomira, Lomira WI
Quad/Graphics, Pewaukee, Pewaukee WI
Quad/Graphics, Sussex, Sussex WI
Quebecor Book Group, New York NY
Quebecor Printing (USA) Corp, Boston MA
Quebecor Printing Inc, Saint Paul MN
Quebecor Printing Inc, Richmond Hill ON
Quebecor Printing/St Cloud, St Cloud MN
RBW Graphics, Owen Sound ON
George Rice & Sons, Los Angeles CA
Rich Printing Company Inc, Nashville TN
Ringier America, Itasca IL
Ringier America Inc, Corinth MS
Ringier America/Brookfield, Brookfield WI
Ringier America/Jonesboro, Jonesboro AR
Ringier America/Phoenix, Phoenix AZ
Ronalds Printing, Vancouver BC
Ronalds Printing, Montreal PQ
S Rosenthal & Company Inc, Cincinnati OH
Royle Printing, Sun Prairie WI
Sanders Printing Company, Garretson SD
The Saratoga Printing Co, Saratoga Springs NY
Schelin Printing Co, St Louis MO
Schumann Printers, Fall River WI
Science Press, Ephrata PA
Sexton Printing, Saint Paul MN
Shea Communications, Oklahoma City OK
Shepard Poorman Communications, Indianapolis IN
The Sheridan Press, Hanover PA
Sheridan Printing Company, Phillipsburg NJ
SLC Graphics, Pittston PA
R E Smith Printing, Fall River MA
Smith-Edwards-Dunlap, Philadelphia PA
Snohomish Publishing, Snohomish WA
Southeastern Printing Company Inc, Stuart FL
Southern California Graphics, Culver City CA
Sowers Printing Company, Lebanon PA

Magazine Printers

Speakes-Hines & Thomas Inc, Lansing MI
St Croix Press Inc, New Richmond WI
St Joseph Printing Ltd, Concord ON
Standard Publishing, Cincinnati OH
Stephenson Lithograph Inc, Alexandria VA
Strathmore Printing, Geneva IL
Straus Printing Company, Madison WI
The Studley Press, Dalton MA
Sullivan Graphics Inc, Brentwood TN
Sundance Press, Tucson AZ
Tapco Inc, Pemberton NJ
Times Litho Inc, Forest Grove OR
Times Printing, Random Lake WI
Todd Publications Inc, Smithville TX
Tracor Publications, Austin TX
Trade Litho Inc, Miami FL
Trend Offset Printing Service Inc, Los Alamitos CA
Tri-Graphic Printing, Ottawa ON
TSO General Corporation, Long Island City NY
United Lithographic Services Inc, Falls Church VA
Universal Press, Providence RI
Universal Printing Company, Miami FL
Universal Printing Company, Saint Louis MO
Valco Graphics, Seattle WA
Waldman Graphics, Pennsauken NJ
Waldon Press Inc, New York NY
Watkins Printing Company, Columbus OH
Watt/Peterson, Plymouth MN
Waverly Press, Linthicum MD
Webcrafters, Madison WI
F A Weber & Sons, Park Falls WI
The Wimmer Companies, Memphis TN
Wisconsin Color Press Inc, Milwaukee WI
Wood & Jones Printers, Pasadena CA
World Color Press Inc, New York NY
Henry Wurst Inc, North Kansas City MO
Ye Olde Genealogie Shoppe, Indianapolis IN

Magazine Short-Run Printers

The following magazine printers specialize in printing magazines in runs from 10,000 on up to 100,000.

Adviser Graphics, Red Deer AB
Alan Lithograph Inc, Inglewood CA

Americomp, Brattleboro VT
The Argus Press Inc, Niles IL
Arizona Lithographers Inc, Tucson AZ
Associated Printers, Grafton ND
Bay Port Press, Chula Vista CA
Beacon Press, Seattle WA
R L Bryan Company, Columbia SC
Capital Printing Company Inc, Austin TX
The College Press, Collegedale TN
Community Press, Provo UT
Dataco A WBC Company, Albuquerque NM
Delmar Printing & Publishing, Charlotte NC
Democrat Printing & Litho, Little Rock AR
Dharma Press, Oakland CA
Diamond Graphics Tech Data, Milwaukee WI
Dragon Press, Delta Junction AK
EBSCO Media, Birmingham AL
Graphic Ways Inc, Buena Park CA
Graphics Ltd, Indianapolis IN
GRIT Commercial Printing Services, Williamsport PA
Guynes Printing Company, Albuquerque NM
Hawkes Publishing, Salt Lake City UT
IPC Publishing Services, Saint Joseph MI
Jersey Printing Company, Bayonne NJ
Johnson Graphics, East Dubuque IL
Jostens Printing & Publishing, Visalia CA
Jostens Printing & Publishing, Clarksville TN
Jostens Printing & Publishing, Topeka KS
Jostens Printing & Publishing, State College PA
Jostens Printing & Publishing, Minneapolis MN
Kaufman Press Printing, Syracuse NY
La Crosse Graphics, LaCrosse WI
Lancaster Press, Lancaster PA
Lithocolor Press, Westchester IL
Lithoid Printing Corporation, East Brunswick NJ
Marrakech Express, Tarpon Springs FL
MBP Lithographics, Hillsboro KS
McClain Printing Company, Parsons WV
The McFarland Press, Harrisburg PA
Louis Neibauer Co Inc, Warminster PA
Neyenesch Printers, San Diego CA
Nimrod Press Inc, Boston MA
Nystrom Publishing Company, Osseo MN
PennWell Printing, Tulsa OK
Port Publications, Port Washington WI
Printing Corporation of America, Pompano Beach FL
The Saratoga Printing Co, Saratoga Springs NY
Science Press, Ephrata PA
Sexton Printing, St Paul MN

Magazine Short-Run

The Sheridan Press, Hanover PA
Sheridan Printing Company, Phillipsburg NJ
R E Smith Printing, Fall River MA
Snohomish Publishing, Snohomish WA
Southeastern Printing Company Inc, Stuart FL
St Croix Press Inc, New Richmond WI
The Studley Press, Dalton MA
Sundance Press, Tucson AZ
Tracor Publications, Austin TX
TSO General Corporation, Long Island City NY
Waldman Graphics, Pennsauken NJ
F A Weber & Sons, Park Falls WI
The Wimmer Companies, Memphis TN
Wood & Jones Printers, Pasadena CA

Ultra-Short-Run Book Printers

The following printers specialize in printing books in runs of less than 1,000 copies.

aBCD, Seattle WA
Advanced Data Reproductions, Wichita KS
Advanced Duplicating & Printing, Minneapolis MN
Americomp, Brattleboro VT
BookMasters Inc, Ashland OH
Braun-Brumfield Inc, Ann Arbor MI
Brunswick Publishing Corp, Lawrenceville VA
C & M Press, Denver CO
Camelot Publishing Co, Ormond Beach FL
Champagne Offset Co Inc, Natick MA
Coach House Press, Toronto ON
Coneco Litho Graphics, Glen Falls NY
Country Press Inc, Lakeville MA
Crane Duplicating Service, West Barnstable MA
The Everton Publishers Inc, Logan UT
Faculty Press Inc, Brooklyn NY
Graphic Litho Corporation, Lawrence MA
GRT Book Printing, Oakland CA
Hawkes Publishing, Salt Lake City UT
JK Creative Printers, Quincy IL
McDowell Publications, Utica KY
Morgan Printing, Austin TX
National Lithographers Inc, Miami FL
National Reproductions Corp, Madison Heights MI
OMNIPRESS, Madison WI

PFP Printing Corporation, Gaithersburg MD
Publishers Press, Salt Lake City UT
Quebecor Book Group, New York NY
Sanders Printing Company, Garretson SD
Stewart Publishing & Printing Co, Marble Hill MO
Stinehour Press, Lunenburg VT
Van Volumes, Palmer MA
Walter's Publishing, Waseca MN
Weber Printing Company, Lincoln NE
Westview Press, Boulder CO
Wickersham Printing Co Inc, Lancaster PA

Short-Run Book Printers

The following book printers report an optimum print run somewhere between 1,001 copies and 10,000 copies.

Academy Books, Rutland VT
Adviser Graphics, Red Deer AB
Baker Johnson Inc, Dexter MI
Bay Port Press, Chula Vista CA
Blake Printery, San Luis Obispo CA
Blue Dolphin Press Inc, Grass Valley CA
BookCrafters, Chelsea MI
BookCrafters/Fredericksburg, Fredericksburg VA
Boyd Printing Co Inc, Albany NY
Brennan Printing, Deep River IA
Capital City Press, Montpelier VT
Columbus Bookbinders & Printers, Columbus GA
Community Press, Provo UT
Cushing-Malloy, Ann Arbor MI
Daamen Printing Company, West Rutland VT
Dickinson Press Inc, Grand Rapids MI
R R Donnelley & Sons Co, Crawfordsville IN
Edwards Brothers Inc, Ann Arbor, MI
Edwards Brothers Inc, Lillington NC
Evangel Press, Nappanee IN
The Forms Man Inc, Deer Park NY
Friesen Printers, Altona MB
Fundcraft Publishing, Collierville TN
Gilliland Printing Inc, Arkansas City KS
Graphic Printing, New Carlisle OH
Griffin Printing and Lithograph Company, Sacramento CA
Harlo Printing Co, Detroit MI
Heritage Printers Inc, Charlotte NC
Hignell Printing, Winnepeg MB

Short-Run Book Printers

Rae Horowitz Book Manufacturers, Fairfield NJ
Jostens Printing & Publishing, Visalia CA
Jostens Printing & Publishing, Clarksville TN
Jostens Printing & Publishing, Topeka KS
Jostens Printing & Publishing, State College PA
Kimberly Press, Santa Barbara CA
Kirby Lithographic Co Inc, Arlington VA
KNI Incorporated, Anaheim CA
La Crosse Graphics, LaCrosse WI
Les Editions Marquis, Montmagny PQ
Mackintosh Typography Inc, Santa Barbara CA
Malloy Lithographing, Ann Arbor MI
Maple-Vail Book Manufacturing Group, York PA
Marrakech Express, Tarpon Springs FL
MBP Lithographics, Hillsboro KS
McNaughton & Gunn Inc, Saline MI
Monument Printers & Lithographer Inc, Verplanck NY
Naturegraph Publishers, Happy Camp CA
Optic Graphics Inc, Glen Burnie MD
Patterson Printing Company, Benton Harbor MI
Port City Press Inc, Baltimore MD
Princeton Academic Press, Lawrenceville NJ
Prinit Press, Dublin IN
Rose Printing Company, Tallahassee FL
The Saratoga Printing Co, Saratoga Springs NY
Science Press, Ephrata PA
Service Printing Company, San Leandro CA
Snohomish Publishing, Snohomish WA
The Studley Press, Dalton MA
TEACH Services, Brushton NY
Thomson-Shore, Dexter MI
Versa Press Inc, East Peoria IL
Vicks Lithograph & Printing, Yorkville NY
Victor Graphics, Baltimore MD
Walsworth Publishing Co Inc, Marceline MO
Webcom Ltd, Scarborough ON

Index: Other Printed Items

The following indexes list some of the printers who are capable of printing various items, from annual reports to software manuals, from calendars to posters. These indexes, like all the other indexes, are derived from the surveys returned by printers (plus other sources describing the capabilities of those printers who did not respond to the survey).

Annual Reports

Annual Reports

Electronic Printing, Brentwood NY
EU Services, Rockville MD
Faculty Press Inc, Brooklyn NY
First Impressions & Lithographers Inc,
    Elk Grove IL
Fisher Printing Co, Galion OH
Fleming Printing Co, St Louis MO
Flower City Printing Inc, Rochester NY
Focus Direct, San Antonio TX
Fort Orange Press Inc, Albany NY
Foster Printing Service, Michigan City IN
Frye & Smith, Costa Mesa CA
Futura Printing Inc, Boynton Beach FL
Gateway Press Inc, Louisville KY
Gaylord Printing, Detroit MI
George Lithograph, San Francisco CA
Glundal Color Service, East Syracuse NY
Graphic Arts Center, Portland OR
Graphic Litho Corporation, Lawrence MA
Graphic Ways Inc, Buena Park CA
Graphics Ltd, Indianapolis IN
Gray Printing, Fostoria OH
Great Lakes Lithograph Co, Cleveland
    OH
Great Northern/Design Printing, Skokie
    IL
Gregath Publishing Company, Wyan-
    dotte OK
Griffin Printing & Lithograph, Glendale
    CA
GRIT Commercial Printing Services, Wil-
    liamsport PA
Gulf Printing, Houston TX
Guynes Printing Company, Albuquerque
    NM
Harlo Printing Co, Detroit MI
Harmony Printing, Liberty MO
Heartland Press Inc, Spencer IA
Hennegan Company, Cincinnati OH
Hinz Lithographing Company, Mt
    Prospect IL
A B Hirschfeld Press, Denver CO
Imperial Litho/Graphics Inc, Phoenix AZ
Inland Printing Company, Syosset NY
Intelligencer Printing, Lancaster PA
Interform Corporation, Bridgeville PA
Interstate Printing Company, Omaha NE
Jersey Printing Company, Bayonne NJ
JK Creative Printers, Quincy IL
The Johnson & Hardin Company, Cin-
    cinnati OH
Johnson Graphics, East Dubuque IL
Jostens Printing & Publishing, Winston-
    Salem NC
Jostens Printing & Publishing, Min-
    neapolis MN
Julin Printing, Monticello IA
K-B Offset Printing, State College PA
Kaufman Press Printing, Syracuse NY
Kolor View Press, Aurora MO

The C J Krehbiel Company, Cincinnati
    OH
La Crosse Graphics, LaCrosse WI
Lasky Company, Milburn NJ
Lehigh Press Inc, Cherry Hill NJ
Les Editions Marquis, Montmagny PQ
Lighthouse Press Inc, Manchester NH
Litho Productions Inc, Madison WI
Litho Specialties, St Paul MN
Lithograph Printing Company, Memphis
    TN
Lithoid Printing Corporation, East
    Brunswick NJ
The Little River Press Inc, Miami FL
John D Lucas Printing Company, Bal-
    timore MD
Marathon Communications Group,
    Wausau WI
Marrakech Express, Tarpon Springs FL
Maverick Publications Inc, Bend OR
MBP Lithographics, Hillsboro KS
McClain Printing Company, Parsons WV
The McFarland Press, Harrisburg PA
McGill Jensen, St Paul MN
McGrew Color Graphics, Kansas City
    MO
McQuiddy Printing Company, Nashville
    TN
Meaker the Printer, Phoenix AZ
Meehan-Tooker & Company, East
    Rutherford NJ
Mercury Printing Company, Memphis
    TN
Mitchell Press, Vancouver BC
Moebius Printing, Milwaukee WI
Moran Printing Company, Orlando FL
Moran Printing Inc, Baton Rouge LA
Morgan Press, Dobbs Ferry NY
Motheral Printing, Fort Worth TX
William H Muller Printing, Santa Clara
    CA
MultiPrint Co Inc, Skokie IL
Murphy's Printing Company, Campbell
    CA
National Lithographers Inc, Miami FL
Nationwide Printing, Burlington KY
Louis Neibauer Co Inc, Warminster PA
Neyenesch Printers, San Diego CA
Nielsen Lithographing Co, Cincinnati OH
Nies/Artcraft Printing Companies, St
    Louis MO
Nimrod Press Inc, Boston MA
Northlight Studio Press, Barre VT
Nystrom Publishing Company, Osseo MN
Oaks Printing Company, Bethlehem PA
Omaha Printing Company, Omaha NE
The Ovid Bell Press Inc, Fulton MO
The Oxford Group, Norway ME
Padgett Printing Corporation, Dallas TX
Panorama Press Inc, Clifton NJ
Paraclete Press, Brewster MA
Paust Inc, Richmond IN

Annual Reports

Pearl-Pressman-Liberty Communications Group, Philadelphia PA
Peconic Companies, Mattituck NY
The Perlmuter Printing Co, Cleveland OH
Perry Printing Corp, Waterloo WI
Perry/Baraboo, Baraboo WI
Pine Hill Press Inc, Freeman SD
Plain Talk Printing Company, Des Moines IA
Port Publications, Port Washington WI
Practical Graphics, New York NY
The Press, Chanhassen MN
Print Northwest, Tacoma WA
The Printer Inc, Osseo MN
The Printing Company, Indianapolis IN
Printing Corporation of America, Pompano Beach FL
Printright International, Palmdale CA
Printwest, Regina SK
Progress Printing, Lynchburg VA
ProLitho, Provo UT
Publishers Press, Salt Lake City UT
Quad/Graphics Inc, Saratoga Springs NY
Quad/Graphics, Lomira, Lomira WI
Quad/Graphics, Pewaukee, Pewaukee WI
Quad/Graphics, Sussex, Sussex WI
Quebecor Book Group, New York NY
Quebecor Printing (USA) Corp, Boston MA
Quebecor Printing Inc, Richmond Hill ON
Repro-Tech, Fairfield CT
George Rice & Sons, Los Angeles CA
Ringier America/Brookfield, Brookfield WI
Ringier America/Jonesboro, Jonesboro AR
Ringier America/Phoenix, Phoenix AZ
John Roberts Company, Minneapolis MN
Ronalds Printing, Montreal PQ
Royle Printing, Sun Prairie WI
RPP Enterprises Inc, Libertyville IL
Sanders Printing Company, Garretson SD
The Saratoga Printing Co, Saratoga Springs NY
Schelin Printing Co, St Louis MO
Schiff Printing, Pittsburgh PA
Science Press, Ephrata PA
Sexton Printing, Saint Paul MN
Shepard Poorman Communications, Indianapolis IN
R E Smith Printing, Fall River MA
Smith-Edwards-Dunlap, Philadelphia PA
Snohomish Publishing, Snohomish WA
Southeastern Printing Company Inc, Stuart FL
Southern California Graphics, Culver City CA

Sowers Printing Company, Lebanon PA
Spencer Press, Wells ME
John Stark Printing Co Inc, St Louis MO
Stephenson Lithograph Inc, Alexandria VA
Stinehour Press, Lunenburg VT
Straus Printing Company, Madison WI
The Studley Press, Dalton MA
John S Swift Company Inc, St Louis MO
John S Swift/Chicago, Chicago IL
John S Swift/Cincinnati, Cincinnati OH
John S Swift/Teterboro, Teterboro NJ
TAN Books & Publishers Inc, Rockford IL
Taylor Publishing Company, Dallas TX
Times Printing, Random Lake WI
Tracor Publications, Austin TX
Trade Litho Inc, Miami FL
Triangle Booklet Company, Nashville TN
TSO General Corporation, Long Island City NY
Universal Press, Providence RI
Universal Printing Company, Saint Louis MO
US Press Inc, Valdosta GA
Van Volumes, Palmer MA
Vogue Printers, North Chicago IL
Wagners Printers, Davenport IA
Waldman Graphics, Pennsauken NJ
Waldon Press Inc, New York NY
Wallace Press, Hillside IL
Watkins Printing Company, Columbus OH
Watt/Peterson, Plymouth MN
F A Weber & Sons, Park Falls WI
Westview Press, Boulder CO
White Arts Inc, Indianapolis IN
Wolfer Printing, Commerce CA
Wood & Jones Printers, Pasadena CA
World Color Press Inc, New York NY
Henry Wurst Inc, North Kansas City MO
Y/Z Printing Co, Elizabethtown PA

Brochures

A-1 Business Service Inc, St Paul MN
Academy Books, Rutland VT
Accurate Printing, Post Falls ID
Action Printing, Fond Du Lac WI
Alcom Printing Group Inc, Bethlehem PA
AM Graphics & Printing, San Marcos CA
American Litho, Hayward CA
American Printers & Litho, Niles IL
American Signature, Memphis Division, Olive Branch MS
American Signature Graphics, Dallas TX
American Signature/Foote, Atlanta GA
American Signature, Lincoln NE
Amidon & Associates Inc, St Paul MN

Brochures

Amos Press, Sidney OH
Anundsen Publishing Company, Decorah IA
Apollo Graphics Ltd, Southampton PA
Arandell Corporation, Menomonee Falls WI
The Argus Press Inc, Niles IL
Arizona Lithographers Inc, Tucson AZ
Artcraft Press, Waterloo WI
Associated Printers, Grafton ND
Austin Printing Co Inc, Akron OH
Automated Graphic Systems, White Plains MD
Bang Printing Company, Brainerd MN
Barton Press Inc, West Orange NJ
Bay Port Press, Chula Vista CA
Bay State Press, Framingham MA
Beacon Press Inc, Seattle WA
Ben-Wal Printing, Pomona CA
Bertelsmann Printing & Manufacturing, New York NY
Blake Printery, San Luis Obispo CA
Blue Dolphin Press Inc, Grass Valley CA
Bolger Publications/Creative Printing, Minneapolis MN
Booklet Publishing Company Inc, Elk Grove Village IL
Bradley Printing Company, Des Plaines IL
Brennan Printing, Deep River IA
Brookshore Lithographers Inc, Elk Grove Village IL
Brunswick Publishing Corp, Lawrenceville VA
R L Bryan Company, Columbia SC
BSC Litho, Harrisburg PA
Buse Printing & Advertising, Phoenix AZ
Cal Central Press, Sacramento CA
Caldwell Printers, Arcadia CA
California Offset Printers, Glendale CA
Calsonic Miura Graphics, Irvine CA
Canterbury Press, Rome NY
Capital Printing Company Inc, Austin TX
Catalog King, Clifton NJ
Champagne Offset Co Inc, Natick MA
Chicago Color Express, Chicago IL
Chicago Press Corp, Chicago IL
Coach House Press, Toronto ON
The College Press, Collegedale TN
Colonial Graphics, Paterson NJ
Color Express, Commerce CA
Color Express, Irvine CA
Color Express, Minneapolis MN
Color Express, Atlanta GA
Color World Printers, Bozeman MT
ColorGraphics Printing Corp, Tulsa OK
Colorlith Corporation, Johnston RI
Colortone Press, Landover MD
Colotone Riverside Inc, Dallas TX

Columbus Bookbinders & Printers, Columbus GA
Comfort Printing, St Louis MO
Community Press, Provo UT
Concord Litho Company Inc, Concord NH
Coneco Litho Graphics, Glen Falls NY
Continental Web, Itasca IL
Cooley Printers & Office Supply, Monroe LA
Country Press, Mohawk NY
Crane Duplicating Service, West Barnstable MA
Crusader Printing Co, East Saint Louis IL
Danbury Printing & Litho Inc, Danbury CT
Dartmouth Printing, Hanover NH
Dataco A WBC Company, Albuquerque NM
Dayal Graphics, Lemont IL
Dellas Graphics, Syracuse NY
Delmar Printing & Publishing, Charlotte NC
Desaulniers Printing Company, Milan IL
Deven Lithographers, Brooklyn NY
Dharma Press, Oakland CA
Diamond Graphics Tech Data, Milwaukee WI
Direct Graphics Inc, Sidney OH
Direct Press/Modern Litho, Huntington Station NY
Dittler Brothers, Atlanta GA
Dollco Printing, Ottawa ON
Donihe Graphics Inc, Kingsport TN
Dragon Press, Delta Junction AK
Dynagraphics, Carlsbad CA
E & D Web Inc, Cicero IL
Eagle Web Press, Salem OR
Eastwood Printing Co, Denver CO
EBSCO Media, Birmingham AL
Edison Lithographing, North Bergen NJ
Editors Press Inc, Hyattsville MD
Electronic Printing, Brentwood NY
EU Services, Rockville MD
Eureka Printing Co Inc, Eureka CA
Eusey Press Inc, Leominster MA
Eva-Tone Inc, Clearwater FL
Evangel Press, Nappanee IN
Faculty Press Inc, Brooklyn NY
Federated Lithographers, Providence RI
Fetter Printing, Louisville KY
First Impressions & Lithographers Inc, Elk Grove IL
Fisher Printers Inc, Cedar Rapids IA
Fisher Printing Co, Galion OH
Fleming Printing Co, St Louis MO
Flower City Printing Inc, Rochester NY
Focus Direct, San Antonio TX
The Forms Man Inc, Deer Park NY
Fort Orange Press Inc, Albany NY
Foster Printing Service, Michigan City IN

Brochures

Friesen Printers, Altona MB
Frye & Smith, Costa Mesa CA
Futura Printing Inc, Boynton Beach FL
Gateway Press Inc, Louisville KY
Gaylord Printing, Detroit MI
George Lithograph, San Francisco CA
Geryon Press Limited, Tunnel NY
Glundal Color Service, East Syracuse NY
Golden Belt Printing Inc, Great Bend KS
Gowe Printing Company, Medina OH
Graphic Arts Center, Portland OR
Graphic Litho Corporation, Lawrence MA
Graphic Printing, New Carlisle OH
Graphic Production Center, Norcross GA
Graphic Ways Inc, Buena Park CA
Graphics Ltd, Indianapolis IN
Gray Printing, Fostoria OH
Great Impressions Printing, Dallas TX
Great Lakes Lithograph Co, Cleveland
    OH
Great Northern/Design Printing, Skokie
    IL
Gregath Publishing Co, Wyandotte OK
Griffin Printing & Lithograph, Glendale
    CA
GRIT Commercial Printing Services, Wil-
    liamsport PA
Gulf Printing, Houston TX
Guynes Printing Company, Albuquerque
    NM
Harlo Printing Co, Detroit MI
Harmony Printing, Liberty MO
Hart Graphics, Austin TX
Hawkes Publishing, Salt Lake City UT
Heartland Press Inc, Spencer IA
Henington Publishing Company, Wolfe
    City TX
Hennegan Company, Cincinnati OH
Hi-Tech Color House, Chicago IL
Hinz Lithographing Company, Mt
    Prospect IL
A B Hirschfeld Press, Denver CO
Imperial Litho/Graphics Inc, Phoenix AZ
Independent Printing Company, New
    York NY
Independent Publishing Company,
    Sarasota FL
Instant Web, Chanhassen MN
Intelligencer Printing, Lancaster PA
Interform Corporation, Bridgeville PA
Interstate Printing Company, Omaha NE
Ivy Hill Packaging, Louisville KY
Japs-Olson Company, Minneapolis MN
Jersey Printing Company, Bayonne NJ
JK Creative Printers, Quincy IL
The Job Shop, Woods Hole MA
The Johnson & Hardin Company, Cin-
    cinnati OH
Johnson Graphics, East Dubuque IL
Jostens Printing & Publishing, Visalia CA

Jostens Printing & Publishing, Winston-
    Salem NC
Jostens Printing & Publishing, Min-
    neapolis MN
Julin Printing, Monticello IA
K-B Offset Printing, State College PA
Kaufman Press Printing, Syracuse NY
Keys Printing Co Inc, Greenville SC
Kirby Lithographic Co Inc, Arlington VA
Kolor View Press, Aurora MO
Kordet Graphics Inc, Oceanside NY
The C J Krehbiel Company, Cincinnati
    OH
La Crosse Graphics, LaCrosse WI
Lasky Company, Milburn NJ
Lehigh Press Inc, Cherry Hill NJ
Les Editions Marquis, Montmagny PQ
Lighthouse Press Inc, Manchester NH
Litho Productions Inc, Madison WI
Litho Specialties, St Paul MN
Lithograph Printing Company, Memphis
    TN
Lithoid Printing Corporation, East
    Brunswick NJ
The Little River Press Inc, Miami FL
John D Lucas Printing Company, Bal-
    timore MD
Mackintosh Typography Inc, Santa Bar-
    bara CA
Marathon Communications Group,
    Wausau WI
Marrakech Express, Tarpon Springs FL
Mars Graphic Services, Westville NJ
The Mazer Corporation, Dayton OH
MBP Lithographics, Hillsboro KS
McArdle Printing Co, Upper Marlboro
    MD
McClain Printing Company, Parsons WV
The McFarland Press, Harrisburg PA
McGill Jensen, St Paul MN
McGrew Color Graphics, Kansas City
    MO
McKay Printing Services, Michigan City
    IN
McQuiddy Printing Company, Nashville
    TN
Meaker the Printer, Phoenix AZ
Media Printing, Miami FL
Media Publications, Santa Clara CA
Meehan-Tooker & Company, East
    Rutherford NJ
Mercury Printing Company, Memphis
    TN
Mitchell Graphics, Petosky MI
Mitchell Press, Vancouver BC
Moebius Printing, Milwaukee WI
Moran Printing Company, Orlando FL
Moran Printing Inc, Baton Rouge LA
Morgan Press, Dobbs Ferry NY
Morgan Printing, Austin TX
Morningrise Printing, Costa Mesa CA
Motheral Printing, Fort Worth TX

Brochures

William H Muller Printing, Santa Clara
    CA
MultiPrint Co Inc, Skokie IL
Murphy's Printing Co, Campbell CA
National Graphics Corp, Columbus OH
National Lithographers Inc, Miami FL
National Reproductions Corp, Madison
    Heights MI
Nationwide Printing, Burlington KY
Louis Neibauer Co Inc, Warminster PA
Nielsen Lithographing Co, Cincinnati OH
Nimrod Press Inc, Boston MA
Northlight Studio Press, Barre VT
Nystrom Publishing Company, Osseo MN
Oaks Printing Company, Bethlehem PA
Ohio Valley Litho Color, Florence KY
Omaha Printing Company, Omaha NE
Outstanding Graphics, Kenosha WI
The Oxford Group, Norway ME
Padgett Printing Corporation, Dallas TX
Panorama Press Inc, Clifton NJ
Paraclete Press, Brewster MA
Parker Graphics, Fuquay-Varina NC
Paust Inc, Richmond IN
Pearl-Pressman-Liberty Communica-
    tions Group, Philadelphia PA
Peconic Companies, Mattituck NY
PennWell Printing, Tulsa OK
Pentagram Press, Minneapolis MN
Penton Press Publishing, Berea OH
The Perlmuter Printing Co, Cleveland
    OH
Perry Printing Corp, Waterloo WI
Perry/Baraboo, Baraboo WI
Petty Co Inc, Effingham IL
PFP Printing Corporation, Gaithersburg
    MD
Phillips Brothers Printers Inc,
    Springfield IL
Pine Hill Press Inc, Freeman SD
Pioneer Press, Wilmette IL
Plain Talk Printing Company, Des
    Moines IA
Plus Communications, Saint Louis MO
Port Publications, Port Washington WI
Practical Graphics, New York NY
The Press, Chanhassen MN
Prinit Press, Dublin IN
Print Northwest, Tacoma WA
The Printing Company, Indianapolis IN
Printing Corporation of America, Pom-
    pano Beach FL
Printright International, Palmdale CA
Printwest, Regina SK
Progress Printing, Lynchburg VA
ProLitho, Provo UT
Promotional Printing Corporation, Hous-
    ton TX
Publishers Express Press, Ladysmith WI
Quad/Graphics Inc, Saratoga Springs NY

Quad/Graphics, Lomira, Lomira WI
Quad/Graphics, Pewaukee, Pewaukee WI
Quad/Graphics, Sussex, Sussex WI
Quebecor Printing (USA) Corp, Boston
    MA
Quebecor Printing Inc, Saint Paul MN
Quebecor Printing Inc, Richmond Hill
    ON
Quebecor/Semline, Braintree MA
Rapidocolor Corporation, West Chester
    PA
Repro-Tech, Fairfield CT
Ringier America/Brookfield, Brookfield
    WI
Ringier America/Jonesboro, Jonesboro
    AR
Ringier America/Phoenix, Phoenix AZ
John Roberts Company, Minneapolis MN
Ronalds Printing, Vancouver BC
Ronalds Printing, Montreal PQ
Rose Printing Company, Tallahassee FL
S Rosenthal & Company Inc, Cincinnati
    OH
RPP Enterprises Inc, Libertyville IL
Sanders Printing Company, Garretson
    SD
The Saratoga Printing Co, Saratoga
    Springs NY
Schelin Printing Co, St Louis MO
Schiff Printing, Pittsburgh PA
Science Press, Ephrata PA
Service Web Offset Corporation, Chicago
    IL
Sexton Printing, Saint Paul MN
Shea Communications, Oklahoma City
    OK
Shepard Poorman Communications, In-
    dianapolis IN
Sheridan Printing Company, Phil-
    lipsburg NJ
Sinclair Printing Co, Los Angeles CA
R E Smith Printing, Fall River MA
Smith-Edwards-Dunlap, Philadelphia PA
Snohomish Publishing, Snohomish WA
Southeastern Printing Company Inc,
    Stuart FL
Southern California Graphics, Culver
    City CA
Sowers Printing Company, Lebanon PA
Speakes-Hines & Thomas Inc, Lansing
    MI
Spencer Press, Wells ME
St Croix Press Inc, New Richmond WI
Standard Publishing, Cincinnati OH
John Stark Printing Co Inc, St Louis MO
Stephenson Lithograph Inc, Alexandria
    VA
Stewart Publishing & Printing Co,
    Marble Hill MO
Straus Printing Company, Madison WI
The Studley Press, Dalton MA
Sun Graphics, Parsons KS

Brochures

The Sun Hill Press, North Brookfield MA
John S Swift Company Inc, St Louis MO
John S Swift/Chicago, Chicago IL
John S Swift/Cincinnati, Cincinnati OH
John S Swift/Teterboro, Teterboro NJ
TAN Books & Publishers Inc, Rockford IL
Tapco Inc, Pemberton NJ
TEACH Services, Brushton NY
Times Litho Inc, Forest Grove OR
Times Printing, Random Lake WI
Todd Publications Inc, Smithville TX
Trade Litho Inc, Miami FL
Tri-Graphic Printing, Ottawa ON
Triangle Booklet Company, Nashville TN
TSO General Corporation, Long Island City NY
Ultra-Color Corporation, Saint Louis MO
Universal Press, Providence RI
Universal Printing Company, Saint Louis MO
US Press Inc, Valdosta GA
Valco Graphics, Seattle WA
Vogue Printers, North Chicago IL
Wagners Printers, Davenport IA
Waldman Graphics, Pennsauken NJ
Waldon Press Inc, New York NY
Wallace Press, Hillside IL
Watt/Peterson, Plymouth MN
Web Specialties, Wheeling IL
Webcraft Technologies, North Brunswick NJ
F A Weber & Sons, Park Falls WI
Weber Printing Company, Lincoln NE
The Wessel Company Inc, Elk Grove Village IL
White Arts Inc, Indianapolis IN
The Wimmer Companies, Memphis TN
Wolfer Printing, Commerce CA
Wood & Jones Printers, Pasadena CA
The Wood Press, Paterson NJ
Henry Wurst Inc, North Kansas City MO
Y/Z Printing Co, Elizabethtown PA
Ye Olde Genealogie Shoppe, Indianapolis IN

Calendars

Accurate Printing, Post Falls ID
Action Printing, Fond Du Lac WI
Adviser Graphics, Red Deer AB
AM Graphics & Printing, San Marcos CA
Apollo Graphics Ltd, Southampton PA
Arandell Corporation, Menomonee Falls WI
The Argus Press Inc, Niles IL
Austin Printing Co Inc, Akron OH

Bang Printing Company, Brainerd MN
Bay Port Press, Chula Vista CA
Bay State Press, Framingham MA
Beacon Press Inc, Seattle WA
Bertelsmann Printing & Manufacturing, New York NY
Blake Printery, San Luis Obispo CA
Bolger Publications/Creative Printing, Minneapolis MN
Bradley Printing Company, Des Plaines IL
R L Bryan Company, Columbia SC
BSC Litho, Harrisburg PA
Buse Printing & Advertising, Phoenix AZ
Cal Central Press, Sacramento CA
California Offset Printers, Glendale CA
Calsonic Miura Graphics, Irvine CA
Capital Printing Company Inc, Austin TX
Champagne Offset Co Inc, Natick MA
Colonial Graphics, Paterson NJ
Color World Printers, Bozeman MT
ColorCorp Inc, Littleton CO
Colorlith Corporation, Johnston RI
Colortone Press, Landover MD
Colotone Riverside Inc, Dallas TX
Columbus Bookbinders & Printers, Columbus GA
Community Press, Provo UT
Concord Litho Company Inc, Concord NH
Coneco Litho Graphics, Glen Falls NY
Country Press, Mohawk NY
Crusader Printing Co, East Saint Louis IL
Danner Press Corporation, Canton OH
Dataco A WBC Company, Albuquerque NM
Dellas Graphics, Syracuse NY
Delmar Printing & Publishing, Charlotte NC
Des Plaines Publishing Co, Des Plaines IL
Desaulniers Printing Company, Milan IL
Deven Lithographers, Brooklyn NY
Dharma Press, Oakland CA
Diamond Graphics Tech Data, Milwaukee WI
Dickinson Press Inc, Grand Rapids MI
Dollco Printing, Ottawa ON
Donihe Graphics Inc, Kingsport TN
Dragon Press, Delta Junction AK
Dynagraphics, Carlsbad CA
Eastwood Printing Co, Denver CO
EU Services, Rockville MD
Eusey Press Inc, Leominster MA
Faculty Press Inc, Brooklyn NY
Federated Lithographers, Providence RI
Fisher Printing Co, Galion OH
Fleming Printing Co, St Louis MO
Flower City Printing Inc, Rochester NY
Focus Direct, San Antonio TX
The Forms Man Inc, Deer Park NY

Calendars

Fort Orange Press Inc, Albany NY
Friesen Printers, Altona MB
Futura Printing Inc, Boynton Beach FL
Geiger Brothers, Lewiston ME
Glundal Color Service, East Syracuse NY
Golden Belt Printing Inc, Great Bend KS
Graphic Litho Corporation, Lawrence MA
Graphic Ways Inc, Buena Park CA
Great Impressions Printing & Graphics,
    Dallas TX
Great Lakes Lithograph Co, Cleveland
    OH
Great Northern/Design Printing, Skokie
    IL
Gregath Publishing Co, Wyandotte OK
Griffin Printing & Lithograph, Glendale
    CA
GRIT Commercial Printing Services, Wil-
    liamsport PA
Guynes Printing Company, Albuquerque
    NM
Harlo Printing Co, Detroit MI
Harmony Printing, Liberty MO
Heartland Press Inc, Spencer IA
Henington Publishing Company, Wolfe
    City TX
Hinz Lithographing Company, Mt
    Prospect IL
A B Hirschfeld Press, Denver CO
Intelligencer Printing, Lancaster PA
Interform Corporation, Bridgeville PA
Interstate Printing Company, Omaha NE
Jersey Printing Company, Bayonne NJ
Johns Byrne Co, Niles IL
Jostens Printing & Publishing, Visalia CA
Jostens Printing & Publishing,
    Clarksville TN
Jostens Printing & Publishing, Winston-
    Salem NC
Jostens Printing & Publishing, Topeka
    KS
Jostens Printing & Publishing, State Col-
    lege PA
Jostens Printing & Publishing, Min-
    neapolis MN
Julin Printing, Monticello IA
Kaufman Press Printing, Syracuse NY
Kolor View Press, Aurora MO
The C J Krehbiel Company, Cincinnati
    OH
La Crosse Graphics, LaCrosse WI
Lighthouse Press Inc, Manchester NH
Litho Productions Inc, Madison WI
Lithograph Printing Company, Memphis
    TN
Lithoid Printing Corporation, East
    Brunswick NJ
John D Lucas Printing Company, Bal-
    timore MD

Marathon Communications Group,
    Wausau WI
McClain Printing Company, Parsons WV
The McFarland Press, Harrisburg PA
McGrew Color Graphics, Kansas City
    MO
McQuiddy Printing Company, Nashville
    TN
Mercury Printing Company, Memphis
    TN
Moebius Printing, Milwaukee WI
Moran Printing Company, Orlando FL
Morgan Press, Dobbs Ferry NY
William H Muller Printing, Santa Clara
    CA
MultiPrint Co Inc, Skokie IL
National Graphics Corp, Columbus OH
National Lithographers Inc, Miami FL
Nationwide Printing, Burlington KY
Louis Neibauer Co Inc, Warminster PA
Oaks Printing Company, Bethlehem PA
Omaha Printing Company, Omaha NE
The Oxford Group, Norway ME
Padgett Printing Corporation, Dallas TX
Panel Prints Inc, Old Forge PA
Panorama Press Inc, Clifton NJ
Paraclete Press, Brewster MA
Paust Inc, Richmond IN
Pearl-Pressman-Liberty Communica-
    tions Group, Philadelphia PA
PennWell Printing, Tulsa OK
Plain Talk Printing Company, Des
    Moines IA
Plus Communications, Saint Louis MO
Prinit Press, Dublin IN
Print Northwest, Tacoma WA
Printing Corporation of America, Pom-
    pano Beach FL
Printright International, Palmdale CA
Printwest, Regina SK
Progress Printing, Lynchburg VA
Quebecor Book Group, New York NY
Quebecor Printing (USA) Corp, Boston
    MA
Quebecor/Semline, Braintree MA
RBW Graphics, Owen Sound ON
Ringier America/Brookfield, Brookfield
    WI
Ringier America/Jonesboro, Jonesboro
    AR
Ringier America/Phoenix, Phoenix AZ
Ronalds Printing, Vancouver BC
S Rosenthal & Company Inc, Cincinnati
    OH
Royle Printing, Sun Prairie WI
RPP Enterprises Inc, Libertyville IL
The Saratoga Printing Co, Saratoga
    Springs NY
Schelin Printing Co, St Louis MO
Schiff Printing, Pittsburgh PA
Science Press, Ephrata PA

Calendars

Shea Communications, Oklahoma City OK
Shepard Poorman Communications, Indianapolis IN
R E Smith Printing, Fall River MA
Snohomish Publishing, Snohomish WA
Southeastern Printing Company Inc, Stuart FL
Spencer Press, Wells ME
St Croix Press Inc, New Richmond WI
St Joseph Printing Ltd, Concord ON
Standard Publishing, Cincinnati OH
Stinehour Press, Lunenburg VT
Straus Printing Company, Madison WI
Sun Graphix, Lewiston ME
John S Swift/Chicago, Chicago IL
John S Swift/Cincinnati, Cincinnati OH
John S Swift/Teterboro, Teterboro NJ
Times Printing, Random Lake WI
Trade Litho Inc, Miami FL
Tri-Graphic Printing, Ottawa ON
Viking Press, Eden Prairie MN
Wagners Printers, Davenport IA
Waldman Graphics, Pennsauken NJ
Wallace Press, Hillside IL
Watt/Peterson, Plymouth MN
F A Weber & Sons, Park Falls WI
Weber Printing Company, Lincoln NE
The Wessel Company Inc, Elk Grove Village IL
White Arts Inc, Indianapolis IN
Wolfer Printing, Commerce CA
World Color Press Inc, New York NY
Henry Wurst Inc, North Kansas City MO

Comic Books

Adviser Graphics, Red Deer AB
Americomp, Brattleboro VT
Amidon & Associates Inc, Saint Paul MN
Amos Press Inc, Sidney OH
Arizona Lithographers Inc, Tucson AZ
Associated Printers, Grafton ND
Cal Central Press, Sacramento CA
California Offset Printers, Glendale CA
Capital Printing Company Inc, Austin TX
Danner Press Corporation, Canton OH
Des Plaines Publishing Co, Des Plaines IL
Deven Lithographers, Brooklyn NY
Diversified Printing, Brea CA
Dollco Printing, Ottawa ON
Donihe Graphics Inc, Kingsport TN
EU Services, Rockville MD
Gowe Printing Company, Medina OH
Graphic Litho Corporation, Lawrence MA

Griffin Printing & Lithograph, Glendale CA
Interstate Printing Company, Omaha NE
Marrakech Express, Tarpon Springs FL
Mercury Printing Company, Memphis TN
Louis Neibauer Co Inc, Warminster PA
The Oxford Group, Norway ME
Port Publications, Port Washington WI
Quebecor Printing (USA) Corp, Boston MA
Quebecor Printing Inc, Richmond Hill ON
Ronalds Printing, Montreal PQ
Shea Communications, Oklahoma City OK
Standard Publishing, Cincinnati OH
Stevens Graphics Inc, Birmingham AL
Times Litho Inc, Forest Grove OR
Times Printing, Random Lake WI
Trend Offset Printing Service Inc, Los Alamitos CA
Valco Graphics, Seattle WA
Western Publishing Co Inc, Racine WI
World Color Press Inc, New York NY

Computer Documentation

aBCD, Seattle WA
Action Printing, Fond Du Lac WI
Advanced Duplicating & Printing, Minneapolis MN
Adviser Graphics, Red Deer AB
AM Graphics & Printing, San Marcos CA
Americomp, Brattleboro VT
Apollo Graphics Ltd, Southampton PA
Arcata Graphics/Fairfield, Fairfield PA
The Argus Press Inc, Niles IL
Austin Printing Co Inc, Akron OH
Automated Graphic Systems, White Plains MD
Bang Printing Company, Brainerd MN
Banta Company, Menasha WI
Banta-Harrisonburg, Harrisonburg VA
Barton Press Inc, West Orange NJ
Bawden Printing Inc, Eldridge IA
Bay Port Press, Chula Vista CA
Bay State Press, Framingham MA
Beacon Press Inc, Seattle WA
Ben-Wal Printing, Pomona CA
Bertelsmann Printing & Manufacturing, New York NY
Best Gagne Book Manufacturers, Toronto ON
Bolger Publications/Creative Printing, Minneapolis MN

Computer Documentation

The Book Press Inc, Brattleboro VT
Book-Mart Press Inc, North Bergen NJ
BookCrafters, Chelsea MI
BookCrafters/Fredericksburg, Fredericksburg VA
Booklet Publishing Company Inc, Elk Grove Village IL
BookMasters Inc, Ashland OH
Boyd Printing Co Inc, Albany NY
Braceland Brothers Inc, Philadelphia PA
Bradley Printing Company, Des Plaines IL
Buse Printing & Advertising, Phoenix AZ
C & M Press, Denver CO
Cal Central Press, Sacramento CA
Caldwell Printers, Arcadia CA
California Offset Printers, Glendale CA
Camelot Publishing Co, Ormond Beach FL
Capital City Press, Montpelier VT
Capital Printing Company Inc, Austin TX
Carpenter Lithographic Co, Springfield OH
Champagne Offset Co Inc, Natick MA
Chicago Press Corp, Chicago IL
Clarkwood Corporation, Totowa NJ
The College Press, Collegedale TN
Colonial Graphics, Paterson NJ
Color World Printers, Bozeman MT
Colotone Riverside Inc, Dallas TX
Columbus Bookbinders & Printers, Columbus GA
Comfort Printing, St Louis MO
Commercial Documentation Services, Medford OR
Coneco Litho Graphics, Glen Falls NY
Consolidated Printers Inc, Berkeley CA
Cookbook Publishers Inc, Lenexa KS
Cookbooks by Mom's Press, Kearney NE
Corley Printing Company, Earth City MO
Country Press, Mohawk NY
Courier-Kendallville Corp, Kendallville IN
Courier-Westford Inc, Westford MA
Crane Duplicating Service, West Barnstable MA
Crusader Printing Co, East Saint Louis IL
Daamen Printing Company, West Rutland VT
Data Reproductions Corp, Rochester Hills MI
Dataco A WBC Company, Albuquerque NM
Dellas Graphics, Syracuse NY
Delta Lithograph Company, Valencia CA
Des Plaines Publishing Co, Des Plaines IL
Dharma Press, Oakland CA

Diamond Graphics Tech Data, Milwaukee WI
Diversified Printing, Brea CA
Dollco Printing, Ottawa ON
R R Donnelley & Sons Company Inc, Crawfordsville IN
R R Donnelley & Sons Company Inc, Harrisonburg VA
R R Donnelley & Sons Company Inc, Willard OH
Eastwood Printing Co, Denver CO
EBSCO Media, Birmingham AL
Edwards Brothers, Ann Arbor MI
Edwards Brothers Inc, Lillington NC
Electronic Printing, Brentwood NY
EP Graphics, Berne IN
EU Services, Rockville MD
Eva-Tone Inc, Clearwater FL
Faculty Press Inc, Brooklyn NY
Federated Lithographers, Providence RI
First Impressions Printer & Lithographers Inc, Elk Grove IL
Flower City Printing Inc, Rochester NY
Fort Orange Press Inc, Albany NY
Gateway Press Inc, Louisville KY
George Lithograph, San Francisco CA
Goodway Graphics of Virginia, Springfield VA
Graphic Printing, New Carlisle OH
Graphics Ltd, Indianapolis IN
Gray Printing, Fostoria OH
Great Impressions Printing & Graphics, Dallas TX
Great Lakes Lithograph Co, Cleveland OH
Gregath Publishing Co, Wyandotte OK
Griffin Printing & Lithograph Company, Glendale CA
Griffin Printing and Lit, Sacramento CA
GRT Book Printing, Oakland CA
Guynes Printing Company, Albuquerque NM
Haddon Craftsmen, Scranton PA
Harlo Printing Co, Detroit MI
Harmony Printing, Liberty MO
Hawkes Publishing, Salt Lake City UT
A B Hirschfeld Press, Denver CO
Imperial Litho/Graphics Inc, Phoenix AZ
Interform Corporation, Bridgeville PA
IPC Publishing Services, Saint Joseph MI
JK Creative Printers, Quincy IL
The Job Shop, Woods Hole MA
The Johnson & Hardin Company, Cincinnati OH
Jostens Printing & Publishing, Visalia CA
Jostens Printing & Publishing, Winston-Salem NC
Kaufman Press Printing, Syracuse NY
Keys Printing Co Inc, Greenville SC
Kirby Lithographic Co Inc, Arlington VA
KNI Incorporated, Anaheim CA

Computer Documention

The C J Krehbiel Company, Cincinnati OH
Les Editions Marquis, Montmagny PQ
Lighthouse Press Inc, Manchester NH
Lithocolor Press, Westchester IL
Lithoid Printing Corporation, East Brunswick NJ
Mack Printing Group Inc, Easton PA
Malloy Lithographing, Ann Arbor MI
Maple-Vail Book Group, York PA
Marathon Communications Group, Wausau WI
Mark IV Press, Hauppauge NY
Marrakech Express, Tarpon Springs FL
Maverick Publications Inc, Bend OR
The Mazer Corporation, Dayton OH
The McFarland Press, Harrisburg PA
McNaughton & Gunn Inc, Saline MI
Media Publications, Santa Clara CA
Mercury Printing Company, Memphis TN
Merrill Corporation, Los Angeles CA
Metromail Corporation, Lincoln NE
Moran Printing Company, Orlando FL
Moran Printing Inc, Baton Rouge LA
Morgan Printing, Austin TX
Morningrise Printing, Costa Mesa CA
William H Muller Printing, Santa Clara CA
National Reproductions Corp, Madison Heights MI
Nationwide Printing, Burlington KY
Louis Neibauer Co Inc, Warminster PA
Nimrod Press Inc, Boston MA
Northwest Web, Eugene OR
Oaks Printing Company, Bethlehem PA
Odyssey Press Inc, Dover NH
OMNIPRESS, Madison WI
Optic Graphics Inc, Glen Burnie MD
Original Copy Centers, Cleveland OH
Outstanding Graphics, Kenosha WI
The Ovid Bell Press Inc, Fulton MO
The Oxford Group, Norway ME
Padgett Printing Corporation, Dallas TX
Page Litho Inc, Detroit MI
Park Press, South Holland IL
Patterson Printing Company, Benton Harbor MI
Paust Inc, Richmond IN
Peconic Companies, Mattituck NY
PFP Printing Corporation, Gaithersburg MD
Phillips Brothers Printers Inc, Springfield IL
Pine Hill Press Inc, Freeman SD
Plain Talk Printing Company, Des Moines IA
Plus Communications, Saint Louis MO
Port City Press Inc, Baltimore MD
Print Northwest, Tacoma WA

Printing Corporation of America, Pompano Beach FL
Progress Printing, Lynchburg VA
ProLitho, Provo UT
Publishers Express, Ladysmith WI
Publishers Press Press, Salt Lake City UT
Quebecor Book Group, New York NY
Quebecor Printing (USA) Corp, Boston MA
Quebecor Printing Inc, Richmond Hill ON
Quebecor/Semline, Braintree MA
Repro-Tech, Fairfield CT
Ronalds Printing, Vancouver BC
Ronalds Printing, Montreal PQ
Rose Printing Company, Tallahassee FL
Sanders Printing Company, Garretson SD
Schelin Printing Co, St Louis MO
Science Press, Ephrata PA
Service Printing Company, San Leandro CA
Sexton Printing, Saint Paul MN
Shepard Poorman Communications, Indianapolis IN
Sheridan Printing Company, Phillipsburg NJ
R E Smith Printing, Fall River MA
Snohomish Publishing, Snohomish WA
Sowers Printing Company, Lebanon PA
Spencer Press, Wells ME
Standard Publishing, Cincinnati OH
John Stark Printing Co Inc, St Louis MO
Strathmore Printing, Geneva IL
Straus Printing Company, Madison WI
John S Swift Company Inc, St Louis MO
TEACH Services, Brushton NY
Technical Communication Services, North Kansas City MO
Thomson-Shore, Dexter MI
Times Printing, Random Lake WI
Todd Publications Inc, Smithville TX
Trade Litho Inc, Miami FL
Trend Offset Printing Service, Los Alamitos CA
Tri-Graphic Printing, Ottawa ON
TSO General Corporation, Long Island City NY
Tweddle Litho Co, Clinton Township MI
Valco Graphics, Seattle WA
Van Volumes, Palmer MA
Vicks Lithograph & Printing, Yorkville NY
Victor Graphics, Baltimore MD
Waldman Graphics, Pennsauken NJ
Waldon Press Inc, New York NY
Wallace Press, Hillside IL
Webcom Ltd, Scarborough ON
Webcrafters, Madison WI
Weber Printing Company, Lincoln NE
Western Publishing Co Inc, Racine WI
White Arts Inc, Indianapolis IN

Computer Documentation

Whitehall Printing Company, Naples FL
Wickersham Printing Co Inc, Lancaster
    PA
Windsor Associates, San Diego CA
Wolfer Printing, Commerce CA
Wood & Jones Printers, Pasadena CA
Worzalla Publishing Company, Stevens
    Point WI

Cookbooks

aBCD, Seattle WA
Academy Books, Rutland VT
Accurate Printing, Post Falls ID
Action Printing, Fond Du Lac WI
Adams Press, Chicago IL
Adviser Graphics, Red Deer AB
AM Graphics & Printing, San Marcos CA
American Litho, Hayward CA
American Printers & Litho, Niles IL
Americomp, Brattleboro VT
Amos Press Inc, Sidney OH
Anundsen Publishing Company, Decorah
    IA
Apollo Graphics Ltd, Southampton PA
Arandell Corporation, Menomonee Falls
    WI
Arcata Graphics/Fairfield, Fairfield PA
The Argus Press Inc, Niles IL
Austin Printing Co Inc, Akron OH
Automated Graphic Systems, White
    Plains MD
Bang Printing Company, Brainerd MN
Banta Company, Menasha WI
Barton Press Inc, West Orange NJ
Bay Port Press, Chula Vista CA
Bay State Press, Framingham MA
Beacon Press Inc, Seattle WA
Berryville Graphics, Berryville VA
Bertelsmann Printing & Manufacturing,
    New York NY
Best Gagne Book Manufacturers, Toron-
    to ON
Blake Printery, San Luis Obispo CA
Blue Dolphin Press Inc, Grass Valley CA
Bolger Publications/Creative Printing,
    Minneapolis MN
The Book Press Inc, Brattleboro VT
Book-Mart Press Inc, North Bergen NJ
BookCrafters, Chelsea MI
BookCrafters/Fredericksburg, Frederick-
    sburg VA
BookMasters Inc, Ashland OH
Braun-Brumfield Inc, Ann Arbor MI
Brennan Printing, Deep River IA
Brunswick Publishing Corp, Lawren-
    ceville VA

R L Bryan Company, Columbia SC
BSC Litho, Harrisburg PA
Buse Printing & Advertising, Phoenix AZ
C & M Press, Denver CO
Cal Central Press, Sacramento CA
Camelot Publishing CO, Ormond Beach
    FL
Capital City Press, Montpelier VT
Capital Printing Company Inc, Austin TX
Carpenter Lithographic Co, Springfield
    OH
The College Press, Collegedale TN
Colonial Graphics, Paterson NJ
Color World Printers, Bozeman MT
ColorCorp Inc, Littleton CO
Colorlith Corporation, Johnston RI
Colortone Press, Landover MD
Colotone Riverside Inc, Dallas TX
Columbus Bookbinders & Printers,
    Columbus GA
Comfort Printing, St Louis MO
Community Press, Provo UT
Coneco Litho Graphics, Glen Falls NY
Consolidated Printers Inc, Berkeley CA
Cookbook Publishers Inc, Lenexa KS
Cookbooks by Mom's Press, Kearney NE
Cooley Printers & Office Supply, Monroe
    LA
Corley Printing Company, Earth City MO
Country Press, Mohawk NY
Country Press Inc, Lakeville MA
Courier-Kendallville Corp, Kendallville
    IN
Courier-Westford Inc, Westford MA
Crane Duplicating Service, West
    Barnstable MA
Crusader Printing Co, East Saint Louis
    IL
Cushing-Malloy, Ann Arbor MI
Daamen Printing Company, West Rut-
    land VT
Danner Press Corporation, Canton OH
Data Reproductions Corp, Rochester
    Hills MI
Dellas Graphics, Syracuse NY
Delmar Printing & Publishing, Charlotte
    NC
Des Plaines Publishing Co, Des Plaines
    IL
Diamond Graphics Tech Data, Mil-
    waukee WI
Dickinson Press Inc, Grand Rapids MI
Dollco Printing, Ottawa ON
R R Donnelley & Sons Company Inc,
    Crawfordsville IN
R R Donnelley & Sons Company Inc,
    Harrisonburg, Harrisonburg VA
R R Donnelley & Sons Company Inc,
    Willard, Willard OH
Dragon Press, Delta Junction AK
Eastwood Printing Co, Denver CO
EBSCO Media, Birmingham AL

Cookbooks

Edwards Brothers Inc, Ann Arbor MI
Edwards Brothers Inc, Lillington NC
Eerdmans Printing Co, Grand Rapids MI
Electronic Printing, Brentwood NY
EP Graphics, Berne IN
Eureka Printing Co Inc, Eureka CA
Eusey Press Inc, Leominster MA
Eva-Tone Inc, Clearwater FL
Evangel Press, Nappanee IN
Faculty Press Inc, Brooklyn NY
Federated Lithographers, Providence RI
Fleming Printing Co, St Louis MO
Flower City Printing Inc, Rochester NY
Fort Orange Press Inc, Albany NY
Friesen Printers, Altona MB
Fundcraft Publishing, Collierville TN
Futura Printing Inc, Boynton Beach FL
Gateway Press Inc, Louisville KY
George Lithograph, San Francisco CA
Geryon Press Limited, Tunnel NY
Golden Belt Printing Inc, Great Bend KS
Goodway Graphics of Virginia,
    Springfield VA
Gorham Printing, Rochester WA
Graphic Ways Inc, Buena Park CA
Graphics Ltd, Indianapolis IN
Great Impressions Printing & Graphics,
    Dallas TX
Great Lakes Lithograph Co, Cleveland
    OH
Gregath Publishing Co, Wyandotte OK
Griffin Printing & Lithograph, Glendale
    CA
Griffin Printing and Lithograph Com-
    pany, Sacramento CA
GRT Book Printing, Oakland CA
Guynes Printing Company, Albuquerque
    NM
Haddon Craftsmen, Scranton PA
Harlo Printing Co, Detroit MI
Harmony Printing, Liberty MO
Hawkes Publishing, Salt Lake City UT
Heartland Press Inc, Spencer IA
Henington Publishing Company, Wolfe
    City TX
Heritage Printers Inc, Charlotte NC
A B Hirschfeld Press, Denver CO
Rae Horowitz Book Manufacturers, Fair-
    field NJ
Interform Corporation, Bridgeville PA
IPC Publishing Services, Saint Joseph MI
The Johnson & Hardin Company, Cin-
    cinnati OH
Jostens Printing & Publishing, Visalia CA
Jostens Printing & Publishing,
    Clarksville TN
Jostens Printing & Publishing, Winston-
    Salem NC
Jostens Printing & Publishing, Topeka
    KS

Jostens Printing & Publishing, State Col-
    lege PA
Jostens Printing & Publishing, Min-
    neapolis MN
Julin Printing, Monticello IA
Keys Printing Co Inc, Greenville SC
KNI Incorporated, Anaheim CA
Kolor View Press, Aurora MO
The C J Krehbiel Company, Cincinnati
    OH
La Crosse Graphics, LaCrosse WI
Les Editions Marquis, Montmagny PQ
Litho Specialties, St Paul MN
Lithograph Printing Company, Memphis
    TN
Lithoid Printing Corporation, East
    Brunswick NJ
Malloy Lithographing, Ann Arbor MI
Maple-Vail Book Manufacturing Group,
    York PA
Marathon Communications Group,
    Wausau WI
Mark IV Press, Hauppauge NY
Marrakech Express, Tarpon Springs FL
Maverick Publications Inc, Bend OR
MBP Lithographics, Hillsboro KS
McArdle Printing Co, Upper Marlboro
    MD
McClain Printing Company, Parsons WV
McNaughton & Gunn Inc, Saline MI
Meaker the Printer, Phoenix AZ
Media Publications, Santa Clara CA
Mercury Printing Company, Memphis
    TN
Moran Printing Company, Orlando FL
Moran Printing Inc, Baton Rouge LA
Morgan Printing, Austin TX
MultiPrint Co Inc, Skokie IL
National Graphics Corp, Columbus OH
National Reproductions Corp, Madison
    Heights MI
Louis Neibauer Co Inc, Warminster PA
Nimrod Press Inc, Boston MA
Northlight Studio Press, Barre VT
Oaks Printing Company, Bethlehem PA
Odyssey Press Inc, Dover NH
Omaha Printing Company, Omaha NE
Optic Graphics Inc, Glen Burnie MD
Outstanding Graphics, Kenosha WI
The Ovid Bell Press Inc, Fulton MO
The Oxford Group, Norway ME
Paraclete Press, Brewster MA
Parker Graphics, Fuquay-Varina NC
Patterson Printing Company, Benton
    Harbor MI
Pearl-Pressman-Liberty Communica-
    tions Group, Philadelphia PA
Peconic Companies, Mattituck NY
Pentagram Press, Minneapolis MN
PFP Printing Corporation, Gaithersburg
    MD
Pine Hill Press Inc, Freeman SD

Cookbooks

Plain Talk Printing Company, Des Moines IA
Port City Press Inc, Baltimore MD
Prinit Press, Dublin IN
Print Northwest, Tacoma WA
Printing Corporation of America, Pompano Beach FL
Printright International, Palmdale CA
Printwest, Regina SK
Progress Printing, Lynchburg VA
Publishers Express Press, Ladysmith WI
Publishers Press, Salt Lake City UT
Quebecor Book Group, New York NY
Quebecor Printing (USA) Corp, Boston MA
Quebecor Printing Inc, Richmond Hill ON
Quebecor/Semline, Braintree MA
Quinn-Woodbine Inc, Woodbine NJ
Repro-Tech, Fairfield CT
Ringier America, Itasca IL
Ronalds Printing, Montreal PQ
Rose Printing Company, Tallahassee FL
Roxbury Publishing Company, Los Angeles CA
Sanders Printing Company, Garretson SD
The Saratoga Printing Co, Saratoga Springs NY
Schiff Printing, Pittsburgh PA
Science Press, Ephrata PA
Sexton Printing, Saint Paul MN
Shepard Poorman Communications, Indianapolis IN
Sinclair Printing Co, Los Angeles CA
Snohomish Publishing, Snohomish WA
Southeastern Printing Company Inc, Stuart FL
Spencer Press, Wells ME
Starr Toof Printing, Memphis TN
Stewart Publishing & Printing Co, Marble Hill MO
The Stinehour Press, Lunenburg VT
Straus Printing Company, Madison WI
John S Swift Company Inc, St Louis MO
TAN Books & Publishers Inc, Rockford IL
Taylor Publishing Company, Dallas TX
TEACH Services, Brushton NY
Technical Communication Services, North Kansas City MO
Thomson-Shore, Dexter MI
Times Printing, Random Lake WI
Town House Press, Pittsboro NC
Tracor Publications, Austin TX
Trade Litho Inc, Miami FL
Tri-Graphic Printing, Ottawa ON
Triangle Booklet Company, Nashville TN
Tweddle Litho Co, Clinton Township MI
Universal Printers, Winnipeg MB

Universal Printing Company, Miami FL
Vail Ballou Press, Binghamton NY
Valco Graphics, Seattle WA
Van Volumes, Palmer MA
Vogue Printers, North Chicago IL
Walsworth Publishing Co Inc, Marceline MO
Walter's Publishing, Waseca MN
Web Specialties, Wheeling IL
Webcom Ltd, Scarborough ON
F A Weber & Sons, Park Falls WI
Weber Printing Company, Lincoln NE
Westview Press, Boulder CO
Whitehall Printing Company, Naples FL
Wickersham Printing Co Inc, Lancaster PA
The Wimmer Companies, Memphis TN
Windsor Associates, San Diego CA
Worzalla Publishing Company, Stevens Point WI
Y/Z Printing Co, Elizabethtown PA

Directories

A-1 Business Service Inc, St Paul MN
aBCD, Seattle WA
Accurate Web, Central Islip NY
Action Printing, Fond Du Lac WI
Ad Infinitum Press, Mount Vernon NY
Adams Press, Chicago IL
Advanced Duplicating & Printing, Minneapolis MN
Adviser Graphics, Red Deer AB
Alcom Printing Group Inc, Bethlehem PA
AM Graphics & Printing, San Marcos CA
American Litho, Hayward CA
American Signature, Lincoln NE
Apollo Graphics Ltd, Southampton PA
Arcata Graphics/Fairfield, Fairfield PA
The Argus Press Inc, Niles IL
Artcraft Press, Waterloo WI
Austin Printing Co Inc, Akron OH
Automated Graphic Systems, White Plains MD
Bang Printing Company, Brainerd MN
Banta Company, Menasha WI
Banta-Harrisonburg, Harrisonburg VA
Bawden Printing Inc, Eldridge IA
Bay Port Press, Chula Vista CA
Bay State Press, Framingham MA
Beacon Press Inc, Seattle WA
Ben-Wal Printing, Pomona CA
Bertelsmann Printing & Manufacturing, New York NY
Best Gagne Book Manufacturers, Toronto ON
Blue Dolphin Press Inc, Grass Valley CA
Bolger Publications/Creative Printing, Minneapolis MN

Directories

The Book Press Inc, Brattleboro VT
Book-Mart Press Inc, North Bergen NJ
BookCrafters, Chelsea MI
BookCrafters/Fredericksburg, Fredericksburg VA
Booklet Publishing Company Inc, Elk Grove IL
BookMasters Inc, Ashland OH
Braceland Brothers Inc, Philadelphia PA
Bradley Printing Company, Des Plaines IL
Braun-Brumfield Inc, Ann Arbor MI
Brennan Printing, Deep River IA
Brown Printing, Waseca MN
R L Bryan Company, Columbia SC
BSC Litho, Harrisburg PA
Buse Printing & Advertising, Phoenix AZ
The William Byrd Press, Richmond VA
C & M Press, Denver CO
Cal Central Press, Sacramento CA
Caldwell Printers, Arcadia CA
California Offset Printers, Glendale CA
Camelot Publishing Co, Ormond Beach FL
Canterbury Press, Rome NY
Capital City Press, Montpelier VT
Capital Printing Company Inc, Austin TX
Carpenter Lithographic Co, Springfield OH
Champagne Offset Co Inc, Natick MA
Chicago Press Corp, Chicago IL
Citizen Printing, Beaver Dam WI
The College Press, Collegedale TN
Colonial Graphics, Paterson NJ
Color World Printers, Bozeman MT
ColorCorp Inc, Littleton CO
Colotone Riverside Inc, Dallas TX
Columbus Bookbinders & Printers, Columbus GA
Comfort Printing, St Louis MO
Community Press, Provo UT
Coneco Litho Graphics, Glen Falls NY
Consolidated Printers Inc, Berkeley CA
Cookbooks by Mom's Press, Kearney NE
Corley Printing Company, Earth City MO
Country Press, Mohawk NY
Country Press Inc, Lakeville MA
Courier Corporation, Lowell MA
Courier Stoughton Inc, Stoughton MA
Courier-Kendallville Corp, Kendallville IN
Courier-Westford Inc, Westford MA
Crane Duplicating Service, West Barnstable MA
Crusader Printing Co, East Saint Louis IL
Cushing-Malloy, Ann Arbor MI
Daamen Printing Company, West Rutland VT

Data Reproductions Corp, Rochester Hills MI
Dataco A WBC Company, Albuquerque NM
Delmar Printing & Publishing, Charlotte NC
Delta Lithograph Company, Valencia CA
Des Plaines Publishing Co, Des Plaines IL
Desaulniers Printing Company, Milan IL
Dharma Press, Oakland CA
Diamond Graphics Tech Data, Milwaukee WI
Dickinson Press Inc, Grand Rapids MI
Directories America, Chicago IL
Diversified Printing, Brea CA
Dollco Printing, Ottawa ON
R R Donnelley & Sons Company, Crawfordsville IN
R R Donnelley & Sons Company, Harrisonburg, Harrisonburg VA
R R Donnelley & Sons Company, Willard, Willard OH
Dragon Press, Delta Junction AK
Dynagraphics, Carlsbad CA
Eastwood Printing Co, Denver CO
EBSCO Media, Birmingham AL
Editors Press Inc, Hyattsville MD
Edwards Brothers Inc, Ann Arbor MI
Edwards Brothers Inc, Lillington NC
Eerdmans Printing Co, Grand Rapids MI
Electronic Printing, Brentwood NY
EP Graphics, Berne IN
Eureka Printing Co Inc, Eureka CA
Eva-Tone Inc, Clearwater FL
The Everton Publishers Inc, Logan UT
Faculty Press Inc, Brooklyn NY
Federated Lithographers, Providence RI
First Impressions & Lithographers Inc, Elk Grove IL
Fisher Printers Inc, Cedar Rapids IA
Fisher Printing Co, Galion OH
Fleming Printing Co, St Louis MO
Flower City Printing Inc, Rochester NY
Focus Direct, San Antonio TX
The Forms Man Inc, Deer Park NY
Fort Orange Press Inc, Albany NY
Futura Printing Inc, Boynton Beach FL
Gateway Press Inc, Louisville KY
George Lithograph, San Francisco CA
Gilliland Printing Inc, Arkansas City KS
Glundal Color Service, East Syracuse NY
Golden Belt Printing Inc, Great Bend KS
Goodway Graphics of Virginia, Springfield VA
Gorham Printing, Rochester WA
Gowe Printing Company, Medina OH
Graftek Press, Woodstock IL
Graphic Litho Corporation, Lawrence MA
Graphic Printing, New Carlisle OH
Graphics Ltd, Indianapolis IN
Gray Printing, Fostoria OH

Directories

Great Impressions Printing & Graphics, Dallas TX
Great Lakes Lithograph Co, Cleveland OH
Gregath Publishing Co, Wyandotte OK
Griffin Printing & Lithograph, Glendale CA
Griffin Printing and Lithograph Company, Sacramento CA
GRT Book Printing, Oakland CA
GTE Directories Corp, Mt Prospect IL
GTE Directories Corp, Des Plaines IL
Gulf Printing, Houston TX
Guynes Printing Company, Albuquerque NM
Haddon Craftsmen, Scranton PA
Harlo Printing Co, Detroit MI
Harmony Printing, Liberty MO
Heartland Press Inc, Spencer IA
Henington Publishing Company, Wolfe City TX
Hennegan Company, Cincinnati OH
D B Hess Company, Woodstock IL
Hignell Printing, Winnepeg MB
Holladay-Tyler Printing Corp, Glenn Dale MD
Rae HorowitzBook Manufacturers, Fairfield NJ
Imperial Litho/Graphics Inc, Phoenix AZ
Independent Publishing Company, Sarasota FL
Inland Printing Company, Syosset NY
Interform Corporation, Bridgeville PA
IPC Publishing Services, Saint Joseph MI
Jersey Printing Company, Bayonne NJ
The Job Shop, Woods Hole MA
The Johnson & Hardin Company, Cincinnati OH
Johnson Graphics, East Dubuque IL
Jostens Printing & Publishing, Visalia CA
Jostens Printing & Publishing, Winston-Salem NC
Jostens Printing & Publishing, Minneapolis MN
Julin Printing, Monticello IA
Kirby Lithographic Co Inc, Arlington VA
KNI Incorporated, Anaheim CA
Kolor View Press, Aurora MO
The C J Krehbiel Company, Cincinnati OH
La Crosse Graphics, LaCrosse WI
Les Editions Marquis, Montmagny PQ
Lighthouse Press Inc, Manchester NH
Litho Productions Inc, Madison WI
Lithocolor Press, Westchester IL
Lithoid Printing Corporation, East Brunswick NJ
The Little River Press Inc, Miami FL
Long Island Web Printing, Jericho NY

John D Lucas Printing Company, Baltimore MD
Mack Printing Group Inc, Easton PA
Mail-O-Graph, Kewanee IL
Malloy Lithographing, Ann Arbor MI
Maquoketa Web Printing, Maquoketa IA
Mark IV Press, Hauppauge NY
Marrakech Express, Tarpon Springs FL
Maverick Publications Inc, Bend OR
The Mazer Corporation, Dayton OH
MBP Lithographics, Hillsboro KS
McArdle Printing Co, Upper Marlboro MD
McClain Printing Company, Parsons WV
McDowell Publications, Utica KY
The McFarland Press, Harrisburg PA
McGill Jensen, St Paul MN
McGrew Color Graphics, Kansas City MO
McNaughton & Gunn Inc, Saline MI
McQuiddy Printing Company, Nashville TN
Meaker the Printer, Phoenix AZ
Media Publications, Santa Clara CA
Mercury Printing Company, Memphis TN
Merrill Corporation, Los Angeles CA
Metromail Corporation, Lincoln NE
Metroweb, Erlanger KY
Mitchell Press, Vancouver BC
MMI Press, Harrisville NH
Monument Printers & Lithographer Inc, Verplanck NY
Moran Printing Company, Orlando FL
Moran Printing Inc, Baton Rouge LA
Morgan Printing, Austin TX
MultiPrint Co Inc, Skokie IL
National Lithographers Inc, Miami FL
National Publishing Company, Philadelphia PA
National Reproductions Corp, Madison Heights MI
Louis Neibauer Co Inc, Warminster PA
Newsfoto Publishing Company, San Angelo TX
Neyenesch Printers, San Diego CA
Nimrod Press Inc, Boston MA
Noll Printing Company, Huntington IN
Northlight Studio Press, Barre VT
Nystrom Publishing Company, Osseo MN
Oaks Printing Company, Bethlehem PA
Odyssey Press Inc, Dover NH
OMNIPRESS, Madison WI
Optic Graphics Inc, Glen Burnie MD
Original Copy Centers, Cleveland OH
Outstanding Graphics, Kenosha WI
The Ovid Bell Press Inc, Fulton MO
The Oxford Group, Norway ME
Padgett Printing Corporation, Dallas TX
Page Litho Inc, Detroit MI
Panorama Press Inc, Clifton NJ
Pantagraph Printing, Bloomington IL

Directories

Paraclete Press, Brewster MA
Park Press, South Holland IL
Parker Graphics, Fuquay-Varina NC
Patterson Printing Company, Benton
    Harbor MI
Paust Inc, Richmond IN
Peconic Companies, Mattituck NY
Pendell Printing, Midland MI
Perry Printing Corp, Waterloo WI
Perry/Baraboo, Baraboo WI
PFP Printing Corporation, Gaithersburg
    MD
Pine Hill Press Inc, Freeman SD
Pioneer Press, Wilmette IL
Plain Talk Printing Company, Des
    Moines IA
Plus Communications, Saint Louis MO
Port City Press Inc, Baltimore MD
Practical Graphics, New York NY
Princeton Academic Press, Lawrenceville
    NJ
Prinit Press, Dublin IN
Print Northwest, Tacoma WA
The Printing Company, Indianapolis IN
Printing Corporation of America, Pom-
    pano Beach FL
Printright International, Palmdale CA
Printwest, Regina SK
Progress Printing, Lynchburg VA
ProLitho, Provo UT
Publishers Express Press, Ladysmith WI
Publishers Press, Salt Lake City UT
Quebecor Book Group, New York NY
Quebecor Printing (USA) Corp, Boston
    MA
Quebecor Printing Inc, Richmond Hill
    ON
Quebecor Printing/St Cloud, Saint Cloud
    MN
Quebecor/Semline, Braintree MA
Quinn-Woodbine Inc, Woodbine NJ
RBW Graphics, Owen Sound ON
Repro-Tech, Fairfield CT
Rich Printing Company Inc, Nashville
    TN
Ringier America, Itasca IL
Ronalds Printing, Vancouver BC
Ronalds Printing, Montreal PQ
Rose Printing Company, Tallahassee FL
S Rosenthal & Company, Cincinnati OH
Roxbury Publishing Company, Los An-
    geles CA
RPP Enterprises Inc, Libertyville IL
Sanders Printing Company, Garretson
    SD
The Saratoga Printing Co, Saratoga
    Springs NY
Schiff Printing, Pittsburgh PA
Science Press, Ephrata PA

Service Printing Company, San Leandro
    CA
Sexton Printing, Saint Paul MN
Shepard Poorman Communications, In-
    dianapolis IN
Sheridan Printing Company, Phil-
    lipsburg NJ
Sinclair Printing Co, Los Angeles CA
R E Smith Printing, Fall River MA
Snohomish Publishing, Snohomish WA
Sowers Printing Company, Lebanon PA
Speakes-Hines & Thomas Inc, Lansing
    MI
St Croix Press Inc, New Richmond WI
Standard Printing Service, Chicago IL
John Stark Printing Co Inc, St Louis MO
Stephenson Lithograph Inc, Alexandria
    VA
Stevens Graphics, Atlanta GA
Stevens Graphics Inc, Birmingham AL
Stewart Publishing & Printing Co,
    Marble Hill MO
The Stinehour Press, Lunenburg VT
Strathmore Printing, Geneva IL
Straus Printing Company, Madison WI
John S Swift Company Inc, St Louis MO
TAN Books & Publishers Inc, Rockford
    IL
Taylor Publishing Company, Dallas TX
TEACH Services, Brushton NY
Technical Communication Services,
    North Kansas City MO
Thomson-Shore, Dexter MI
Times Litho Inc, Forest Grove OR
Times Printing, Random Lake WI
Todd Publications Inc, Smithville TX
Town House Press, Pittsboro NC
Tracor Publications, Austin TX
Trade Litho Inc, Miami FL
Trend Offset Printing Service Inc, Los
    Alamitos CA
Tri-Graphic Printing, Ottawa ON
Triangle Booklet Company, Nashville TN
TSO General Corporation, Long Island
    City NY
Tweddle Litho Co, Clinton Township MI
Universal Printers, Winnipeg MB
Universal Printing Company, Miami FL
Universal Printing Company, Saint
    Louis MO
Valco Graphics, Seattle WA
Van Volumes, Palmer MA
Victor Graphics, Baltimore MD
Viking Press, Eden Prairie MN
Vogue Printers, North Chicago IL
Wagners Printers, Davenport IA
Waldman Graphics, Pennsauken NJ
Waldon Press Inc, New York NY
Wallace Press, Hillside IL
Walsworth Publishing Co Inc, Marceline
    MO
Watt/Peterson, Plymouth MN

Directories

Webcom Ltd, Scarborough ON
Webcrafters, Madison WI
F A Weber & Sons, Park Falls WI
Weber Printing Company, Lincoln NE
Western Web Printing, Sioux Falls SD
Westview Press, Boulder CO
Whitehall Printing Company, Naples FL
Wickersham Printing Co Inc, Lancaster PA
Windsor Associates, San Diego CA
Wolfer Printing, Commerce CA
Wood & Jones Printers, Pasadena CA
World Color Press Inc, New York NY
Worzalla Publishing Company, Stevens Point WI
Henry Wurst Inc, North Kansas City MO
Y/Z Printing Co, Elizabethtown PA

Galley Copies

Ad Infinitum Press, Mount Vernon NY
Adviser Graphics, Red Deer AB
AM Graphics & Printing, San Marcos CA
Americomp, Brattleboro VT
Apollo Graphics Ltd, Southampton PA
The Argus Press Inc, Niles IL
Automated Graphic Systems, White Plains MD
Bang Printing Company, Brainerd MN
Ben-Wal Printing, Pomona CA
Bertelsmann Printing & Manufacturing, New York NY
Blue Dolphin Press Inc, Grass Valley CA
Buse Printing & Advertising, Phoenix AZ
Cal Central Press, Sacramento CA
Capital Printing Company Inc, Austin TX
Champagne Offset Co Inc, Natick MA
Comfort Printing, St Louis MO
Coneco Litho Graphics, Glen Falls NY
Country Press, Mohawk NY
Country Press Inc, Lakeville MA
Crane Duplicating Service, West Barnstable MA
Daamen Printing Company, West Rutland VT
R R Donnelley & Sons Company, Crawfordsville IN
Dragon Press, Delta Junction AK
Electronic Printing, Brentwood NY
Eva-Tone Inc, Clearwater FL
Fleming Printing Co, St Louis MO
Foster Printing Service, Michigan City IN
Graphic Printing, New Carlisle OH
Graphics Ltd, Indianapolis IN
Haddon Craftsmen, Scranton PA
Independent Printing Company, New York NY

Interform Corporation, Bridgeville PA
Jostens Printing & Publishing, Winston-Salem NC
Kimberly Press, Santa Barbara CA
The C J Krehbiel Company, Cincinnati OH
Lithoid Printing Corporation, East Brunswick NJ
Mark IV Press, Hauppauge NY
The Mazer Corporation, Dayton OH
MBP Lithographics, Hillsboro KS
Monument Printers & Lithographer Inc, Verplanck NY
National Reproductions Corp, Madison Heights MI
Louis Neibauer Co Inc, Warminster PA
Northlight Studio Press, Barre VT
Odyssey Press Inc, Dover NH
OMNIPRESS, Madison WI
Peconic Companies, Mattituck NY
PFP Printing Corporation, Gaithersburg MD
Publishers Press, Salt Lake City UT
Quebecor Printing (USA) Corp, Boston MA
Repro-Tech, Fairfield CT
Sanders Printing Company, Garretson SD
Times Printing, Random Lake WI
Trade Litho Inc, Miami FL
Van Volumes, Palmer MA
Westview Press, Boulder CO
Wickersham Printing Co Inc, Lancaster PA
Y/Z Printing Co, Elizabethtown PA

Genealogies

The following small companies specialize in printing genealogies and family histories for individuals. Of course, many of the other printers who manufacture books can also print genealogies.

The Everton Publishers Inc, Logan UT
Gateway Press, Baltimore MD
McDowell Publications, Utica KY
Ye Olde Genealogie Shoppe, Indianapolis IN

Journals

Academy Books, Rutland VT
Action Printing, Fond Du Lac WI
Ad Infinitum Press, Mount Vernon NY
Adviser Graphics, Red Deer AB
Allen Press Inc, Lawrence KS
AM Graphics & Printing, San Marcos CA

Journals

American Web Inc, Denver CO
Americomp, Brattleboro VT
Apollo Graphics Ltd, Southampton PA
The Argus Press Inc, Niles IL
Automated Graphic Systems, White
    Plains MD
Baker Johnson Inc, Dexter MI
Bang Printing Company, Brainerd MN
Banta Publications Group, Hinsdale IL
Bawden Printing Inc, Eldridge IA
Bay Port Press, Chula Vista CA
Bay State Press, Framingham MA
Beacon Press Inc, Richmond VA
Blake Printery, San Luis Obispo CA
Bolger Publications/Creative Printing,
    Minneapolis MN
Book-Mart Press Inc, North Bergen NJ
BookCrafters, Chelsea MI
BookCrafters/Fredericksburg, Frederick-
    sburg VA
BookMasters Inc, Ashland OH
Boyd Printing Co Inc, Albany NY
Braun-Brumfield Inc, Ann Arbor MI
Brown Printing, Waseca MN
Brown Printing/East Greenville Division,
    East Greenville PA
R L Bryan Company, Columbia SC
Buse Printing & Advertising, Phoenix AZ
Cal Central Press, Sacramento CA
California Offset Printers, Glendale CA
Canterbury Press, Rome NY
Capital City Press Inc, Montpelier VT
Capital Printing Company Inc, Austin TX
Champagne Offset Co Inc, Natick MA
Coach House Press, Toronto ON
Colonial Graphics, Paterson NJ
Color World Printers, Bozeman MT
Columbus Bookbinders & Printers,
    Columbus GA
Combined Communication Services,
    Mendota IL
Comfort Printing, St Louis MO
Community Press, Provo UT
Coneco Litho Graphics, Glen Falls NY
Cookbooks by Mom's Press, Kearney NE
Country Press, Mohawk NY
Crane Duplicating Service, West
    Barnstable MA
Lew A Cummings Printing Company
    Inc, Manchester NH
Cushing-Malloy, Ann Arbor MI
Daamen Printing Company, West Rut-
    land VT
Danner Press Corporation, Canton OH
Data Reproductions Corp, Rochester
    Hills MI
Dataco A WBC Company, Albuquerque
    NM
Des Plaines Publishing Co, Des Plaines
    IL

Desaulniers Printing Company, Milan IL
Dharma Press, Oakland CA
Directories America, Chicago IL
Dollco Printing, Ottawa ON
R R Donnelley & Sons Company, Craw-
    fordsville IN
Dragon Press, Delta Junction AK
EBSCO Media, Birmingham AL
Editors Press Inc, Hyattsville MD
Edwards Brothers Inc, Ann Arbor MI
Edwards Brothers Inc, Lillington NC
Electronic Printing, Brentwood NY
Evangel Press, Nappanee IN
Faculty Press Inc, Brooklyn NY
First Impressions & Lithographers Inc,
    Elk Grove IL
Fisher Printers Inc, Cedar Rapids IA
Fleming Printing Co, St Louis MO
Fort Orange Press Inc, Albany NY
Friesen Printers, Altona MB
Gateway Press Inc, Louisville KY
Goodway Graphics of Virginia,
    Springfield VA
Graftek Press, Woodstock IL
Graphic Printing, New Carlisle OH
Graphics Ltd, Indianapolis IN
Gray Printing, Fostoria OH
Great Lakes Lithograph Co, Cleveland
    OH
Gregath Publishing Co, Wyandotte OK
Griffin Printing & Lithograph, Glendale
    CA
GRT Book Printing, Oakland CA
Harmony Printing, Liberty MO
Hawkes Publishing, Salt Lake City UT
Heartland Press Inc, Spencer IA
Hennegan Company, Cincinnati OH
Heritage Printers Inc, Charlotte NC
Hignell Printing, Winnepeg MB
A B Hirschfeld Press, Denver CO
Holladay-Tyler Printing Corp, Glenn
    Dale MD
Independent Printing Company, New
    York NY
Independent Publishing Company,
    Sarasota FL
Interform Corporation, Bridgeville PA
IPC Publishing Services, Saint Joseph MI
Jersey Printing Company, Bayonne NJ
The Johnson & Hardin Company, Cin-
    cinnati OH
Jostens Printing & Publishing, Visalia CA
Jostens Printing & Publishing, Winston-
    Salem NC
Jostens Printing & Publishing, Min-
    neapolis MN
Kaufman Press Printing, Syracuse NY
Kimberly Press, Santa Barbara CA
Kirby Lithographic Co Inc, Arlington VA
KNI Incorporated, Anaheim CA
The C J Krehbiel Company, Cincinnati
    OH

Journals

La Crosse Graphics, LaCrosse WI
Lancaster Press, Lancaster PA
The Lane Press, Burlington VT
Lighthouse Press Inc, Manchester NH
Lithoid Printing Corporation, East
    Brunswick NJ
The Little River Press Inc, Miami FL
John D Lucas Printing Company, Bal-
    timore MD
Mack Printing Group Inc, Easton PA
Mackintosh Typography Inc, Santa Bar-
    bara CA
The Mad Printers, Mattituck NY
Mail-O-Graph, Kewanee IL
Malloy Lithographing, Ann Arbor MI
Maquoketa Web Printing, Maquoketa IA
Marrakech Express, Tarpon Springs FL
The Mazer Corporation, Dayton OH
MBP Lithographics, Hillsboro KS
McArdle Printing Co, Upper Marlboro
    MD
McClain Printing Company, Parsons WV
The McFarland Press, Harrisburg PA
McNaughton & Gunn Inc, Saline MI
Meaker the Printer, Phoenix AZ
Merrill Corporation, Los Angeles CA
Metroweb, Erlanger KY
Mitchell Press, Vancouver BC
Monument Printers & Lithographer Inc,
    Verplanck NY
Moran Printing Company, Orlando FL
Moran Printing Inc, Baton Rouge LA
Morgan Press, Dobbs Ferry NY
Morgan Printing, Austin TX
Motheral Printing, Fort Worth TX
William H Muller Printing, Santa Clara
    CA
National Graphics Corp, Columbus OH
National Reproductions C, Madison
    Heights MI
Nationwide Printing, Burlington KY
Louis Neibauer Co Inc, Warminster PA
Neyenesch Printers, San Diego CA
Nimrod Press Inc, Boston MA
Noll Printing Company, Huntington IN
Nystrom Publishing Company, Osseo MN
Oaks Printing Company, Bethlehem PA
Odyssey Press Inc, Dover NH
OMNIPRESS, Madison WI
Optic Graphics Inc, Glen Burnie MD
The Ovid Bell Press Inc, Fulton MO
The Oxford Group, Norway ME
Pantagraph Printing, Bloomington IL
Paraclete Press, Brewster MA
Park Press, South Holland IL
Patterson Printing Company, Benton
    Harbor MI
Peconic Companies, Mattituck NY
Pendell Printing, Midland MI
PennWell Printing, Tulsa OK

Perry Printing Corp, Waterloo WI
Perry/Baraboo, Baraboo WI
PFP Printing Corporation, Gaithersburg
    MD
Pine Hill Press Inc, Freeman SD
Plain Talk Printing Company, Des
    Moines IA
Plus Communications, Saint Louis MO
Port City Press Inc, Baltimore MD
Port Publications, Port Washington WI
Practical Graphics, New York NY
Princeton Academic Press, Lawrenceville
    NJ
The Printing Company, Indianapolis IN

Printing Corporation of America, Pom-
    pano Beach FL
Printwest, Regina SK
Progress Printing, Lynchburg VA
Publishers Press, Salt Lake City UT
Quebecor Printing (USA) Corp, Boston
    MA
Quebecor Printing Inc, Saint Paul MN
Quebecor Printing Inc, Richmond Hill
    ON
Quebecor Printing/St Cloud, Saint Cloud
    MN
Quinn-Woodbine Inc, Woodbine NJ
Ringier America, Itasca IL
Ronalds Printing, Vancouver BC
Ronalds Printing, Montreal PQ
Rose Printing Company, Tallahassee FL
S Rosenthal & Company Inc, Cincinnati
    OH
Royle Printing, Sun Prairie WI
Sanders Printing Company, Garretson
    SD
Schumann Printers, Fall River WI
Science Press, Ephrata PA
Shepard Poorman Communications, In-
    dianapolis IN
The Sheridan Press, Hanover PA
Sheridan Printing Company, Phil-
    lipsburg NJ
Smith-Edwards-Dunlap, Philadelphia PA
Southeastern Printing Company Inc,
    Stuart FL
Sowers Printing Company, Lebanon PA
St Croix Press Inc, New Richmond WI
Standard Publishing, Cincinnati OH
Stephenson Lithograph Inc, Alexandria
    VA
Stewart Publishing & Printing Co,
    Marble Hill MO
The Stinehour Press, Lunenburg VT
The Studley Press, Dalton MA
John S Swift Company Inc, St Louis MO
Technical Communication Services,
    North Kansas City MO
Thomson-Shore, Dexter MI
Times Litho Inc, Forest Grove OR
Times Printing, Random Lake WI

Journals

Town House Press, Pittsboro NC
Tracor Publications, Austin TX
Trend Offset Printing Service Inc, Los Alamitos CA
Tri-Graphic Printing, Ottawa ON
Universal Printing Company, Miami FL
Universal Printing Company, Saint Louis MO
Valco Graphics, Seattle WA
Van Volumes, Palmer MA
Vicks Lithograph & Printing, Yorkville NY
Victor Graphics, Baltimore MD
Waldman Graphics, Pennsauken NJ
Waldon Press Inc, New York NY
Waverly Press, Linthicum MD
Webcom Ltd, Scarborough ON
Whitehall Printing Company, Naples FL
Wickersham Printing Co Inc, Lancaster PA
Y/Z Printing Co, Elizabethtown PA

Maps

Adviser Graphics, Red Deer AB
Alcom Printing Group Inc, Bethlehem PA
AM Graphics & Printing, San Marcos CA
American Litho, Hayward CA
American Printers & Litho, Niles IL
Amos Press, Sidney OH
Apollo Graphics Ltd, Southampton PA
The Argus Press Inc, Niles IL
Bang Printing Company, Brainerd MN
Barton Press Inc, West Orange NJ
Bay State Press, Framingham MA
Beacon Press Inc, Seattle WA
Bradley Printing Company, Des Plaines IL
R L Bryan Company, Columbia SC
Buse Printing & Advertising, Phoenix AZ
Cal Central Press, Sacramento CA
Calsonic Miura Graphics, Irvine CA
Capital Printing Company Inc, Austin TX
Colonial Graphics, Paterson NJ
Colorlith Corporation, Johnston RI
Colotone Riverside Inc, Dallas TX
Continental Web, Itasca IL
Corley Printing Company, Earth City MO
Dellas Graphics, Syracuse NY
Deven Lithographers, Brooklyn NY
Dharma Press, Oakland CA
Dollco Printing, Ottawa ON
R R Donnelley & Sons Company, Crawfordsville IN
Dragon Press, Delta Junction AK
Eastwood Printing Co, Denver CO
Editors Press Inc, Hyattsville MD

Faculty Press Inc, Brooklyn NY
Federated Lithographers, Providence RI
Fisher Printers Inc, Cedar Rapids IA
The Forms Man Inc, Deer Park NY
Fort Orange Press Inc, Albany NY
Futura Printing Inc, Boynton Beach FL
Gateway Press Inc, Louisville KY
Gaylord Printing, Detroit MI
Graphic Litho Corporation, Lawrence MA
Great Lakes Lithograph Co, Cleveland OH
Gregath Publishing Co, Wyandotte OK
Harmony Printing, Liberty MO
Hinz Lithographing Company, Mt Prospect IL
A B Hirschfeld Press, Denver CO
Interstate Printing Company, Omaha NE
Julin Printing, Monticello IA
Kaufman Press Printing, Syracuse NY
Keys Printing Co Inc, Greenville SC
Kolor View Press, Aurora MO
The C J Krehbiel Company, Cincinnati OH
La Crosse Graphics, LaCrosse WI
Lehigh Press Inc, Cherry Hill NJ
Les Editions Marquis, Montmagny PQ
Marathon Communications Group, Wausau WI
The Mazer Corporation, Dayton OH
The McFarland Press, Harrisburg PA
McGill Jensen, St Paul MN
Meehan-Tooker & Company, East Rutherford NJ
Mercury Printing Company, Memphis TN
National Lithographers Inc, Miami FL
Neyenesch Printers, San Diego CA
Nystrom Publishing Company, Osseo MN
The Oxford Group, Norway ME
Padgett Printing Corporation, Dallas TX
Panel Prints Inc, Old Forge PA
Paraclete Press, Brewster MA
Paust Inc, Richmond IN
The Perlmuter Printing Co, Cleveland OH
The Press, Chanhassen MN
Print Northwest, Tacoma WA
Printing Corporation of America, Pompano Beach FL
Progress Printing, Lynchburg VA
Quad/Graphics Inc, Saratoga Springs NY
Quad/Graphics, Lomira, Lomira WI
Quad/Graphics, Pewaukee, Pewaukee WI
Quad/Graphics, Sussex, Sussex WI
Quebecor Book Group, New York NY
Quebecor/Semline, Braintree MA
Ringier America Inc, Corinth MS
Ronalds Printing, Vancouver BC
Royle Printing, Sun Prairie WI
The Saratoga Printing Co, Saratoga Springs NY

Maps

Shea Communications, Oklahoma City OK
Spencer Press, Wells ME
Stephenson Lithograph Inc, Alexandria VA
Straus Printing Company, Madison WI
John S Swift Company Inc, St Louis MO
John S Swift/Chicago, Chicago IL
John S Swift/Cincinnati, Cincinnati OH
John S Swift/Teterboro, Teterboro NJ
Times Printing, Random Lake WI
Trade Litho Inc, Miami FL
Valco Graphics, Seattle WA
Viking Press, Eden Prairie MN
F A Weber & Sons, Park Falls WI
The Wessel Company Inc, Elk Grove Village IL
The Wimmer Companies, Memphis TN
Ye Olde Genealogie Shoppe, Indianapolis IN

Mass-Market Paperbacks

Adams Press, Chicago IL
Banta Company, Menasha WI
Bertelsmann Printing & Manufacturing, New York NY
Best Gagne Book Manufacturers, Toronto ON
BookMasters Inc, Ashland OH
Buse Printing & Advertising, Phoenix AZ
C & M Press, Denver CO
Cal Central Press, Sacramento CA
Comfort Printing, St Louis MO
Cookbooks by Mom's Press, Kearney NE
Country Press, Mohawk NY
Crane Duplicating Service, West Barnstable MA
Crusader Printing Co, East Saint Louis IL
Cushing-Malloy, Ann Arbor MI
Des Plaines Publishing Co, Des Plaines IL
Dickinson Press Inc, Grand Rapids MI
R R Donnelley & Sons Company, Crawfordsville IN
Edwards Brothers Inc, Ann Arbor MI
Fleming Printing Co, St Louis MO
Golden Horn Press, Berkeley CA
Guynes Printing Company, Albuquerque NM
Harmony Printing, Liberty MO
The C J Krehbiel Company, Cincinnati OH

Lithocolor Press, Westchester IL
The Little River Press Inc, Miami FL
Mark IV Press, Hauppauge NY
The Mazer Corporation, Dayton OH
MBP Lithographics, Hillsboro KS
McClain Printing Company, Parsons WV
Merrill Corporation, Los Angeles CA
MMI Press, Harrisville NH
Moebius Printing, Milwaukee WI
Noll Printing Company, Huntington IN
Odyssey Press Inc, Dover NH
Offset Paperback Manufacturers, Dallas PA
Paraclete Press, Brewster MA
Publishers Press, Salt Lake City UT
Quad Graphics - Thomaston Press, Thomaston GA
Quebecor Book Group, New York NY
Quebecor Printing (USA) Corp, Boston MA
Quebecor Printing Inc, Richmond Hill ON
Ringier America, Itasca IL
Ringier America/Dresden, Dresden TN
Ronalds Printing, Montreal PQ
S Rosenthal & Company Inc, Cincinnati OH
Roxbury Publishing Company, Los Angeles CA
The Saratoga Printing Co, Saratoga Springs NY
Standard Publishing, Cincinnati OH
Stewart Publishing & Printing, Marble Hill MO
TEACH Services, Brushton NY
Thomson-Shore, Dexter MI
Todd Publications Inc, Smithville TX
Trade Litho Inc, Miami FL
Tweddle Litho Co, Clinton Township MI
Universal Printers Ltd, Winnipeg MB
Universal Printing Company, Miami FL
Webcom Ltd, Scarborough ON
Weber Printing Company, Lincoln NE

Postcards

A-1 Business Service Inc, St Paul MN
Accurate Printing, Post Falls ID
Adviser Graphics, Red Deer AB
Alcom Printing Group Inc, Bethlehem PA
American Litho, Hayward CA
Amidon & Associates Inc, Saint Paul MN
Apollo Graphics Ltd, Southampton PA
The Argus Press Inc, Niles IL
Arizona Lithographers Inc, Tucson AZ
Artcraft Press, Waterloo WI
Bay Port Press, Chula Vista CA
Beacon Press Inc, Seattle WA
Ben-Wal Printing, Pomona CA

Postcards

Blake Printery, San Luis Obispo CA
Blue Dolphin Press Inc, Grass Valley CA
Bradley Printing Company, Des Plaines IL
Brookshore Lithographers Inc, Elk Grove Village IL
BSC Litho, Harrisburg PA
Buse Printing & Advertising, Phoenix AZ
Cal Central Press, Sacramento CA
Caldwell Printers, Arcadia CA
Calsonic Miura Graphics, Irvine CA
Champagne Offset Co Inc, Natick MA
Chicago Color Express, Chicago IL
Coach House Press, Toronto ON
Color World Printers, Bozeman MT
Colotone Riverside Inc, Dallas TX
Comfort Printing, St Louis MO
Coneco Litho Graphics, Glen Falls NY
Continental Web, Itasca IL
Cooley Printers & Office Supply, Monroe LA
Country Press, Mohawk NY
Dataco A WBC Company, Albuquerque NM
Dellas Graphics, Syracuse NY
Dharma Press, Oakland CA
Diamond Graphics Tech Data, Milwaukee WI
Direct Graphics Inc, Sidney OH
Direct Press/Modern Litho, Huntington Station NY
Dollco Printing, Ottawa ON
Dragon Press, Delta Junction AK
E & D Web Inc, Cicero IL
Eagle Web Press, Salem OR
EU Services, Rockville MD
Eureka Printing Co Inc, Eureka CA
Eva-Tone Inc, Clearwater FL
First Impressions & Lithographers Inc, Elk Grove IL
Flower City Printing Inc, Rochester NY
Focus Direct, San Antonio TX
Futura Printing Inc, Boynton Beach FL
Geryon Press Limited, Tunnel NY
Golden Belt Printing Inc, Great Bend KS
Graphic Litho Corporation, Lawrence MA
Graphic Ways Inc, Buena Park CA
Gregath Publishing Co, Wyandotte OK
Guynes Printing Company, Albuquerque NM
Harlo Printing Co, Detroit MI
Hennegan Company, Cincinnati OH
A B Hirschfeld Press, Denver CO
Interform Corporation, Bridgeville PA
Interstate Printing Company, Omaha NE
Jersey Printing Company, Bayonne NJ
JK Creative Printers, Quincy IL
The Job Shop, Woods Hole MA
Jostens Printing & Publishing, Visalia CA

Jostens Printing & Publishing, Winston-Salem NC
Julin Printing, Monticello IA
Kaufman Press Printing, Syracuse NY
Kolor View Press, Aurora MO
La Crosse Graphics, LaCrosse WI
Lehigh Press Inc, Cherry Hill NJ
Lighthouse Press Inc, Manchester NH
Lithograph Printing Company, Memphis TN
John D Lucas Printing Company, Baltimore MD
Mackintosh Typography Inc, Santa Barbara CA
MBP Lithographics, Hillsboro KS
McClain Printing Company, Parsons WV
McGrew Color Graphics, Kansas City MO
Meaker the Printer, Phoenix AZ
Mercury Printing Company, Memphis TN
Moebius Printing, Milwaukee WI
Morgan Press, Dobbs Ferry NY
MultiPrint Co Inc, Skokie IL
Murphy's Printing Company, Campbell CA
National Lithographers Inc, Miami FL
Nationwide Printing, Burlington KY
Omaha Printing Company, Omaha NE
Outstanding Graphics, Kenosha WI
The Oxford Group, Norway ME
Padgett Printing Corporation, Dallas TX
Panorama Press Inc, Clifton NJ
Paraclete Press, Brewster MA
Paust Inc, Richmond IN
Pearl-Pressman-Liberty Communications Group, Philadelphia PA
Peconic Companies, Mattituck NY
Plain Talk Printing Company, Des Moines IA
Practical Graphics, New York NY
The Press, Chanhassen MN
Prinit Press, Dublin IN
Print Northwest, Tacoma WA
The Printing Company, Indianapolis IN
Printing Corporation of America, Pompano Beach FL
Printright International, Palmdale CA
Progress Printing, Lynchburg VA
ProLitho, Provo UT
Quebecor Book Group, New York NY
Quebecor/Semline, Braintree MA
Rapidocolor Corporation, West Chester PA
Sanders Printing Company, Garretson SD
The Saratoga Printing Co, Saratoga Springs NY
Schelin Printing Co, St Louis MO
Schiff Printing, Pittsburgh PA
R E Smith Printing, Fall River MA
The Stinehour Press, Lunenburg VT

Postcards

Straus Printing Company, Madison WI
John S Swift Company Inc, St Louis MO
John S Swift/Chicago, Chicago IL
John S Swift/Cincinnati, Cincinnati OH
John S Swift/Teterboro, Teterboro NJ
Times Printing, Random Lake WI
Triangle Booklet Company, Nashville TN
TSO General Corporation, Long Island
    City NY
US Press Inc, Valdosta GA
Vogue Printers, North Chicago IL
Waldon Press Inc, New York NY
Web Specialties, Wheeling IL
F A Weber & Sons, Park Falls WI
Weber Printing Company, Lincoln NE
The Wessel Company Inc, Elk Grove Vil-
    lage IL
Wood & Jones Printers, Pasadena CA
Y/Z Printing Co, Elizabethtown PA

Posters

Alan Lithograph Inc, Inglewood CA
AM Graphics & Printing, San Marcos CA
American Litho, Hayward CA
American Printers & Litho, Niles IL
American Signature, Memphis Division,
    Olive Branch MS
American Signature Graphics, Dallas TX
Apollo Graphics Ltd, Southampton PA
The Argus Press Inc, Niles IL
Artcraft Press, Waterloo WI
Bang Printing Company, Brainerd MN
Barton Press Inc, West Orange NJ
Bay State Press, Framingham MA
Beacon Press Inc, Seattle WA
Bertelsmann Printing & Manufacturing,
    New York NY
Blake Printery, San Luis Obispo CA
Blue Dolphin Press Inc, Grass Valley CA
Bolger Publications/Creative Printing,
    Minneapolis MN
Bradley Printing Company, Des Plaines
    IL
Brookshore Lithographers Inc, Elk
    Grove Village IL
R L Bryan Company, Columbia SC
Buse Printing & Advertising, Phoenix AZ
Cal Central Press, Sacramento CA
Calsonic Miura Graphics, Irvine CA
Capital Printing Company Inc, Austin TX
Champagne Offset Co Inc, Natick MA
Chicago Color Express, Chicago IL
Coach House Press, Toronto ON
Colonial Graphics, Paterson NJ
Color Express, Commerce CA
Color Express, Minneapolis MN

Color Express, Atlanta GA
Color World Printers, Bozeman MT
ColorGraphics Printing Corp, Tulsa OK
Colortone Press, Landover MD
Colotone Riverside Inc, Dallas TX
Comfort Printing, St Louis MO
Community Press, Provo UT
Concord Litho Company Inc, Concord
    NH
Coneco Litho Graphics, Glen Falls NY
Continental Web, Itasca IL
Country Press, Mohawk NY
Crusader Printing Co, East Saint Louis
    IL
Danbury Printing & Litho Inc, Danbury
    CT
Dataco A WBC Company, Albuquerque
    NM
Dellas Graphics, Syracuse NY
Delmar Printing & Publishing, Charlotte
    NC
Des Plaines Publishing Co, Des Plaines
    IL
Desaulniers Printing Company, Milan IL
Deven Lithographers, Brooklyn NY
Dharma Press, Oakland CA
Diamond Graphics Tech Data, Mil-
    waukee WI
Direct Press/Modern Litho, Huntington
    Station NY
Dollco Printing, Ottawa ON
Dragon Press, Delta Junction AK
E & D Web Inc, Cicero IL
Eagle Web Press, Salem OR
Edison Lithographing, North Bergen NJ
Editors Press Inc, Hyattsville MD
EU Services, Rockville MD
Eureka Printing Co Inc, Eureka CA
Faculty Press Inc, Brooklyn NY
Federated Lithographers, Providence RI
First Impressions & Lithographers Inc,
    Elk Grove IL
Fisher Printers Inc, Cedar Rapids IA
Fleming Printing Co, St Louis MO
Flower City Printing Inc, Rochester NY
Focus Direct, San Antonio TX
The Forms Man Inc, Deer Park NY
Fort Orange Press Inc, Albany NY
Foster Printing Service, Michigan City IN
Frye & Smith, Costa Mesa CA
Futura Printing Inc, Boynton Beach FL
Gateway Press Inc, Louisville KY
Gaylord Printing, Detroit MI
Geryon Press Limited, Tunnel NY
Glundal Color Service, East Syracuse NY
Graphic Litho Corporation, Lawrence MA
Graphic Ways Inc, Buena Park CA
Gray Printing, Fostoria OH
Great Lakes Lithograph Co, Cleveland
    OH
Great Northern/Design Printing, Skokie
    IL

Posters

Gregath Publishing Co, Wyandotte OK
GRIT Commercial Printing Services, Williamsport PA
Gulf Printing, Houston TX
Guynes Printing Company, Albuquerque NM
Harlo Printing Co, Detroit MI
Hennegan Company, Cincinnati OH
Hi-Tech Color House, Chicago IL
Hinz Lithographing Company, Mt Prospect IL
A B Hirschfeld Press, Denver CO
Intelligencer Printing, Lancaster PA
Interform Corporation, Bridgeville PA
Interstate Printing Company, Omaha NE
Jersey Printing Company, Bayonne NJ
JK Creative Printers, Quincy IL
Johns Byrne Co, Niles IL

Posters

Jostens Printing & Publishing, Visalia CA
Jostens Printing & Publishing, Clarksville TN
Jostens Printing & Publishing, Winston-Salem NC
Jostens Printing & Publishing, Topeka KS
Jostens Printing & Publishing, State College PA
Jostens Printing & Publishing, Minneapolis MN
Julin Printing, Monticello IA
Kaufman Press Printing, Syracuse NY
Keys Printing Co Inc, Greenville SC
Kolor View Press, Aurora MO
The C J Krehbiel Company, Cincinnati OH
La Crosse Graphics, LaCrosse WI
Lasky Company, Milburn NJ
Lehigh Press Inc, Cherry Hill NJ
Les Editions Marquis, Montmagny PQ
Lighthouse Press Inc, Manchester NH
Litho Specialties, St Paul MN
Lithograph Printing Company, Memphis TN
John D Lucas Printing Company, Baltimore MD
Mackintosh Typography Inc, Santa Barbara CA
Marathon Communications Group, Wausau WI
The Mazer Corporation, Dayton OH
McClain Printing Company, Parsons WV
The McFarland Press, Harrisburg PA
McGill Jensen, St Paul MN
McGrew Color Graphics, Kansas City MO
McQuiddy Printing Company, Nashville TN
Meaker the Printer, Phoenix AZ

Meehan-Tooker & Company, East Rutherford NJ
Mercury Printing Company, Memphis TN
Mitchell Graphics, Petosky MI
Moebius Printing, Milwaukee WI
Moran Printing Company, Orlando FL
Moran Printing Inc, Baton Rouge LA
Morgan Press, Dobbs Ferry NY
Motheral Printing, Fort Worth TX
William H Muller Printing, Santa Clara CA
MultiPrint Co Inc, Skokie IL
Murphy's Printing Co, Campbell CA
National Graphics Corp, Columbus OH
National Lithographers Inc, Miami FL
Nationwide Printing, Burlington KY
Louis Neibauer Co Inc, Warminster PA
Neyenesch Printers, San Diego CA
Noll Printing Company, Huntington IN
Oaks Printing Company, Bethlehem PA
Ohio Valley Litho Color, Florence KY
Omaha Printing Company, Omaha NE
Optic Graphics Inc, Glen Burnie MD
The Oxford Group, Norway ME
Padgett Printing Corporation, Dallas TX
Panel Prints Inc, Old Forge PA
Panorama Press Inc, Clifton NJ
Paraclete Press, Brewster MA
Paust Inc, Richmond IN
Pearl-Pressman-Liberty Communications Group, Philadelphia PA
Peconic Companies, Mattituck NY
The Perlmuter Printing Co, Cleveland OH
Perry Printing Corp, Waterloo WI
Perry/Baraboo, Baraboo WI
Pine Hill Press Inc, Freeman SD
Plain Talk Printing Company, Des Moines IA
Port Publications, Port Washington WI
Practical Graphics, New York NY
The Press, Chanhassen MN
Prinit Press, Dublin IN
Print Northwest, Tacoma WA
The Printing Company, Indianapolis IN
Printing Corporation of America, Pompano Beach FL
Printright International, Palmdale CA
Progress Printing, Lynchburg VA
ProLitho, Provo UT
Publishers Express Press, Ladysmith WI
Quebecor Book Group, New York NY
Quebecor Printing Inc, Richmond Hill ON
Quebecor/Semline, Braintree MA
Rapidocolor Corporation, West Chester PA
Ringier America Inc, Corinth MS
John Roberts Company, Minneapolis MN
Ronalds Printing, Vancouver BC
Ronalds Printing, Montreal PQ

Posters

S Rosenthal & Company, Cincinnati OH
Royle Printing, Sun Prairie WI
Sanders Printing Company, Garretson SD
The Saratoga Printing Co, Saratoga Springs NY
Schelin Printing Co, St Louis MO
Schiff Printing, Pittsburgh PA
R E Smith Printing, Fall River MA
Smith-Edwards-Dunlap, Philadelphia PA
Southeastern Printing Company Inc, Stuart FL
Standard Publishing, Cincinnati OH
John Stark Printing Co Inc, St Louis MO
Stephenson Lithograph Inc, Alexandria VA
Stewart Publishing & Printing Co, Marble Hill MO
The Stinehour Press, Lunenburg VT
Straus Printing Company, Madison WI
John S Swift Company Inc, St Louis MO
John S Swift/Chicago, Chicago IL
John S Swift/Cincinnati, Cincinnati OH
John S Swift/Teterboro, Teterboro NJ
Times Printing, Random Lake WI
Todd Publications Inc, Smithville TX
Trade Litho Inc, Miami FL
Universal Press, Providence RI
US Press Inc, Valdosta GA
Van Volumes, Palmer MA
Vogue Printers, North Chicago IL
Waldman Graphics, Pennsauken NJ
Waldon Press Inc, New York NY
Watt/Peterson, Plymouth MN
F A Weber & Sons, Park Falls WI
Weber Printing Company, Lincoln NE
Western Publishing Co Inc, Racine WI
White Arts Inc, Indianapolis IN
The Wood Press, Paterson NJ
Y/Z Printing Co, Elizabethtown PA

Textbooks

aBCD, Seattle WA
Action Printing, Fond Du Lac WI
Ad Infinitum Press, Mount Vernon NY
Adams Press, Chicago IL
Advanced Duplicating & Printing, Minneapolis MN
Adviser Graphics, Red Deer AB
Alcom Printing Group Inc, Bethlehem PA
AM Graphics & Printing, San Marcos CA
Apollo Graphics Ltd, Southampton PA
Arcata Graphics/Fairfield, Fairfield PA
The Argus Press Inc, Niles IL
Automated Graphic Systems, White Plains MD

Baker Johnson Inc, Dexter MI
Banta Company, Menasha WI
Bay Port Press, Chula Vista CA
Ben-Wal Printing, Pomona CA
Bertelsmann Printing & Manufacturing, New York NY
Best Gagne Book Manufacturers, Toronto ON
Bolger Publications/Creative Printing, Minneapolis MN
The Book Press Inc, Brattleboro VT
Book-Mart Press Inc, North Bergen NJ
BookCrafters, Chelsea MI
BookCrafters/Fredericksburg, Fredericksburg VA
BookMasters Inc, Ashland OH
Braceland Brothers Inc, Philadelphia PA
Braun-Brumfield Inc, Ann Arbor MI
Brunswick Publishing Corporation, Lawrenceville VA
Buse Printing & Advertising, Phoenix AZ
C & M Press, Denver CO
Cal Central Press, Sacramento CA
Camelot Publishing Co, Ormond Beach FL
Capital City Press, Montpelier VT
Capital Printing Company Inc, Austin TX
Carpenter Lithographic Co, Springfield OH
The College Press, Collegedale TN
Color World Printers, Bozeman MT
Columbus Bookbinders & Printers, Columbus GA
Comfort Printing, St Louis MO
Community Press, Provo UT
Coneco Litho Graphics, Glen Falls NY
Consolidated Printers Inc, Berkeley CA
Cookbooks by Mom's Press, Kearney NE
Country Press, Mohawk NY
Country Press Inc, Lakeville MA
Courier Stoughton Inc, Stoughton MA
Courier-Kendallville Corp, Kendallville IN
Courier-Westford Inc, Westford MA
Crane Duplicating Service, West Barnstable MA
Crusader Printing Co, East Saint Louis IL
Cushing-Malloy, Ann Arbor MI
Daamen Printing Company, West Rutland VT
Danner Press Corporation, Canton OH
Data Reproductions Corp, Rochester Hills MI
Delmar Printing & Publishing, Charlotte NC
Des Plaines Publishing Co, Des Plaines IL
Diamond Graphics Tech Data, Milwaukee WI
Dickinson Press Inc, Grand Rapids MI
Diversified Printing, Brea CA

Textbooks

Dollco Printing, Ottawa ON
R R Donnelley & Sons Company, Crawfordsville IN
R R Donnelley & Sons Company Inc, Harrisonburg, Harrisonburg VA
R R Donnelley & Sons Company Inc, Willard, Willard OH
Dragon Press, Delta Junction AK
Eastwood Printing Co, Denver CO
Edwards Brothers Inc, Ann Arbor MI
Edwards Brothers Inc, Lillington NC
Electronic Printing, Brentwood NY
Eureka Printing Co Inc, Eureka CA
Federated Lithographers, Providence RI
First Impressions & Lithographers Inc, Elk Grove IL
Fleming Printing Co, St Louis MO
Flower City Printing Inc, Rochester NY
Friesen Printers, Altona MB
Gateway Press Inc, Louisville KY
Gorham Printing, Rochester WA
Graphic Printing, New Carlisle OH
Great Lakes Lithograph Co, Cleveland OH
Gregath Publishing Co, Wyandotte OK
Griffin Printing and Lithograph Company, Sacramento CA
GRT Book Printing, Oakland CA
Haddon Craftsmen, Scranton PA
Harlo Printing Co, Detroit MI
Hawkes Publishing, Salt Lake City UT
Hennegan Company, Cincinnati OH
D B Hess Company, Woodstock IL
Hignell Printing, Winnepeg MB
Inland Printing Company, Syosset NY
IPC Publishing Services, Saint Joseph MI
JK Creative Printers, Quincy IL
The Johnson & Hardin Company, Cincinnati OH
Jostens Printing & Publishing, Winston-Salem NC
Jostens Printing & Publishing, Minneapolis MN
KNI Incorporated, Anaheim CA
The C J Krehbiel Company, Cincinnati OH
Les Editions Marquis, Montmagny PQ

John D Lucas Printing Company, Baltimore MD
Malloy Lithographing, Ann Arbor MI
Maple-Vail Book Manufacturing Group, York PA
Marrakech Express, Tarpon Springs FL
Maverick Publications Inc, Bend OR
The Mazer Corporation, Dayton OH
McArdle Printing Co, Upper Marlboro MD
McClain Printing Company, Parsons WV
The McFarland Press, Harrisburg PA

McNaughton & Gunn Inc, Saline MI
Media Publications, Santa Clara CA
Mercury Printing Company, Memphis TN
Merrill Corporation, Los Angeles CA
MMI Press, Harrisville NH
Moran Printing Company, Orlando FL
National Publishing Company, Philadelphia PA
National Reproductions Corp, Madison Heights MI
Nationwide Printing, Burlington KY
Louis Neibauer Co Inc, Warminster PA
Nimrod Press Inc, Boston MA
Oaks Printing Company, Bethlehem PA
OMNIPRESS, Madison WI
Optic Graphics Inc, Glen Burnie MD
The Ovid Bell Press Inc, Fulton MO
The Oxford Group, Norway ME
Padgett Printing Corporation, Dallas TX
Pantagraph Printing, Bloomington IL
Paraclete Press, Brewster MA
Peconic Companies, Mattituck NY
PFP Printing Corporation, Gaithersburg MD
Pine Hill Press Inc, Freeman SD
Plus Communications, Saint Louis MO
Port City Press Inc, Baltimore MD
The Press of Ohio, Kent OH
Princeton Academic Press, Lawrenceville NJ
Printing Corporation of America, Pompano Beach FL
Publishers Express Press, Ladysmith WI
Publishers Press, Salt Lake City UT
Quebecor Book Group, New York NY
Quebecor Printing (USA) Corp, Boston MA
Quebecor/Semline, Braintree MA
Rich Printing Company Inc, Nashville TN
Ringier America, Itasca IL
Ringier America/Brookfield, Brookfield WI
Ringier America/Jonesboro, Jonesboro AR
Ringier America/Phoenix, Phoenix AZ
Ronalds Printing, Vancouver BC
Rose Printing Company, Tallahassee FL
Roxbury Publishing Company, Los Angeles CA
Sanders Printing Company, Garretson SD
The Saratoga Printing Co, Saratoga Springs NY
Science Press, Ephrata PA
Shepard Poorman Communications, Indianapolis IN
Sheridan Printing Company, Phillipsburg NJ
Sowers Printing Company, Lebanon PA
Standard Printing Service, Chicago IL

Textbooks

Standard Publishing, Cincinnati OH
The Stinehour Press, Lunenburg VT
Straus Printing Company, Madison WI
John S Swift Company Inc, St Louis MO
TAN Books & Publishers Inc, Rockford
    IL
TEACH Services, Brushton NY
Technical Communication Services,
    North Kansas City MO
Todd Publications Inc, Smithville TX
Town House Press, Pittsboro NC
Tri-Graphic Printing, Ottawa ON
Tweddle Litho Co, Clinton Township MI
Valco Graphics, Seattle WA
Versa Press Inc, East Peoria IL
Vicks Lithograph & Printing, Yorkville
    NY
Waldon Press Inc, New York NY
Walsworth Publishing Co Inc, Marceline
    MO
Webcom Ltd, Scarborough ON
Webcrafters, Madison WI
Westview Press, Boulder CO
Whitehall Printing Company, Naples FL
Windsor Associates, San Diego CA
Worzalla Publishing Company, Stevens
    Point WI
Y/Z Printing Co, Elizabethtown PA

Workbooks

aBCD, Seattle WA
Academy Books, Rutland VT
Action Printing, Fond Du Lac WI
Ad Infinitum Press, Mount Vernon NY
Advanced Duplicating & Printing, Min-
    neapolis MN
Adviser Graphics, Red Deer AB
Alcom Printing Group Inc, Bethlehem PA
All-Star Printing, Lansing MI
Alonzo Printing, Hayward CA
AM Graphics & Printing, San Marcos CA
American Signature, Lincoln NE
Americomp, Brattleboro VT
Amidon & Associates Inc, Saint Paul MN
Amos Press Inc, Sidney OH
Apollo Graphics Ltd, Southampton PA
The Argus Press Inc, Niles IL
Arizona Lithographers Inc, Tucson AZ
Artcraft Press, Waterloo WI
Automated Graphic Systems, White
    Plains MD
Baker Johnson Inc, Dexter MI
Bang Printing Company, Brainerd MN
Banta Company, Menasha WI
Banta-Harrisonburg, Harrisonburg VA
Barton Press Inc, West Orange NJ
Bawden Printing Inc, Eldridge IA

Bay Port Press, Chula Vista CA
Bay State Press, Framingham MA
Beacon Press Inc, Seattle WA
Ben-Wal Printing, Pomona CA
Bertelsmann Printing & Manufacturing,
    New York NY
Best Gagne Book Manufacturers, Toron-
    to ON
Blue Dolphin Press Inc, Grass Valley CA
Bolger Publications/Creative Printing,
    Minneapolis MN
The Book Press Inc, Brattleboro VT
Book-Mart Press Inc, North Bergen NJ
BookCrafters, Chelsea MI
BookCrafters/Fredericksburg, Frederick-
    sburg VA
Booklet Publishing Company Inc, Elk
    Grove Village IL
BookMasters Inc, Ashland OH
Boyd Printing Co Inc, Albany NY
Braceland Brothers Inc, Philadelphia PA
Braun-Brumfield Inc, Ann Arbor MI
Brennan Printing, Deep River IA
Buse Printing & Advertising, Phoenix AZ
C & M Press, Denver CO
Cal Central Press, Sacramento CA
Caldwell Printers, Arcadia CA
Canterbury Press, Rome NY
Capital City Press, Montpelier VT
Capital Printing Company Inc, Austin TX
Carpenter Lithographic Co, Springfield
    OH
Champagne Offset Co Inc, Natick MA
Chicago Press Corp, Chicago IL
The College Press, Collegedale TN
Color World Printers, Bozeman MT
Colortone Press, Landover MD
Columbus Bookbinders & Printers,
    Columbus GA
Comfort Printing, St Louis MO
Community Press, Provo UT
Coneco Litho Graphics, Glen Falls NY
Cookbook Publishers Inc, Lenexa KS
Cookbooks by Mom's Press, Kearney NE
Corley Printing Company, Earth City MO
Country Press, Mohawk NY
Country Press Inc, Lakeville MA
Courier-Kendallville Corporation, Ken-
    dallville IN
Courier-Westford Inc, Westford MA
Crane Duplicating Service, West
    Barnstable MA
Crusader Printing Co, East Saint Louis
    IL
Daamen Printing Company, West Rut-
    land VT
Danner Press Corporation, Canton OH
Data Reproductions Corp, Rochester
    Hills MI
Delmar Printing & Publishing, Charlotte
    NC

Workbooks

Des Plaines Publishing Co, Des Plaines IL
Desaulniers Printing Company, Milan IL
Deven Lithographers, Brooklyn NY
Dharma Press, Oakland CA
Diamond Graphics Tech Data, Milwaukee WI
Dickinson Press Inc, Grand Rapids MI
Diversified Printing, Brea CA
Dollco Printing, Ottawa ON
R R Donnelley & Sons Company Inc, Crawfordsville IN
R R Donnelley & Sons Company Inc, Harrisonburg, Harrisonburg VA
R R Donnelley & Sons Company Inc, Willard, Willard OH
Dragon Press, Delta Junction AK
Eastwood Printing Co, Denver CO
EBSCO Media, Birmingham AL
Edwards Brothers Inc, Ann Arbor MI
Edwards Brothers Inc, Lillington NC
Eerdmans Printing Co, Grand Rapids MI
Electronic Printing, Brentwood NY
EP Graphics, Berne IN
EU Services, Rockville MD
Eureka Printing Co Inc, Eureka CA
Eva-Tone Inc, Clearwater FL
Evangel Press, Nappanee IN
Faculty Press Inc, Brooklyn NY
Federated Lithographers, Providence RI
First Impressions & Lithographers Inc, Elk Grove IL
Fisher Printers Inc, Cedar Rapids IA
Fleming Printing Co, St Louis MO
Flower City Printing Inc, Rochester NY
Fort Orange Press Inc, Albany NY
Friesen Printers, Altona MB
Gateway Press Inc, Louisville KY
Gaylord Printing, Detroit MI
Golden Belt Printing Inc, Great Bend KS
Goodway Graphics of Virginia, Springfield VA
Gorham Printing, Rochester WA
Graphic Litho Corporation, Lawrence MA
Graphic Printing, New Carlisle OH
Gray Printing, Fostoria OH
Great Impressions Printing & Graphics, Dallas TX
Great Lakes Lithograph Co, Cleveland OH
Gregath Publishing Co, Wyandotte OK
Griffin Printing & Lithograph, Glendale CA
Griffin Printing and Litograph Co, Sacramento CA
GRT Book Printing, Oakland CA
Haddon Craftsmen, Scranton PA
Harlo Printing Co, Detroit MI
Hawkes Publishing, Salt Lake City UT
Hennegan Company, Cincinnati OH

D B Hess Company, Woodstock IL
Hignell Printing, Winnepeg MB
Hoechstetter Printing, Pittsburgh PA
Holladay-Tyler Printing, Glenn Dale MD
Interform Corporation, Bridgeville PA
Interstate Printing Company, Omaha NE
IPC Publishing Services, Saint Joseph MI
JK Creative Printers, Quincy IL
The Job Shop, Woods Hole MA
The Johnson & Hardin Company, Cincinnati OH
Johnson Graphics, East Dubuque IL
Jostens Printing & Publishing, Visalia CA
Jostens Printing & Publishing, Clarksville TN
Jostens Printing & Publishing, Winston-Salem NC
Jostens Printing & Publishing, Topeka KS
Jostens Printing & Publishing, State College PA
Jostens Printing & Publishing, Minneapolis MN
Kirby Lithographic Co Inc, Arlington VA
KNI Incorporated, Anaheim CA
The C J Krehbiel Company, Cincinnati OH
Les Editions Marquis, Montmagny PQ
Lighthouse Press Inc, Manchester NH
Lithocolor Press, Westchester IL
Lithograph Printing Company, Memphis TN
Lithoid Printing Corporation, East Brunswick NJ
The Little River Press Inc, Miami FL
John D Lucas Printing Company, Baltimore MD
Malloy Lithographing, Ann Arbor MI
Maple-Vail Book Manufacturing Group, York PA
Maquoketa Web Printing, Maquoketa IA
Marrakech Express, Tarpon Springs FL
Maverick Publications Inc, Bend OR
The Mazer Corporation, Dayton OH
MBP Lithographics, Hillsboro KS
McClain Printing Company, Parsons WV
McGill Jensen, St Paul MN
McKay Printing Services, Michigan City IN
McNaughton & Gunn Inc, Saline MI
Meaker the Printer, Phoenix AZ
Media Publications, Santa Clara CA
Mercury Printing Company, Memphis TN
Merrill Corporation, Los Angeles CA
Metromail Corporation, Lincoln NE
Mitchell Press, Vancouver BC
MMI Press, Harrisville NH
Moran Printing Company, Orlando FL
Morgan Printing, Austin TX
National Reproductions Corp, Madison Heights MI

Workbooks

Nationwide Printing, Burlington KY
Louis Neibauer Co Inc, Warminster PA
Neyenesch Printers, San Diego CA
Nimrod Press Inc, Boston MA
Oaks Printing Company, Bethlehem PA
Odyssey Press Inc, Dover NH
OMNIPRESS, Madison WI
Optic Graphics Inc, Glen Burnie MD
Original Copy Centers, Cleveland OH
Outstanding Graphics, Kenosha WI
The Ovid Bell Press Inc, Fulton MO
The Oxford Group, Norway ME
Padgett Printing Corporation, Dallas TX
Pantagraph Printing, Bloomington IL
Paraclete Press, Brewster MA
Park Press, South Holland IL
Parker Graphics, Fuquay-Varina NC
Patterson Printing Company, Benton
    Harbor MI
Paust Inc, Richmond IN
Perry Printing Corp, Waterloo WI
Perry/Baraboo, Baraboo WI
PFP Printing Corporation, Gaithersburg
    MD
Pine Hill Press Inc, Freeman SD
Plus Communications, Saint Louis MO
Port City Press Inc, Baltimore MD
The Press of Ohio, Kent OH
The Press, Chanhassen MN
Princeton Academic Press, Lawrenceville
    NJ
Printing Corporation of America, Pom-
    pano Beach FL
Printright International, Palmdale CA
Progress Printing, Lynchburg VA
ProLitho, Provo UT
Publishers Express Press, Ladysmith WI
Publishers Press, Salt Lake City UT
Quebecor Book Group, New York NY
Quebecor Printing (USA) Corp, Boston
    MA
Quebecor/Semline, Braintree MA
RBW Graphics, Owen Sound ON
Repro-Tech, Fairfield CT
Rich Printing Company, Nashville TN
John Roberts Company, Minneapolis MN
Ronalds Printing, Vancouver BC
Rose Printing Company, Tallahassee FL
S Rosenthal & Company Inc, Cincinnati
    OH
Roxbury Publishing Company, Los An-
    geles CA
Royle Printing, Sun Prairie WI
Sanders Printing Company, Garretson
    SD
The Saratoga Printing Co, Saratoga
    Springs NY
Schelin Printing Co, St Louis MO
Schiff Printing, Pittsburgh PA
Science Press, Ephrata PA

Service Printing Company, San Leandro
    CA
Sexton Printing, Saint Paul MN
Shepard Poorman Communications, In-
    dianapolis IN
Sheridan Printing Company, Phil-
    lipsburg NJ
Sinclair Printing Co, Los Angeles CA
R E Smith Printing, Fall River MA
Smith-Edwards-Dunlap, Philadelphia PA
Snohomish Publishing, Snohomish WA
Sowers Printing Company, Lebanon PA
Speakes-Hines & Thomas Inc, Lansing
    MI
Standard Publishing, Cincinnati OH
Stephenson Lithograph Inc, Alexandria
    VA
Stevens Graphics Inc, Birmingham AL
Straus Printing Company, Madison WI
John S Swift Company Inc, St Louis MO
TAN Books & Publishers Inc, Rockford
    IL
Tapco Inc, Pemberton NJ
TEACH Services, Brushton NY
Times Printing, Random Lake WI
Todd Publications Inc, Smithville TX
Town House Press, Pittsboro NC
Tracor Publications, Austin TX
Trade Litho Inc, Miami FL
Trend Offset Printing Service Inc, Los
    Alamitos CA
Tri-Graphic Printing, Ottawa ON
Triangle Booklet Company, Nashville TN
TSO General Corporation, Long Island
    City NY
Tweddle Litho Co, Clinton Township MI
Universal Printing Company, Miami FL
Universal Printing Company, Saint
    Louis MO
Valco Graphics, Seattle WA
Van Volumes, Palmer MA
Versa Press Inc, East Peoria IL
Vicks Lithograph & Printing, Yorkville
    NY
Victor Graphics, Baltimore MD
Vogue Printers, North Chicago IL
Waldon Press Inc, New York NY
Wallace Press, Hillside IL
Watt/Peterson, Plymouth MN
Webcom Ltd, Scarborough ON
Webcrafters, Madison WI
Weber Printing Company, Lincoln NE
Western Publishing Co Inc, Racine WI
Westview Press, Boulder CO
White Arts Inc, Indianapolis IN
Whitehall Printing Company, Naples FL
Wickersham Printing Co Inc, Lancaster
    PA
Winter Sun Inc, Phoenix AZ
Worzalla Publishing Company, Stevens
    Point WI
Y/Z Printing Co, Elizabethtown PA

Workbooks

Ye Olde Genealogie Shoppe, Indianapolis IN

Yearbooks

aBCD, Seattle WA
Adams Press, Chicago IL
Adviser Graphics, Red Deer AB
Americomp, Brattleboro VT
Apollo Graphics Ltd, Southampton PA
Automated Graphic Systems, White Plains MD
Barton Press Inc, West Orange NJ
Bay State Press, Framingham MA
Beacon Press Inc, Seattle WA
Bolger Publications/Creative Printing, Minneapolis MN
Book-Mart Press Inc, North Bergen NJ
Buse Printing & Advertising, Phoenix AZ
Cal Central Press, Sacramento CA
Capital Printing Company Inc, Austin TX
Carpenter Lithographic Co, Springfield OH
Comfort Printing, St Louis MO
Community Press, Provo UT
Coneco Litho Graphics, Glen Falls NY
Crane Duplicating Service, West Barnstable MA
Crusader Printing Co, East Saint Louis IL
Daamen Printing Company, West Rutland VT
Danner Press Corporation, Canton OH
Delmar Printing & Publishing, Charlotte NC
Diamond Graphics Tech Data, Milwaukee WI
Dollco Printing, Ottawa ON
R R Donnelley & Sons Company, Crawfordsville IN
Eastwood Printing Co, Denver CO
Eerdmans Printing Co, Grand Rapids MI
EU Services, Rockville MD
Eureka Printing Co Inc, Eureka CA
Faculty Press Inc, Brooklyn NY
Flower City Printing Inc, Rochester NY
Friesen Printers, Altona MB
Futura Printing Inc, Boynton Beach FL
Great Lakes Lithograph Co, Cleveland OH
Gregath Publishing Co, Wyandotte OK
Harlo Printing Co, Detroit MI
Henington Publishing Company, Wolfe City TX
Jostens Printing & Publishing, Visalia CA
Jostens Printing & Publishing, Clarksville TN
Jostens Printing & Publishing, Winston-Salem NC
Jostens Printing & Publishing, Topeka KS
Jostens Printing & Publishing, State College PA
Jostens Printing & Publishing, Minneapolis MN
The C J Krehbiel Company, Cincinnati OH
Les Editions Marquis, Montmagny PQ
Malloy Lithographing, Ann Arbor MI
Marrakech Express, Tarpon Springs FL
Maverick Publications Inc, Bend OR
The Mazer Corporation, Dayton OH
McClain Printing Company, Parsons WV
Mercury Printing Company, Memphis TN
Moran Printing Company, Orlando FL
Louis Neibauer Co Inc, Warminster PA
Newsfoto Publishing Company, San Angelo TX
Outstanding Graphics, Kenosha WI
Princeton Academic Press, Lawrenceville NJ
Printing Corporation of America, Pompano Beach FL
Printright International, Palmdale CA
Publishers Press, Salt Lake City UT
Quebecor Printing (USA) Corp, Boston MA
Rose Printing Company, Tallahassee FL
The Saratoga Printing Co, Saratoga Springs NY
Sexton Printing, Saint Paul MN
Shepard Poorman Communications, Indianapolis IN
Sheridan Printing Company, Phillipsburg NJ
Straus Printing Company, Madison WI
John S Swift Company Inc, St Louis MO
Tracor Publications, Austin TX
Trade Litho Inc, Miami FL
Valco Graphics, Seattle WA
Waldon Press Inc, New York NY
Walsworth Publishing Co Inc, Marceline MO
F A Weber & Sons, Park Falls WI
Windsor Associates, San Diego CA

Pop-ups

The following foreign printers specialize in producing pop-ups and other novelty books that require lots of intricate handwork.

Carvajal S.A. (Columbia)
Dominie Press (Singapore)
Ediciones Lerner (Columbia)
Saik Wah Press (Singapore)
Tien Wah Press (Singapore)

Index: Binding Capabilities

These indexes list those printers who report that they have in-house binding capabilities for the following types of books.

Casebound Books

Academy Books, Rutland VT
Action Printing, Fond Du Lac WI
Adams Press, Chicago IL
Adviser Graphics, Red Deer AB
Alan Lithograph Inc, Inglewood CA
All-Star Printing, Lansing MI
American Litho, Hayward CA
American Printers & Litho, Niles IL
Anundsen Publishing Company, Decorah IA
Apollo Graphics Ltd, Southampton PA
Arandell Corporation, Menomonee Falls WI
Arcata Graphics/Fairfield, Fairfield PA
The Argus Press Inc, Niles IL
Arizona Lithographers, Tucson AZ
Associated Printers, Grafton ND
Austin Printing Co Inc, Akron OH
Bang Printing Co, Brainerd MN
Barton Press Inc, West Orange NJ
Bay Port Press, Chula Vista CA
Beacon Press, Seattle WA
Berryville Graphics, Berryville VA
Bertelsmann Printing & Manufacturing, New York NY
Blake Printery, San Luis Obispo CA
Bolger Publications/Creative Printing Minneapolis MN
The Book Press Inc, Brattleboro VT
Book-Mart Press Inc, North Bergen NJ
BookCrafters, Chelsea MI
BookCrafters/Frederick, Fredericksburg VA
BookMasters Inc, Ashland OH
Bradley Printing Company, Des Plaines IL
Braun-Brumfield Inc, Ann Arbor MI
Brown Printing, Waseca MN
R L Bryan Company, Columbia SC
BSC Litho, Harrisburg PA
The William Byrd Press, Richmond VA
C & M Press, Denver CO
Caldwell Printers, Arcadia CA
Calsonic Miura Graphics, Irvine CA

Canterbury Press, Rome NY
Capital City Press, Montpelier VT
Capital Printing Company Inc, Austin TX
Chicago Color Express, Chicago IL
Coach House Press, Toronto ON
The College Press, Collegedale TN
Color World Printers, Bozeman MT
ColorCorp Inc, Littleton CO
Colortone Press, Landover MD
Colotone Riverside Inc, Dallas TX
Comfort Printing, St Louis MO
Community Press, Provo UT
Concord Litho Company Inc, Concord NH
Coneco Litho Graphics, Glen Falls NY
Consolidated Printers Inc, Berkeley CA
Cookbook Publishers Inc, Lenexa KS
Cookbooks by Mom's Press, Kearney NE
Corley Printing Company, Earth City MO
Country Press, Mohawk NY
Courier Stoughton Inc, Stoughton MA
Courier-Kendallville Corp, Kendallville IN
Courier-Westford Inc, Westford MA
Crusader Printing Co, East Saint Louis IL
Cushing-Malloy, Ann Arbor MI
Daamen Printing Company, West Rutland VT
Danbury Printing & Litho Inc, Danbury CT
Dataco A WBC Company, Albuquerque NM
Delmar Printing & Publishing, Charlotte NC
Delta Lithograph Company, Valencia CA
Desaulniers Printing Company, Milan IL
Diamond Graphics Tech Data, Milwaukee WI
Dickinson Press Inc, Grand Rapids MI
Direct Press/Modern Litho, Huntington Station NY
Dollco Printing, Ottawa ON
R R Donnelley & Sons Company Inc, Crawfordsville IN
R R Donnelley & Sons Company Inc, Harrisonburg, Harrisonburg VA
R R Donnelley & Sons Company Inc, Willard, Willard OH
Dragon Press, Delta Junction AK
Edwards Brothers Inc, Ann Arbor MI

Casebound Books

Edwards Brothers Inc, Lillington NC
Eerdmans Printing Co, Grand Rapids MI
Electronic Printing, Brentwood NY
Eureka Printing Co Inc, Eureka CA
Evangel Press, Nappanee IN
Faculty Press Inc, Brooklyn NY
Federated Lithographers, Providence RI
First Impressions Printer & Lithographers Inc, Elk Grove IL
Fleming Printing Co, St Louis MO
Flower City Printing Inc, Rochester NY
Focus Direct, San Antonio TX
Friesen Printers, Altona MB
Futura Printing Inc, Boynton Beach FL
Gateway Press Inc, Louisville KY
Geryon Press Limited, Tunnel NY
Golden Belt Printing Inc, Great Bend KS
Gorham Printing, Rochester WA
Graftek Press, Woodstock IL
Graphic Arts Center, Portland OR
Graphic Printing, New Carlisle OH
Graphic Ways Inc, Buena Park CA
Graphics Ltd, Indianapolis IN
Great Impressions Printing & Graphics, Dallas TX
Great Lakes Lithograph Co, Cleveland OH
Gregath Publishing Co, Wyandotte OK
Griffin Printing & Lithograph, Glendale CA
Griffin Printing and Litograph Company, Sacramento CA
GRIT Commercial Printing Services, Williamsport PA
Guynes Printing Company, Albuquerque NM
Haddon Craftsmen, Scranton PA
Harlo Printing Co, Detroit MI
Harmony Printing, Liberty MO
Henington Publishing Company, Wolfe City TX
Heritage Printers Inc, Charlotte NC
Hignell Printing, Winnepeg MB
Hinz Lithographing Company, Mt Prospect IL
A B Hirschfeld Press, Denver CO
Holladay-Tyler Printing Corp, Glenn Dale MD
Rae Horowitz Book Manufacturers, Fairfield CT
Interform Corporation, Bridgeville PA
The Job Shop, Woods Hole MA
The Johnson & Hardin Company, Cincinnati OH
Jostens Printing & Publishing, Visalia CA
Jostens Printing & Publishing, Clarksville TN
Jostens Printing & Publishing, Winston-Salem NC
Jostens Printing & Publishing, Topeka KS
Jostens Printing & Publishing, State College PA
Jostens Printing & Publishing, Minneapolis MN
Julin Printing, Monticello IA
Kaufman Press Printing, Syracuse NY
Kimberly Press, Santa Barbara CA
Kirby Lithographic Co Inc, Arlington VA
KNI Incorporated, Anaheim CA
The C J Krehbiel Company, Cincinnati OH
La Crosse Graphics, LaCrosse WI
Les Editions Marquis, Montmagny PQ
Lighthouse Press Inc, Manchester NH
Litho Specialties, St Paul MN
Lithocolor Press, Westchester IL
Lithograph Printing Company, Memphis TN
Lithoid Printing Corporation, East Brunswick NJ
The Little River Press Inc, Miami FL
John D Lucas Printing Company, Baltimore MD
Mackintosh Typography Inc, Santa Barbara CA
Malloy Lithographing, Ann Arbor MI
Maple-Vail Book Manufacturing Group, York PA
Marathon Communications Group, Wausau WI
Marrakech Express, Tarpon Springs FL
McArdle Printing Co, Upper Marlboro MD
McClain Printing Company, Parsons WV
McDowell Publications, Utica KY
McNaughton & Gunn Inc, Saline MI
McQuiddy Printing Company, Nashville TN
Meaker the Printer, Phoenix AZ
Mercury Printing Company, Memphis TN
Merrill Corporation, Los Angeles CA
Metromail Corporation, Lincoln NE
MMI Press, Harrisville NH
Moran Printing Company, Orlando FL
Moran Printing Inc, Baton Rouge LA
Morgan Printing, Austin TX
Morningrise Printing, Costa Mesa CA
William H Muller Printing, Santa Clara CA
MultiPrint Co Inc, Skokie IL
National Graphics Corp, Columbus OH
National Lithographers Inc, Miami FL
National Publishing Company, Philadelphia PA
Newsfoto Publishing Company, San Angelo TX
Nimrod Press Inc, Boston MA
Northlight Studio Press, Barre VT
Northwest Web, Eugene OR

Comb-Bound Books

Comb-Bound Books

Advanced Duplicating & Printing, Edina MN
Alan Lithograph Inc, Inglewood CA
Alcom Printing Group Inc, Bethlehem PA
AM Graphics & Printing, San Marcos CA
American Litho, Hayward CA
American Printers & Litho, Niles IL
Anundsen Publishing Company, Decorah IA
Apollo Graphics Ltd, Southampton PA
The Argus Press Inc, Niles IL
Austin Printing Co Inc, Akron OH
Automated Graphic Systems, White Plains MD
Bang Printing Co, Brainerd MN
Banta Company, Menasha WI
Barton Press Inc, West Orange NJ
Bawden Printing Inc, Eldridge IA
Bay Port Press, Chula Vista CA
Beacon Press, Seattle WA
Ben-Wal Printing, Pomona CA
Berryville Graphics, Berryville VA
Bertelsmann Printing & Manufacturing, New York NY
Best Gagne Book Manufacturers, Toronto ON
Blake Printery, San Luis Obispo CA
Blue Dolphin Press Inc, Grass Valley CA
Bolger Publications/Creative Printing, Minneapolis MN
Book-Mart Press Inc, North Bergen NJ
BookCrafters, Chelsea MI
BookCrafters/Fredericksburg, Fredericksburg VA
BookMasters Inc, Ashland OH
Bradley Printing Company, Des Plaines IL
Braun-Brumfield Inc, Ann Arbor MI
Brennan Printing, Deep River IA
Brown Printing, Waseca MN
Brunswick Publishing Corp, Lawrenceville VA
R L Bryan Company, Columbia SC
BSC Litho, Harrisburg PA
Buse Printing & Advertising, Phoenix AZ
C & M Press, Denver CO
Cal Central Press, Sacramento CA
Caldwell Printers, Arcadia CA
Calsonic Miura Graphics, Irvine CA
Camelot Publishing Company, Ormond Beach FL
Canterbury Press, Rome NY
Capital City Press, Montpelier VT
Capital Printing Company Inc, Austin TX
Chicago Color Express, Chicago IL
The College Press, Collegedale TN
Colonial Graphics, Paterson NJ
Color World Printers, Bozeman MT
Colortone Press, Landover MD
Colotone Riverside Inc, Dallas TX

Columbus Bookbinders & Printers, Columbus GA
Comfort Printing, St Louis MO
Community Press, Provo UT
Concord Litho Company Inc, Concord NH
Coneco Litho Graphics, Glen Falls NY
Cookbook Publishers Inc, Lenexa KS
Cookbooks by Mom's Press, Kearney NE
Cooley Printers & Office Supply, Monroe LA
Corley Printing Company, Earth City MO
Country Press, Mohawk NY
Courier Stoughton Inc, Stoughton MA
Crane Duplicating Service, West Barnstable MA
Crusader Printing Co, East Saint Louis IL
Cushing-Malloy, Ann Arbor MI
Daamen Printing Company, West Rutland VT
Data Reproductions Corp, Rochester Hills MI
Dataco A WBC Company, Albuquerque NM
Delmar Printing & Publishing, Charlotte NC
Delta Lithograph Company, Valencia CA
Desaulniers Printing Company, Milan IL
Deven Lithographers, Brooklyn NY
Diamond Graphics Tech Data, Milwaukee WI
Dickinson Press Inc, Grand Rapids MI
Direct Press/Modern Litho, Huntington Station NY
Diversified Printing, Brea CA
Dollco Printing, Ottawa ON
R R Donnelley & Sons Co, Crawfordsville IN
Dragon Press, Delta Junction AK
Edwards Brothers Inc, Ann Arbor MI
Eerdmans Printing Co, Grand Rapids MI
Electronic Printing, Brentwood NY
Eureka Printing Co Inc, Eureka CA
Evangel Press, Nappanee IN
Faculty Press Inc, Brooklyn NY
Federated Lithographers, Providence RI
First Impressions Printer & Lithographers Inc, Elk Grove IL
Fleming Printing Co, St Louis MO
Flower City Printing Inc, Rochester NY
Focus Direct, San Antonio TX
Fort Orange Press Inc, Albany NY
Friesen Printers, Altona MB
Fundcraft Publishing, Collierville TN
Futura Printing Inc, Boynton Beach FL
Gateway Press Inc, Louisville KY
George Lithograph, San Francisco CA
Glundal Color Service, East Syracuse NY
Golden Belt Printing Inc, Great Bend KS
Goodway Graphics of Virginia, Springfield VA

Comb-bound Books

Quebecor/Semline, Braintree MA
Repro-Tech, Fairfield CT
Ringier America, Itasca IL
Ronalds Printing, Vancouver BC
Ronalds Printing, Montreal PQ
Rose Printing Company, Tallahassee FL
S Rosenthal & Company Inc, Cincinnati OH
Roxbury Publishing Company, Los Angeles CA
Royle Printing, Sun Prairie WI
Sanders Printing Company, Garretson SD
The Saratoga Printing Co, Saratoga Springs NY
Schiff Printing, Pittsburgh PA
Service Printing Company, San Leandro CA
Sexton Printing, Saint Paul MN
The Sheridan Press, Hanover PA
Sheridan Printing Company, Phillipsburg NJ
Sinclair Printing Co, Los Angeles CA
R E Smith Printing, Fall River MA
Southeastern Printing Company Inc, Stuart FL
Sowers Printing Company, Lebanon PA
Speakes-Hines & Thomas Inc, Lansing MI
John Stark Printing Co Inc, St Louis MO
Starr Toof Printing, Memphis TN
Stewart Publishing & Printing Co, Marble Hill MO
The Stinehour Press, Lunenburg VT
Straus Printing Company, Madison WI
The Studley Press, Dalton MA
Sun Graphix, Lewiston ME
John S Swift Company Inc, St Louis MO
John S Swift/Chicago, Chicago IL
John S Swift/Cincinnati, Cincinnati OH
John S Swift/Teterboro, Teterboro NJ
TAN Books & Publishers Inc, Rockford IL
TEACH Services, Brushton NY
Technical Communication Services, North Kansas City MO
Thomson-Shore, Dexter MI
Times Litho Inc, Forest Grove OR
Times Printing, Random Lake WI
Town House Press, Pittsboro NC
Tracor Publications, Austin TX
Trade Litho Inc, Miami FL
Triangle Booklet Company, Nashville TN
TSO General Corporation, Long Island City NY
Tweddle Litho Co, Clinton Township MI
Vail Ballou Press, Binghamton NY
Valco Graphics, Seattle WA
Van Volumes, Palmer MA
Versa Press Inc, East Peoria IL

Vicks Lithograph & Printng, Yorkville NY
Victor Graphics, Baltimore MD
Vogue Printers, North Chicago IL
Waldman Graphics, Pennsauken NJ
Waldon Press Inc, New York NY
Walsworth Publishing Co Inc, Marceline MO
Walter's Publishing, Waseca MN
Watkins Printing Company, Columbus OH
Watt/Peterson, Plymouth MN
Waverly Press, Linthicum MD
F A Weber & Sons, Park Falls WI
Westview Press, Boulder CO
Wickersham Printing Co Inc, Lancaster PA
The Wimmer Companies, Memphis TN
Windsor Associates, San Diego CA
Winter Sun Inc, Phoenix AZ
Wolfer Printing, Commerce CA
World Color Press Inc, New York NY
Worzalla Publishing Company, Stevens Point WI
Y/Z Printing Co, Elizabethtown PA
Ye Olde Genealogie Shoppe, Indianapolis IN

Loose-Leaf Binding

aBCD, Seattle WA
Accurate Printing, Post Falls ID
Action Printing, Fond Du Lac WI
Advanced Duplicating & Printing, Minnerapolis MN
Adviser Graphics, Red Deer AB
Alan Lithograph Inc, Inglewood CA
Alcom Printing Group Inc, Bethlehem PA
All-Star Printing, Lansing MI
AM Graphics & Printing, San Marcos CA
American Litho, Hayward CA
American Printers & Litho, Niles IL
Anundsen Publishing Company, Decorah IA
Apollo Graphics Ltd, Southampton PA
The Argus Press Inc, Niles IL
Austin Printing Co Inc, Akron OH
Automated Graphic Systems, White Plains MD
Banta Company, Menasha WI
Barton Press Inc, West Orange NJ
Bawden Printing Inc, Eldridge IA
Bay Port Press, Chula Vista CA
Bay State Press, Framingham MA
Beacon Press, Seattle WA
Ben-Wal Printing, Pomona CA

Loose-Leaf Binding

Bertelsmann Printing & Manufacturing, New York NY
Best Gagne Book Manufacturers, Toronto ON
Blue Dolphin Press Inc, Grass Valley CA
Bolger Publications/Creative Printing, Minneapolis MN
BookCrafters, Chelsea MI
BookCrafters/Fredericksburg, Fredericksburg VA
BookMasters Inc, Ashland OH
Braceland Brothers Inc, Philadelphia PA
Bradley Printing Company, Des Plaines IL
Braun-Brumfield Inc, Ann Arbor MI
Brennan Printing, Deep River IA
Brown Printing, Waseca MN
Brunswick Publishing Corp, Lawrenceville VA
R L Bryan Company, Columbia SC
BSC Litho, Harrisburg PA
Buse Printing & Advertising, Phoenix AZ
C & M Press, Denver CO
Cal Central Press, Sacramento CA
Caldwell Printers, Arcadia CA
Calsonic Miura Graphics, Irvine CA
Capital City Press, Montpelier VT
Capital Printing Company Inc, Austin TX
Carpenter Lithographic Co, Springfield OH
Champagne Offset Co Inc, Natick MA
Chicago Color Express, Chicago IL
Clarkwood Corporation, Totowa NJ
The College Press, Collegedale TN
Color World Printers, Bozeman MT
Colotone Riverside Inc, Dallas TX
Comfort Printing, St Louis MO
Concord Litho Company Inc, Concord NH
Coneco Litho Graphics, Glen Falls NY
Cookbook Publishers Inc, Lenexa KS
Cookbooks by Mom's Press, Kearney NE
Cooley Printers & Office Supply, Monroe LA
Corley Printing Company, Earth City MO
Country Press, Mohawk NY
Country Press Inc, Lakeville MA
Crusader Printing Co, East Saint Louis IL
Daamen Printing Company, West Rutland VT
Data Reproductions Corp, Rochester Hills MI
Dataco A WBC Company, Albuquerque NM
Dharma Press, Oakland CA
Diamond Graphics Tech Data, Milwaukee WI
Dickinson Press Inc, Grand Rapids MI

Direct Press/Modern Litho, Huntington Station NY
Dollco Printing, Ottawa ON
R R Donnelley & Sons Company Inc, Crawfordsville IN
Dragon Press, Delta Junction AK
Eastwood Printing Co, Denver CO
EBSCO Media, Birmingham AL
Edwards Brothers Inc, Lillington NC
Eerdmans Printing Co, Grand Rapids MI
Electronic Printing, Brentwood NY
Eureka Printing Co Inc, Eureka CA
Eva-Tone Inc, Clearwater FL
Faculty Press Inc, Brooklyn NY
Federated Lithographers, Providence RI
First Impressions Printer & Lithographers Inc, Elk Grove IL
Fleming Printing Co, St Louis MO
Flower City Printing Inc, Rochester NY
Focus Direct, San Antonio TX
The Forms Man Inc, Deer Park NY
Futura Printing Inc, Boynton Beach FL
Gateway Press Inc, Louisville KY
Geiger Brothers, Lewiston ME
George Lithograph, San Francisco CA
Glundal Color Service, East Syracuse NY
Golden Belt Printing Inc, Great Bend KS
Gorham Printing, Rochester WA
Graphic Printing, New Carlisle OH
Graphic Ways Inc, Buena Park CA
Graphics Ltd, Indianapolis IN
Great Impressions Printing & Graphics, Dallas TX
Great Lakes Lithograph Co, Cleveland OH
Gregath Publishing Co, Wyandotte OK
Griffin Printing and Lithograph Company, Sacramento CA
GRIT Commercial Printing Services, Williamsport PA
GRT Book Printing, Oakland CA
Guynes Printing Company, Albuquerque NM
Harlo Printing Co, Detroit MI
Hignell Printing, Winnepeg MB
A B Hirschfeld Press, Denver CO
Independent Printing Company, New York NY
Interform Corporation, Bridgeville PA
IPC Publishing Services, Saint Joseph MI
The Job Shop, Woods Hole MA
Jostens Printing & Publishing, Visalia CA
Jostens Printing & Publishing, Winston-Salem NC
Julin Printing, Monticello IA
Kaufman Press Printing, Syracuse NY
Keys Printing Co Inc, Greenville SC
Kirby Lithographic Co Inc, Arlington VA
KNI Incorporated, Anaheim CA
La Crosse Graphics, LaCrosse WI
Les Editions Marquis, Montmagny PQ
Lighthouse Press Inc, Manchester NH

Loose-Leaf Binding

Lithograph Printing Company, Memphis TN

Lithoid Printing Corporation, East Brunswick NJ

Malloy Lithographing, Ann Arbor MI

Maquoketa Web Printing, Maquoketa IA

Marathon Communications Group, Wausau WI

Mark IV Press, Hauppauge NY

Maverick Publications Inc, Bend OR

The Mazer Corporation, Dayton OH

McArdle Printing Co, Upper Marlboro MD

McClain Printing Company, Parsons WV

McQuiddy Printing Company, Nashville TN

Media Publications, Santa Clara CA

Mercury Printing Company, Memphis TN

Metromail Corporation, Lincoln NE

Moran Printing Company, Orlando FL

Moran Printing Inc, Baton Rouge LA

William H Muller Printing, Santa Clara CA

MultiPrint Co Inc, Skokie IL

National Graphics Corp, Columbus OH

Northlight Studio Press, Barre VT

Northwest Web, Eugene OR

Odyssey Press Inc, Dover NH

Omaha Printing Co, Omaha NE

OMNIPRESS, Madison WI

Optic Graphics, Glen Burnie MD

Original Copy Centers, Cleveland OH

Outstanding Graphics, Kenosha WI

The Oxford Group, Norway ME

Padgett Printing Corporation, Dallas TX

Page Litho Inc, Detroit MI

Parker Graphics, Fuquay-Varina NC

Patterson Printing Company, Benton Harbor MI

Paust Inc, Richmond IN

PFP Printing Corporation, Gaithersburg MD

Pine Hill Press Inc, Freeman SD

Plus Communications, Saint Louis MO

Port City Press Inc, Baltimore MD

Practical Graphics, New York NY

The Printing Company, Indianapolis IN

Printing Corporation of America, Pompano Beach FL

Printright International, Palmdale CA

Printwest, Regina SK

Progress Printing, Lynchburg VA

Publishers Express Press, Ladysmith WI

Publishers Press, Salt Lake City UT

Quebecor Printing (USA) Corp, Boston MA

Quebecor Printing Inc, Richmond Hill ON

Quebecor/Semline, Braintree MA

Repro-Tech, Fairfield CT

Ronalds Printing, Vancouver BC

Ronalds Printing, Montreal PQ

Roxbury Publishing Company, Los Angeles CA

Royle Printing, Sun Prairie WI

Sanders Printing Company, Garretson SD

The Saratoga Printing Co, Saratoga Springs NY

Schiff Printing, Pittsburgh PA

Science Press, Ephrata PA

Sexton Printing, Saint Paul MN

Sheridan Printing Company, Phillipsburg NJ

Sinclair Printing Co, Los Angeles CA

R E Smith Printing, Fall River MA

Sowers Printing Company, Lebanon PA

Speakes-Hines & Thomas Inc, Lansing MI

John Stark Printing Co Inc, St Louis MO

Stewart Publishing & Printing Co, Marble Hill MO

The Stinehour Press, Lunenburg VT

Straus Printing Company, Madison WI

Sun Graphix, Lewiston ME

TEACH Services, Brushton NY

Times Litho Inc, Forest Grove OR

Times Printing, Random Lake WI

Trade Litho Inc, Miami FL

Tri-Graphic Printing, Ottawa ON

Triangle Booklet Company, Nashville TN

TSO General Corporation, Long Island City NY

Tweddle Litho Co, Clinton Township MI

Valco Graphics, Seattle WA

Victor Graphics, Baltimore MD

Waldman Graphics, Pennsauken NJ

Waldon Press Inc, New York NY

Watt/Peterson, Plymouth MN

Waverly Press, Linthicum MD

Webcom Ltd, Scarborough ON

F A Weber & Sons, Park Falls WI

Westview Press, Boulder CO

Wickersham Printing Co Inc, Lancaster PA

The Wimmer Companies, Memphis TN

Wolfer Printing, Commerce CA

World Color Press Inc, New York NY

Y/Z Printing Co, Elizabethtown PA

Ye Olde Genealogie Shoppe, Indianapolis IN

Perfectbound Books

aBCD, Seattle WA
Academy Books, Rutland VT
Accurate Printing, Post Falls ID
Accurate Web, Central Islip NY
Action Printing, Fond Du Lac WI
Ad Infinitum Press, Mount Vernon NY
Adams Press, Chicago IL
Advanced Data Reproductions, Wichita KS
Advanced Duplicating & Printing, Minneapolis MN
Adviser Graphics, Red Deer AB
Allen Press Inc, Lawrence KS
American Printers & Litho, Niles IL
American Signature, Memphis Division, Olive Branch MS
American Signature/Foote, Atlanta GA
American Signature, Lincoln NE
American Signature Graphics, Dallas TX
American Web Inc, Denver CO
Americomp, Brattleboro VT
Anundsen Publishing Company, Decorah IA
Apollo Graphics Ltd, Southampton PA
The Argus Press Inc, Niles IL
Arizona Lithographers Inc, Tucson AZ
Artcraft Press, Waterloo WI
Associated Printers, Grafton ND
Automated Graphic Systems, White Plains MD
Baker Johnson Inc, Dexter MI
Bang Printing Co, Brainerd MN
Banta Company, Menasha WI
Banta Publications Group, Hinsdale IL
Banta-Harrisonburg, Harrisonburg VA
Barton Press Inc, West Orange NJ
Bawden Printing Inc, Eldridge IA
Beacon Press Inc, Richmond VA
Beacon Press, Seattle WA
Ben-Wal Printing, Pomona CA
Berryville Graphics, Berryville VA
Bertelsmann Printing & Manufacturing, New York NY
Best Gagne Book Manufacturers, Toronto ON
Blake Printery, San Luis Obispo CA
Blue Dolphin Press Inc, Grass Valley CA
Bolger Publications/Creative Printing, Minneapolis MN
The Book Press Inc, Brattleboro VT
Book-Mart Press Inc, North Bergen NJ
BookCrafters, Chelsea MI
BookCrafters/Frederick,sburg Fredericksburg VA
Booklet Publishing Company Inc, Elk Grove Village IL

BookMasters Inc, Ashland OH
Boyd Printing Co Inc, Albany NY
Braceland Brothers Inc, Philadelphia PA
Bradley Printing Company, Des Plaines IL
Braun-Brumfield Inc, Ann Arbor MI
Brennan Printing, Deep River IA
Brown Printing, Waseca MN
Brown Printing / Franklin, Franklin KY
Brown Printing/East Greenville, East Greenville PA
Brunswick Publishing Corp, Lawrenceville VA
R L Bryan Company, Columbia SC
BSC Litho, Harrisburg PA
Buse Printing & Advertising, Phoenix AZ
The William Byrd Press, Richmond VA
C & M Press, Denver CO
Cal Central Press, Sacramento CA
Calsonic Miura Graphics, Irvine CA
Camelot Publishing Company, Ormond Beach FL
Canterbury Press, Rome NY
Capital City Press, Montpelier VT
Capital Printing Company Inc, Austin TX
Carpenter Lithographic Co, Springfield OH
Champagne Offset Co Inc, Natick MA
Chicago Color Express, Chicago IL
Coach House Press, Toronto ON
The College Press, Collegedale TN
Color World Printers, Bozeman MT
ColorCorp Inc, Littleton CO
ColorGraphics Printing Corp, Tulsa OK
Colortone Press, Landover MD
Colotone Riverside Inc, Dallas TX
Columbus Bookbinders & Printers, Columbus GA
Combined Communication Services, Mendota IL
Comfort Printing, St Louis MO
Commercial Documentation Services, Medford OR
Community Press, Provo UT
Concord Litho Company Inc, Concord NH
Coneco Litho Graphics, Glen Falls NY
Consolidated Printers Inc, Berkeley CA
Cookbook Publishers Inc, Lenexa KS
Cookbooks by Mom's Press, Kearney NE
Corley Printing Company, Earth City MO
Country Press, Mohawk NY
Country Press Inc, Lakeville MA
Courier Corporation, Lowell MA
Courier Stoughton Inc, Stoughton MA
Courier-Kendallville Corp, Kendallville IN
Courier-Westford Inc, Westford MA
The Craftsman Press Inc, Seattle WA
Craftsman Printing, Charlotte NC

Perfectbound Books

Crane Duplicating Service, West
 Barnstable MA
Crusader Printing Co, East Saint Louis
 IL
Lew A Cummings Company Inc,
 Manchester NH
Cushing-Malloy, Ann Arbor MI
Daamen Printing Company, West Rut-
 land VT
Danbury Printing & Litho Inc, Danbury
 CT
Danner Press Corporation, Canton OH
Dartmouth Printing, Hanover NH
Data Reproductions Corp, Rochester
 Hills MI
Dataco A WBC Company, Albuquerque
 NM
Dellas Graphics, Syracuse NY
Delmar Printing & Publishing, Charlotte
 NC
Delta Lithograph Company, Valencia CA
Des Plaines Publishing Co, Des Plaines
 IL
Desaulniers Printing Company, Milan IL
Deven Lithographers, Brooklyn NY
Dharma Press, Oakland CA
Diamond Graphics Tech Data, Mil-
 waukee WI
Dickinson Press Inc, Grand Rapids MI
Direct Press/Modern Litho, Huntington
 Station NY
Directories America, Chicago IL
Dittler Brothers, Atlanta GA
Diversified Printing, Brea CA
Dollco Printing, Ottawa ON
R R Donnelley & Sons Company Inc,
 Crawfordsville IN
R R Donnelley & Sons Company Inc,
 Harrisonburg, Harrisonburg VA
R R Donnelley & Sons Company Inc,
 Willard, Willard OH
Dragon Press, Delta Junction AK
Dynagraphics, Carlsbad CA
Eastwood Printing Co, Denver CO
EBSCO Media, Birmingham AL
Edison Lithographing, North Bergen NJ
Editors Press Inc, Hyattsville MD
Edwards Brothers Inc, Ann Arbor MI
Edwards Brothers Inc, Lillington NC
Eerdmans Printing Co, Grand Rapids MI
Electronic Printing, Brentwood NY
EP Graphics, Berne IN
Eureka Printing Co Inc, Eureka CA
Eusey Press Inc, Leominster MA
Eva-Tone Inc, Clearwater FL
Evangel Press, Nappanee IN
Faculty Press Inc, Brooklyn NY
Federated Lithographers, Providence RI
First Impressions Printer & Lithog-
 raphers Inc, Elk Grove IL

Fleming Printing Co, St Louis MO
Flower City Printing Inc, Rochester NY
Focus Direct, San Antonio TX
The Forms Man Inc, Deer Park NY
Fort Orange Press Inc, Albany NY
Foster Printing Service, Michigan City IN
Friesen Printers, Altona MB
Frye & Smith, Costa Mesa CA
Futura Printing Inc, Boynton Beach FL
Gateway Press Inc, Louisville KY
George Lithograph, San Francisco CA
Geryon Press Limited, Tunnel NY
Gilliland Printing Inc, Arkansas City KS
Glundal Color Service, East Syracuse NY
Golden Belt Printing Inc, Great Bend KS
Golden Horn Press, Berkeley CA
Goodway Graphics of Virginia,
 Springfield VA
Gorham Printing, Rochester WA
Gowe Printing Company, Medina OH
Graftek Press, Woodstock IL
Graphic Arts Center, Portland OR
Graphic Litho Corporation, Lawrence MA
Graphic Printing, New Carlisle OH
Graphics Ltd, Indianapolis IN
Gray Printing, Fostoria OH
Great Impressions Printing & Graphics,
 Dallas TX
Greenfield Printing & Publishing, Green-
 field OH
Griffin Printing & Lithography, Glendale
 CA
Griffin Printing and Lithograph Com-
 pany, Sacramento CA
GRT Book Printing, Oakland CA
GTE Directories Printing Corp, Mt
 Prospect IL
GTE Directories Corp, Des Plaines IL
Gulf Printing, Houston TX
Guynes Printing Company, Albuquerque
 NM
Haddon Craftsmen, Scranton PA
Harlo Printing Co, Detroit MI
Harmony Printing, Liberty MO
Hart Graphics, Austin TX
Hart Graphics, Simpsonville SC
Hart Press, Long Prairie MN
Hawkes Publishing, Salt Lake City UT
Heartland Press Inc, Spencer IA
Henington Publishing Company, Wolfe
 City TX
Hennegan Company, Cincinnati OH
Heritage Printers Inc, Charlotte NC
D B Hess Company, Woodstock IL
Hignell Printing, Winnepeg MB
Hinz Lithographing Company, Mt
 Prospect IL
A B Hirschfeld Press, Denver CO
Hoechstetter Printing, Pittsburgh PA
Holladay-Tyler Printing Corp, Glenn
 Dale MD

Perfectbound Books

Rae Horowitz Book Manufacturers, Fairfield CT
Imperial Litho/Graphics Inc, Phoenix AZ
Independent Printing Company, New York NY
Independent Publishing Co, Sarasota FL
Intelligencer Printing, Lancaster PA
Interform Corporation, Bridgeville PA
IPC Publishing Services, Saint Joseph MI
Jersey Printing Company, Bayonne NJ
JK Creative Printers, Quincy IL
The Job Shop, Woods Hole MA
The Johnson & Hardin Company, Cincinnati OH
Jostens Printing & Publishing, Visalia CA
Jostens Printing & Publishing, Clarksville TN
Jostens Printing & Publishing, Winston-Salem NC
Jostens Printing & Publishing, Topeka KS
Jostens Printing & Publishing, State College PA
Jostens Printing & Publishing, Minneapolis MN
Judd's Incorporated, Washington DC
Judd's Incorporated, Strasburg VA
Julin Printing, Monticello IA
Kaufman Press Printing, Syracuse NY
Kimberly Press, Santa Barbara CA
Kirby Lithographic Co Inc, Arlington VA
KNI Incorporated, Anaheim CA
The C J Krehbiel Company, Cincinnati OH
La Crosse Graphics, LaCrosse WI
Lancaster Press, Lancaster PA
The Lane Press, Burlington VT
Lasky Company, Milburn NJ
Les Editions Marquis, Montmagny PQ
Litho Specialties, St Paul MN
Lithocolor Press, Westchester IL
Lithograph Printing Company, Memphis TN
Lithoid Printing Corporation, East Brunswick NJ
The Little River Press Inc, Miami FL
John D Lucas Printing Company, Baltimore MD
Mack Printing Group Inc, Easton PA
Mackintosh Typography Inc, Santa Barbara CA
The Mad Printers, Mattituck NY
Malloy Lithographing, Ann Arbor MI
Maple-Vail Book Manufacturing Group, York PA
Maquoketa Web Printing, Maquoketa IA
Marathon Communications Group, Wausau WI
Mark IV Press, Hauppauge NY
Marrakech Express, Tarpon Springs FL

Maverick Publications Inc, Bend OR
The Mazer Corporation, Dayton OH
MBP Lithographics, Hillsboro KS
McClain Printing Company, Parsons WV
McDowell Publications, Utica KY
The McFarland Press, Harrisburg PA
McKay Printing Services, Michigan City IN
McNaughton & Gunn Inc, Saline MI
McQuiddy Printing Company, Nashville TN
Meaker the Printer, Phoenix AZ
Media Publications, Santa Clara CA
Meehan-Tooker & Company, East Rutherford NJ
Mercury Printing Company, Memphis TN
Merrill Corporation, Los Angeles CA
Metromail Corporation, Lincoln NE
Mitchell Press, Vancouver BC
MMI Press, Harrisville NH
Monument Printers & Lithographer Inc, Verplanck NY
Moran Printing Inc, Baton Rouge LA
Morgan Printing, Austin TX
Morningrise Printing, Costa Mesa CA
William H Muller Printing, Santa Clara CA
MultiPrint Co Inc, Skokie IL
Murphy's Printing Co, Campbell CA
National Graphics Corp, Columbus OH
National Lithographers Inc, Miami FL
National Publishing Company, Philadelphia PA
National Reproductions Corp, Madison Heights MI
Naturegraph Publishers, Happy Camp CA
Neyenesch Printers, San Diego CA
Louis Neibauer Co Inc, Warminster PA
The Nielsen Lithographing Co, Cincinnati OH
Nimrod Press Inc, Boston MA
Noll Printing Company, Huntington IN
Northlight Studio Press, Barre VT
Northwest Web, Eugene OR
Nystrom Publishing Company, Osseo MN
Odyssey Press Inc, Dover NH
Offset Paperback Manufacturers, Dallas PA
Omaha Printing Co, Omaha NE
OMNIPRESS, Madison WI
Optic Graphics Inc, Glen Burnie MD
Outstanding Graphics, Kenosha WI
The Ovid Bell Press Inc, Fulton MO
The Oxford Group, Norway ME
Padgett Printing Corporation, Dallas TX
Page Litho Inc, Detroit MI
Panorama Press Inc, Clifton NJ
Pantagraph Printing, Bloomington IL
Paraclete Press, Brewster MA
Parker Graphics, Fuquay-Varina NC

Perfectbound Books

Patterson Printing Company, Benton Harbor MI
Paust Inc, Richmond IN
Peconic Companies, Mattituck NY
Pendell Printing, Midland MI
PennWell Printing, Tulsa OK
Penton Press Publishing, Berea OH
The Perlmuter Printing Co, Cleveland OH
Perry Printing Corp, Waterloo WI
Perry/Baraboo, Baraboo WI
PFP Printing Corporation, Gaithersburg MD
Phillips Brothers Printers Inc, Springfield IL
Pine Hill Press Inc, Freeman SD
Plus Communications, Saint Louis MO
Port City Press Inc, Baltimore MD
Port Publications, Port Washington WI
Practical Graphics, New York NY
The Press of Ohio, Kent OH
Princeton Academic Press, Lawrenceville NJ
Prinit Press, Dublin IN
Print Northwest, Tacoma WA
The Printer Inc, Osseo MN
The Printing Company, Indianapolis IN
Printing Corporation of America, Pompano Beach FL
Printright International, Palmdale CA
Printwest, Regina SK
Progress Printing, Lynchburg VA
Publishers Express Press, Ladysmith WI
Publishers Press, Salt Lake City UT
Publishers Printing Co, Shepherdsville KY
Quad Graphics — Thomaston Press, Thomaston GA
Quad/Graphics Inc, Saratoga Springs NY
Quad/Graphics, Lomira, Lomira WI
Quad/Graphics, Pewaukee, Pewaukee WI
Quad/Graphics, Sussex, Sussex WI
Quebecor Book Group, New York NY
Quebecor Printing (USA) Corp, Boston MA
Quebecor Printing Inc, Saint Paul MN
Quebecor Printing Inc, Richmond Hill ON
Quebecor Printing/St Cloud, Saint Cloud MN
Quebecor/Dallas, Dallas TX
Quebecor/Semline, Braintree MA
Quinn-Woodbine Inc, Woodbine NJ
RBW Graphics, Owen Sound ON
Repro-Tech, Fairfield CT
Rich Printing Company Inc, Nashville TN
Ringier America, Itasca IL
Ringier America/Brookfield, Brookfield WI

Ringier America/Dresden, Dresden TN
Ronalds Printing, Vancouver BC
Ronalds Printing, Montreal PQ
Rose Printing Company, Tallahassee FL
S Rosenthal & Company Inc, Cincinnati OH
Roxbury Publishing Company, Los Angeles CA
Royle Printing, Sun Prairie WI
The Saratoga Printing Co, Saratoga Springs NY
Schiff Printing, Pittsburgh PA
Schumann Printers, Fall River WI
Science Press, Ephrata PA
Service Printing Company, San Leandro CA
Sexton Printing, Saint Paul MN
Shepard Poorman Communications, Indianapolis IN
The Sheridan Press, Hanover PA
Sheridan Printing Company, Phillipsburg NJ
Sinclair Printing Co, Los Angeles CA
SLC Graphics, Pittston PA
R E Smith Printing, Fall River MA
Snohomish Publishing, Snohomish WA
Southeastern Printing Company Inc, Stuart FL
Sowers Printing Company, Lebanon PA
Speakes-Hines & Thomas Inc, Lansing MI
St Croix Press Inc, New Richmond WI
Standard Publishing, Cincinnati OH
Stephenson Lithograph Inc, Alexandria VA
Stevens Graphics, Atlanta GA
Stevens Graphics Inc, Birmingham AL
Stewart Publishing & Printing Co, Marble Hill MO
The Stinehour Press, Lunenburg VT
Strathmore Printing, Geneva IL
Straus Printing Company, Madison WI
The Studley Press, Dalton MA
The Sun Hill Press, North Brookfield MA
John S Swift Company Inc, St Louis MO
John S Swift/Chicago, Chicago IL
John S Swift/Cincinnati, Cincinnati OH
John S Swift/Teterboro, Teterboro NJ
TAN Books & Publishers Inc, Rockford IL
Taylor Publishing Company, Dallas TX
TEACH Services, Brushton NY
Technical Communication, North Kansas City MO
Thomson-Shore, Dexter MI
Times Litho Inc, Forest Grove OR
Times Printing, Random Lake WI
Todd Publications Inc, Smithville TX
Tracor Publications, Austin TX
Trade Litho Inc, Miami FL
Trend Offset Printing Service Inc, Los Alamitos CA

Perfectbound Books

Tri-Graphic Printing, Ottawa ON
Triangle Booklet Company, Nashville TN
TSO General Corporation, Long Island
    City NY
Tweddle Litho Co, Clinton Township MI
Universal Press, Providence RI
Universal Printers Ltd, Winnipeg MB
Universal Printing Company, Miami FL
Universal Printing Company, Saint
    Louis MO
Vail Ballou Press, Binghamton NY
Valco Graphics, Seattle WA
Van Volumes, Palmer MA
Vicks Lithograph & Printing, Yorkville
    NY
Victor Graphics, Baltimore MD
Viking Press, Eden Prairie MN
Vogue Printers, North Chicago IL
Waldman Graphics, Pennsauken NJ
Waldon Press Inc, New York NY
Wallace Press, Hillside IL
Walsworth Publishing Co Inc, Marceline
    MO
Watt/Peterson, Plymouth MN
Waverly Press, Linthicum MD
Webcom Ltd, Scarborough ON
Webcrafters, Madison WI
F A Weber & Sons, Park Falls WI
Weber Printing Company, Lincoln NE
Western Publishing Co Inc, Racine WI
Western Web Printing, Sioux Falls SD
Westview Press, Boulder CO
Whitehall Printing Company, Naples FL
Wickersham Printing Co Inc, Lancaster
    PA
The Wimmer Companies, Memphis TN
Windsor Associates, San Diego CA
Wisconsin Color Press Inc, Milwaukee WI
Wolfer Printing, Commerce CA
World Color Press Inc, New York NY
Worzalla Publishing Company, Stevens
    Point WI
Y/Z Printing Co, Elizabethtown PA

Spiral Bound Books

A-1 Business Service Inc, St Paul MN
aBCD, Seattle WA
Academy Books, Rutland VT
Accurate Printing, Post Falls ID
Action Printing, Fond Du Lac WI
Adams Press, Chicago IL
Advanced Data Reproductions, Wichita
    KS

Advanced Duplicating & Printing, Min-
    neapolis MN
Adviser Graphics, Red Deer AB
Apollo Graphics Ltd, Southampton PA
Automated Graphic Systems, White
    Plains MD
Bang Printing Co, Brainerd MN
Banta Company, Menasha WI
Bawden Printing Inc, Eldridge IA
Ben-Wal Printing, Pomona CA
Berryville Graphics, Berryville VA
Blue Dolphin Press Inc, Grass Valley CA
BookCrafters, Chelsea MI
Braun-Brumfield Inc, Ann Arbor MI
Brunswick Publishing Corp, Lawren-
    ceville VA
C & M Press, Denver CO
Caldwell Printers, Arcadia CA
Canterbury Press, Rome NY
Champagne Offset Co Inc, Natick MA
Chicago Color Express, Chicago IL
Colonial Graphics, Paterson NJ
ColorCorp Inc, Littleton CO
Colortone Press, Landover MD
Columbus Bookbinders & Printers,
    Columbus GA
Commercial Documentation Services,
    Medford OR
Community Press, Provo UT
Cookbooks by Mom's Press, Kearney NE
Cooley Printers & Officer Supply, Mon-
    roe LA
Courier Stoughton Inc, Stoughton MA
Craftsman Printing Co, Charlotte NC
Crane Duplicating Service, West
    Barnstable MA
Cushing-Malloy, Ann Arbor MI
Danner Press Corporation, Canton OH
Data Reproductions Corp, Rochester
    Hills MI
Dataco A WBC Company, Albuquerque
    NM
Dellas Graphics, Syracuse NY
Des Plaines Publishing Co, Des Plaines
    IL
Deven Lithographers, Brooklyn NY
Dharma Press, Oakland CA
Diamond Graphics Tech Data, Mil-
    waukee WI
Direct Press/Modern Litho, Huntington
    Station NY
Diversified Printing, Brea CA
R R Donnelley & Sons Company Inc,
    Crawfordsville IN
Eastwood Printing Co, Denver CO
Edwards Brothers Inc, Ann Arbor MI
Edwards Brothers Inc, Lillington NC
Eerdmans Printing Co, Grand Rapids MI
Evangel Press, Nappanee IN
Flower City Printing Inc, Rochester NY
Friesen Printers, Altona MB
Geiger Brothers, Lewiston ME

Spiral Bound Books

Glundal Color Service, East Syracuse NY
Graftek Press, Woodstock IL
Graphic Arts Center, Portland OR
Graphic Litho Corporation, Lawrence MA
Griffin Printing & Lithography, Glendale CA
GRT Book Printing, Oakland CA
Harmony Printing, Liberty MO
Henington Publishing Company, Wolfe City TX
Heritage Printers Inc, Charlotte NC
D B Hess Company, Woodstock IL
Hinz Lithographing Company, Mt Prospect IL
Rae Horowitz Book Manufacturers, Fairfield CT
Imperial Litho/Graphics Inc, Phoenix AZ
IPC Publishing Services, Saint Joseph MI
JK Creative Printers, Quincy IL
The Johnson & Hardin Company, Cincinnati OH
Jostens Printing & Publishing, Clarksville TN
Jostens Printing & Publishing, Topeka KS
Jostens Printing & Publishing, State College PA
Keys Printing Co Inc, Greenville SC
Kolor View Press, Aurora MO
The C J Krehbiel Company, Cincinnati OH
Litho Specialties, St Paul MN
Lithocolor Press, Westchester IL
Lithoid Printing Corporation, East Brunswick NJ
Mark IV Press, Hauppauge NY
Marrakech Express, Tarpon Springs FL
Maverick Publications Inc, Bend OR
The Mazer Corporation, Dayton OH
MBP Lithographics, Hillsboro KS
McGrew Color Graphics, Kansas City MO
McNaughton & Gunn Inc, Saline MI
Media Publications, Santa Clara CA
Merrill Corporation, Los Angeles CA
Monument Printers & Lithographer Inc, Verplanck NY
Moran Printing Company, Orlando FL
Morgan Printing, Austin TX
MultiPrint Co Inc, Skokie IL
National Lithographers Inc, Miami FL
Nimrod Press Inc, Boston MA
Nystrom Publishing Company, Osseo MN
Omaha Printing Co, Omaha NE
Outstanding Graphics, Kenosha WI
The Ovid Bell Press Inc, Fulton MO
Panorama Press Inc, Clifton NJ
Paraclete Press, Brewster MA
Pine Hill Press Inc, Freeman SD
Port Publications, Port Washington WI

The Press of Ohio, Kent OH
Princeton Academic Press, Lawrenceville NJ
Printing Corporation of America, Pompano Beach FL
Printwest, Regina SK
Publishers Express Press, Ladysmith WI
Publishers Press, Salt Lake City UT
Quebecor/Semline, Braintree MA
Quinn-Woodbine Inc, Woodbine NJ
Rose Printing Company, Tallahassee FL
S Rosenthal & Company Inc, Cincinnati OH
The Saratoga Printing Co, Saratoga Springs NY
Science Press, Ephrata PA
Service Printing Company, San Leandro CA
Shepard Poorman Communications, Indianapolis IN
The Sheridan Press, Hanover PA
R E Smith Printing, Fall River MA
Sowers Printing Company, Lebanon PA
John Stark Printing Co Inc, St Louis MO
The Studley Press, Dalton MA
Sun Graphix, Lewiston ME
John S Swift Company Inc, St Louis MO
John S Swift/Chicago, Chicago IL
John S Swift/Cincinnati, Cincinnati OH
John S Swift/Teterboro, Teterboro NJ
TAN Books & Publishers Inc, Rockford IL
Trade Litho Inc, Miami FL
Tri-Graphic Printing, Ottawa ON
Triangle Booklet Company, Nashville TN
TSO General Corporation, Long Island City NY
Vail Ballou Press, Binghamton NY
Van Volumes, Palmer MA
Versa Press Inc, East Peoria IL
Vicks Lithograph & Printing, Yorkville NY
Victor Graphics, Baltimore MD
Vogue Printers, North Chicago IL
Wagners Printers, Davenport IA
Waldon Press Inc, New York NY
Walsworth Publishing Co Inc, Marceline MO
Webcom Ltd, Scarborough ON
The Wimmer Companies, Memphis TN
Wolfer Printing, Commerce CA
World Color Press Inc, New York NY
Y/Z Printing Co, Elizabethtown PA
Ye Olde Genealogie Shoppe, Indianapolis IN

Wire-O Bound Books

A-1 Business Service Inc, St Paul MN
Action Printing, Fond Du Lac WI
Adams Press, Chicago IL
Advanced Data Reproductions, Wichita KS
Advanced Duplicating & Printing, Edina MN
AM Graphics & Printing, San Marcos CA
Apollo Graphics Ltd, Southampton PA
Banta Company, Menasha WI
Bawden Printing Inc, Eldridge IA
Book-Mart Press Inc, North Bergen NJ
BookMasters Inc, Ashland OH
Brennan Printing, Deep River IA
Canterbury Press, Rome NY
Champagne Offset Co Inc, Natick MA
Chicago Color Express, Chicago IL
Colonial Graphics, Paterson NJ
ColorCorp Inc, Littleton CO
Columbus Bookbinders & Printers, Columbus GA
Commercial Documentation Services, Medford OR
Coneco Litho Graphics, Glen Falls NY
Consolidated Printers Inc, Berkeley CA
Cookbook Publishers Inc, Lenexa KS
Courier Stoughton Inc, Stoughton MA
Cushing-Malloy, Ann Arbor MI
Dataco A WBC Company, Albuquerque NM
Delmar Printing & Publishing, Charlotte NC
Des Plaines Publishing Co, Des Plaines IL
Deven Lithographers, Brooklyn NY
Dharma Press, Oakland CA
Direct Press/Modern Litho, Huntington Station NY
R R Donnelley & Sons Company Inc, Crawfordsville IN
Eastwood Printing Co, Denver CO
Edwards Brothers Inc, Ann Arbor MI
Eva-Tone Inc, Clearwater FL
Flower City Printing Inc, Rochester NY
Friesen Printers, Altona MB
George Lithograph, San Francisco CA
Graphic Arts Center, Portland OR
Graphic Litho Corporation, Lawrence MA
Gray Printing, Fostoria OH
Griffin Printing & Lithograph, Glendale CA
Harmony Printing, Liberty MO
Hennegan Company, Cincinnati OH
Rae Horowitz Book Manufacturers, Fairfield CT
Imperial Litho/Graphics, Phoenix AZ

Independent Printing Company, New York NY
IPC Publishing Services, Saint Joseph MI
Kolor View Press, Aurora MO
The C J Krehbiel Company, Cincinnati OH
Lighthouse Press Inc, Manchester NH
Lithocolor Press, Westchester IL
Lithoid Printing Corporation, East Brunswick NJ
Marrakech Express, Tarpon Springs FL
The McFarland Press, Harrisburg PA
McGrew Color Graphics, Kansas City MO
McNaughton & Gunn Inc, Saline MI
Media Publications, Santa Clara CA
Merrill Corporation, Los Angeles CA
Moran Printing Company, Orlando FL
Morgan Printing, Austin TX
MultiPrint Co Inc, Skokie IL
National Lithographers Inc, Miami FL
Nystrom Publishing Company, Osseo MN
Odyssey Press Inc, Dover NH
Omaha Printing Co, Omaha NE
OMNIPRESS, Madison WI
Panorama Press Inc, Clifton NJ
Paraclete Press, Brewster MA
Patterson Printing Company, Benton Harbor MI
Pine Hill Press Inc, Freeman SD
The Press of Ohio, Kent OH
Princeton Academic Press, Lawrenceville NJ
Print Northwest, Tacoma WA
Quinn-Woodbine Inc, Woodbine NJ
Science Press, Ephrata PA
Shepard Poorman Communications, Indianapolis IN
Southeastern Printing Company Inc, Stuart FL
Sowers Printing Company, Lebanon PA
Stewart Publishing & Printing Co, Marble Hill MO
John S Swift Company Inc, St Louis MO
TAN Books & Publishers Inc, Rockford IL
Technical Communication Services, North Kansas City MO
Trade Litho Inc, Miami FL
Versa Press Inc, East Peoria IL
Waldon Press Inc, New York NY
Walsworth Publishing Co Inc, Marceline MO
The Wimmer Companies, Memphis TN
Wolfer Printing, Commerce CA
World Color Press Inc, New York NY
Y/Z Printing Co, Elizabethtown PA

Otabind

This newer method of binding is available through the following printers.

aBCD, Seattle WA
Action Printing, Fond Du Lac WI
Alan Lithograph Inc, Inglewood CA
Alcom Printing Group Inc, Bethlehem PA
All-Star Printing, Lansing MI
AM Graphics & Printing, San Marcos CA
American Litho, Hayward CA
Apollo Graphics Ltd, Southampton PA
Arcata Graphics/Fairfield, Fairfield PA
Austin Printing Co Inc, Akron OH
Bang Printing Co, Brainerd MN
Barton Press Inc, West Orange NJ
Bay Port Press, Chula Vista CA
Best Gagne Book Manufactures, Toronto ON
Brown Printing, Waseca MN
R L Bryan Company, Columbia SC
Buse Printing & Advertising, Phoenix AZ
The William Byrd Press, Richmond VA
Cal Central Press, Sacramento CA
Caldwell Printers, Arcadia CA
Capital City Press, Montpelier VT
Chicago Color Express, Chicago IL
Colotone Riverside Inc, Dallas TX
Comfort Printing, St Louis MO
Concord Litho Company Inc, Concord NH
Coneco Litho Graphics, Glen Falls NY
Delta Lithograph Company, Valencia CA
Desaulniers Printing Company, Milan IL
Dickinson Press Inc, Grand Rapids MI
Direct Press/Modern Litho, Huntington Station NY
Dollco Printing, Ottawa ON
Electronic Printing, Brentwood NY
Eureka Printing Co Inc, Eureka CA
Eusey Press Inc, Leominster MA
Eva-Tone Inc, Clearwater FL
Evangel Press, Nappanee IN
Faculty Press Inc, Brooklyn NY
Fleming Printing Co, St Louis MO
Flower City Printing Inc, Rochester NY
Gateway Press Inc, Louisville KY
Gorham Printing, Rochester WA
Gowe Printing Company, Medina OH
Graphic Ways Inc, Buena Park CA
Great Impressions Printing & Graphics, Dallas TX
Great Lakes Lithograph Co, Cleveland OH
Gregath Publishing Co, Wyandotte OK
GRIT Commercial Printing Services, Williamsport PA
Haddon Craftsmen, Scranton PA
Heritage Printers Inc, Charlotte NC
Interform Corporation, Bridgeville PA

IPC Publishing Services, Saint Joseph MI
Jostens Printing & Publishing, Minneapolis MN
The C J Krehbiel Company, Cincinnati OH
Litho Specialties, St Paul MN
Maquoketa Web Printing, Maquoketa IA
McArdle Printing Co, Upper Marlboro MD
McNaughton & Gunn Inc, Saline MI
McQuiddy Printing Company, Nashville TN
Media Publications, Santa Clara CA
MultiPrint Co Inc, Skokie IL
National Lithographers Inc, Miami FL
Northwest Web, Eugene OR
The Ovid Bell Press Inc, Fulton MO
The Oxford Group, Norway ME
Padgett Printing Corporation, Dallas TX
The Printing Company, Indianapolis IN
Quebecor Book Group, New York NY
Quebecor Printing (USA) Corp, Boston MA
Quebecor/Semline, Braintree MA
Rose Printing Company, Tallahassee FL
S Rosenthal & Company Inc, Cincinnati OH
Sanders Printing Company, Garretson SD
Sexton Printing, St Paul MN
Sheridan Printing Company, Phillipsburg NJ
Sinclair Printing Co, Los Angeles CA
Sowers Printing Company, Lebanon PA
Speakes-Hines & Thomas Inc, Lansing MI
The Stinehour Press, Lunenburg VT
Straus Printing Company, Madison WI
TEACH Services, Brushton NY
Times Litho Inc, Forest Grove OR
Trade Litho Inc, Miami FL
Tweddle Litho Co, Clinton Township MI
Versa Press Inc, East Peoria IL
Victor Graphics, Baltimore MD
Viking Press, Eden Prairie MN
Waldman Graphics, Pennsauken NJ
Waldon Press Inc, New York NY
Watt/Peterson, Plymouth MN
Westview Press, Boulder CO
Whitehall Printing Company, Naples FL
Wickersham Printing Co Inc, Lancaster PA
Wolfer Printing, Commerce CA
World Color Press Inc, New York NY
Worzalla Publishing Company, Stevens Point WI

Index: Printing Equipment

This index lists those printers which have the following printing equipment in-house. Since not all printers responded to our survey forms, this index lists only those printers which could be verified as having the printing equipment indexed here.

Cameron Belt Press Printers

Arcata Graphics/Fairfield, Fairfield PA
Banta Company, Menasha WI
Banta-Harrisonburg, Harrisonburg VA
The Book Press, Brattleboro VT
BookCrafters, Chelsea MI
BookCrafters/Frederick, Fredericksburg VA
Columbus Bookbinders & Printers, Columbus GA
R R Donnelley & Sons Inc Harrisonburg, Harrisonburg VA
Interstate Printing Company, Omaha NE
Long Island Web Printing, Jericho NY
Marathon Communications Group, Wausau WI
MMI Press, Harrisville NH
Paraclete Press, Brewster MA
Practical Graphics, New York NY
Quebecor Book Group, New York NY
Quebecor Printing (USA) Corp, Boston MA
Quebecor/Semline, Braintree MA
Western Publishing Co Inc, Racine WI

Gravure Printers

3PI Prepress, Herndon VA
Action Nicholson Color, Cleveland OH
American Color, Phoenix AZ
Applied Graphics Technologies, New York NY
Black Dot Graphics, Crystal Lake IL
Blanks Color Imaging, Dallas TX
Brown Printing, Waseca MN
Brown Printing / Franklin, Franklin KY
Capitol Engraving, Nashville TN
Carey Color, Medina OH
CGI, Burlington MA

Characters, Houston TX
Chroma-Graphics, Kansas City MO
Classic Litho Arts, Oak Park IL
Colorbrite, Minneapolis MN
Colorhouse, Minneapolis MN
Colotone Group, Branford CT
Designer Color Systems, St Louis MO
Dittler Brothers, Atlanta GA
Enteron Group, Chicago IL
Flying Color Graphics, Pontiac IL
Gamma One, North Haven CT
Graphic Art Service, Marietta GA
Imaging Systems, Collingswood NJ
Intaglio Vivi-Color, Schiller Park IL
Kreber Graphics, Columbus OH
Kwik International Color, New York NY
Lanman Companies, Washington DC
Magna Graphic, Lexington KY
MCP, Milwaukee WI
Midwest Litho Arts, Des Plaines IL
NEC, Nashville TN
One Color Communications, Emeryville CA
Phototype Color Graphics, Merchantville NJ
Potomac Graphic Industries, New York NY
Prepsat, Franklin KY
Quad/Graphics, Lomira, Lomira WI
Quality House of Graphic, Long Island City NY
Quebecor Printing (USA) Corp, Boston MA
Quebecor/Dallas, Dallas TX
Ringier America, Itasca IL
Ringier America Inc, Corinth MS
Stevenson Photo Color, Cincinnati OH
TSI Graphics, St Louis MO
Tukaiz Innovative, Franklin Park IL
Wace USA, Chicago IL

Letterpress Printers

A-1 Business Service Inc, St Paul MN
Ad Infinitum Press, Mount Vernon NY
Adviser Graphics, Red Deer AB
American Printers & Litho, Niles IL
Anundsen Publishing Company, Decorah IA
Austin Printing Co Inc, Akron OH
Bang Printing Co, Brainerd MN
Barton Press Inc, West Orange NJ
Bay State Press, Framingham MA
Beacon Press, Seattle WA
Berryville Graphics, Berryville VA
Blue Dolphin Press Inc, Grass Valley CA
Bradley Printing Company, Des Plaines IL
R L Bryan Company, Columbia SC
BSC Litho, Harrisburg PA
Buse Printing & Advertising, Phoenix AZ
Caldwell Printers, Arcadia CA
Capital Printing Company Inc, Austin TX
Champagne Offset Co Inc, Natick MA
Coach House Press, Toronto ON
Colotone Riverside Inc, Dallas TX
Columbus Bookbinders & Printers, Columbus GA
Comfort Printing, St Louis MO
Cooley Printers & Office Supply, Monroe LA
Country Press Inc, Lakeville MA
Crusader Printing Co, East Saint Louis IL
Delmar Printing & Publishing, Charlotte NC
Delprint Inc, Mt Prospect IL
Dollco Printing, Ottawa ON
EBSCO Media, Birmingham AL
Eureka Printing Co Inc, Eureka CA
Fisher Printing Co, Galion OH
Fleming Printing Co, St Louis MO
Focus Direct, San Antonio TX
Fort Orange Press Inc, Albany NY
Foster Printing Service, Michigan City IN
Geiger Brothers, Lewiston ME
Geryon Press Limited, Tunnel NY
Hart Graphics, Austin TX
Henington Publishing Company, Wolfe City TX
Heritage Printers Inc, Charlotte NC
Interform Corporation, Bridgeville PA
Interstate Printing Company, Omaha NE
Jersey Printing Company, Bayonne NJ
JK Creative Printers, Quincy IL
The Johnson & Hardin Company, Cincinnati OH
Kaufman Press Printing, Syracuse NY
Kolor View Press, Aurora MO
La Crosse Graphics, LaCrosse WI

Long Island Web Printing, Jericho NY
Mackintosh Typography Inc, Santa Barbara CA
MBP Lithographics, Hillsboro KS
McGill Jensen, St Paul MN
Mercury Printing Company, Memphis TN
Moran Printing Company, Orlando FL
Moran Printing Inc, Baton Rouge LA
Morgan Press, Dobbs Ferry NY
National Graphics Corp, Columbus OH
National Publishing Company, Philadelphia PA
Neyenesch Printers, San Diego CA
Omaha Printing Co, Omaha NE
Padgett Printing Corporation, Dallas TX
Paraclete Press, Brewster MA
Pentagram Press, Minneapolis MN
Pine Hill Press Inc, Freeman SD
Plain Talk Printing Co, Des Moines IA
Print Northwest, Tacoma WA
Printwest, Regina SK
Progress Printing, Lynchburg VA
Quebecor Printing (USA) Corp, Boston MA
RBW Graphics, Owen Sound ON
John Roberts Company, Minneapolis MN
Ronalds Printing, Vancouver BC
RPP Enterprises Inc, Libertyville IL
Sanders Printing Company, Garretson SD
The Saratoga Printing Co, Saratoga Springs NY
Sexton Printing, Saint Paul MN
R E Smith Printing, Fall River MA
Smith-Edwards-Dunlap, Philadelphia PA
Stevens Graphics, Atlanta GA
Stevens Graphics Inc, Birmingham AL
Stinehour Press, Lunenburg VT
Sullivan Graphics Inc, Brentwood TN
Sun Graphix, Lewiston ME
The Sun Hill Press, North Brookfield MA
Triangle Booklet Company, Nashville TN
Universal Printing Company, Saint Louis MO
Western Publishing Co Inc, Racine WI
White Arts Inc, Indianapolis IN
Wood & Jones Printers, Pasadena CA
Y/Z Printing Co, Elizabethtown PA

Sheetfed Offset Printers

3PI Prepress, Herndon VA
A-1 Business Service Inc, St Paul MN
Accu-Color, St Louis MO
Accurate Printing, Post Falls ID
Action Nicholson Color, Cleveland OH

Sheetfed Offset Printers

Action Printing, Fond Du Lac WI
Ad Infinitum Press, Mount Vernon NY
Adams Press, Chicago IL
Advanced Data Reproductions, Wichita KS
Advanced Duplicating & Printing, Minneapolis MN
Adviser Graphics, Red Deer AB
Alan Lithograph Inc, Inglewood CA
Alcom Printing Group Inc, Bethlehem PA
All-Star Printing, Lansing MI
Allen Press Inc, Lawrence KS
Allied Color, Guilford CT
AM Graphics & Printing, San Marcos CA
American Color, Phoenix AZ
American Litho, Hayward CA
American Printers & Litho, Niles IL
American Signature Memphis Division, Olive Branch MS
American Signature Graphics, Dallas TX
American Signature/Foote, Atlanta GA
Americomp, Brattleboro VT
Anundsen Publishing Company, Decorah IA
Apollo Graphics Ltd, Southampton PA
Applied Graphics Technologies, New York NY
The Argus Press Inc, Niles IL
Arizona Lithographers Inc, Tucson AZ
Artcraft Press, Waterloo WI
Associated Printers, Grafton ND
Austin Printing Co Inc, Akron OH
Automated Graphic Systems, White Plains MD
Autumn House Graphics, Hagerstown MD
Baker Johnson Inc, Dexter MI
Balzer/Shopes, Brisbane CA
Banta Company, Menasha WI
Banta Publications Group, Hinsdale IL
Banta-Harrisonburg, Harrisonburg VA
Barton Press Inc, West Orange NJ
Bawden Printing Inc, Eldridge IA
Bay Port Press, Chula Vista CA
Bay State Press, Framingham MA
Beacon Press Inc, Richmond VA
Beacon Press, Seattle WA
Berryville Graphics, Berryville VA
Bertelsmann Printing & Manufacturing, New York NY
Best Gagne Book Manufacturers, Toronto ON
Black Dot Graphics, Crystal Lake IL
Blanks Color Imaging, Dallas TX
Blue Dolphin Press Inc, Grass Valley CA
Bolger Publications/Creative Printing, Minneapolis MN
The Book Press Inc, Brattleboro VT
Book-Mart Press Inc, North Bergen NJ
BookCrafters, Chelsea MI

BookCrafters/Frederick, Fredericksburg VA
Booklet Publishing Company, Elk Grove Village IL
BookMasters Inc, Ashland OH
Boyd Printing Co Inc, Albany NY
Bradley Printing Company, Des Plaines IL
Braun-Brumfield Inc, Ann Arbor MI
Brennan Printing, Deep River IA
Brunswick Publishing Corp, Lawrenceville VA
R L Bryan Company, Columbia SC
BSC Litho, Harrisburg PA
Buse Printing & Advertising, Phoenix AZ
C & M Press, Denver CO
Cal Central Press, Sacramento CA
Caldwell Printers, Arcadia CA
Calsonic Miura Graphics, Irvine CA
Canterbury Press, Rome NY
Capital City Press, Montpelier VT
Capital Printing Company Inc, Austin TX
Capitol Engraving, Nashville TN
Carey Color, Medina OH
Carpenter Lithographic Company, Springfield OH
CGI, Burlington MA
Champagne Offset Co Inc, Natick MA
Characters, Houston TX
Chicago Color Express, Chicago IL
Chicago Color Express, Westmont IL
Chicago Press Corp, Chicago IL
Chroma-Graphics, Kansas City MO
Clarkwood Corporation, Totowa NJ
Classic Litho Arts, Oak Park IL
The College Press, Collegedale TN
Colonial Graphics, Paterson NJ
Color Express, Commerce CA
Color Express, Minneapolis MN
Color Express, Atlanta GA
Color Express, Irvine CA
Color Service, Monterey Park CA
Color World Printers, Bozeman MT
Colorbrite, Minneapolis MN
ColorCorp Inc, Littleton CO
ColorGraphics Printing Corp, Tulsa OK
Colorhouse, Minneapolis MN
Colorlith Corporation, Johnston RI
Colotone Group, Branford CT
Colotone Riverside Inc, Dallas TX
Columbus Bookbinders & Printers, Columbus GA
Combined Communication Services, Mendota IL
Comfort Printing, St Louis MO
Commercial Documentation Services, Medford OR
Community Press, Provo UT
Concord Litho Company Inc, Concord NH
Coneco Litho Graphics, Glen Falls NY
Consolidated Printers Inc, Berkeley CA

Sheetfed Offset Printers

Cookbook Publishers Inc, Lenexa KS
Cookbooks by Mom's Press, Kearney NE
Cooley Printers & Office Supply, Monroe LA
Corley Printing Company, Earth City MO
Country Press, Mohawk NY
Country Press Inc, Lakeville MA
Courier Stoughton Inc, Stoughton MA
Courier-Kendallville Corp, Kendallville IN
Courier-Westford Inc, Westford MA
The Craftsman Press Inc, Seattle WA
Craftsman Printing Co, Charlotte NC
Crane Duplicating Service, West Barnstable MA
Crusader Printing Co, East Saint Louis IL
Lew A Cummings Company, Manchester NH
Cushing-Malloy, Ann Arbor MI
Daamen Printing Company, West Rutland VT
Danbury Printing & Litho Inc, Danbury CT
Danner Press Corporation, Canton OH
Dartmouth Printing, Hanover NH
Data Reproductions Corp, Rochester Hills MI
Dataco A WBC Company, Albuquerque NM
Dellas Graphics, Syracuse NY
Delmar Printing & Publishing, Charlotte NC
Delprint Inc, Mt Prospect IL
Delta Lithograph Company, Valencia CA
Desaulniers Printing Company, Milan IL
Designer Color Systems, St Louis MO
Dharma Press, Oakland CA
Diamond Graphics Tech Data, Milwaukee WI
Dickinson Press Inc, Grand Rapids MI
Direct Press/Modern Litho, Huntington Station NY
Directories America, Chicago IL
Dittler Brothers, Atlanta GA
Diversified Printing, Brea CA
Dollco Printing, Ottawa ON
R R Donnelley & Sons Company Inc, Crawfordsville IN
R R Donnelley & Sons Company Inc Harrisonburg, Harrisonburg VA
R R Donnelley & Sons Company Inc Willard, Willard OH
Dragon Press, Delta Junction AK
Eastern Rainbow, Derry NH
Eastwood Printing Co, Denver CO
EBSCO Media, Birmingham AL
Edison Lithographing, North Bergen NJ
Editors Press Inc, Hyattsville MD
Edwards Brothers Inc, Ann Arbor MI

Eerdmans Printing Co, Grand Rapids MI
Electronic Printing, Brentwood NY
Enteron Group, Chicago IL
EP Graphics, Berne IN
EU Services, Rockville MD
Eureka Printing Co Inc, Eureka CA
Eusey Press Inc, Leominster MA
Eva-Tone Inc, Clearwater FL
Evangel Press, Nappanee IN
Faculty Press Inc, Brooklyn NY
Federated Lithographers, Providence RI
Fetter Printing, Louisville KY
First Impressions Printer & Lithographers Inc, Elk Grove IL
Fisher Printing Co, Galion OH
Fleming Printing Co, St Louis MO
Flower City Printing Inc, Rochester NY
Flying Color Graphics, Pontiac IL
Focus Direct, San Antonio TX
The Forms Man Inc, Deer Park NY
Fort Orange Press Inc, Albany NY
Foster Printing Service, Michigan City IN
Friesen Printers, Altona MB
Frye & Smith, Costa Mesa CA
Futura Printing Inc, Boynton Beach FL
Gamma One, North Haven CT
Gateway Press Inc, Louisville KY
Gaylord Printing, Detroit MI
Geiger Brothers, Lewiston ME
George Lithograph, San Francisco CA
Gilliland Printing Inc, Arkansas City KS
Glundal Color Service, East Syracuse NY
Golden Belt Printing Inc, Great Bend KS
Goodway Graphics of Virginia, Springfield VA
Gorham Printing, Rochester WA
Graftek Press, Woodstock IL
Graphic Art Service, Marietta GA
Graphic Arts Center, Portland OR
Graphic Arts Publishing, Boise ID
Graphic Litho Corporation, Lawrence MA
Graphic Printing, New Carlisle OH
Graphic Production Ctr, Norcross GA
Graphic Technology, Baltimore MD
Graphic Ways Inc, Buena Park CA
Graphics Express, Boston MA
Graphics Ltd, Indianapolis IN
Gray Printing, Fostoria OH
Great Impressions Printing & Graphics, Dallas TX
Great Lakes Lithograph Co, Cleveland OH
Greenfield Printing & Publishing, Greenfield OH
Gregath Publishing Co, Wyandotte OK
Griffin Printing & Lithograph, Glendale CA
Griffin Printing and Lithograph Company, Sacramento CA
GRIT Commercial Printing Services, Williamsport PA
GRT Book Printing, Oakland CA

Sheetfed Offset Printers

Gulf Printing, Houston TX
Guynes Printing Company, Albuquerque NM
Harlo Printing Co, Detroit MI
Harmony Printing, Liberty MO
Hart Graphics, Austin TX
Hawkes Publishing, Salt Lake City UT
Henington Publishing Company, Wolfe City TX
Hennegan Company, Cincinnati OH
D B Hess Company, Woodstock IL
Hi-Tech Color House, Chicago IL
Hignell Printing, Winnepeg MB
Hinz Lithographing Company, Mt Prospect IL
A B Hirschfeld Press, Denver CO
Hoechstetter Printing, Pittsburgh PA
Rae Horowitz Book Manufacturers, Fairfield NJ
Imaging Systems, Collingswood NJ
Independent Publishing Company, Sarasota FL
Intaglio Vivi-Color, Schiller Park IL
Intelligencer Printing, Lancaster PA
Interform Corporation, Bridgeville PA
Interstate Printing Company, Omaha NE
IPC Publishing Services, Saint Joseph MI
Ivy Hill Packaging, Louisville KY
Japs-Olson Company, Minneapolis MN
Jersey Printing Company, Bayonne NJ
JK Creative Printers, Quincy IL
The Job Shop, Woods Hole MA
Johns Byrne Co, Niles IL
The Johnson & Hardin Company, Cincinnati OH
Johnson Graphics, East Dubuque IL
Jostens Printing & Publishing, Visalia CA
Jostens Printing & Publishing, Clarksville TN
Jostens Printing & Publishing, Winston-Salem NC
Jostens Printing & Publishing, Topeka KS
Jostens Printing & Publishing, State College PA
Jostens Printing & Publishing, Minneapolis MN
Julin Printing, Monticello IA
Kaufman Press Printing, Syracuse NY
Keys Printing Co Inc, Greenville SC
Kirby Lithographic Co Inc, Arlington VA
KNI Incorporated, Anaheim CA
Kolor View Press, Aurora MO
Kordet Graphics Inc, Oceanside NY
Kreber Graphics, Columbus OH
The C J Krehbiel Company, Cincinnati OH
Kwik International Color, New York NY
La Crosse Graphics, LaCrosse WI
Lancaster Press, Lancaster PA

The Lane Press, Burlington VT
Lanman Companies, Washington DC
Lasky Company, Milburn NJ
Les Editions Marquis, Montmagny PQ
Lighthouse Press Inc, Manchester NH
Litho Productions Inc, Madison WI
Litho Specialties, St Paul MN
Lithocolor Press, Westchester IL
Lithograph Printing Company, Memphis TN
Lithoid Printing Corporation, East Brunswick NJ
John D Lucas Printing Company, Baltimore MD
Mack Printing Group Inc, Easton PA
Mackintosh Typography Inc, Santa Barbara CA
The Mad Printers, Mattituck NY
Magna Graphic, Lexington KY
Malloy Lithographing, Ann Arbor MI
Maple-Vail Book Manufacturing Group, York PA
Maquoketa Web Printing, Maquoketa IA
Marathon Communications Group, Wausau WI
Mark IV Press, Hauppauge NY
Marrakech Express, Tarpon Springs FL
Maverick Publications Inc, Bend OR
The Mazer Corporation, Dayton OH
MBP Lithographics, Hillsboro KS
McArdle Printing Co, Upper Marlboro MD
McClain Printing Company, Parsons WV
The McFarland Press, Harrisburg PA
McGill Jensen, St Paul MN
McKay Printing Services, Michigan City IN
McNaughton & Gunn Inc, Saline MI
MCP, Milwaukee WI
McQuiddy Printing Company, Nashville TN
Media Publications, Santa Clara CA
Meehan-Tooker & Company, East Rutherford NJ
Mercury Printing Company, Memphis TN
Merrill Corporation, Los Angeles CA
Metromail Corporation, Lincoln NE
Midwest Litho Arts, Des Plaines IL
Mitchell Graphics, Petosky MI
Mitchell Press, Vancouver BC
Moebius Printing, Milwaukee WI
Monument Printers & Lithographers Inc, Verplanck NY
Moran Printing Company, Orlando FL
Moran Printing Inc, Baton Rouge LA
Morgan Press, Dobbs Ferry NY
Morgan Printing, Austin TX
Morningrise Printing, Costa Mesa CA
Motheral Printing, Fort Worth TX
William H Muller Printing, Santa Clara CA

Sheetfed Offset Printers

Multi-Ad Services, Peoria IL
MultiPrint Co Inc, Skokie IL
National Graphics Corp, Columbus OH
National Lithographers Inc, Miami FL
National Reproductions Corp, Madison Heights MI
Nationwide Printing, Burlington KY
Naturegraph Publishers, Happy Camp CA
NEC, Nashville TN
Louis Neibauer Co Inc, Warminster PA
Newsfoto Publishing Company, San Angelo TX
Neyenesch Printers, San Diego CA
The Nielsen Lithographing Co, Cincinnati OH
Nies/Artcraft Printing Companies, Saint Louis MO
Nimrod Press Inc, Boston MA
Northeast Graphics, New York NY
Northlight Studio Press, Barre VT
Nystrom Publishing Company, Osseo MN
Oaks Printing Company, Bethlehem PA
Odyssey Press Inc, Dover NH
Offset Paperback Manufacturers, Dallas PA
Omaha Printing Co, Omaha NE
Optic Graphics Inc, Glen Burnie MD
Outstanding Graphics, Kenosha WI
The Ovid Bell Press Inc, Fulton MO
The Oxford Group, Norway ME
Padgett Printing Corporation, Dallas TX
Page Litho Inc, Detroit MI
Panel Prints Inc, Old Forge PA
Panorama Press Inc, Clifton NJ
Pantagraph Printing, Bloomington IL
Paraclete Press, Brewster MA
Parker Graphics, Fuquay-Varina NC
Paust Inc, Richmond IN
Pearl-Pressman-Liberty Communications Group, Philadelphia PA
Peconic Companies, Mattituck NY
Peerless Engraving, Little Rock AR
Pendell Printing, Midland MI
Penn Colour Graphics, Huntington Valley PA
PennWell Printing, Tulsa OK
Penton Press Publishing, Berea OH
The Perlmuter Printing Co, Cleveland OH
Petty Co Inc, Effingham IL
PFP Printing Corporation, Gaithersburg MD
Phillips Brothers Printers Inc, Springfield IL
Phototype Color Graphics, Merchantville NJ
Pinecliffe Printers, Shawnee OK
Pioneer Press, Wilmette IL
Plain Talk Printing Co, Des Moines IA

Plus Communications, Saint Louis MO
Port City Press Inc, Baltimore MD
Port Publications, Port Washington WI
Potomac Graphic Industries, New York NY
Practical Graphics, New York NY
Prepsat, Franklin KY
The Press of Ohio, Kent OH
The Press, Chanhassen MN
Princeton Academic Press, Lawrenceville NJ
Prinit Press, Dublin IN
Print Northwest, Tacoma WA
The Printing Company, Indianapolis IN
Printing Corporation of America, Pompano Beach FL
Printwest, Regina SK
Progress Printing, Lynchburg VA
ProLitho, Provo UT
Publishers Choice Reprints, The Woodlands TX
Publishers Express Press, Ladysmith WI
Publishers Press, Salt Lake City UT
Publishers Printing Co, Shepherdsville KY
Quality House of Graphic, Long Island City NY
Quebecor Book Group, New York NY
Quebecor Printing (USA) Corp, Boston MA
Quebecor Printing Inc, Saint Paul MN
Quebecor Printing Inc, Richmond Hill ON
Quebecor Printing/St Cloud, Saint Cloud MN
Quebecor/Semline, Braintree MA
Quinn-Woodbine Inc, Woodbine NJ
Rapidocolor Corporation, West Chester PA
RBW Graphics, Owen Sound ON
Repro-Tech, Fairfield CT
Rich Printing Company Inc, Nashville TN
Ringier America, Itasca IL
John Roberts Company, Minneapolis MN
Ronalds Printing, Vancouver BC
Ronalds Printing, Montreal PQ
Rose Printing Company, Tallahassee FL
S Rosenthal & Company Inc, Cincinnati OH
Roxbury Publishing Company, Los Angeles CA
Royle Printing, Sun Prairie WI
RPP Enterprises Inc, Libertyville IL
Sanders Printing Company, Garretson SD
The Saratoga Printing Co, Saratoga Springs NY
Schelin Printing Co, St Louis MO
Schiff Printing, Pittsburgh PA
Science Press, Ephrata PA

Sheetfed Offset Printers

Service Printing Company, San Leandro CA
Sexton Printing, Saint Paul MN
Shepard Poorman Communications, Indianapolis IN
The Sheridan Press, Hanover PA
Sheridan Printing Company, Phillipsburg NJ
Sinclair Printing Co, Los Angeles CA
R E Smith Printing, Fall River MA
Smith-Edwards-Dunlap, Philadelphia PA
Snohomish Publishing, Snohomish WA
Southeastern Printing Company Inc, Stuart FL
Sowers Printing Company, Lebanon PA
Speakes-Hines & Thomas,, Lansing MI
St Croix Press Inc, New Richmond WI
Standard Printing Service, Chicago IL
Standard Publishing, Cincinnati OH
John Stark Printing Co Inc, St Louis MO
Stephenson Lithograph Inc, Alexandria VA
Stevens Graphics, Atlanta GA
Stevens Graphics Inc, Birmingham AL
Stevenson Photo Color, Cincinnati OH
Stewart Publishing & Printing Co, Marble Hill MO
The Stinehour Press, Lunenburg VT
Straus Printing Company, Madison WI
The Studley Press, Dalton MA
Sundance Press, Tucson AZ
John S Swift Company Inc, St Louis MO
John S Swift/Chicago, Chicago IL
John S Swift/Cincinnati, Cincinnati OH
John S Swift/Teterboro, Teterboro NJ
TAN Books & Publishers Inc, Rockford IL
Tapco Inc, Pemberton NJ
Taylor Publishing Company, Dallas TX
TEACH Services, Brushton NY
Technical Communication Services, North Kansas City MO
Thomson-Shore, Dexter MI
Times Printing, Random Lake WI
Todd Publications Inc, Smithville TX
Tracor Publications, Austin TX
Trade Litho Inc, Miami FL
Tri-Graphic Printing, Ottawa ON
Triangle Booklet Company, Nashville TN
TSI Graphics, St Louis MO
TSO General Corporation, Long Island City NY
Tukaiz Innovative, Franklin Park IL
Tweddle Litho Co, Clinton Township MI
Universal Press, Providence RI
Universal Printing Company, Saint Louis MO
US Press Inc, Valdosta GA
Vail Ballou Press, Binghamton NY
Valco Graphics, Seattle WA

Van Volumes, Palmer MA
Versa Press Inc, East Peoria IL
Vicks Lithograph & Printing, Yorkville NY
Victor Graphics, Baltimore MD
Viking Press, Eden Prairie MN
Vogue Printers, North Chicago IL
Wace USA, Chicago IL
Wagners Printers, Davenport IA
Waldman Graphics, Pennsauken NJ
Waldon Press Inc, New York NY
Wallace Press, Hillside IL
Walsworth Publishing Co Inc, Marceline MO
Watkins Printing Company, Columbus OH
Watt/Peterson, Plymouth MN
Waverly Press, Linthicum MD
Webcom Ltd, Scarborough ON
Webcrafters, Madison WI
F A Weber & Sons, Park Falls WI
Weber Printing Company, Lincoln NE
Western Publishing Co Inc, Racine WI
Western Web Printing, Sioux Falls SD
Westview Press, Boulder CO
White Arts Inc, Indianapolis IN
Whitehall Printing Company, Naples FL
Wickersham Printing Co Inc, Lancaster PA
Windsor Associates, San Diego CA
Winter Sun Inc, Phoenix AZ
Wood & Jones Printers, Pasadena CA
The Wood Press, Paterson NJ
World Color Press Inc, New York NY
Worzalla Publishing Company, Stevens Point WI
Y/Z Printing Co, Elizabethtown PA

Web Offset Printers

3PI Prepress, Herndon VA
Accu-Color, St Louis MO
Accurate Web, Central Islip NY
Action Nicholson Color, Cleveland OH
Action Printing, Fond Du Lac WI
Advanced Duplicating & Printing, Minneapolis MN
Adviser Graphics, Red Deer AB
Alan Lithograph Inc, Inglewood CA
Alcom Printing Group Inc, Bethlehem PA
Alden Press, Elk Grove Village IL
Alden Press/Wm Feathers Printers, Oberlin OH
Allied Color, Guilford CT
Alonzo Printing, Hayward CA
American Color, Phoenix AZ

Web Offset Printers

American Litho, Hayward CA
American Signature Memphis Division, Olive Branch MS
American Signature Graphics, Dallas TX
American Signature/Foote, Atlanta GA
American Signature, Lincoln NE
American Web Inc, Denver CO
Amidon Associated Inc, Saint Paul MN
Applied Graphics Technologies, New York NY
Arandell Corporation, Menomonee Falls WI
Arcata Graphics/Fairfield, Fairfield PA
Artcraft Press, Waterloo WI
Associated Printers, Grafton ND
B & W Press Inc, Georgetown MA
Balzer/Shopes, Brisbane CA
Baniff Direct Mail Printers, Norwalk CT
Banta Company, Menasha WI
Banta Publications Group, Hinsdale IL
Banta-Harrisonburg, Harrisonburg VA
Barton Press Inc, West Orange NJ
Bawden Printing Inc, Eldridge IA
Bay Port Press, Chula Vista CA
Bay State Press, Framingham MA
Beacon Press Inc, Richmond VA
Ben-Wal Printing, Pomona CA
Berryville Graphics, Berryville VA
Bertelsmann Printing & Manufacturing, New York NY
Best Gagne Book Manufacturers, Toronto ON
Black Dot Graphics, Crystal Lake IL
Blanks Color Imaging, Dallas TX
The Book Press, Brattleboro VT
BookCrafters, Chelsea MI
Boyd Printing Co Inc, Albany NY
Braceland Brothers Inc, Philadelphia PA
Bradley Printing Company, Des Plaines IL
Brown Printing, Waseca MN
Brown Printing/East Greenville Division, East Greenville PA
BSC Litho, Harrisburg PA
Buse Printing & Advertising, Phoenix AZ
The William Byrd Press, Richmond VA
Cal Central Press, Sacramento CA
California Offset Printers, Glendale CA
Capitol Engraving, Nashville TN
Carey Color, Medina OH
Carlson Color Graphics, Ocala FL
Carpenter Lithographic Co, Springfield OH
CGI, Burlington MA
Champion Printing, Cincinnati OH
Characters, Houston TX
Chroma-Graphics, Kansas City MO
Citizen Printing, Beaver Dam WI
Clarkwood Corporation, Totowa NJ
Classic Litho Arts, Oak Park IL

Colonial Graphics, Paterson NJ
Color Service, Monterey Park CA
Colorbrite, Minneapolis MN
ColorGraphics Printing Corp, Tulsa OK
Colorhouse, Minneapolis MN
Colorlith Corporation, Johnston RI
Colotone Group, Branford CT
Columbus Bookbinders & Printers, Columbus GA
Combined Communication Services, Mendota IL
Comfort Printing, St Louis MO
Commercial Documention Services, Medford OR
Communicolor, Newark OH
Concord Litho Company Inc, Concord NH
Consolidated Printers Inc, Berkeley CA
Continental Web, Itasca IL
Cookbook Publishers Inc, Lenexa KS
Cookbooks by Mom's Press, Kearney NE
Corley Printing Company, Earth CityMO
Courier Stoughton Inc, Stoughton MA
Courier-Kendallville Corp, Kendallville IN
Courier-Westford Inc, Westford MA
The Craftsman Press Inc, Seattle WA
Crusader Printing Co, East Saint Louis IL
Danbury Printing & Litho Inc, Danbury CT
Danner Press Corporation, Canton OH
Dartmouth Printing, Hanover NH
Dataco A WBC Company, Albuquerque NM
Delta Lithograph Company, Valencia CA
Des Plaines Publishing Co, Des Plaines IL
Desaulniers Printing Company, Milan IL
Designer Color Systems, St Louis MO
Deven Lithographers, Brooklyn NY
Dickinson Press Inc, Grand Rapids MI
The Dingley Press, Lisbon ME
Direct Graphics Inc, Sidney OH
Direct Press/Modern Litho, Huntington Station NY
Directories America, Chicago IL
Dittler Brothers, Atlanta GA
Diversified Printing, Brea CA
Dollco Printing, Ottawa ON
Donihe Graphics Inc, Kingsport TN
R R Donnelley & Sons Company Inc, Crawfordsville IN
R R Donnelley & Sons Company Inc Harrisonburg, Harrisonburg VA
R R Donnelley & Sons Company Inc Willard, Willard OH
Dynagraphics, Carlsbad CA
E & D Web Inc, Cicero IL
Eagle Web Press, Salem OR
Eastern Rainbow, Derry NH
Eastwood Printing Co, Denver CO

Web Offset Printers

Editors Press Inc, Hyattsville MD
Edwards Brothers Inc, Ann Arbor MI
Eerdmans Printing Co, Grand Rapids MI
Enteron Group, Chicago IL
EP Graphics, Berne IN
EU Services, Rockville MD
Eusey Press Inc, Leominster MA
Faculty Press Inc, Brooklyn NY
Fetter Printing, Louisville KY
First Impressions Printer & Lithographers Inc, Elk Grove IL
Fisher Printers Inc, Cedar Rapids IA
Fisher Printing Co, Galion OH
Fleming Printing Co, St Louis MO
Flying Color Graphics, Pontiac IL
Focus Direct, San Antonio TX
The Forms Man Inc, Deer Park NY
Frye & Smith, Costa Mesa CA
Gamma One, North Haven CT
Gateway Press Inc, Louisville KY
Gaylord Printing, Detroit MI
Geiger Brothers, Lewiston ME
George Lithograph, San Francisco CA
Glundal Color Service, East Syracuse NY
Goodway Graphics of Virginia, Springfield VA
Gowe Printing Company, Medina OH
Graftek Press, Woodstock IL
Graphic Art Service, Marietta GA
Graphic Arts Center, Portland OR
Graphic Arts Publishing, Boise ID
Graphic Technology, Baltimore MD
Graphics Express, Boston MA
Gray Printing, Fostoria OH
Great Impressions Printing & Graphics, Dallas TX
Great Lakes Lithograph Co, Cleveland OH
Greenfield Printing & Publishing, Greenfield OH
Gregath Publishing Co, Wyandotte OK
Griffin Printing & Lithograph, Glendale CA
Griffin Printing and Lithograph Company, Sacramento CA
GRIT Commercial Printing Services, Williamsport PA
GRT Book Printing, Oakland CA
GTE Directories Printing Corp, Mt Prospect IL
GTE Directories Corp, Des Plaines IL
Gulf Printing, Houston TX
Haddon Craftsmen, Scranton PA
Harlo Printing Co, Detroit MI
Harmony Printing, Liberty MO
Hart Graphics, Austin TX
Hart Graphics, Simpsonville SC
Hart Press, Long Prairie MN
Hawkes Publishing, Salt Lake City UT
Heartland Press Inc, Spencer IA

Hennegan Company, Cincinnati OH
D B Hess Company, Woodstock IL
Hinz Lithographing Company, Mt Prospect IL
A B Hirschfeld Press, Denver CO
Holladay-Tyler Printing Corp, Glenn Dale MD
Horan Engraving, New York NY
Rae Horowitz Book Manufacturers, Fairfield NJ
Imaging Systems, Collingswood NJ
Imperial Litho/Graphics Inc, Phoenix AZ
Instant Web, Chanhassen MN
Intaglio Vivi-Color, Schiller Park IL
Intelligencer Printing, Lancaster PA
Interstate Printing Company, Omaha NE
IPC Publishing Services, Saint Joseph MI
Japs-Olson Company, Minneapolis MN
The Johnson & Hardin Company, Cincinnati OH
Kaufman Press Printing, Syracuse NY
KNI Incorporated, Anaheim CA
Kreber Graphics, Columbus OH
The C J Krehbiel Company, Cincinnati OH
Kwik International Color, New York NY
Lancaster Press, Lancaster PA
The Lane Press, Burlington VT
Lanman Companies, Washington DC
Lasky Company, Milburn NJ
Lehigh Press Inc, Cherry Hill NJ
Lehigh Press/Cadillac, Broadview IL
Lighthouse Press Inc, Manchester NH
Litho Specialties, St Paul MN
Lithocolor Press, Westchester IL
Lithograph Printing, Memphis TN
Lithoid Printing Corporation, East Brunswick NJ
Long Island Web Printing, Jericho NY
John D Lucas Printing Company, Baltimore MD
Mack Printing Group Inc, Easton PA
Magna Graphic, Lexington KY
Mail-O-Graph, Kewanee IL
Malloy Lithographing, Ann Arbor MI
Maple-Vail Book ManufacturingGroup, York PA
Maquoketa Web Printing, Maquoketa IA
Mars Graphic Services, Westville NJ
The Mazer Corporation, Dayton OH
McArdle Printing Co, Upper Marlboro MD
McGill Jensen, St Paul MN
McKay Printing Services, Michigan City IN
McNaughton & Gunn Inc, Saline MI
MCP, Milwaukee WI
McQuiddy Printing Company, Nashville TN
Media Printing, Miami FL
Media Publications, Santa Clara CA

Web Offset Printers

Meehan-Tooker & Company, East Rutherford NJ
Merrill Corporation, Los Angeles CA
Metromail Corporation, Lincoln NE
Metroweb, Erlanger KY
Midwest Litho Arts, Des Plaines IL
Mitchell Press, Vancouver BC
Moebius Printing, Milwaukee WI
Motheral Printing, Fort Worth TX
William H Muller Printing, Santa Clara CA
Multi-Ad Services, Peoria IL
National Graphics Corp, Columbus OH
National Publishing Company, Philadelphia PA
NEC, Nashville TN
Neyenesch Printers, San Diego CA
The Nielsen Lithographing Co, Cincinnati OH
Nies/Artcraft Printing Companies, Saint Louis MO
Noll Printing Company, Huntington IN
Northeast Graphics, New York NY
Northeast Web Printing Inc, Farmingdale NY
Northprint International Inc, Grand Rapids MN
Northwest Web, Eugene OR
Offset Paperback Manufacturers, Dallas PA
Omaha Printing Co, Omaha NE
One Color Communications, Emeryville CA
Optic Graphics Inc, Glen Burnie MD
The Oxford Group, Norway ME
Padgett Printing Corporation, Dallas TX
Page Litho Inc, Detroit MI
Pantagraph Printing, Bloomington IL
Park Press, South Holland IL
Parker Graphics, Fuquay-Varina NC
Patterson Printing Company, Benton Harbor MI
Peerless Engraving, Little Rock AR
Pendell Printing, Midland MI
PennWell Printing, Tulsa OK
Penton Press Publishing, Berea OH
The Perlmuter Printing Co, Cleveland OH
Perry Printing Corp, Waterloo WI
Perry/Baraboo, Baraboo WI
Petty Co Inc, Effingham IL
PFP Printing Corporation, Gaithersburg MD
Phillips Brothers Printers Inc, Springfield IL
Phototype Color Graphics, Merchantville NJ
Pioneer Press, Wilmette IL
Plus Communications, Saint Louis MO
Port City Press Inc, Baltimore MD

Port Publications, Port Washington WI
Potomac Graphic Industries, New York NY
Practical Graphics, New York NY
Prepsat, Franklin KY
The Press of Ohio, Kent OH
The Press, Chanhassen MN
Print Northwest, Tacoma WA
The Printer Inc, Osseo MN
Printing Corporation of America, Pompano Beach FL
Printwest, Regina SK
Progress Printing, Lynchburg VA
ProLitho, Provo UT
Promotional Printing Corporation, Houston TX
Publishers Press, Salt Lake City UT
Quad Graphics—Thomaston, Thomaston GA
Quad/Graphics Inc, Saratoga Springs NY
Quad/Graphics, Lomira, Lomira WI
Quad/Graphics, Pewaukee, Pewaukee WI
Quad/Graphics, Sussex, Sussex WI
Quality House of Graphic, Long Island City NY
Quebecor Book Group, New York NY
Quebecor Printing (USA) Corp, Boston MA
Quebecor Printing Inc, Saint Paul MN
Quebecor Printing Inc, Richmond Hill ON
Quebecor Printing/St Cloud, Saint Cloud MN
Quebecor/Dallas, Dallas TX
Quebecor/Semline, Braintree MA
RBW Graphics, Owen Sound ON
Rich Printing Company Inc, Nashville TN
Ringier America, Itasca IL
Ringier America Inc, Corinth MS
Ringier America/Brookfield, Brookfield WI
Ringier America/Jonesboro, Jonesboro AR
Ringier America/Phoenix, Phoenix AZ
John Roberts Company, Minneapolis MN
Ronalds Printing, Vancouver BC
Ronalds Printing, Montreal PQ
Rose Printing Company, Tallahassee FL
S Rosenthal & Company Inc, Cincinnati OH
Roxbury Publishing Company, Los Angeles CA
Royle Printing, Sun Prairie WI
RPP Enterprises Inc, Libertyville IL
Schumann Printers, Fall River WI
Science Press, Ephrata PA
Service Web Offset Corporation, Chicago IL
Shepard Poorman Communications, Indianapolis IN
Sinclair Printing Co, Los Angeles CA

Web Offset Printers

SLC Graphics, Pittston PA
R E Smith Printing, Fall River MA
Smith-Edwards-Dunlap, Philadelphia PA
Snohomish Publishing, Snohomish WA
Sowers Printing Company, Lebanon PA
Speakes-Hines & Thomas, Lansing MI
Spencer Press, Wells ME
St Croix Press Inc, New Richmond WI
Standard Printing Service, Chicago IL
Standard Publishing, Cincinnati OH
John Stark Printing Co Inc, St Louis MO
Stephenson Lithograph Inc, Alexandria
     VA
Stevens Graphics, Atlanta GA
Stevenson Photo Color, Cincinnati OH
Strathmore Printing, Geneva IL
Straus Printing Company, Madison WI
Sullivan Graphics Inc, Brentwood TN
Sun Graphix, Lewiston ME
John S Swift Company Inc, St Louis MO
John S Swift/Chicago, Chicago IL
John S Swift/Cincinnati, Cincinnati OH
John S Swift/Teterboro, Teterboro NJ
Tapco Inc, Pemberton NJ
TEACH Services, Brushton NY
Times Litho Inc, Forest Grove OR
Times Printing, Random Lake WI
Todd Publications Inc, Smithville TX
Trend Offset Printing Service Inc, Los
     Alamitos CA
Tri-Graphic Printing, Ottawa ON
Triangle Booklet Company, Nashville TN
TSI Graphics, St Louis MO
Tukaiz Innovative, Franklin Park IL
Tweddle Litho Co, Clinton Township MI
Universal Press, Providence RI
Universal Printing Company, Miami FL
Universal Printing Company, Saint
     Louis MO
Vail Ballou Press, Binghamton NY
Valco Graphics, Seattle WA
Vicks Lithograph & Printing, Yorkville
     NY
Victor Graphics, Baltimore MD
Viking Press, Eden Prairie MN
Wace USA, Chicago IL
Wagners Printers, Davenport IA
Waldon Press Inc, New York NY
Wallace Press, Hillside IL
Watkins Printing Company, Columbus
     OH
Waverly Press, Linthicum MD
Web Specialties, Wheeling IL
Webcom Ltd, Scarborough ON
Webcraft Technologies, North Brunswick
     NJ
Webcrafters, Madison WI
The Wessel Company Inc, Elk Grove Vil-
     lage IL
Western Publishing Co Inc, Racine WI
Western Web Printing, Sioux Falls SD
Whitehall Printing Company, Naples FL
Wickersham Printing Co Inc, Lancaster
     PA
The Wimmer Companies, Memphis TN
Windsor Associates, San Diego CA
Wisconsin Color Press Inc, Milwaukee WI
Wolfer Printing, Commerce CA
The Wood Press, Paterson NJ
World Color Press, New York NY
Worzalla Publishing Company, Stevens
     Point WI
Henry Wurst Inc, North Kansas City MO

Index: Printing Services

These indexes list those printers who provide additional services such as color separations, typesetting, warehousing, etc. Separate indexes note those printers who stock acid free and/or recycled paper.

4-Color Printers

Action Printing, Fond Du Lac WI
Ad Infinitum Press, Mount Vernon NY
Adams Press, Chicago IL
Adviser Graphics, Red Deer AB
Alan Lithograph Inc, Inglewood CA
Alcom Printing Group Inc, Bethlehem PA
Alden Press, Elk Grove Village IL
Alden Press/Wm Feathers Printer, Oberlin OH
All-Star Printing, Lansing MI
Allen Press Inc, Lawrence KS
Alonzo Printing, Hayward CA
AM Graphics & Printing, San Marcos CA
American Litho, Hayward CA
American Printers & Litho, Niles IL
American Signature Memphis Division, Olive Branch MS
American Signature Graphics, Dallas TX
American Signature/Foote, Atlanta GA
American Signature, Lincoln NE
American Web Inc, Denver CO
Amidon & Associates Inc, St Paul MN
Anundsen Publishing Company, Decorah IA
Apollo Graphics Ltd, Southampton PA
Arandell Corporation, Menomonee Falls WI
The Argus Press Inc, Niles IL
Arizona Lithographers Inc, Tucson AZ
Artcraft Press, Waterloo WI
Associated Printers, Grafton ND
Austin Printing Co Inc, Akron OH
Automated Graphic System, White Plains MD
B & W Press Inc, Georgetown MA
Baker Johnson Inc, Dexter MI
Bang Printing Co, Brainerd MN
Banta Company, Menasha WI
Banta Publications Group, Hinsdale IL
Banta-Harrisonburg, Harrisonburg VA
Barton Press Inc, West Orange NJ
Bay Port Press, Chula Vista CA
Bay State Press, Framingham MA
Beacon Press Inc, Richmond VA
Beacon Press, Seattle WA
Ben-Wal Printing, Pomona CA
Berryville Graphics, Berryville VA
Bertelsmann Printing & Manufacturing, New York NY
Blake Printery, San Luis Obispo CA
Blue Dolphin Press Inc, Grass Valley CA
Bolger Publications/Creative Printing, Minneapolis MN
Book-Mart Press Inc, North Bergen NJ
BookCrafters, Chelsea MI
BookCrafters/Frederick, Fredericksburg VA
BookMasters Inc, Ashland OH
Bradley Printing Company, Des Plaines IL
Brookshore Lithographers Inc, Elk Grove Village IL
Brown Printing, Waseca MN
Brown Printing/Franklin, Franklin KY
Brown Printing/East Greenville Division, East Greenville PA
R L Bryan Company, Columbia SC
BSC Litho, Harrisburg PA
Buse Printing & Advertising, Phoenix AZ
The William Byrd Press, Richmond VA
Cal Central Press, Sacramento CA
Caldwell Printers, Arcadia CA
California Offset Printers, Glendale CA
Calsonic Miura Graphics, Irvine CA
Canterbury Press, Rome NY
Capital City Press, Montpelier VT
Capital Printing Company Inc, Austin TX
Carlson Color Graphics, Ocala FL
Carpenter Lithographic Co, Springfield OH
Catalog King, Clifton NJ
Champagne Offset Co Inc, Natick MA
Chicago Color Express, Chicago IL
Chicago Press Corp, Chicago IL
Citizen Printing, Beaver Dam WI
Coach House Press, Toronto ON
The College Press, Collegedale TN
Color Express, Commerce CA
Color Express, Minneapolis MN
Color Express, Atlanta GA
Color Express, Irvine CA
Color World Printers, Bozeman MT

4-Color Printers

ColorCorp Inc, Littleton CO
ColorGraphics Printing Corp, Tulsa OK
Colorlith Corporation, Johnston RI
Colortone Press, Landover MD
Colotone Riverside Inc, Dallas TX
Columbus Bookbinders & Printers,
    Columbus GA
Combined Communication Services,
    Mendota IL
Comfort Printing, St Louis MO
Commercial Documentation Services,
    Medford OR
Communicolor, Newark OH
Community Press, Provo UT
Concord Litho Company Inc, Concord
    NH
Coneco Litho Graphics, Glen Falls NY
Continental Web, Itasca IL
Cookbooks by Mom's Press, Kearney NE
Cooley Printers & Office Supply, Monroe
    LA
Coral Graphic Services, Plainview NY
Corley Printing Company, Earth City MO
Country Press, Mohawk NY
Country Press Inc, Lakeville MA
The Craftsman Press Inc, Seattle WA
Crane Duplicating Service, West
    Barnstable MA
Crusader Printing Co, East Saint Louis
    IL
Cushing-Malloy, Ann Arbor MI
Danbury Printing & Litho Inc, Danbury
    CT
Danner Press Corporation, Canton OH
Dartmouth Printing, Hanover NH
Data Reproductions Corp, Rochester
    Hills MI
Dataco A WBC Company, Albuquerque
    NM
Dayal Graphics, Lemont IL
Dellas Graphics, Syracuse NY
Delmar Printing & Publishing, Charlotte
    NC
Delta Lithograph Company, Valencia CA
Des Plaines Publishing Co, Des Plaines
    IL
Desaulniers Printing Company, Milan IL
Deven Lithographers, Brooklyn NY
Dharma Press, Oakland CA
Diamond Graphics Tech Data, Mil-
    waukee WI
Dickinson Press Inc, Grand Rapids MI
The Dingley Press, Lisbon ME
Direct Graphics Inc, Sidney OH
Direct Press/Modern Litho, Huntington
    Station NY
Directories America, Chicago IL
Dollco Printing, Ottawa ON
Donihe Graphics Inc, Kingsport TN

R R Donnelley & Sons Company Inc,
    Crawfordsville IN
R R Donnelley & Sons Company Inc,
    Harrisonburg, Harrisonburg VA
R R Donnelley & Sons Company Inc,
    Willard, Willard OH
Dragon Press, Delta Junction AK
Dynagraphics, Carlsbad CA
E & D Web, Cicero IL
Eastwood Printing Co, Denver CO
EBSCO Media, Birmingham AL
Editors Press Inc, Hyattsville MD
Edwards Brothers Inc, Lillington NC
Eerdmans Printing Co, Grand Rapids MI
EP Graphics, Berne IN
EU Services, Rockville MD
Eureka Printing Co Inc, Eureka CA
Eusey Press Inc, Leominster MA
Eva-Tone Inc, Clearwater FL
Faculty Press Inc, Brooklyn NY
Federated Lithographers, Providence RI
Fetter Printing, Louisville KY
First Impressions Printer & Lithog-
    raphers Inc, Elk Grove IL
Fisher Printers Inc, Cedar Rapids IA
Fisher Printing Co, Galion OH
Fleming Printing Co, St Louis MO
Flower City Printing Inc, Rochester NY
Focus Direct, San Antonio TX
The Forms Man Inc, Deer Park NY
Fort Orange Press Inc, Albany NY
Foster Printing Service, Michigan City IN
Friesen Printers, Altona MB
Frye & Smith, Costa Mesa CA
Fundcraft Publishing, Collierville TN
Futura Printing Inc, Boynton Beach FL
Gateway Press Inc, Louisville KY
Gaylord Printing, Detroit MI
George Lithograph, San Francisco CA
Glundal Color Service, East Syracuse NY
Golden Belt Printing Inc, Great Bend KS
Gowe Printing Company, Medina OH
Graftek Press, Woodstock IL
Graphic Arts Center, Portland OR
Graphic Litho Corporation, Lawrence MA
Graphic Printing, New Carlisle OH
Graphic Production Center, Norcross GA
Graphic Ways Inc, Buena Park CA
Graphics Ltd, Indianapolis IN
Gray Printing, Fostoria OH
Great Impressions Printing & Graphics
    and Graphics, Dallas TX
Great Lakes Lithograph Co, Cleveland
    OH
Great Northern/Design Printing, Skokie
    IL
Greenfield Printing & Publishing, Green-
    field OH
Gregath Publishing Co, Wyandotte OK
GRIT Commercial Printing Services, Wil-
    liamsport PA
Gulf Printing, Houston TX

4-Color Printers

Guynes Printing Company, Albuquerque NM
Harlo Printing Co, Detroit MI
Harmony Printing, Liberty MO
Hart Graphics, Austin TX
Hart Graphics, Simpsonville SC
Hart Press, Long Prairie MN
Heartland Press Inc, Spencer IA
Henington Publishing, Wolfe City TX
Hennegan Company, Cincinnati OH
Heritage Printers Inc, Charlotte NC
D B Hess Company, Woodstock IL
Hi-Tech Color House, Chicago IL
Hignell Printing, Winnepeg MB
Hinz Lithographing Company, Mt Prospect IL
A B Hirschfeld Press, Denver CO
Hoechstetter Printing, Pittsburgh PA
Holladay-Tyler Printing Corp, Glenn Dale MD
Rae Horowitz Book Manufacturers, Fairfield NJ
Imperial Litho/Graphics, Phoenix AZ
Independent Publishing Company, Sarasota FL
Instant Web, Chanhassen MN
Intelligencer Printing, Lancaster PA
Interform Corporation, Bridgeville PA
Interstate Printing Company, Omaha NE
IPC Publishing Services, Saint Joseph MI
Japs-Olson Company, Minneapolis MN
Jersey Printing Company, Bayonne NJ
JK Creative Printers, Quincy IL
Johns Byrne Co, Niles IL
The Johnson & Hardin Company, Cincinnati OH
Jostens Printing & Publishing, Visalia CA
Jostens Printing & Publishing, Clarksville TN
Jostens Printing & Publishing, Winston-Salem NC
Jostens Printing & Publishing, Topeka KS
Jostens Printing & Publishing, State College PA
Jostens Printing & Publishing, Minneapolis MN
Julin Printing, Monticello IA
K-B Offset Printing, State College PA
Kaufman Press Printing, Syracuse NY
Keys Printing Co Inc, Greenville SC
Kimberly Press, Santa Barbara CA
KNI Incorporated, Anaheim CA
Kolor View Press, Aurora MO
Kordet Graphics Inc, Oceanside NY
The C J Krehbiel Company, Cincinnati OH
La Crosse Graphics, LaCrosse WI
Lancaster Press, Lancaster PA
The Lane Press, Burlington VT

Lasky Company, Milburn NJ
Lehigh Press Inc, Cherry Hill NJ
Lehigh Press/Cadillac, Broadview IL
Les Editions Marquis, Montmagny PQ
Lighthouse Press Inc, Manchester NH
Litho Productions Inc, Madison WI
Litho Specialties, St Paul MN
Lithograph Printing Company, Memphis TN
Lithoid Printing Corporation, East Brunswick NJ
The Little River Press Inc, Miami FL
John D Lucas Printing Company, Baltimore MD
Mack Printing Group Inc, Easton PA
Mackintosh Typography Inc, Santa Barbara CA
Mail-O-Graph, Kewanee IL
Maquoketa Web Printing, Maquoketa IA
Marathon Communications Group, Wausau WI
Mark IV Press, Hauppauge NY
Marrakech Express, Tarpon Springs FL
Mars Graphic Services, Westville NJ
Maverick Publications Inc, Bend OR
The Mazer Corporation, Dayton OH
MBP Lithographics, Hillsboro KS
McClain Printing Company, Parsons WV
The McFarland Press, Harrisburg PA
McGill Jensen, St Paul MN
McGrew Color Graphics, Kansas City MO
McQuiddy Printing Company, Nashville TN
Meaker the Printer, Phoenix AZ
Media Printing, Miami FL
Media Publications, Santa Clara CA
Meehan-Tooker & Company, East Rutherford NJ
Mercury Printing Company, Memphis TN
Merrill Corporation, Los Angeles CA
Metroweb, Erlanger KY
Mitchell Graphics, Petosky MI
Mitchell Press, Vancouver BC
MMI Press, Harrisville NH
Moebius Printing, Milwaukee WI
Moran Printing Company, Orlando FL
Moran Printing Inc, Baton Rouge LA
Morgan Press, Dobbs Ferry NY
Morgan Printing, Austin TX
Motheral Printing, Fort Worth TX
William H Muller Printing Co, Santa Clara CA
MultiPrint Co Inc, Skokie IL
Murphy's Printing Co, Campbell CA
National Graphics Corp, Columbus OH
National Lithographers Inc, Miami FL
National Publishing Company, Philadelphia PA
National Reproductions Corp, Madison Heights MI

4-Color Printers

Nationwide Printing, Burlington KY
Naturegraph Publishers, Happy Camp CA
Louis Neibauer Co Inc, Warminster PA
Newsfoto Publishing Company, San Angelo TX
Neyenesch Printers, San Diego CA
The Nielsen Lithographing Co, Cincinnati OH
Nies/Artcraft Printing Companies, Saint Louis MO
Nimrod Press Inc, Boston MA
Noll Printing Company, Huntington IN
Northeast Graphics, New York NY
Northeast Web Printing Inc, Farmingdale NY
Northlight Studio Press, Barre VT
Northwest Web, Eugene OR
Nystrom Publishing Company, Osseo MN
Oaks Printing Company, Bethlehem PA
Offset Paperback Manufacturers, Dallas PA
Ohio Valley Litho Color, Florence KY
Omaha Printing Co, Omaha NE
Optic Graphics Inc, Glen Burnie MD
Outstanding Graphics, Kenosha WI
The Ovid Bell Press Inc, Fulton MO
The Oxford Group, Norway ME
Padgett Printing Corporation, Dallas TX
Page Litho Inc, Detroit MI
Panel Prints Inc, Old Forge PA
Panorama Press Inc, Clifton NJ
Paraclete Press, Brewster MA
Park Press, South Holland IL
Parker Graphics, Fuquay-Varina NC
Patterson Printing Company, Benton Harbor MI
Paust Inc, Richmond IN
Pearl-Pressman-Liberty Communications Group, Philadelphia PA
Peconic Companies, Mattituck NY
Pendell Printing, Midland MI
Penn Colour Graphics, Huntington Valley PA
PennWell Printing, Tulsa OK
Penton Press Publishing, Berea OH
The Perlmuter Printing Co, Cleveland OH
Perry Printing Corp, Waterloo WI
Perry/Baraboo, Baraboo WI
PFP Printing Corporation, Gaithersburg MD
Phillips Brothers Printers Inc, Springfield IL
Pine Hill Press Inc, Freeman SD
Pinecliffe Printers, Shawnee OK
Pioneer Press, Wilmette IL
Plain Talk Printing Co, Des Moines IA
Plus Communications, Saint Louis MO
Port City Press Inc, Baltimore MD

Port Publications, Port Washington WI
Practical Graphics, New York NY
The Press of Ohio, Kent OH
The Press, Chanhassen MN
Princeton Academic Press, Lawrenceville NJ
Prinit Press, Dublin IN
Print Northwest, Tacoma WA
The Printer Inc, Osseo MN
The Printing Company, Indianapolis IN
Printing Corporation of America, Pompano Beach FL
Printright International, Palmdale CA
Printwest, Regina SK
Progress Printing, Lynchburg VA
ProLitho, Provo UT
Promotional Printing Corporation, Houston TX
Publishers Choice Reprints, The Woodlands TX
Publishers Express Press, Ladysmith WI
Publishers Press, Salt Lake City UT
Publishers Printing Co, Shepherdsville KY
Quad Graphics — Thomaston, Thomaston GA
Quad/Graphics Inc, Saratoga Springs NY
Quad/Graphics, Lomira, Lomira WI
Quad/Graphics, Pewaukee, Pewaukee WI
Quad/Graphics, Sussex, Sussex WI
Quebecor Book Group, New York NY
Quebecor Printing (USA) Corp, Boston MA
Quebecor Printing Inc, Saint Paul MN
Quebecor Printing Inc, Richmond Hill ON
Quebecor Printing/St Cloud, Saint Cloud MN
Quebecor/Dallas, Dallas TX
Quebecor/Semline, Braintree MA
Rapidocolor Corporation, West Chester PA
RBW Graphics, Owen Sound ON
Ringier America, Itasca IL
Ringier America Inc, Corinth MS
Ringier America/Brookfield, Brookfield WI
Ringier America/Jonesboro, Jonesboro AR
Ringier America/Phoenix, Phoenix AZ
John Roberts Company, Minneapolis MN
Ronalds Printing, Vancouver BC
Ronalds Printing, Montreal PQ
Rose Printing Company, Tallahassee FL
S Rosenthal & Company Inc, Cincinnati OH
Roxbury Publishing Company, Los Angeles CA
Royle Printing, Sun Prairie WI
RPP Enterprises Inc, Libertyville IL
Sanders Printing Company, Garretson SD

4-Color Printers

The Saratoga Printing Co, Saratoga
 Springs NY
Schelin Printing Co, St Louis MO
Schiff Printing, Pittsburgh PA
Schumann Printers, Fall River WI
Science Press, Ephrata PA
Service Web Offset Corporation, Chicago
 IL
Sexton Printing, Saint Paul MN
Shea Communications, Oklahoma City
 OK
Shepard Poorman Communications, In-
 dianapolis IN
The Sheridan Press, Hanover PA
Sheridan Printing Company, Phil-
 lipsburg NJ
Sinclair Printing Co, Los Angeles CA
SLC Graphics, Pittston PA
R E Smith Printing, Fall River MA
Smith-Edwards-Dunlap, Philadelphia PA
Snohomish Publishing, Snohomish WA
Southeastern Printing Company Inc,
 Stuart FL
Sowers Printing Company, Lebanon PA
Speakes-Hines & Thomas Inc, Lansing
 MI
Spencer Press, Wells ME
St Croix Press Inc, New Richmond WI
Standard Publishing, Cincinnati OH
John Stark Printing Co Inc, St Louis MO
Stephenson Lithograph Inc, Alexandria
 VA
Stevens Graphics, Atlanta GA
The Stinehour Press, Lunenburg VT
Strathmore Printing, Geneva IL
Straus Printing Company, Madison WI
The Studley Press, Dalton MA
Sullivan Graphics Inc, Brentwood TN
Sun Graphics, Parsons KS
John S Swift Company Inc, St Louis MO
John S Swift/Chicago, Chicago IL
John S Swift/Cincinnati, Cincinnati OH
John S Swift/Teterboro, Teterboro NJ
TAN Books & Publishers Inc, Rockford
 IL
Tapco Inc, Pemberton NJ
Tapemark, West St Paul MN
Taylor Publishing Company, Dallas TX
TEACH Services, Brushton NY
Thomson-Shore, Dexter MI
Times Litho Inc, Forest Grove OR
Times Printing, Random Lake WI
Todd Publications Inc, Smithville TX
Town House Press, Pittsboro NC
Tracor Publications, Austin TX
Trade Litho Inc, Miami FL
Trend Offset Printing Service Inc, Los
 Alamitos CA
Tri-Graphic Printing, Ottawa ON
Tweddle Litho Co, Clinton Township MI

Ultra-Color Corporation, Saint Louis MO
United Lithographic Services Inc, Falls
 Church VA
Universal Press, Providence RI
Universal Printing Company, Miami FL
Universal Printing Company, Saint
 Louis MO
US Press Inc, Valdosta GA
Valco Graphics, Seattle WA
Vicks Lithograph & Printing, Yorkville
 NY
Vogue Printers, North Chicago IL
Wagners Printers, Davenport IA
Waldman Graphics, Pennsauken NJ
Waldon Press Inc, New York NY
Wallace Press, Hillside IL
Walsworth Publishing Co Inc, Marceline
 MO
Watkins Printing Company, Columbus
 OH
Watt/Peterson Inc, Plymouth MN
Waverly Press, Linthicum MD
Web Specialties, Wheeling IL
Webcom Ltd, Scarborough ON
Webcraft Technologies, North Brunswick
 NJ
Webcrafters, Madison WI
F A Weber & Sons, Park Falls WI
Weber Printing Company, Lincoln NE
The Wessel Company Inc, Elk Grove Vil-
 lage IL
Western Publishing Co Inc, Racine WI
Western Web Printing, Sioux Falls SD
White Arts Inc, Indianapolis IN
Whitehall Printing Company, Naples FL
The Wimmer Companies, Memphis TN
Windsor Associates, San Diego CA
Wisconsin Color Press Inc, Milwaukee WI
Wolfer Printing, Commerce CA
Wood & Jones Printers, Pasadena CA
The Wood Press, Paterson NJ
World Color Press Inc, New York NY
Worzalla Publishing Company, Stevens
 Point WI
Henry Wurst Inc, North Kansas City MO
Y/Z Printing Co, Elizabethtown PA

Color Separations

Action Printing, Fond Du Lac WI
Ad Infinitum Press, Mount Vernon NY
Alan Lithograph Inc, Inglewood CA
Alcom Printing Group Inc, Bethlehem PA
American Litho, Hayward CA
American Signature Memphis Division,
 Olive Branch MS

Color Separations

American Signature Graphics, Dallas TX
American Signature, Atlanta GA
American Web Inc, Denver CO
Amidon & Associates Inc, St Paul MN
Apollo Graphics Ltd, Southampton PA
The Argus Press Inc, Niles IL
Austin Printing Co Inc, Akron OH
Automated Graphic Systems, White
    Plains MD
Baker Johnson Inc, Dexter MI
Bang Printing Co, Brainerd MN
Banta Company, Menasha WI
Banta Publications Group, Hinsdale IL
Banta-Harrisonburg, Harrisonburg VA
Barton Press Inc, West Orange NJ
Bay State Press, Framingham MA
Ben-Wal Printing, Pomona CA
Bertelsmann Printing & Manufacturing,
    New York NY
Blue Dolphin Press Inc, Grass Valley CA
Bolger Publications/Creative Printing,
    Minneapolis MN
BookCrafters, Chelsea MI
Bradley Printing Company, Des Plaines
    IL
Brown Printing, Waseca MN
Brown Printing/Franklin, Franklin KY
R L Bryan Company, Columbia SC
Buse Printing & Advertising, Phoenix AZ
The William Byrd Press, Richmond VA
Cal Central Press, Sacramento CA
California Offset Printers, Glendale CA
Calsonic Miura Graphics, Irvine CA
Capital City Press, Montpelier VT
Catalog King, Clifton NJ
Champagne Offset Co Inc, Natick MA
Citizen Printing, Beaver Dam WI
The College Press, Collegedale TN
Color Express, Commerce CA
Color Express, Minneapolis MN
Color Express, Atlanta GA
Color Express, Irvine CA
ColorCorp Inc, Littleton CO
Colorlith Corporation, Johnston RI
Colotone Riverside Inc, Dallas TX
Columbus Bookbinders & Printers,
    Columbus GA
Combined Communication Services,
    Mendota IL
Commercial Documentation Services,
    Medford OR
Concord Litho Company Inc, Concord
    NH
Coneco Litho Graphics, Glen Falls NY
Continental Web, Itasca IL
Cookbooks by Mom's Press, Kearney NE
Coral Graphic Services, Plainview NY
Country Press Inc, Lakeville MA
The Craftsman Press Inc, Seattle WA

Crane Duplicating Service, West
    Barnstable MA
Cushing-Malloy, Ann Arbor MI
Danbury Printing & Litho Inc, Danbury
    CT
Danner Press Corporation, Canton OH
Dartmouth Printing, Hanover NH
Data Reproductions Corp, Rochester
    Hills MI
Dayal Graphics, Lemont IL
Dellas Graphics, Syracuse NY
Delmar Printing & Publishing, Charlotte
    NC
Delta Lithograph Company, Valencia CA
Desaulniers Printing Company, Milan IL
Direct Press/Modern Litho, Huntington
    Station NY
Directories America, Chicago IL
Dollco Printing, Ottawa ON
Donihe Graphics Inc, Kingsport TN
R R Donnelley & Sons Company Inc, ,
    Crawfordsville IN
R R Donnelley & Sons Company Inc,
    Harrisonburg, Harrisonburg VA
R R Donnelley & Sons Company Inc,
    Willard, Willard OH
Dynagraphics, Carlsbad CA
EP Graphics, Berne IN
EU Services, Rockville MD
Eva-Tone Inc, Clearwater FL
Faculty Press Inc, Brooklyn NY
Fetter Printing, Louisville KY
Fisher Printers Inc, Cedar Rapids IA
Fisher Printing Co, Galion OH
Flower City Printing Inc, Rochester NY
Focus Direct, San Antonio TX
The Forms Man Inc, Deer Park NY
Friesen Printers, Altona MB
Frye & Smith, Costa Mesa CA
Gateway Press Inc, Louisville KY
Gaylord Printing, Detroit MI
George Lithograph, San Francisco CA
Gilliland Printing Inc, Arkansas City KS
Glundal Color Service, East Syracuse NY
Gowe Printing Company, Medina OH
Graftek Press, Woodstock IL
Graphic Arts Center, Portland OR
Graphic Litho Corporation, Lawrence MA
Graphic Production Center, Norcross GA
Graphic Ways Inc, Buena Park CA
Graphics Ltd, Indianapolis IN
Great Lakes Lithograph Co, Cleveland
    OH
Great Northern/Design Printing, Skokie
    IL
Greenfield Printing & Publishing, Green-
    field OH
Gulf Printing, Houston TX
Guynes Printing Company, Albuquerque
    NM
Harmony Printing, Liberty MO
Hart Graphics, Austin TX

Color Separations

Hart Press, Long Prairie MN
Hennegan Company, Cincinnati OH
Hi-Tech Color House, Chicago IL
Hinz Lithographing Company, Mt
    Prospect IL
A B Hirschfeld Press, Denver CO
Rae Horowitz Book Manufacturers, Fair-
    field NJ
Imperial Litho/Graphics, Phoenix AZ
Intelligencer Printing, Lancaster PA
Interform Corporation, Bridgeville PA
IPC Publishing Services, Saint Joseph MI
Jersey Printing Company, Bayonne NJ
Johns Byrne Co, Niles IL
Jostens Printing & Publishing, Visalia CA
Jostens Printing & Publishing,
    Clarksville TN
Jostens Printing & Publishing, Winston-
    Salem NC
Jostens Printing & Publishing, Topeka
    KS
Jostens Printing & Publishing, State Col-
    lege PA
Jostens Printing & Publishing, Min-
    neapolis MN
Julin Printing, Monticello IA
Kaufman Press Printing, Syracuse NY
Keys Printing Co Inc, Greenville SC
Kolor View Press, Aurora MO
Kordet Graphics Inc, Oceanside NY
Lancaster Press, Lancaster PA
The Lane Press, Burlington VT
Lasky Company, Milburn NJ
Lehigh Press Inc, Cherry Hill NJ
Les Editions Marquis, Montmagny PQ
Litho Productions Inc, Madison WI
Litho Specialties, St Paul MN
Lithograph Printing Company, Memphis
    TN
Lithoid Printing Corporation, East
    Brunswick NJ
John D Lucas Printing Company, Bal-
    timore MD
Mackintosh Typography Inc, Santa Bar-
    bara CA
Marathon Communications Group,
    Wausau WI
Mark IV Press, Hauppauge NY
Marrakech Express, Tarpon Springs FL
Maverick Publications Inc, Bend OR
McClain Printing Company, Parsons WV
McGill Jensen, St Paul MN
McGrew Color Graphics, Kansas City
    MO
McNaughton & Gunn Inc, Saline MI
McQuiddy Printing Company, Nashville
    TN
Meehan-Tooker & Company, East
    Rutherford NJ
Merrill Corporation, Los Angeles CA

Mitchell Press, Vancouver BC
Moebius Printing, Milwaukee WI
Moran Printing Company, Orlando FL
William H Muller Printing Co, Santa
    Clara CA
MultiPrint Co Inc, Skokie IL
National Lithographers Inc, Miami FL
National Reproductions Corp, Madison
    Heights MI
Newsfoto Publishing Company, San An-
    gelo TX
Neyenesch Printers, San Diego CA
Nielsen Lithographing Co, Cincinnati OH
Noll Printing Company, Huntington IN
Oaks Printing Company, Bethlehem PA
Offset Paperback Manufacturers, Dallas
    PA
Ohio Valley Litho Color, Florence KY
Omaha Printing Co, Omaha NE
Optic Graphics Inc, Glen Burnie MD
The Oxford Group, Norway ME
Padgett Printing Corporation, Dallas TX
Panel Prints Inc, Old Forge PA
Panorama Press Inc, Clifton NJ
Paust Inc, Richmond IN
Pearl-Pressman-Liberty Communica-
    tions Group, Philadelphia PA
Pendell Printing, Midland MI
Penn Colour Graphics, Huntington Val-
    ley PA
PennWell Printing, Tulsa OK
The Perlmuter Printing Co, Cleveland
    OH
Pinecliffe Printers, Shawnee OK
The Press of Ohio, Kent OH
The Press, Chanhassen MN
Princeton Academic Press, Lawrenceville
    NJ
The Printing Company, Indianapolis IN
Printing Corporation of America, Pom-
    pano Beach FL
Printright International, Palmdale CA
Printwest, Regina SK
Progress Printing, Lynchburg VA
Progressive Typographers Inc, Emigsville
    PA
Promotional Printing Corporation, Hous-
    ton TX
Publishers Express Press, Ladysmith WI
Publishers Printing Co, Shepherdsville
    KY
Quad/Graphics Inc, Saratoga Springs NY
Quad/Graphics, Lomira, Lomira WI
Quad/Graphics, Pewaukee, Pewaukee WI
Quad/Graphics, Sussex, Sussex WI
Quebecor Book Group, New York NY
Quebecor Printing (USA) Corp, Boston
    MA
Quebecor Printing Inc, Saint Paul MN
Quebecor Printing/St Cloud, Saint Cloud
    MN
Quebecor/Dallas, Dallas TX

Color Separations

Ringier America, Itasca IL
Ringier America/Brookfield, Brookfield WI
Ringier America/Jonesboro, Jonesboro AR
Ringier America/Phoenix, Phoenix AZ
John Roberts Company, Minneapolis MN
Roxbury Publishing Company, Los Angeles CA
Royle Printing, Sun Prairie WI
Sanders Printing Company, Garretson SD
Schelin Printing Co, St Louis MO
Schiff Printing, Pittsburgh PA
Schumann Printers, Fall River WI
Science Press, Ephrata PA
Shea Communications, Oklahoma City OK
Shepard Poorman Communications, Indianapolis IN
SLC Graphics, Pittston PA
Southeastern Printing Company Inc, Stuart FL
Speakes-Hines & Thomas Inc, Lansing MI
Spencer Press, Wells ME
St Croix Press, Inc, New Richmond WI
John Stark Printing Co Inc, St Louis MO
Stephenson Lithograph Inc, Alexandria VA
The Stinehour Press, Lunenburg VT
Straus Printing Company, Madison WI
Sullivan Graphics Inc, Brentwood TN
Sun Graphics, Parsons KS
Taylor Publishing Company, Dallas TX
TEACH Services, Brushton NY
Todd Publications Inc, Smithville TX
Town House Press, Pittsboro NC
Trade Litho Inc, Miami FL
Trend Offset Printing Service Inc, Los Alamitos CA
Ultra-Color Corporation, Saint Louis MO
Universal Press, Providence RI
Universal Printing Company, St Louis MO
US Press Inc, Valdosta GA
Versa Press Inc, East Peoria IL
Wagners Printers, Davenport IA
Waldman Graphics, Pennsauken NJ
Waldon Press Inc, New York NY
Wallace Press, Hillside IL
Walsworth Publishing Co Inc, Marceline MO
Watt/Peterson Inc, Plymouth MN
Web Specialties, Wheeling IL
Webcraft Technologies, North Brunswick NJ
F A Weber & Sons, Park Falls WI
Western Publishing Co Inc, Racine WI
Whitehall Printing Company, Naples FL

Windsor Associates, San Diego CA
Wisconsin Color Press Inc, Milwaukee WI
Wolfer Printing, Commerce CA
Wood & Jones Printers, Pasadena CA
The Wood Press, Paterson NJ
World Color Press Inc, New York NY
Henry Wurst Inc, North Kansas City MO

Design and Artwork

A-1 Business Service Inc, St Paul MN
aBCD, Seattle WA
Accurate Printing, Post Falls ID
Action Printing, Fond Du Lac WI
Ad Infinitum Press, Mount Vernon NY
Adviser Graphics, Red Deer AB
AM Graphics & Printing, San Marcos CA
American Litho, Hayward CA
Apollo Graphics Ltd, Southampton PA
Automated Graphic Systems, White Plains MD
Bang Printing Co, Brainerd MN
Bay Port Press, Chula Vista CA
Bay State Press, Framingham MA
Ben-Wal Printing, Pomona CA
Blake Printery, San Luis Obispo CA
Blue Dolphin Press Inc, Grass Valley CA
Bolger Publications/Creative Printing, Minneapolis MN
BookCrafters, Chelsea MI
BookMasters Inc, Ashland OH
Bradley Printing Company, Des Plaines IL
R L Bryan Company, Columbia SC
BSC Litho, Harrisburg PA
Buse Printing & Advertising, Phoenix AZ
Caldwell Printers, Arcadia CA
Calsonic Miura Graphics, Irvine CA
Camelot Publishing Co, Ormond Beach FL
Carlson Color Graphics, Ocala FL
Catalog King, Clifton NJ
Champagne Offset Co Inc, Natick MA
The College Press, Collegedale TN
Color World Printers, Bozeman MT
ColorGraphics Printing Corp, Tulsa OK
Colortone Press, Landover MD
Columbus Bookbinders & Printers, Columbus GA
Community Press, Provo UT
Concord Litho Company Inc, Concord NH
Coneco Litho Graphics, Glen Falls NY
Cookbooks by Mom's Press, Kearney NE
Cooley Printers & Office Supply, Monroe LA

Design

Country Press, Mohawk NY
Crane Duplicating Service, West Barnstable MA
Crusader Printing Co, East Saint Louis IL
Dataco A WBC Company, Albuquerque NM
Delmar Printing & Publishing, Charlotte NC
Delta Lithograph Company, Valencia CA
Des Plaines Publishing Co, Des Plaines IL
Diamond Graphics Tech Data, Milwaukee WI
Direct Graphics Inc, Sidney OH
Dollco Printing, Ottawa ON
Donihe Graphics Inc, Kingsport TN
R R Donnelley & Sons Company Inc, Crawfordsville IN
R R Donnelley & Sons Company Inc, Harrisonburg, Harrisonburg VA
R R Donnelley & Sons Company Inc, Willard, Willard OH
Dragon Press, Delta Junction AK
EP Graphics, Berne IN
Eureka Printing Co Inc, Eureka CA
Eva-Tone Inc, Clearwater FL
Faculty Press Inc, Brooklyn NY
Fetter Printing, Louisville KY
Flower City Printing Inc, Rochester NY
Focus Direct, San Antonio TX
The Forms Man Inc, Deer Park NY
Fort Orange Press Inc, Albany NY
Fundcraft Publishing, Collierville TN
Gateway Press Inc, Louisville KY
Geryon Press Limited, Tunnel NY
Glundal Color Service, East Syracuse NY
Graphic Ways Inc, Buena Park CA
Gray Printing, Fostoria OH
Gregath Publishing Co, Wyandotte OK
GRIT Commercial Printing Services, Williamsport PA
Harlo Printing Co, Detroit MI
Harmony Printing, Liberty MO
A B Hirschfeld Press, Denver CO
Imperial Litho/Graphics, Phoenix AZ
Interform Corporation, Bridgeville PA
IPC Publishing Services, Saint Joseph MI
Jersey Printing Company, Bayonne NJ
JK Creative Printers, Quincy IL
The Job Shop, Woods Hole MA
Jostens Printing & Publishing, Visalia CA
Jostens Printing & Publishing, Clarksville TN
Jostens Printing & Publishing, Winston-Salem NC
Jostens Printing & Publishing, Topeka KS
Jostens Printing & Publishing, State College PA
Jostens Printing & Publishing, Minneapolis MN
Julin Printing, Monticello IA
Kaufman Press Printing, Syracuse NY
Kolor View Press, Aurora MO
Kordet Graphics Inc, Oceanside NY
The C J Krehbiel Company, Cincinnati OH
La Crosse Graphics, LaCrosse WI
Lithoid Printing Corporation, East Brunswick NJ
Long Island Web Printing, Jericho NY
Mack Printing Group Inc, Easton PA
Mackintosh Typography Inc, Santa Barbara CA
Maquoketa Web Printing, Maquoketa IA
Marathon Communications Group, Wausau WI
Mark IV Press, Hauppauge NY
Maverick Publications Inc, Bend OR
The Mazer Corporation, Dayton OH
McGill Jensen, St Paul MN
McGrew Color Graphics, Kansas City MO
Meaker the Printer, Phoenix AZ
Mercury Printing Company, Memphis TN
Mitchell Press, Vancouver BC
Morgan Press, Dobbs Ferry NY
Morningrise Printing, Costa Mesa CA
MultiPrint Co Inc, Skokie IL
National Lithographers Inc, Miami FL
Louis Neibauer Co Inc, Warminster PA
The Nielsen Lithographing Co, Cincinnati OH
Oaks Printing Company, Bethlehem PA
The Oxford Group, Norway ME
Panorama Press Inc, Clifton NJ
Paraclete Press, Brewster MA
Paust Inc, Richmond IN
Pearl-Pressman-Liberty Communications Group, Philadelphia PA
Pendell Printing, Midland MI
Penn Colour Graphics, Huntington Valley PA
PennWell Printing, Tulsa OK
Pentagram Press, Minneapolis MN
The Perlmuter Printing Co, Cleveland OH
Pine Hill Press Inc, Freeman SD
Plus Communications, Saint Louis MO
Practical Graphics, New York NY
Prinit Press, Dublin IN
The Printing Company, Indianapolis IN
Printwest, Regina SK
Progress Printing, Lynchburg VA
Progressive Typographers, Emigsville PA
Publishers Choice Reprints, The Woodlands TX
Quebecor Printing Inc, Saint Paul MN

Design

Quebecor Printing/St Cloud, Saint Cloud MN
Repro-Tech, Fairfield NJ
S Rosenthal & Company Inc, Cincinnati OH
Roxbury Publishing Company, Los Angeles CA
RPP Enterprises Inc, Libertyville IL
Sanders Printing Company, Garretson SD
Schiff Printing, Pittsburgh PA
Sexton Printing, Saint Paul MN
Southeastern Printing Company Inc, Stuart FL
Sowers Printing Company, Lebanon PA
Standard Publishing, Cincinnati OH
Stewart Publishing & Printing Co, Marble Hill MO
The Stinehour Press, Lunenburg VT
Strathmore Printing, Geneva IL
Straus Printing Company, Madison WI
The Studley Press, Dalton MA
Sullivan Graphics Inc, Brentwood TN
The Sun Hill Press, North Brookfield MA
Tapemark, West St Paul MN
TEACH Services, Brushton NY
Times Printing, Random Lake WI
Town House Press, Pittsboro NC
Tracor Publications, Austin TX
Trade Litho Inc, Miami FL
Trend Offset Printing Service Inc, Los Alamitos CA
Tweddle Litho Co, Clinton Township MI
Ultra-Color Corporation, Saint Louis MO
Vogue Printers, North Chicago IL
Waldon Press Inc, New York NY
Wallace Press, Hillside IL
Walsworth Publishing Co Inc, Marceline MO
Waverly Press, Linthicum MD
Web Specialties, Wheeling IL
F A Weber & Sons, Park Falls WI
Weber Printing Company, Lincoln NE
Western Publishing Co Inc, Racine WI
Western Web Printing, Sioux Falls SD
Windsor Associates, San Diego CA
Winter Sun Inc, Phoenix AZ
Wood & Jones Printers, Pasadena CA
Y/Z Printing Co, Elizabethtown PA

Fulfillment and Mailing Services

Adviser Graphics, Red Deer AB
Alcom Printing Group Inc, Bethlehem PA
All-Star Printing, Lansing MI

Allen Press Inc, Lawrence KS
Alonzo Printing, Hayward CA
American Signature Memphis Division, Olive Branch MS
American Signature Graphics, Dallas TX
American Signature, Atlanta GA
American Signature, Lincoln NE
Apollo Graphics Ltd, Southampton PA
Arandell Corporation, Menomonee Falls WI
The Argus Press Inc, Niles IL
Artcraft Press, Waterloo WI
Austin Printing Co Inc, Akron OH
Automated Graphic Systems, White Plains MD
Bang Printing Co, Brainerd MN
Banta Company, Menasha WI
Banta Publications Group, Hinsdale IL
Banta-Harrisonburg, Harrisonburg VA
Barton Press Inc, West Orange NJ
Bawden Printing Inc, Eldridge IA
Bay State Press, Framingham MA
Beacon Press, Seattle WA
Blue Dolphin Press Inc, Grass Valley CA
Bolger Publications/Creative Printing, Minneapolis MN
The Book Press, Brattleboro VT
Book-Mart Press Inc, North Bergen NJ
BookCrafters, Chelsea MI
BookCrafters/Fredericksburg, Fredericksburg VA
BookMasters Inc, Ashland OH
Boyd Printing Co Inc, Albany NY
Bradley Printing Company, Des Plaines IL
Braun-Brumfield Inc, Ann Arbor MI
Brown Printing / Franklin, Franklin KY
R L Bryan Company, Columbia SC
Buse Printing & Advertising, Phoenix AZ
The William Byrd Press, Richmond VA
Cal Central Press, Sacramento CA
Calsonic Miura Graphics, Irvine CA
Capital City Press, Montpelier VT
Carlson Color Graphics, Ocala FL
Carpenter Lithographic Co, Springfield OH
Champagne Offset Co Inc, Natick MA
Clarkwood Corporation, Totowa NJ
Colonial Graphics, Paterson NJ
Color World Printers, Bozeman MT
ColorGraphics Printing Corp, Tulsa OK
Colotone Riverside Inc, Dallas TX
Columbus Bookbinders & Printers, Columbus GA
Combined Communication Services, Mendota IL
Comfort Printing, St Louis MO
Commercial Documentation Services, Medford OR
Concord Litho Company Inc, Concord NH
The Craftsman Press Inc, Seattle WA

Fulfillment

Crusader Printing Co, East Saint Louis
IL
Lew ACummings Printing Company Inc,
Manchester NH
Danbury Printing & Litho Inc, Danbury
CT
Dartmouth Printing, Hanover NH
Dataco A WBC Company, Albuquerque
NM
Dellas Graphics, Syracuse NY
Delmar Printing & Publishing, Charlotte
NC
Delta Lithograph Company, Valencia CA
Des Plaines Publishing Co, Des Plaines
IL
Desaulniers Printing Company, Milan IL
Dharma Press, Oakland CA
Diamond Graphics Tech Data, Mil-
waukee WI
Directories America, Chicago IL
Diversified Printing, Brea CA
Donihe Graphics Inc, Kingsport TN
R R Donnelley & Sons Company Inc,
Crawfordsville IN
R R Donnelley & Sons Company Inc,
Harrisonburg, Harrisonburg VA
R R Donnelley & Sons Company Inc,
Willard, Willard OH
Edwards Brothers Inc, Ann Arbor MI
Edwards Brothers Inc, Lillington NC
EU Services, Rockville MD
Eva-Tone Inc, Clearwater FL
Fetter Printing, Louisville KY
First Impressions Printer & Lithog-
raphers Inc, Elk Grove IL
Fisher Printers Inc, Cedar Rapids IA
Fleming Printing Co, St Louis MO
Flower City Printing Inc, Rochester NY
Focus Direct, San Antonio TX
Friesen Printers, Altona MB
Frye & Smith, Costa Mesa CA
Gateway Press Inc, Louisville KY
George Lithograph, San Francisco CA
Golden Belt Printing Inc, Great Bend KS
Gowe Printing Company, Medina OH
Graphic Arts Center, Portland OR
Great Impressions Printing & Graphics
and Graphics, Dallas TX
Great Lakes Lithograph Co, Cleveland
OH
Greenfield Printing & Publishing, Green-
field OH
Griffin Printing & Lithograph, Glendale
CA
Gulf Printing, Houston TX
Guynes Printing Company, Albuquerque
NM
Haddon Craftsmen, Scranton PA
Harlo Printing Co, Detroit MI
Harmony Printing, Liberty MO

Hart Graphics, Austin TX
Hart Graphics, Simpsonville SC
Hart Press, Long Prairie MN
Heartland Press Inc, Spencer IA
D B Hess Company, Woodstock IL
A B Hirschfeld Press, Denver CO
Rae Horowitz Book Manufacturers, Fair-
field NJ
Imperial Litho/Graphics Inc, Phoenix AZ
Instant Web, Chanhassen MN
Interform Corporation, Bridgeville PA
Interstate Printing Company, Omaha NE
IPC Publishing Services, Saint Joseph MI
Japs-Olson Company, Minneapolis MN
Jersey Printing Company, Bayonne NJ
Johnson Graphics, East Dubuque IL
Kaufman Press Printing, Syracuse NY
Keys Printing Co Inc, Greenville SC
Kimberly Press, Santa Barbara CA
Kolor View Press, Aurora MO
The C J Krehbiel Company, Cincinnati
OH
Lancaster Press, Lancaster PA
Lithograph Printing Company, Memphis
TN
Lithoid Printing Corporation, East
Brunswick NJ
Long Island Web Printing, Jericho NY
Mack Printing Group Inc, Easton PA
Malloy Lithographing, Ann Arbor MI
Maple-Vail Book Manufacturing Group,
York PA
Marathon Communications Group,
Wausau WI
Mark IV Press, Hauppauge NY
Marrakech Express, Tarpon Springs FL
The Mazer Corporation, Dayton OH
MBP Lithographics, Hillsboro KS
McClain Printing Company, Parsons WV
McGill Jensen, St Paul MN
McKay Printing Services, Michigan City
IN
Media Printing, Miami FL
Mercury Printing Company, Memphis
TN
Merrill Corporation, Los Angeles CA
Moebius Printing, Milwaukee WI
Moran Printing Company, Orlando FL
Moran Printing Inc, Baton Rouge LA
National Lithographers Inc, Miami FL
National Publishing Company, Philadel-
phia PA
Louis Neibauer Co Inc, Warminster PA
Noll Printing Company, Huntington IN
Nystrom Publishing Company, Osseo MN
Oaks Printing Company, Bethlehem PA
Odyssey Press Inc, Dover NH
Offset Paperback Manufacturers, Dallas
PA
OMNIPRESS, Madison WI
The Ovid Bell Press Inc, Fulton MO
The Oxford Group, Norway ME

Fulfillment

Padgett Printing Corporation, Dallas TX
Panorama Press Inc, Clifton NJ
Patterson Printing Company, Benton Harbor MI
The Perlmuter Printing Co, Cleveland OH
Perry Printing Corp, Waterloo WI
Perry/Baraboo, Baraboo WI
Petty Co Inc, Effingham IL
PFP Printing Corporation, Gaithersburg MD
Plus Communications, Saint Louis MO
Port City Press Inc, Baltimore MD
The Press, Chanhassen MN
The Printer Inc, Osseo MN
Printing Corporation of America, Pompano Beach FL
Printwest, Regina SK
Progress Printing, Lynchburg VA
Publishers Express Press, Ladysmith WI
Publishers Press, Salt Lake City UT
Quad Graphics — Thomaston, Thomaston GA
Quad/Graphics Inc, Saratoga Springs NY
Quad/Graphics, Lomira, Lomira WI
Quad/Graphics, Pewaukee, Pewaukee WI
Quad/Graphics, Sussex, Sussex WI
Quebecor Printing (USA) Corp, Boston MA
Quebecor Printing Inc, Saint Paul MN
Quebecor Printing Inc, Richmond Hill ON
Quebecor Printing/St Cloud, St Cloud MN
Quebecor/Dallas, Dallas TX
Quinn-Woodbine Inc, Woodbine NJ
RBW Graphics, Owen Sound ON
Ringier America, Itasca IL
Ringier America Inc, Corinth MS
Ronalds Printing, Vancouver BC
Ronalds Printing, Montreal PQ
Rose Printing Company, Tallahassee FL
S Rosenthal & Company Inc, Cincinnati OH
Roxbury Publishing Company, Los Angeles CA
Schelin Printing Co, St Louis MO
Schiff Printing, Pittsburgh PA
Schumann Printers, Fall River WI
Science Press, Ephrata PA
Sexton Printing, Saint Paul MN
Shea Communications, Oklahoma City OK
Shepard Poorman Communications, Indianapolis IN
The Sheridan Press, Hanover PA
Sheridan Printing Company, Phillipsburg NJ
SLC Graphics, Pittston PA
Smith-Edwards-Dunlap, Philadelphia PA

Southeastern Printing Company Inc, Stuart FL
Sowers Printing Company, Lebanon PA
Standard Publishing, Cincinnati OH
John Stark Printing Co Inc, St Louis MO
Stephenson Lithograph, Alexandria VA
The Stinehour Press, Lunenburg VT
Strathmore Printing, Geneva IL
Straus Printing Company, Madison WI
Sullivan Graphics Inc, Brentwood TN
Taylor Publishing Company, Dallas TX
TEACH Services, Brushton NY
Technical Communication Services, North Kansas City MO
Times Printing, Random Lake WI
Todd Publications Inc, Smithville TX
Trade Litho Inc, Miami FL
Tweddle Litho Co, Clinton Township MI
Universal Press, Providence RI
Vail Ballou Press, Binghamton NY
Victor Graphics, Baltimore MD
Waldon Press Inc, New York NY
Wallace Press, Hillside IL
Walsworth Publishing Co Inc, Marceline MO
Watkins Printing Company, Columbus OH
Waverly Press, Linthicum MD
Webcom Ltd, Scarborough ON
Western Publishing Co Inc, Racine WI
Western Web Printing, Sioux Falls SD
Westview Press, Boulder CO
Wickersham Printing Co Inc, Lancaster PA
The Wimmer Companies, Memphis TN
Winter Sun Inc, Phoenix AZ
World Color Press Inc, New York NY
Worzalla Publishing Company, Stevens Point WI
Henry Wurst Inc, North Kansas City MO

List Maintenance

Adviser Graphics, Red Deer AB
Allen Press Inc, Lawrence KS
Apollo Graphics Ltd, Southampton PA
Arandell Corporation, Menomonee Falls WI
Associated Printers, Grafton ND
Banta Publications Group, Hinsdale IL
Bay State Press, Framingham MA
Beacon Press, Seattle WA
Bolger Publications/Creative Printing, Minneapolis MN
BookMasters Inc, Ashland OH
Boyd Printing Co Inc, Albany NY
Brown Printing, Waseca MN
Buse Printing & Advertising, Phoenix AZ
Capital City Press, Montpelier VT

List Maintenance

Citizen Printing, Beaver Dam WI
Clarkwood Corporation, Totowa NJ
Cooley Printers & Office Supply, Monroe LA
Crusader Printing Co, East Saint Louis IL
Dataco A WBC Company, Albuquerque NM
Dellas Graphics, Syracuse NY
Des Plaines Publishing Co, Des Plaines IL
Desaulniers Printing Company, Milan IL
Diamond Graphics Tech Data, Milwaukee WI
R R Donnelley & Sons Company Inc, Crawfordsville IN
R R Donnelley & Sons Company Inc, Harrisonburg, Harrisonburg VA
R R Donnelley & Sons Company Inc, Willard, Willard OH
EU Services, Rockville MD
Eva-Tone Inc, Clearwater FL
Fleming Printing Co, St Louis MO
Focus Direct, San Antonio TX
Gowe Printing Company, Medina OH
Great Lakes Lithograph Co, Cleveland OH
Gregath Publishing Co, Wyandotte OK
Haddon Craftsmen, Scranton PA
Hart Press, Long Prairie MN
Interform Corporation, Bridgeville PA
Keys Printing Co Inc, Greenville SC
Kolor View Press, Aurora MO
Lehigh Press Inc, Cherry Hill NJ
Lithoid Printing Corporation, East Brunswick NJ
Mack Printing Group Inc, Easton PA
Marathon Communications Group, Wausau WI
Marrakech Express, Tarpon Springs FL
MBP Lithographics, Hillsboro KS
Noll Printing Company, Huntington IN
Optic Graphics Inc, Glen Burnie MD
Paust Inc, Richmond IN
Pendell Printing, Midland MI
PennWell Printing, Tulsa OK
Port City Press Inc, Baltimore MD
Practical Graphics, New York NY
Publishers Express Press, Ladysmith WI
Quebecor Printing (USA) Corp, Boston MA
Repro-Tech, Fairfield CT
Rose Printing Company, Tallahassee FL
Royle Printing, Sun Prairie WI
Sexton Printing, Saint Paul MN
The Sheridan Press, Hanover PA
Spencer Press, Wells ME
Standard Publishing, Cincinnati OH
The Stinehour Press, Lunenburg VT
Strathmore Printing, Geneva IL

TEACH Services, Brushton NY
Trend Offset Printing Service Inc, Los Alamitos CA
Tweddle Litho Co, Clinton Township MI
Wallace Press, Hillside IL
Waverly Press, Linthicum MD
Western Web Printing, Sioux Falls SD
World Color Press Inc, New York NY

Opti-Copy Systems

The Argus Press Inc, Niles IL
Austin Printing Co Inc, Akron OH
Bang Printing Co, Brainerd MN
Banta Company, Menasha WI
Banta Publications Group, Hinsdale IL
Banta-Harrisonburg, Harrisonburg VA
Bawden Printing Inc, Eldridge IA
Berryville Graphics, Berryville VA
Bertelsmann Printing & Manufacturing, New York NY
Best Gagne Book Manufacturers, Toronto ON
Blue Dolphin Press Inc, Grass Valley CA
Bolger Publications/Creative Printing, Minneapolis MN
The Book Press, Brattleboro VT
Braceland Brothers Inc, Philadelphia PA
Braun-Brumfield Inc, Ann Arbor MI
Brown Printing, Waseca MN
Buse Printing & Advertising, Phoenix AZ
The William Byrd Press, Richmond VA
Cal Central Press, Sacramento CA
California Offset Printers, Glendale CA
Capital Printing Company Inc, Austin TX
Champagne Offset Co Inc, Natick MA
Chicago Press Corp, Chicago IL
Citizen Printing, Beaver Dam WI
Colonial Graphics, Paterson NJ
Colotone Riverside Inc, Dallas TX
Commercial Documentation Services, Medford OR
Consolidated Printers Inc, Berkeley CA
Country Press, Mohawk NY
Courier-Westford Inc, Westford MA
Danner Press Corporation, Canton OH
Data Reproductions Corp, Rochester Hills MI
Delta Lithograph Company, Valencia CA
The Dingley Press, Lisbon ME
Dollco Printing, Ottawa ON
R R Donnelley & Sons Company Inc, , Crawfordsville IN
EBSCO Media, Birmingham AL
Editors Press Inc, Hyattsville MD
Edwards Brothers Inc, Ann Arbor MI

Opti-Copy Systems

Flower City Printing Inc, Rochester NY
Gateway Press Inc, Louisville KY
George Lithograph, San Francisco CA
Gilliland Printing Inc, Arkansas City KS
Gowe Printing Company, Medina OH
Graphic Litho Corporation, Lawrence MA
Great Lakes Lithograph Co, Cleveland
    OH
Griffin Printing & Lithograph, Glendale
    CA
Griffin Printing and Lithograph Com-
    pany, Sacramento CA
IPC Publishing Services, Saint Joseph MI
JK Creative Printers, Quincy IL
Keys Printing Co Inc, Greenville SC
Kirby Lithographic Co Inc, Arlington VA
KNI Incorporated, Anaheim CA
Kolor View Press, Aurora MO
The C J Krehbiel Company, Cincinnati
    OH
Lithocolor Press, Westchester IL
Lithoid Printing Corporation, East
    Brunswick NJ
Malloy Lithographing, Ann Arbor MI
Maple-Vail Book Manufacturing Group,
    York PA
The Mazer Corporation, Dayton OH
McArdle Printing Co, Upper Marlboro
    MD
McGrew Color Graphics, Kansas City
    MO
McQuiddy Printing Company, Nashville
    TN
Metroweb, Erlanger KY
Optic Graphics Inc, Glen Burnie MD
The Ovid Bell Press Inc, Fulton MO
Panorama Press Inc, Clifton NJ
Pendell Printing, Midland MI
The Perlmuter Printing Co, Cleveland
    OH
Petty Co Inc, Effingham IL
Port City Press Inc, Baltimore MD
The Press of Ohio, Kent OH
Princeton Academic Press, Lawrenceville
    NJ
Printwest, Regina SK
Progress Printing, Lynchburg VA
Progressive Typographers Inc, Emigsville
    PA
Publishers Press, Salt Lake City UT
Publishers Printing Co, Shepherdsville
    KY
Quebecor Book Group, New York NY
Quebecor Printing (USA) Corp, Boston
    MA
Quebecor/Semline, Braintree MA
Quinn-Woodbine Inc, Woodbine NJ
Rose Printing Company, Tallahassee FL
S Rosenthal & Company Inc, Cincinnati
    OH

Royle Printing, Sun Prairie WI
Science Press, Ephrata PA
Sexton Printing, Saint Paul MN
Shepard Poorman Communications, In-
    dianapolis IN
The Sheridan Press, Hanover PA
Sinclair Printing Co, Los Angeles CA
Southeastern Printing Company Inc,
    Stuart FL
Speakes-Hines & Thomas Inc, Lansing
    MI
St Croix Press Inc, New Richmond WI
Stevens Graphics, Atlanta GA
Strathmore Printing, Geneva IL
Todd Publications Inc, Smithville TX
Trend Offset Printing Service Inc, Los
    Alamitos CA
Tri-Graphic Printing, Ottawa ON
Universal Printing Company, Miami FL
Victor Graphics, Baltimore MD
Wagners Printers, Davenport IA
Waldman Graphics, Pennsauken NJ
Wallace Press, Hillside IL
Walsworth Publishing Co Inc, Marceline
    MO
Webcom Ltd, Scarborough ON
Whitehall Printing Company, Naples FL
Wisconsin Color Press Inc, Milwaukee WI
Wolfer Printing, Commerce CA
World Color Press Inc, New York NY
Worzalla Publishing Company, Stevens
    Point WI

Rachwal System

Alcom Printing Group Inc, Bethlehem PA
American Litho, Hayward CA
Automated Graphic System, White
    Plains MD
Berryville Graphics, Berryville VA
Bertelsmann Printing & Manufacturing,
    New York NY
Best Gagne Book Manufacturers, Toron-
    to ON
Boyd Printing Co Inc, Albany NY
Braun-Brumfield Inc, Ann Arbor MI
Buse Printing & Advertising, Phoenix AZ
Cal Central Press, Sacramento CA
Champagne Offset Co Inc, Natick MA
Consolidated Printers Inc, Berkeley CA
Corley Printing Company, Earth City MO
Courier Stoughton Inc, Stoughton MA
Danner Press Corporation, Canton OH
Des Plaines Publishing Co, Des Plaines
    IL
Dickinson Press Inc, Grand Rapids MI
R R Donnelley & Sons Company Inc, ,
    Crawfordsville IN
Haddon Craftsmen, Scranton PA

Rachwal Systems

Malloy Lithographing, Ann Arbor MI
McNaughton & Gunn Inc, Saline MI
Metromail Corporation, Lincoln NE
MultiPrint Co Inc, Skokie IL
The Ovid Bell Press Inc, Fulton MO
Patterson Printing Company, Benton
    Harbor MI
Peconic Companies, Mattituck NY
Port City Press Inc, Baltimore MD
Science Press, Ephrata PA
Stevens Graphics, Atlanta GA
John S Swift Company Inc, St Louis MO
Thomson-Shore, Dexter MI
Wickersham Printing Co Inc, Lancaster
    PA
World Color Press Inc, New York NY
Worzalla Publishing Company, Stevens
    Point WI

Stock Acid-
Free Paper

Academy Books, Rutland VT
Adams Press, Chicago IL
Adviser Graphics, Red Deer AB
Alcom Printing Group Inc, Bethlehem PA
Allen Press Inc, Lawrence KS
AM Graphics & Printing, San Marcos CA
Anundsen Publishing Company, Decorah
    IA
Apollo Graphics Ltd, Southampton PA
Arcata Graphics/Fairfield, Fairfield PA
The Argus Press Inc, Niles IL
Automated Graphic System, White
    Plains MD
Baker Johnson Inc, Dexter MI
Banta Company, Menasha WI
Banta-Harrisonburg, Harrisonburg VA
Bawden Printing Inc, Eldridge IA
Bay Port Press, Chula Vista CA
Bay State Press, Framingham MA
Beacon Press, Seattle WA
Ben-Wal Printing, Pomona CA
Berryville Graphics, Berryville VA
Bertelsmann Printing & Manufacturing,
    New York NY
Best Gagne Book Manufacturers, Toron-
    to ON
Blue Dolphin Press Inc, Grass Valley CA
Bolger Publications/Creative Printing,
    Minneapolis MN
Book-Mart Press Inc, North Bergen NJ
BookCrafters, Chelsea MI
BookMasters Inc, Ashland OH
Boyd Printing Co Inc, Albany NY

Braun-Brumfield Inc, Ann Arbor MI
Brunswick Publishing Corp, Lawren-
    ceville VA
Buse Printing & Advertising, Phoenix AZ
The William Byrd Press, Richmond VA
Camelot Publishing Co, Ormond Beach
    FL
Canterbury Press, Rome NY
Capital City Press, Montpelier VT
Champagne Offset Co Inc, Natick MA
Color World Printers, Bozeman MT
ColorCorp Inc, Littleton CO
Colorlith Corporation, Johnston RI
Columbus Bookbinders & Printers,
    Columbus GA
Coneco Litho Graphics, Glen Falls NY
Consolidated Printers Inc, Berkeley CA
Continental Web, Itasca IL
Cookbooks by Mom's Press, Kearney NE
Corley Printing Company, Earth City MO
Country Press Inc, Lakeville MA
Courier-Kendallville Corp, Kendallville
    IN
Courier-Westford Inc, Westford MA
Crane Duplicating Service, West
    Barnstable MA
Cushing-Malloy, Ann Arbor MI
Daamen Printing Company, West Rut-
    land VT
Danbury Printing & Litho Inc, Danbury
    CT
Data Reproductions Corp, Rochester
    Hills MI
Dellas Graphics, Syracuse NY
Delmar Printing & Publishing, Charlotte
    NC
Delta Lithograph Company, Valencia CA
Dharma Press, Oakland CA
Dickinson Press Inc, Grand Rapids MI
Dollco Printing, Ottawa ON
R R Donnelley & Sons Company Inc,
    Crawfordsville IN
Dragon Press, Delta Junction AK
Edwards Brothers Inc, Ann Arbor MI
Edwards Brothers Inc, Lillington NC
Eerdmans Printing Co, Grand Rapids MI
EU Services, Rockville MD
Eureka Printing Co Inc, Eureka CA
Evangel Press, Nappanee IN
Faculty Press Inc, Brooklyn NY
Friesen Printers, Altona MB
Gateway Press Inc, Louisville KY
Gilliland Printing Inc, Arkansas City KS
Gowe Printing Company, Medina OH
Graphic Litho Corporation, Lawrence MA
Great Impressions Printing & Graphics,
    Dallas TX
Great Lakes Lithograph Co, Cleveland
    OH
Gregath Publishing Co, Wyandotte OK
Guynes Printing Company, Albuquerque
    NM

Acid-Free Paper

Harlo Printing Co, Detroit MI
Henington Publishing Company, Wolfe City TX
Hignell Printing, Winnepeg MB
Rae Horowitz Book Manufacturers, Fairfield NJ
Imperial Litho/Graphics Inc, Phoenix AZ
Interform Corporation, Bridgeville PA
JK Creative Printers, Quincy IL
Jostens Printing & Publishing, Minneapolis MN
Julin Printing, Monticello IA
Keys Printing Co Inc, Greenville SC
Kirby Lithographic Co Inc, Arlington VA
Kolor View Press, Aurora MO
The C J Krehbiel Company, Cincinnati OH
La Crosse Graphics, LaCrosse WI
Les Editions Marquis, Montmagny PQ
Litho Productions Inc, Madison WI
Lithoid Printing Corporation, East Brunswick NJ
Malloy Lithographing, Ann Arbor MI
Maple-Vail Book Manufacturing Group, York PA
Marathon Communications Group, Wausau WI
Marrakech Express, Tarpon Springs FL
Maverick Publications Inc, Bend OR
MBP Lithographics, Hillsboro KS
McArdle Printing Co, Upper Marlboro MD
McClain Printing Company, Parsons WV
McNaughton & Gunn Inc, Saline MI
Mercury Printing Company, Memphis TN
Merrill Corporation, Los Angeles CA
Moran Printing Company, Orlando FL
Morgan Printing, Austin TX
National Lithographers Inc, Miami FL
Noll Printing Company, Huntington IN
Odyssey Press Inc, Dover NH
The Ovid Bell Press Inc, Fulton MO
Paraclete Press, Brewster MA
Patterson Printing Company, Benton Harbor MI
Paust Inc, Richmond IN
Peconic Companies, Mattituck NY
Pendell Printing, Midland MI
PennWell Printing, Tulsa OK
Pentagram Press, Minneapolis MN
PFP Printing Corporation, Gaithersburg MD
Pine Hill Press Inc, Freeman SD
Port City Press Inc, Baltimore MD
Princeton Academic Press, Lawrenceville NJ
Printright International, Palmdale CA
Progress Printing, Lynchburg VA
Publishers Press, Salt Lake City UT

Quebecor Book Group, New York NY
Quebecor Printing (USA) Corp, Boston MA
Quinn-Woodbine Inc, Woodbine NJ
Rose Printing Company, Tallahassee FL
S Rosenthal & Company Inc, Cincinnati OH
Roxbury Publishing Company, Los Angeles CA
The Saratoga Printing Co, Saratoga Springs NY
Science Press, Ephrata PA
Shepard Poorman Communications, Indianapolis IN
The Sheridan Press, Hanover PA
Sheridan Printing Company, Phillipsburg NJ
Standard Publishing, Cincinnati OH
The Stinehour Press, Lunenburg VT
John S Swift Company Inc, St Louis MO
Technical Communication Services, North Kansas City MO
Thomson-Shore, Dexter MI
Times Printing, Random Lake WI
Town House Press, Pittsboro NC
Trade Litho Inc, Miami FL
Tri-Graphic Printing, Ottawa ON
Versa Press Inc, East Peoria IL
Vicks Lithograph & Printing, Yorkville NY
Victor Graphics, Baltimore MD
Viking Press, Eden Prairie MN
Walsworth Publishing Co Inc, Marceline MO
Waverly Press, Linthicum MD
Webcom Ltd, Scarborough ON
F A Weber & Sons, Park Falls WI
Weber Printing Company, Lincoln NE
Westview Press, Boulder CO
White Arts Inc, Indianapolis IN
Whitehall Printing Company, Naples FL
Wickersham Printing Co Inc, Lancaster PA
Wood & Jones Printers, Pasadena CA
Y/Z Printing Co, Elizabethtown PA

Stock Recycled Paper

A-1 Business Service Inc, St Paul MN
Accurate Printing, Post Falls ID
Action Printing, Fond Du Lac WI
Adams Press, Chicago IL
Alan Lithograph Inc, Inglewood CA
Alcom Printing Group Inc, Bethlehem PA
All-Star Printing, Lansing MI
AM Graphics & Printing, San Marcos CA

Stock Recycled Paper

American Litho, Hayward CA
Apollo Graphics Ltd, Southampton PA
Arcata Graphics/Fairfield, Fairfield PA
The Argus Press Inc, Niles IL
Associated Printers, Grafton ND
Austin Printing Co Inc, Akron OH
Automated Graphic System, White
    Plains MD
Baker Johnson Inc, Dexter MI
Bang Printing Co, Brainerd MN
Banta Company, Menasha WI
Banta Publications Group, Hinsdale IL
Banta-Harrisonburg, Harrisonburg VA
Bawden Printing Inc, Eldridge IA
Bay State Press, Framingham MA
Beacon Press, Seattle WA
Ben-Wal Printing, Pomona CA
Bertelsmann Printing & Manufacturing,
    New York NY
Best Gagne Book Manufacturers, Toron-
    to ON
Blue Dolphin Press Inc, Grass Valley CA
Bolger Publications/Creative Printing,
    Minneapolis MN
BookCrafters, Chelsea MI
BookMasters Inc, Ashland OH
Boyd Printing Co Inc, Albany NY
Braun-Brumfield Inc, Ann Arbor MI
Brown Printing/East Greenville Division,
    East Greenville PA
Brunswick Publishing Corp, Lawren-
    ceville VA
Buse Printing & Advertising, Phoenix AZ
The William Byrd Press, Richmond VA
Cal Central Press, Sacramento CA
Caldwell Printers, Arcadia CA
California Offset Printers , Glendale CA
Calsonic Miura Graphics, Irvine CA
Canterbury Press, Rome NY
Capital Printing Company Inc, Austin TX
Champagne Offset Co Inc, Natick MA
Chicago Color Express, Chicago IL
Chicago Press Corp, Chicago IL
Citizen Printing, Beaver Dam WI
Colorlith Corporation, Johnston RI
Colotone Riverside Inc, Dallas TX
Coneco Litho Graphics, Glen Falls NY
Cookbook Publishers Inc, Lenexa KS
Cookbooks by Mom's Press, Kearney NE
Cooley Printers & Office Supply, Monroe
    LA
Corley Printing Company, Earth City MO
Country Press, Mohawk NY
Country Press Inc, Lakeville MA
Crane Duplicating Service, West
    Barnstable MA
Crusader Printing Co, East Saint Louis
    IL
Danbury Printing & Litho Inc, Danbury
    CT

Danner Press Corporation, Canton OH
Data Reproductions Corp, Rochester
    Hills MI
Dellas Graphics, Syracuse NY
Delmar Printing & Publishing, Charlotte
    NC
Delta Lithograph Company, Valencia CA
Des Plaines Publishing Co, Des Plaines
    IL
Dharma Press, Oakland CA
The Dingley Press, Lisbon ME
Direct Press/Modern Litho, Huntington
    Station NY
Dollco Printing, Ottawa ON
Donihe Graphics Inc, Kingsport TN
Dragon Press, Delta Junction AK
Eastwood Printing Co, Denver CO
Editors Press Inc, Hyattsville MD
Edwards Brothers Inc, Ann Arbor MI
Eerdmans Printing Co, Grand Rapids MI
EU Services, Rockville MD
Eureka Printing Co Inc, Eureka CA
Faculty Press Inc, Brooklyn NY
First Impressions Printer & Lithog-
    raphers Inc, Elk Grove IL
Fleming Printing Co, St Louis MO
Flower City Printing Inc, Rochester NY
The Forms Man Inc, Deer Park NY
Fort Orange Press Inc, Albany NY
Friesen Printers, Altona MB
Gateway Press Inc, Louisville KY
George Lithograph, San Francisco CA
Gilliland Printing Inc, Arkansas City KS
Golden Belt Printing Inc, Great Bend KS
Gowe Printing Company, Medina OH
Graphic Litho Corporation, Lawrence MA
Great Impressions Printing & Graphics,
    Dallas TX
Great Lakes Lithograph Co, Cleveland
    OH
Gregath Publishing Co, Wyandotte OK
Griffin Printing & Lithograph, Glendale
    CA
GRT Book Printing, Oakland CA
Guynes Printing Company, Albuquerque
    NM
Haddon Craftsmen, Scranton PA
Hignell Printing, Winnepeg MB
A B Hirschfeld Press, Denver CO
Rae Horowitz Book Manufacturers, Fair-
    field NJ
Imperial Litho/Graphics Inc, Phoenix AZ
Interform Corporation, Bridgeville PA
JK Creative Printers, Quincy IL
Jostens Printing & Publishing, Min-
    neapolis MN
Keys Printing Co Inc, Greenville SC
KNI Incorporated, Anaheim CA
Kolor View Press, Aurora MO
The C J Krehbiel Company, Cincinnati
    OH
La Crosse Graphics, LaCrosse WI

Stock Recycled Paper

The Lane Press, Burlington VT
Lighthouse Press Inc, Manchester NH
Litho Productions Inc, Madison WI
Lithograph Printing Company, Memphis TN
Lithoid Printing Corporation, East Brunswick NJ
Long Island Web Printing, Jericho NY
Malloy Lithographing, Ann Arbor MI
Maple-Vail Book Group, York PA
Marathon Communications Group, Wausau WI
Marrakech Express, Tarpon Springs FL
Maverick Publications Inc, Bend OR
The Mazer Corporation, Dayton OH
McArdle Printing Co, Upper Marlboro MD
McNaughton & Gunn Inc, Saline MI
McQuiddy Printing Company, Nashville TN
Metroweb, Erlanger KY
Moran Printing Company, Orlando FL
Moran Printing Inc, Baton Rouge LA
Morgan Printing, Austin TX
National Lithographers Inc, Miami FL
Louis Neibauer Co Inc, Warminster PA
Northlight Studio Press, Barre VT
Nystrom Publishing Company, Osseo MN
Oaks Printing Company, Bethlehem PA
Odyssey Press Inc, Dover NH
OMNIPRESS, Madison WI
Outstanding Graphics, Kenosha WI
The Ovid Bell Press Inc, Fulton MO
Page Litho Inc, Detroit MI
Panorama Press Inc, Clifton NJ
Paraclete Press, Brewster MA
Park Press, South Holland IL
Patterson Printing Company, Benton Harbor MI
Pearl-Pressman-Liberty Communications Group, Philadelphia PA
PennWell Printing, Tulsa OK
The Perlmuter Printing Co, Cleveland OH
Pine Hill Press Inc, Freeman SD
Port City Press Inc, Baltimore MD
Practical Graphics, New York NY
Princeton Academic Press, Lawrenceville NJ
Printing Corporation of America, Pompano Beach FL
Printright International, Palmdale CA
Printwest, Regina SK
Progress Printing, Lynchburg VA
Publishers Printing Co, Shepherdsville KY
Quebecor Printing (USA) Corp, Boston MA
Rich Printing Company, Nashville TN
Rose Printing Company, Tallahassee FL

S Rosenthal & Company Inc, Cincinnati OH
Royle Printing, Sun Prairie WI
Sanders Printing Company, Garretson SD
The Saratoga Printing Co, Saratoga Springs NY
Schelin Printing Co, St Louis MO
Science Press, Ephrata PA
Service Web Offset Corporation, Chicago IL
Sexton Printing, Saint Paul MN
Shepard Poorman Communications, Indianapolis IN
Sheridan Printing Company, Phillipsburg NJ
Southeastern Printing Company Inc, Stuart FL
Speakes-Hines & Thomas Inc, Lansing MI
Spencer Press, Wells ME
St Croix Press Inc, New Richmond WI
Standard Publishing, Cincinnati OH
John Stark Printing Co Inc, St Louis MO
The Stinehour Press, Lunenburg VT
Strathmore Printing, Geneva IL
Straus Printing Company, Madison WI
John S Swift Company Inc, St Louis MO
Taylor Publishing Company, Dallas TX
TEACH Services, Brushton NY
Technical Communication Services, North Kansas City MO
Thomson-Shore, Dexter MI
Trade Litho Inc, Miami FL
Trend Offset Printing Service Inc, Los Alamitos CA
Tri-Graphic Printing, Ottawa ON
Tweddle Litho Co, Clinton Township MI
Universal Printing Company, Miami FL
US Press Inc, Valdosta GA
Versa Press Inc, East Peoria IL
Viking Press, Eden Prairie MN
Wagners Printers, Davenport IA
Waldon Press Inc, New York NY
Wallace Press, Hillside IL
Walsworth Publishing Co Inc, Marceline MO
Webcom Ltd, Scarborough ON
Weber Printing Company, Lincoln NE
Western Web Printing, Sioux Falls SD
Westview Press, Boulder CO
White Arts Inc, Indianapolis IN
Whitehall Printing Company, Naples FL
Wickersham Printing Co Inc, Lancaster PA
Wisconsin Color Press Inc, Milwaukee WI
Wolfer Printing, Commerce CA
Wood & Jones Printers, Pasadena CA
Y/Z Printing Co, Elizabethtown PA

Typesetting

A-1 Business Service Inc, St Paul MN
aBCD, Seattle WA
Accurate Printing, Post Falls ID
Action Printing, Fond Du Lac WI
Ad Infinitum Press, Mount Vernon NY
Adams Press, Chicago IL
Adviser Graphics, Red Deer AB
Alcom Printing Group Inc, Bethlehem PA
All-Star Printing, Lansing MI
Allen Press Inc, Lawrence KS
AM Graphics & Printing, San Marcos CA
American Litho, Hayward CA
American Signature, Atlanta GA
Americomp, Brattleboro VT
Anundsen Publishing Company, Decorah
    IA
Apollo Graphics Ltd, Southampton PA
The Argus Press Inc, Niles IL
Associated Printers, Grafton ND
Austin Printing Co Inc, Akron OH
Automated Graphic Systems, White
    Plains MD
Bang Printing Co, Brainerd MN
Bay Port Press, Chula Vista CA
Bay State Press, Framingham MA
Beacon Press, Seattle WA
Ben-Wal Printing, Pomona CA
Berryville Graphics, Berryville VA
Blake Printery, San Luis Obispo CA
Blue Dolphin Press Inc, Grass Valley CA
Bolger Publications/Creative Printing,
    Minneapolis MN
BookMasters Inc, Ashland OH
Boyd Printing Co Inc, Albany NY
Braun-Brumfield Inc, Ann Arbor MI
Brennan Printing, Deep River IA
Brown Printing/East Greenville Division,
    East Greenville PA
Brunswick Publishing Corp, Lawren-
    ceville VA
R L Bryan Company, Columbia SC
BSC Litho, Harrisburg PA
Buse Printing & Advertising, Phoenix AZ
Cal Central Press, Sacramento CA
Caldwell Printers, Arcadia CA
Calsonic Miura Graphics, Irvine CA
Camelot Publishing Co, Ormond Beach
    FL
Canterbury Press, Rome NY
Capital City Press, Montpelier VT
Carlson Color Graphics, Ocala FL
Catalog King, Clifton NJ
Champagne Offset Co Inc, Natick MA
Citizen Printing, Beaver Dam WI
Coach House Press, Toronto ON
The College Press, Collegedale TN
Colonial Graphics, Paterson NJ
Color World Printers, Bozeman MT

ColorGraphics Printing Corp, Tulsa OK
Colorlith Corporation, Johnston RI
Colortone Press, Landover MD
Colotone Riverside Inc, Dallas TX
Columbus Bookbinders & Printers,
    Columbus GA
Combined Communication Services,
    Mendota IL
Comfort Printing, St Louis MO
Communicolor, Newark OH
Community Press, Provo UT
Concord Litho Company Inc, Concord
    NH
Coneco Litho Graphics, Glen Falls NY
Cookbook Publishers Inc, Lenexa KS
Cookbooks by Mom's Press, Kearney NE
Cooley Printers & Office Supply Monroe
    LA
Corley Printing Company, Earth City MO
Country Press, Mohawk NY
The Craftsman Press Inc, Seattle WA
Crane Duplicating Service, West
    Barnstable MA
Crusader Printing Co, East Saint Louis
    IL
Lew A Cummings Company, Manchester
    NH
Danbury Printing & Litho Inc, Danbury
    CT
Dartmouth Printing, Hanover NH
Data Reproductions Corp, Rochester
    Hills MI
Dataco A WBC Company, Albuquerque
    NM
Dellas Graphics, Syracuse NY
Delmar Printing & Publishing, Charlotte
    NC
Des Plaines Publishing Co, Des Plaines
    IL
Desaulniers Printing Company, Milan IL
Dharma Press, Oakland CA
Diamond Graphics Tech Data, Mil-
    waukee WI
Direct Press/Modern Litho, Huntington
    Station NY
Directories America, Chicago IL
Dollco Printing, Ottawa ON
Donihe Graphics Inc, Kingsport TN
R R Donnelley & Sons Company Inc,
    Crawfordsville IN
Dragon Press, Delta Junction AK
Edison Lithographing, North Bergen NJ
Edwards Brothers Inc, Ann Arbor MI
Edwards Brothers Inc, Lillington NC
Eerdmans Printing Co, Grand Rapids MI
EP Graphics, Berne IN
EU Services, Rockville MD
Eureka Printing Co Inc, Eureka CA
Evangel Press, Nappanee IN
Fetter Printing, Louisville KY
Flower City Printing Inc, Rochester NY
Focus Direct, San Antonio TX

Typesetting

Fort Orange Press Inc, Albany NY
Foster Printing Service, Michigan City IN
Fundcraft Publishing, Collierville TN
Futura Printing Inc, Boynton Beach FL
Gateway Press Inc, Louisville KY
Geryon Press Limited, Tunnel NY
Golden Belt Printing Inc, Great Bend KS
Gorham Printing, Rochester WA
Gowe Printing Company, Medina OH
Graphic Arts Center, Portland OR
Graphics Ltd, Indianapolis IN
Gray Printing, Fostoria OH
Great Impressions Printing & Graphics,
    Dallas TX
Gregath Publishing Co, Wyandotte OK
GRIT Commercial Printing Services, Wil-
    liamsport PA
Guynes Printing Company, Albuquerque
    NM
Haddon Craftsmen, Scranton PA
Harlo Printing Co, Detroit MI
Harmony Printing, Liberty MO
Hart Graphics, Austin TX
Hawkes Publishing, Salt Lake City UT
Heritage Printers Inc, Charlotte NC
Hignell Printing, Winnepeg MB
Hinz Lithographing Company, Mt
    Prospect IL
A B Hirschfeld Press, Denver CO
Independent Publishing Company,
    Sarasota FL
Interform Corporation, Bridgeville PA
Interstate Printing Company, Omaha NE
IPC Publishing Services, Saint Joseph MI
Japs-Olson Company, Minneapolis MN
Jersey Printing Company, Bayonne NJ
JK Creative Printers, Quincy IL
The Job Shop, Woods Hole MA
The Johnson & Hardin Company, Cin-
    cinnati OH
Jostens Printing & Publishing, Visalia CA
Jostens Printing & Publishing,
    Clarksville TN
Jostens Printing & Publishing, Winston-
    Salem NC
Jostens Printing & Publishing, Topeka
    KS
Jostens Printing & Publishing, State Col-
    lege PA
Jostens Printing & Publishing, Min-
    neapolis MN
Kaufman Press Printing, Syracuse NY
Keys Printing Co Inc, Greenville SC
Kolor View Press, Aurora MO
Kordet Graphics Inc, Oceanside NY
La Crosse Graphics, LaCrosse WI
Lancaster Press, Lancaster PA
Lasky Company, Milburn NJ
Les Editions Marquis, Montmagny PQ

Lithograph Printing Company, Memphis
    TN
Lithoid Printing Corporation, East
    Brunswick NJ
Long Island Web Printing, Jericho NY
John D Lucas Printing Company, Bal-
    timore MD
Mack Printing Group Inc, Easton PA
Mackintosh Typography Inc, Santa Bar-
    bara CA
The Mad Printers, Mattituck NY
Mail-O-Graph, Kewanee IL
Maple-Vail Book Manufacturing Group,
    York PA
Maquoketa Web Printing, Maquoketa IA
Marathon Communications Group,
    Wausau WI
Mark IV Press, Hauppauge NY
Maverick Publications Inc, Bend OR
The Mazer Corporation, Dayton OH
MBP Lithographics, Hillsboro KS
McClain Printing Company, Parsons WV
McGill Jensen, St Paul MN
McGrew Color Graphics, Kansas City
    MO
McQuiddy Printing Company, Nashville
    TN
Meaker the Printer, Phoenix AZ
Mercury Printing Company, Memphis
    TN
Mitchell Graphics, Petosky MI
Mitchell Press, Vancouver BC
Monument Printers & Lithographer Inc,
    Verplanck NY
Moran Printing Company, Orlando FL
Moran Printing Inc, Baton Rouge LA
Morgan Press, Dobbs Ferry NY
Morgan Printing, Austin TX
Morningrise Printing, Costa Mesa CA
MultiPrint Co Inc, Skokie IL
National Graphics Corp, Columbus OH
Newsfoto Publishing Company, San An-
    gelo TX
Neyenesch Printers, San Diego CA
Louis Neibauer Co Inc, Warminster PA
The Nielsen Lithographing Co, Cincin-
    nati OH
Nimrod Press Inc, Boston MA
Noll Printing Company, Huntington IN
Northlight Studio Press, Barre VT
Nystrom Publishing Company, Osseo MN
Omaha Printing Co, Omaha NE
Optic Graphics Inc, Glen Burnie MD
Outstanding Graphics, Kenosha WI
The Ovid Bell Press Inc, Fulton MO
The Oxford Group, Norway ME
Pantagraph Printing, Bloomington IL
Paraclete Press, Brewster MA
Parker Graphics, Fuquay-Varina NC
Paust Inc, Richmond IN
Pearl-Pressman-Liberty Communica-
    tions Group, Philadelphia PA

Typesetting

Pendell Printing, Midland MI
PennWell Printing, Tulsa OK
The Perlmuter Printing Co, Cleveland OH
Petty Co Inc, Effingham IL
PFP Printing Corporation, Gaithersburg MD
Pine Hill Press Inc, Freeman SD
Pioneer Press, Wilmette IL
Plus Communications, Saint Louis MO
Port City Press Inc, Baltimore MD
Practical Graphics, New York NY
Prinit Press, Dublin IN
The Printer Inc, Osseo MN
The Printing Company, Indianapolis IN
Printwest, Regina SK
Progress Printing, Lynchburg VA
Progressive Typographers Inc, Emigsville PA
Promotional Printing Corporation, Houston TX
Publishers Choice Reprints, The Woodlands TX
Publishers Printing Co, Shepherdsville KY
Quad Graphics — Thomaston, Thomaston GA
Quad/Graphics Inc, Saratoga Springs NY
Quad/Graphics, Lomira, Lomira WI
Quad/Graphics, Pewaukee, Pewaukee WI
Quad/Graphics, Sussex, Sussex WI
Quebecor Printing Inc, St Paul MN
Quebecor Printing/St Cloud, Saint Cloud MN
RBW Graphics, Owen Sound ON
Repro-Tech, Fairfield CT
Ringier America/Brookfield, Brookfield WI
Ringier America/Jonesboro, Jonesboro AR
Ringier America/Phoenix, Phoenix AZ
Ronalds Printing, Vancouver BC
S Rosenthal & Company Inc, Cincinnati OH
Roxbury Publishing Company, Los Angeles CA
Sanders Printing Company, Garretson SD
Schiff Printing, Pittsburgh PA
Science Press, Ephrata PA
Sexton Printing, Saint Paul MN
Shepard Poorman Communications, Indianapolis IN
The Sheridan Press, Hanover PA
Sheridan Printing Company, Phillipsburg NJ
R E Smith Printing, Fall River MA
Smith-Edwards-Dunlap, Philadelphia PA
Snohomish Publishing, Snohomish WA

Southeastern Printing Company Inc, Stuart FL
Speakes-Hines & Thomas Inc, Lansing MI
St Croix Press Inc, New Richmond WI
Standard Printing Service, Chicago IL
Standard Publishing, Cincinnati OH
Stewart Publishing & Printing Co, Marble Hill MO
The Stinehour Press, Lunenburg VT
Strathmore Printing, Geneva IL
The Studley Press, Dalton MA
Sullivan Graphics Inc, Brentwood TN
Sun Graphics, Parsons KS
The Sun Hill Press, North Brookfield MA
John S Swift Company Inc, St Louis MO
John S Swift/Chicago, Chicago IL
John S Swift/Cincinnati, Cincinnati OH
John S Swift/Teterboro, Teterboro NJ
TAN Books & Publishers Inc, Rockford IL
Tapco Inc, Pemberton NJ
Tapemark, West St Paul MN
Taylor Publishing Company, Dallas TX
TEACH Services, Brushton NY
Times Litho Inc, Forest Grove OR
Times Printing, Random Lake WI
Town House Press, Pittsboro NC
Tracor Publications, Austin TX
Trade Litho Inc, Miami FL
Trend Offset Printing Service Inc, Los Alamitos CA
Tri-Graphic Printing, Ottawa ON
Triangle Booklet Company, Nashville TN
Tweddle Litho Co, Clinton Township MI
Ultra-Color Corporation, Saint Louis MO
United Lithographic Services Inc, Falls Church VA
US Press Inc, Valdosta GA
Vail Ballou Press Binghamton NY
Valco Graphics, Seattle WA
Victor Graphics, Baltimore MD
Vogue Printers, North Chicago IL
Waldman Graphics, Pennsauken NJ
Waldon Press Inc, New York NY
Wallace Press, Hillside IL
Walsworth Publishing Co Inc, Marceline MO
Walter's Publishing, Waseca MN
Watkins Printing Company, Columbus OH
Waverly Press, Linthicum MD
Webcom Ltd, Scarborough ON
F A Weber & Sons, Park Falls WI
Weber Printing Company, Lincoln NE
Western Publishing Co Inc, Racine WI
Western Web Printing, Sioux Falls SD
Westview Press, Boulder CO
Windsor Associates, San Diego CA
Winter Sun Inc, Phoenix AZ
Wisconsin Color Press Inc, Milwaukee WI
Wood & Jones Printers, Pasadena CA

Typesetting

World Color Press Inc, New York NY
Y/Z Printing Co, Elizabethtown PA

Typesetting Via Modem or Disk

aBCD, Seattle WA
Accurate Printing, Post Falls ID
Action Printing, Fond Du Lac WI
Adams Press, Chicago IL
Adviser Graphics, Red Deer AB
Alcom Printing Group Inc, Bethlehem PA
All-Star Printing, Lansing MI
Allen Press Inc, Lawrence KS
AM Graphics & Printing, San Marcos CA
American Litho, Hayward CA
Americomp, Brattleboro VT
Anundsen Publishing Company, Decorah
    IA
Apollo Graphics Ltd, Southampton PA
Arcata Graphics/Fairfield, Fairfield PA
The Argus Press Inc, Niles IL
Associated Printers, Grafton ND
Austin Printing Co Inc, Akron OH
Automated Graphic Systems, White
    Plains MD
Bang Printing Co, Brainerd MN
Banta Company, Menasha WI
Banta Publications Group, Hinsdale IL
Bawden Printing Inc, Eldridge IA
Bay State Press, Framingham MA
Beacon Press, Seattle WA
Ben-Wal Printing, Pomona CA
Berryville Graphics, Berryville VA
Bertelsmann Printing & Manufacturing,
    New York NY
Best Gagne Book Manufacturer, Toronto
    ON
Blue Dolphin Press Inc, Grass Valley CA
Bolger Publications/Creative Printing,
    Minneapolis MN
BookCrafters, Chelsea MI
BookMasters Inc, Ashland OH
Boyd Printing Co Inc, Albany NY
Braceland Brothers Inc, Philadelphia PA
Braun-Brumfield Inc, Ann Arbor MI
Brown Printing, Waseca MN
Brown Printing/East Greenville Division,
    East Greenville PA
Brunswick Publishing Corp, Lawren-
    ceville VA
R L Bryan Company, Columbia SC
Buse Printing & Advertising, Phoenix AZ
The William Byrd Press, Richmond VA
Cal Central Press, Sacramento CA

Caldwell Printers, Arcadia CA
Calsonic Miura Graphics, Irvine CA
Canterbury Press, Rome NY
Capital City Press, Montpelier VT
Carpenter Lithographic Co, Springfield
    OH
Champagne Offset Co Inc, Natick MA
Chicago Press Corp, Chicago IL
Citizen Printing, Beaver Dam WI
The College Press, Collegedale TN
Color World Printers, Bozeman MT
Colorlith Corporation, Johnston RI
Colortone Press, Landover MD
Colotone Riverside Inc, Dallas TX
Columbus Bookbinders & Printers,
    Columbus GA
Comfort Printing, St Louis MO
Concord Litho Company Inc, Concord
    NH
Coneco Litho Graphics, Glen Falls NY
Cooley Printers & Office Supply, Monroe
    LA
Country Press, Mohawk NY
Crane Duplicating Service, West
    Barnstable MA
Crusader Printing Co, East Saint Louis
    IL
Danbury Printing & Litho Inc, Danbury
    CT
Danner Press Corporation, Canton OH
Dartmouth Printing, Hanover NH
Dellas Graphics, Syracuse NY
Delmar Printing & Publishing, Charlotte
    NC
Delta Lithograph Company, Valencia CA
Des Plaines Publishing Co, Des Plaines
    IL
Dharma Press, Oakland CA
Diamond Graphics Tech Data, Mil-
    waukee WI
Dickinson Press Inc, Grand Rapids MI
Direct Press/Modern Litho, Huntington
    Station NY
Directories America, Chicago IL
Dollco Printing, Ottawa ON
Donihe Graphics Inc, Kingsport TN
R R Donnelley & Sons Company Inc,
    Crawfordsville IN
Dragon Press, Delta Junction AK
EBSCO Media, Birmingham AL
Edwards Brothers Inc, Ann Arbor MI
Edwards Brothers Inc, Lillington NC
Eerdmans Printing Co, Grand Rapids MI
EU Services, Rockville MD
Fleming Printing Co, St Louis MO
Flower City Printing Inc, Rochester NY
Fort Orange Press Inc, Albany NY
Gateway Press Inc, Louisville KY
George Lithograph, San Francisco CA
Gilliland Printing Inc, Arkansas City KS
Golden Belt Printing Inc, Great Bend KS
Gorham Printing, Rochester WA

Typesetting via Modem or Disk

Gowe Printing Company, Medina OH
Graphic Litho Corporation, Lawrence MA
Graphics Ltd, Indianapolis IN
Gray Printing, Fostoria OH
Great Impressions Printing & Graphics, Dallas TX
Great Lakes Lithograph Co, Cleveland OH
Gregath Publishing Co, Wyandotte OK
GRIT Commercial Printing Services, Williamsport PA
Guynes Printing Company, Albuquerque NM
Haddon Craftsmen, Scranton PA
Harlo Printing Co, Detroit MI
Harmony Printing, Liberty MO
Hart Press, Long Prairie MN
Hawkes Publishing, Salt Lake City UT
Hignell Printing, Winnepeg MB
A B Hirschfeld Press, Denver CO
Independent Publishing Company, Sarasota FL
Interform Corporation, Bridgeville PA
Interstate Printing Company, Omaha NE
IPC Publishing Services, Saint Joseph MI
Jersey Printing Company, Bayonne NJ
JK Creative Printers, Quincy IL
The Job Shop, Woods Hole MA
Jostens Printing & Publishing, Clarksville TN
Jostens Printing & Publishing, Winston-Salem NC
Jostens Printing & Publishing, Topeka KS
Jostens Printing & Publishing, State College PA
Jostens Printing & Publishing, Minneapolis MN
Keys Printing Co Inc, Greenville SC
Kirby Lithographic Co Inc, Arlington VA
Kolor View Press, Aurora MO
La Crosse Graphics, LaCrosse WI
Lancaster Press, Lancaster PA
The Lane Press, Burlington VT
Les Editions Marquis, Montmagny PQ
Lighthouse Press Inc, Manchester NH
Lithograph Printing Company, Memphis TN
Lithoid Printing Corporation, East Brunswick NJ
Mack Printing Group Inc, Easton PA
Mackintosh Typography Inc, Santa Barbara CA
The Mad Printers, Mattituck NY
Maple-Vail Book Manufacturing Group, York PA
Maquoketa Web Printing, Maquoketa IA
Marathon Communications Group, Wausau WI
Mark IV Press, Hauppauge NY

Marrakech Express, Tarpon Springs FL
Maverick Publications Inc, Bend OR
The Mazer Corporation, Dayton OH
MBP Lithographics, Hillsboro KS
McArdle Printing Co, Upper Marlboro MD
McClain Printing Company, Parsons WV
McGrew Color Graphics, Kansas City MO
McNaughton & Gunn Inc, Saline MI
McQuiddy Printing Company, Nashville TN
Meaker the Printer, Phoenix AZ
Mercury Printing Company, Memphis TN
Metroweb, Erlanger KY
Monument Printers & Lithographer Inc, Verplanck NY
Moran Printing Inc, Baton Rouge LA
Morgan Printing, Austin TX
MultiPrint Co Inc, Skokie IL
National Graphics Corp, Columbus OH
Louis Neibauer Co Inc, Warminster PA
Nimrod Press Inc, Boston MA
Noll Printing Company, Huntington IN
Northlight Studio Press, Barre VT
Nystrom Publishing Company, Osseo MN
Oaks Printing Company, Bethlehem PA
Omaha Printing Co, Omaha NE
OMNIPRESS, Madison WI
Optic Graphics Inc, Glen Burnie MD
The Ovid Bell Press Inc, Fulton MO
The Oxford Group, Norway ME
Padgett Printing Corporation, Dallas TX
Page Litho Inc, Detroit MI
Panorama Press Inc, Clifton NJ
Patterson Printing Company, Benton Harbor MI
Pearl-Pressman-Liberty Communications Group, Philadelphia PA
Peconic Companies, Mattituck NY
Pendell Printing, Midland MI
PennWell Printing, Tulsa OK
The Perlmuter Printing Co, Cleveland OH
Perry Printing Corp, Waterloo WI
Perry/Baraboo, Baraboo WI
PFP Printing Corporation, Gaithersburg MD
Pine Hill Press Inc, Freeman SD
Plus Communications, Saint Louis MO
Port City Press Inc, Baltimore MD
Practical Graphics, New York NY
Princeton Academic Press, Lawrenceville NJ
The Printing Company, Indianapolis IN
Printwest, Regina SK
Progress Printing, Lynchburg VA
Progressive Typographers Inc, Emigsville PA
Publishers Choice Reprints, The Woodlands TX

Typesetting via Modem or Disk

Publishers Printing Co, Shepherdsville KY
Quebecor Book Group, New York NY
Quebecor Printing (USA) Corp, Boston MA
RBW Graphics, Owen Sound ON
Repro-Tech, Fairfield CT
Ronalds Printing, Vancouver BC
Rose Printing Company, Tallahassee FL
S Rosenthal & Company Inc, Cincinnati OH
Roxbury Publishing Company, Los Angeles CA
Royle Printing, Sun Prairie WI
Sanders Printing Company, Garretson SD
Schelin Printing Co, St Louis MO
Schiff Printing, Pittsburgh PA
Science Press, Ephrata PA
Sexton Printing, Saint Paul MN
Shepard Poorman Communications, Indianapolis IN
The Sheridan Press, Hanover PA
Sheridan Printing Company, Phillipsburg NJ
Sinclair Printing Co, Los Angeles CA
Snohomish Publishing, Snohomish WA
Southeastern Printing Company Inc, Stuart FL
Speakes-Hines & Thomas Inc, Lansing MI
St Croix Press Inc, New Richmond WI
Standard Publishing, Cincinnati OH
Stewart Publishing & Printing Co, Marble Hill MO
The Stinehour Press, Lunenburg VT
Strathmore Printing, Geneva IL
The Studley Press, Dalton MA
John S Swift Company Inc, St Louis MO
Tapemark, West St Paul MN
Taylor Publishing Company, Dallas TX
TEACH Services, Brushton NY
Technical Communication, North Kansas City MO
Times Printing, Random Lake WI
Town House Press, Pittsboro NC
Trade Litho Inc, Miami FL
Trend Offset Printing Service Inc, Los Alamitos CA
Tri-Graphic Printing, Ottawa ON
Tweddle Litho Co, Clinton Township MI
United Lithographic Services Inc, Falls Church VA
Vail Ballou Press, Binghamton NY
Valco Graphics, Seattle WA
Victor Graphics, Baltimore MD
Viking Press, Eden Prairie MN
Vogue Printers, North Chicago IL
Waldman Graphics, Pennsauken NJ
Waldon Press Inc, New York NY

Wallace Press, Hillside IL
Walsworth Publishing Co Inc, Marceline MO
Waverly Press, Linthicum MD
Webcom Ltd, Scarborough ON
Weber Printing Company, Lincoln NE
Western Publishing Co Inc, Racine WI
Western Web Printing, Sioux Falls SD
Westview Press, Boulder CO
White Arts Inc, Indianapolis IN
Whitehall Printing Company, Naples FL
Wisconsin Color Press Inc, Milwaukee WI
Wood & Jones Printers, Pasadena CA
World Color Press Inc, New York NY
Worzalla Publishing Company, Stevens Point WI
Y/Z Printing Co, Elizabethtown PA

Printers Who Warehouse

Action Printing, Fond Du Lac WI
Ad Infinitum Press, Mount Vernon NY
Advanced Data Reproductions, Wichita KS
Adviser Graphics, Red Deer AB
Alcom Printing Group Inc, Bethlehem PA
All-Star Printing, Lansing MI
Allen Press Inc, Lawrence KS
Apollo Graphics Ltd, Southampton PA
Arandell Corporation, Menomonee Falls WI
The Argus Press Inc, Niles IL
Bang Printing Co, Brainerd MN
Banta Company, Menasha WI
Banta-Harrisonburg, Harrisonburg VA
Bawden Printing Inc, Eldridge IA
Bay State Press, Framingham MA
Beacon Press, Seattle WA
Berryville Graphics, Berryville VA
Bertelsmann Printing & Manufacturing, New York NY
Bolger Publications/Creative Printing, Minneapolis MN
BookCrafters, Chelsea MI
BookMasters Inc, Ashland OH
Boyd Printing Co Inc, Albany NY
Brennan Printing, Deep River IA
Brown Printing/East Greenville Division, East Greenville PA
R L Bryan Company, Columbia SC
BSC Litho, Harrisburg PA
Buse Printing & Advertising, Phoenix AZ
The William Byrd Press, Richmond VA
Cal Central Press, Sacramento CA
Calsonic Miura Graphics, Irvine CA
Canterbury Press, Rome NY

Printers who Warehouse

Capital City Press, Montpelier VT
Carpenter Lithographic Co, Springfield OH
Champagne Offset Co Inc, Natick MA
Chicago Press Corp, Chicago IL
Citizen Printing, Beaver Dam WI
Colonial Graphics, Paterson NJ
Colorlith Corporation, Johnston RI
Colotone Riverside Inc, Dallas TX
Columbus Bookbinders & Printers, Columbus GA
Comfort Printing, St Louis MO
Community Press, Provo UT
Concord Litho Company Inc, Concord NH
Coneco Litho Graphics, Glen Falls NY
Continental Web, Itasca IL
Cooley Printers & Office Supply, Monroe LA
Dellas Graphics, Syracuse NY
Delmar Printing & Publishing, Charlotte NC
Delta Lithograph Company, Valencia CA
Desaulniers Printing Company, Milan IL
Diamond Graphics Tech Data, Milwaukee WI
The Dingley Press, Lisbon ME
Dollco Printing, Ottawa ON
Donihe Graphics Inc, Kingsport TN
R R Donnelley & Sons Company Inc, , Crawfordsville IN
Editors Press Inc, Hyattsville MD
Edwards Brothers Inc, Ann Arbor MI
Edwards Brothers Inc, Lillington NC
EP Graphics, Berne IN
EU Services, Rockville MD
Fetter Printing, Louisville KY
First Impressions Printer & Lithographers Inc, Elk Grove IL
Fleming Printing Co, St Louis MO
Flower City Printing Inc, Rochester NY
Focus Direct, San Antonio TX
The Forms Man Inc, Deer Park NY
Fort Orange Press Inc, Albany NY
Friesen Printers, Altona MB
Gateway Press Inc, Louisville KY
George Lithograph, San Francisco CA
Gilliland Printing Inc, Arkansas City KS
Golden Belt Printing Inc, Great Bend KS
Great Impressions Printing & Graphics, Dallas TX
Great Lakes Lithograph Co, Cleveland OH
Gregath Publishing Co, Wyandotte OK
GRIT Commercial Printing Services, Williamsport PA
Haddon Craftsmen, Scranton PA
Harlo Printing Co, Detroit MI
A B Hirschfeld Press, Denver CO

Rae Horowitz Book Manufacturers, Fairfield NJ
Imperial Litho/Graphics Inc, Phoenix AZ
Interform Corporation, Bridgeville PA
Interstate Printing Company, Omaha NE
IPC Publishing Services, Saint Joseph MI
Jersey Printing Company, Bayonne NJ
Jostens Printing & Publishing, Winston-Salem NC
Jostens Printing & Publishing, Minneapolis MN
Kaufman Press Printing, Syracuse NY
Keys Printing Co Inc, Greenville SC
Kolor View Press, Aurora MO
The C J Krehbiel Company, Cincinnati OH
The Lane Press, Burlington VT
Lighthouse Press Inc, Manchester NH
Litho Productions Inc, Madison WI
Lithograph Printing Company, Memphis TN
Lithoid Printing Corporation, East Brunswick NJ
Mack Printing Group Inc, Easton PA
Mail-O-Graph, Kewanee IL
Maple-Vail Book Manufacturing Group, York PA
Marathon Communications Group, Wausau WI
Marrakech Express, Tarpon Springs FL
The Mazer Corporation, Dayton OH
MBP Lithographics, Hillsboro KS
McArdle Printing Co, Upper Marlboro MD
McClain Printing Company, Parsons WV
Mercury Printing Company, Memphis TN
Merrill Corporation, Los Angeles CA
Moran Printing Company, Orlando FL
Morgan Press, Dobbs Ferry NY
MultiPrint Co Inc, Skokie IL
National Graphics Corp, Columbus OH
National Lithographers Inc, Miami FL
Louis Neibauer Co Inc, Warminster PA
Nystrom Publishing Company, Osseo MN
Oaks Printing Company, Bethlehem PA
Offset Paperback Manufacturers, Dallas PA
Omaha Printing Co, Omaha NE
The Ovid Bell Press Inc, Fulton MO
The Oxford Group, Norway ME
Padgett Printing Corporation, Dallas TX
Panorama Press Inc, Clifton NJ
Patterson Printing Company, Benton Harbor MI
Pearl-Pressman-Liberty Communications Group, Philadelphia PA
Pendell Printing, Midland MI
PennWell Printing, Tulsa OK
The Perlmuter Printing Co, Cleveland OH
Perry Printing Corp, Waterloo WI

Printers who Warehouse

Perry/Baraboo, Baraboo WI
Plus Communications, Saint Louis MO
Port City Press Inc, Baltimore MD
The Printing Company, Indianapolis IN
Printing Corporation of America, Pompano Beach FL
Printwest, Regina SK
Progress Printing, Lynchburg VA
Publishers Express Press, Ladysmith WI
Publishers Printing Co, Shepherdsville KY
Quebecor Book Group, New York NY
Quebecor Printing (USA) Corp, Boston MA
Quebecor Printing Inc, Richmond Hill ON
Repro-Tech, Fairfield CT
Ringier America/Brookfield, Brookfield WI
Ringier America/Jonesboro, Jonesboro AR
Ringier America/Phoenix, Phoenix AZ
Ronalds Printing, Montreal PQ
Rose Printing Company, Tallahassee FL
Roxbury Publishing Company, Los Angeles CA
Royle Printing, Sun Prairie WI
RPP Enterprises Inc, Libertyville IL
Sanders Printing Company, Garretson SD
The Saratoga Printing Co, Saratoga Springs NY
Schelin Printing Co, St Louis MO
Science Press, Ephrata PA
Shea Communications, Oklahoma City OK
Shepard Poorman Communications, Indianapolis IN
The Sheridan Press, Hanover PA
Sinclair Printing Co, Los Angeles CA
SLC Graphics, Pittston PA
Sowers Printing Company, Lebanon PA
Standard Publishing, Cincinnati OH
The Stinehour Press, Lunenburg VT
Strathmore Printing, Geneva IL
Straus Printing Company, Madison WI
John S Swift Company Inc, St Louis MO
TAN Books & Publishers Inc, Rockford IL
Taylor Publishing Company, Dallas TX
TEACH Services, Brushton NY
Technical Communication Services, North Kansas City MO
Times Litho Inc, Forest Grove OR
Times Printing, Random Lake WI
Todd Publications Inc, Smithville TX
Trade Litho Inc, Miami FL
Trend Offset Printing Service Inc, Los Alamitos CA
Tweddle Litho Co, Clinton Township MI

Universal Printing Company, Miami FL
Valco Graphics, Seattle WA
Wagners Printers, Davenport IA
Waldon Press Inc, New York NY
Walsworth Publishing Co Inc, Marceline MO
Watt/Peterson, Plymouth MN
Waverly Press, Linthicum MD
Webcom Ltd, Scarborough ON
F A Weber & Sons, Park Falls WI
Weber Printing Company, Lincoln NE
The Wessel Company Inc, Elk Grove Village IL
Western Publishing Co Inc, Racine WI
Western Web Printing, Sioux Falls SD
Westview Press, Boulder CO
The Wimmer Companies, Memphis TN
Winter Sun Inc, Phoenix AZ
Wisconsin Color Press Inc, Milwaukee WI
Wood & Jones Printers, Pasadena CA
Worzalla Publishing Company, Stevens Point WI
Y/Z Printing Co, Elizabethtown PA

Union Printers

The following printers all employ union workers.

All-Star Printing, Lansing MI
American Litho, Hayward CA
American Printers & Litho, Niles IL
Arandell Corporation, Menomonee Falls WI
Austin Printing Co Inc, Akron OH
Bagcraft Corp of America, Chicago IL
Bang Printing Co, Brainerd MN
Banta Company, Menasha WI
Banta Publications Group, Hinsdale IL
Barton Press Inc, West Orange NJ
Bawden Printing Inc, Eldridge IA
Bertelsmann Printing & Manufacturing, New York NY
Best Gagne Book Manufacturers, Toronto ON
The Book Press, Brattleboro VT
Boyd Printing Co Inc, Albany NY
Bradley Printing Company, Des Plaines IL
California Offset Printers, Glendale CA
Canterbury Press, Rome NY
Capital City Press, Montpelier VT
Chicago Press Corp, Chicago IL
Comfort Printing, St Louis MO
Consolidated Printers Inc, Berkeley CA
Danner Press Corporation, Canton OH
Des Plaines Publishing Co, Des Plaines IL
Desaulniers Printing Company, Milan IL

Union Printers

Dickinson Press Inc, Grand Rapids MI
Direct Press/Modern Litho, Huntington Station NY
Eastwood Printing Co, Denver CO
Editors Press Inc, Hyattsville MD
Eusey Press Inc, Leominster MA
Faculty Press Inc, Brooklyn NY
Federated Lithographers, Providence RI
Fleming Printing Co, St Louis MO
Fort Orange Press Inc, Albany NY
Foster Printing Service, Michigan City IN
George Lithograph, San Francisco CA
Graphics Ltd, Indianapolis IN
Gray Printing, Fostoria OH
GRIT Commercial Printing Services, Williamsport PA
Haddon Craftsmen, Scranton PA
Hart Press, Long Prairie MN
Hennegan Company, Cincinnati OH
A B Hirschfeld Press, Denver CO
Interform Corporation, Bridgeville PA
Interstate Printing Company, Omaha NE
Jersey Printing Company, Bayonne NJ
JK Creative Printers, Quincy IL
Jostens Printing & Publishing, Minneapolis MN
Kaufman Press Printing, Syracuse NY
Lehigh Press/Cadillac, Broadview IL
Les Editions Marquis, Montmagny PQ
Litho Productions Inc, Madison WI
Mail-O-Graph, Kewanee IL
Maple-Vail Book Manufacturing Group, York PA
McArdle Printing Co, Upper Marlboro MD
The McFarland Press, Harrisburg PA
McGrew Color Graphics, Kansas City MO
Merrill Corporation, Los Angeles CA
Moran Printing Inc, Baton Rouge LA
National Graphics Corp, Columbus OH
Noll Printing Company, Huntington IN
Oaks Printing Company, Bethlehem PA
Offset Paperback Manufacturers, Dallas PA
Pearl-Pressman-Liberty Communications Group, Philadelphia PA
PennWell Printing, Tulsa OK
Perry Printing Corp, Waterloo WI
Perry/Baraboo, Baraboo WI
Phillips Brothers Printers Inc, Springfield IL
Plain Talk Printing Co, Des Moines IA
Port Publications, Port Washington WI
Print Northwest, Tacoma WA
Printwest, Regina SK
Quebecor Book Group, New York NY
Quebecor Printing (USA) Corp, Boston MA
Quebecor Printing Inc, Richmond Hill ON
Quebecor/Semline, Braintree MA
Ronalds Printing, Vancouver BC
Ronalds Printing, Montreal PQ
S Rosenthal & Company Inc, Cincinnati OH
Roxbury Publishing Company, Los Angeles CA
Service Web Offset Corporation, Chicago IL
Sheridan Printing Company, Phillipsburg NJ
SLC Graphics, Pittston PA
Sowers Printing Company, Lebanon PA
Speakes-Hines & Thomas Inc, Lansing MI
St Croix Press Inc, New Richmond WI
Standard Publishing, Cincinnati OH
John Stark Printing Co Inc, St Louis MO
Stevens Graphics, Atlanta GA
Taylor Publishing Company, Dallas TX
Tweddle Litho Co, Clinton Township MI
Universal Printers Ltd, Winnipeg MB
Valco Graphics, Seattle WA
Vicks Lithograph & Printing, Yorkville NY
Waldon Press Inc, New York NY
Wallace Press, Hillside IL
Wickersham Printing Co Inc, Lancaster PA
Wisconsin Color Press Inc, Milwaukee WI
Wolfer Printing, Commerce CA
World Color Press Inc, New York NY

Index: Printers by State

This index lists printers alphabetically by the state in which they are located. Canadian printers are listed at the end of this index.

Naturegraph Publishers, Happy Camp CA
Neyenesch Printers, San Diego CA
One Color Communications, Emeryville CA
Pace Lithographers Inc, LA Puente CA
Penn Lithographics Inc, Cerritos CA
Printing: Campanella & Rome, San Diego CA
Printright International, Palmdale CA
Queen Beach Printers Inc, Long Beach CA
Rice Geoge & Sons, Los Angeles CA
Roxbury Publishing Company, Los Angeles CA
Service Printing Company, San Leandro CA
Sinclair Printing Co, Los Angeles CA
Smith Lithographic Arts Inc, Tustin CA
Southern California Grapics, Culver City CA
Sun Litho Inc, Van Nuys CA
Sung in Printing America Inc, San Mateo CA
Swenson Offset Service, Antioch CA
Trend Offset Printing Service, Los Alamitos CA
Windsor Associates, San Diego CA
Wolfer Printing, Commerce CA
Wood & Jones Printers, Pasadena CA

Colorado

American Web Inc, Denver CO
The William Byrd Press, Denver CO
C & M Press, Denver CO
ColorCorp Inc, Littleton CO
Eastwood Printing Co, Denver CO
Frederic Printing Inc, Aurora CO
A B Hirschfeld Press, Denver CO
RH Communications, Colorado Springs CO
Westview Press, Boulder CO

Connecticut

All-Type Printing Inc, Hamden CT
Allied Color, Guilford CT
Baniff Direct Mail Printers, Norwalk CT
Colotone Group, Branford CT
Danbury Printing & Litho Inc, Danbury CT
Eastern Color Printing Co, Avon CT
Gamma One, North Haven CT
Modern Printing & Lithography Inc, Brookfield CT
Repo-Tech, Fairfield CT
Sweet Printing Co, Glastonbury CT

Delaware

Collated Industries, Bear DE

District of Columbia

Judd's Incorporated, Washington DC
Lanman Companies, Washington DC

Florida

Camelot Publishing Co, Ormond Beach FL
Carlson Color Graphics, Ocala FL
Eva-Tone Inc, Clearwater FL
Futura Printing Inc, Boynton Beach FL
General Printing Inc, Miami FL
Hillsboro Printing Co, Tampa FL
Independent Publishing Company, Sarasota FL
Interprint Inc, Clearwater FL
The Little River Press Inc, Miami FL
Marrakech Express, Tarpon Springs FL
Media Printing, Miami FL
Moran Printing Company, Orlando FL
National Lithographers Inc, Miami FL
Printing Corporation of America, Pompano Beach FL
The Printing House Inc, Quincy FL
Relief Printing East, Madison FL
Rose Printing Company, Tallahassee FL
Southeastern Printing Company Inc, Stuart FL
Trade Litho Inc, Miami FL
Universal Printing Company, Miami FL
Whitehall Printing Company, Naples FL

Georgia

American Signature/Foote, Atlanta GA
Broughton Printing Co, Dalton GA
Color Express, Atlanta GA
Columbus Bookbinders & Printers, Columbus GA
Dittler Brothers, Atlanta GA
Graphic Art Service, Marietta GA
Graphic Production Center, Norcross GA
IPD Printing , Chamblee GA
Quad Graphics — Thomaston, Thomaston GA
Stein Printing Co, Atlanta GA
Stevens Graphics, Atlanta GA
US Press Inc, Valdosta GA

Web Inserts, Duluth GA
Williams Printing Co, Atlanta GA

Idaho

Accurate Printing, Post Falls ID
Caxton Printers Ltd, Caldwell ID
Graphic Arts Publishing, Boise ID
Pend Oreille Printers Inc, Sandpoint ID

Illinois

Adams Press, Chicago IL
AGI Inc, Melrose Park IL
Alden Press, Elk Grove Village IL
American Printers & Litho, Niles IL
The Argus Press Inc, Niles IL
Bagcraft Corp of America, Chicago IL
Banta Publications Group, Hinsdale IL
Black Dot Graphics, Crystal Lake IL
Bockman Printing & Services Co,
     Bellwood IL
Booklet Publishing Company Inc, Elk
     Grove Village IL
Bradley Printing Company, Des Plaines
     IL
Bro's Lithographic Co, Chicago IL
Brookshore Lithographers Inc, Elk
     Grove Village IL
Bruce Offset Co, Elk Grove Village IL
Carqueville Printing Co, Elk Grove Vil-
     lage IL
Chicago Color Express, Westmont IL
Chicago Color Express, Chicago IL
Chicago Press Corp, Chicago IL
Classic Litho Arts, Oak Park IL
Combined Communication Services,
     Mendota IL
Continental Web, Itasca IL
Crusader Printing Co, East Saint Louis
     IL
Dayal Graphics, Lemont IL
Delprint Inc, Mt Prospect IL
Des Plaines Publishing Co, Des Plaines
     IL
Desaulniers Printing Company, Milan IL
Directories America, Chicago IL
E & D Web Inc, Cicero IL
Enteron Group, Chicago IL
First Impression Printer, Elk Grove IL
Flying Color Graphics, Pontiac IL
Ft Dearborn Lithograph Co, Niles IL
Graftek Press, Woodstock IL
Great Northern/Design Printing, Skokie
     IL
GTE Directories Corp, Des Plaines IL
GTE Directories Printing Corp, Mt
     Prospect IL

D B Hess Company, Woodstock IL
Hi-Tech Color House, Chicago IL
Hinz Lithographing Company, Mt
     Prospect IL
Intaglio Vivi-Color, Schiller Park IL
Jet LithoColor Inc, Downers Grove IL
JK Creative Printers, Quincy IL
Johns Byrne Co, Niles IL
Johnson Graphics, East Dubuque IL
The Johnson Press Inc, Pontiac IL
Kinney Printing Co, Chicago IL
Lehigh Press/Cadillac, Broadview IL
Lithocolor Press, Westchester IL
Lithographic Industries Inc, Maywood IL
Mail-O-Graph, Kewanee IL
Merrill Corporation, Chicago IL
Midwest Litho Arts, Des Plaines IL
Multi-Ad Services, Peoria IL
MultiPrint Co Inc, Skokie IL
Original Smith Printing Co Inc,
     Bloomington IL
Pantagraph Printing, Bloomington IL
Park Press, South Holland IL
Perfect Plastic Printing Corp, St Charles
     IL
Petty Co Inc, Effingham IL
Phillips Brothers Printers Inc,
     Springfield IL
Pioneer Press, Wilmette IL
Ringier America, Itasca IL
RPP Enterprises Inc, Libertyville IL
Sakurai USA Inc, Elk Grove Village IL
Saltzman Printers Inc, Maywood IL
Service Web Offset Corporation, Chicago
     IL
Sleepeck Printing Co, Bellwood IL
Spartan Printing, Sparta IL
Standard Printing Service, Chicago IL
Strathmore Printing, Geneva IL
John S Swift/Chicago, Chicago IL
TAN Books & Publishers Inc, Rockford
     IL
Triangle Printers Inc, Skokie IL
Tukaiz Innovative, Franklin Park IL
Unique Printers & Lithographers, Cicero
     IL
Versa Press Inc, East Peoria IL
Vogue Printers, North Chicago IL
Wace USA, Chicago IL
Wallace Press, Hillside IL
Web Specialties, Wheeling IL
The Wessel Company Inc, Elk Grove Vil-
     lage IL
Wicklander Printing Corp, Chicago IL

Indiana

Courier-Kendallville Corp, Kendallville IN
R R Donnelley & Sons Co Inc, Crawfordsville IN
EP Graphics, Berne IN
Evangel Press, Nappanee IN
Fast Print, Fort Wayne IN
Foster Printing Service, Michigan City IN
Graphics Ltd, Indianapolis IN
Lithotone Inc, Elkhart IN
McKay Printing Services, Michigan City IN
Noll Printing Company, Huntington IN
Paust Inc, Richmond IN
Petersen Graphics Group, South Bend IN
Prinit Press, Dublin IN
The Printing Company, Indianapolis IN
Shepard Poorman Communications, Indianapolis IN
Uniform Printing & Supply Inc, Kendallville IN
White Arts Inc, Indianapolis IN
Ye Olde Genealogie Shoppe, Indianapolis IN

Iowa

Anundsen Publishing Company, Decorah IA
Bawden Printing Inc, Eldridge IA
Brennan Printing, Deep River IA
Fisher Printers Inc, Cedar Rapids IA
Heartland Press Inc, Spencer IA
Julin Printing, Monticello IA
Maquoketa Web Printing, Maquoketa IA
Plain Talk Printing Co, Des Moines IA
Wagners Printers, Davenport IA

Kansas

Advanced Data Reproductions, Wichita KS
Allen Press Inc, Lawrence KS
Cookbook Publishers Inc, Lenexa KS
Gilliland Printing Inc, Arkansas City KS
Golden Belt Printing Inc, Great Bend KS
Jostens Printing & Publishing, Topeka KS
MBP Lithographics, Hillsboro KS
Sun Graphics, Parsons KS

Kentucky

Brown Printing / Franklin, Franklin KY
Econocolor, Florence KY
Fetter Printing, Louisville KY
Gateway Press Inc, Louisville KY
Hamilton Printing Co Inc, Louisville KY
Ivy Hill Packaging, Louisville KY
Magna Graphic, Lexington KY
McDowell Publications, Utica KY
Merrick Printing Co Inc, Louisville KY
Metroweb, Erlanger KY
Nationwide Printing, Burlington KY
Ohio Valley Litho Color, Florence KY
Prepsat, Franklin KY
Publishers Press, Louisville KY
Publishers Printing Co, Shepherdsville KY
Rand McNally Media Services Co, Versailles KY

Louisana

Cooley Printers & Office Supply, Monroe LA
Gulf South Printing, Lafayette LA
Moran Printing Inc, Baton Rouge LA

Maine

The Dingley Press, Lisbon ME
Geiger Brothers, Lewiston ME
The Oxford Group, Norway ME
Spencer Press, Wells ME
Sun Graphix, Lewiston ME

Maryland

Art Litho Co, Baltimore MD
Automated Graphic Systems, White Plains MD
Collins Lithographic Co Inc, Baltimore MD
Colortone Press, Landover MD
Editors Press Inc, Hyattsville MD
EU Services, Rockville MD
Gamse Lithographing Co, Baltimore MD
Gateway Press, Baltimore MD
Graphic Technology, Baltimore MD
Holladay-Tyler Printing Corp, Glenn Dale MD
John D Lucas Printing Company, Baltimore MD
McArdle Printing Co, Upper Marlboro MD
Optic Graphics Inc, Glen Burnie MD

Peake Printers Inc, Hyattsville MD
PFP Printing Corporation, Gaithersburg
   MD
Port City Press Inc, Baltimore MD
Victor Graphics, Baltimore MD
Watermark Press, Owings Mills MD
Waverly Press, Linthicum MD

Massachusetts

Acme Printing, Wilmington MA
B & W Press Inc, Georgetown MA
Bay State Press, Framingham MA
Carter Rice, Boston MA
CGI, Burlington MA
Champagne Offset Co Inc, Natick MA
Country Press Inc, Lakeville MA
Courier Corporation, Lowell MA
Courier Stoughton Inc, Stoughton MA
Courier-Westford Inc, Westford MA
Crane Duplicating Service, West
   Barnstable MA
Daniels Printing Co Inc, Everett MA
Emco Printers Inc, Everett MA
Eusey Press Inc, Leominster MA
Graphic Litho Corporation, Lawrence MA
Graphics Express, Boston MA
The Job Shop, Woods Hole MA
Nimrod Press Inc, Boston MA
Northeast Offset Inc, Chelmsford MA
Paraclete Press, Brewster MA
Quebecor Printing (USA) Corp, Boston
   MA
Quebecor/Semline, Braintree MA
Screenprint Inc, Wilmington MA
Smith Print Inc, Norwell MA
R E Smith Printing, Fall River MA
The Studley Press, Dalton MA
Sullivan Brothers Printers Inc, Lowell
   MA
The Sun Hill Press, North Brookfield MA
United Lithograph, Somerville MA
Van Volumes, Palmer MA
Winthrop Printing Co Inc, Boston MA

Michigan

Adair Printing Co Inc, Southfield MI
All-Star Printing, Lansing MI
Baker Johnson Inc, Dexter MI
BookCrafters, Chelsea MI
Braun-Brumfield Inc, Ann Arbor MI
Cushing-Malloy, Ann Arbor MI
Data Reproductions Corp, Rochester
   Hills MI
Dickinson Press Inc, Grand Rapids MI
Edwards Brothers Inc, Ann Arbor MI
Eerdmans Printing Co, Grand Rapids MI

EPI Printers, Livonia MI
Gaylord Printing, Detroit MI
Harlo Printing Co, Detroit MI
Inland Press, Detroit MI
IPC Publishing Services, Saint Joseph MI
Lakeland Litho-Plate Inc, Detroit MI
Malloy Lithographing, Ann Arbor MI
McNaughton & Gunn Inc, Saline MI
Mitchell Graphics, Petosky MI
National Reproductions Corp, Madison
   Heights MI
Page Litho Inc, Detroit MI
Patterson Printing Company, Benton
   Harbor MI
Pendell Printing, Midland MI
Printco Inc, Greenville MI
Q & Q Printing Co, Detroit MI
Speakes-Hines & Thomas Inc, Lansing
   MI
Stylecraft Printing Co, Canton MI
Thomson-Shore, Dexter MI
Tweddle Litho Co, Clinton Township MI
Universal Graphics, Warren MI
Universal Litho Inc, Kalamazoo MI

Minnesota

A-1 Business Service Inc, St Paul MN
Advanced Duplicating & Printing, Min-
   neapolis MN
Advertising Unlimited Inc, Sleepy Eye
   MN
Amidon & Associates Inc, Saint Paul MN
Bang Printing, Saint Paul MN
Bang Printing Co, Brainerd MN
Bolger Publications/Creative Printing,
   Minneapolis MN
Brown Printing, Waseca MN
Color Express, Minneapolis MN
Colorbrite, Minneapolis MN
Colorhouse, Minneapolis MN
Daily Printing Inc, Minneapolis MN
Hart Press, Long Prairie MN
Instant Web, Chanhassen MN
Japs-Olson Company, Minneapolis MN
Johnson Printing & Packaging Co, Frid-
   ley MN
Jostens Printing & Publishing, Min-
   neapolis MN
Kimm Printing, Minneapolis MN
Litho Specialties, St Paul MN
May Printing Co, St Cloud MN
McGill Jensen, St Paul MN
Merrill Corporation, St Paul MN
Miller Printing Inc, Woodbury MN
Northprint International Inc, Grand
   Rapids MN
Nystrom Publishing Company, Osseo MN
Pentagram Press, Minneapolis MN
The Press, Chanhassen MN

The Printer Inc, Osseo MN
Process Displays Co, Minneapolis MN
Quebecor Printing Inc, Saint Paul MN
Quebecor Printing/St Cloud, Saint Cloud MN
John Roberts Company, Minneapolis MN
Schmidt Printing Inc, Byron MN
Sexton Printing, Saint Paul MN
Tapemark, West St Paul MN
Viking Press, Eden Prairie MN
Walter's Publishing, Waseca MN
Watt/Peterson, Plymouth MN
Winslow Printing Co, Minneapolis MN

Mississippi

American Signature Memphis Division, Olive Branch MS
Hederman Brothers Printing, Jackson MS
Ringier America Inc, Corinth MS

Missouri

Accu-Color, St Louis MO
Best Printing & Duplicating Inc, Maryland Heights MO
Buxton & Skinner Printing Co, St Louis MO
Chroma-Graphics, Kansas City MO
Clark Printing Co, Liberty MO
Comfort Printing, St Louis MO
Commercial Lithographing Co, Kansas City MO
Corley Printing Company, Earth City MO
Craftmasters, Sikeston MO
Designer Color Systems, St Louis MO
Eagle Lithographing Co, Kansas City MO
Etling Printing Co Inc, St Louis MO
Fleming Printing Co, St Louis MO
Harmony Printing, Liberty MO
Jefferson/Keeler Printing Co, St Louis MO
Kolor View Press, Aurora MO
McGrew Color Graphics, Kansas City MO
Nies/Artcraft Printing Companies, Saint Louis MO
The Ovid Bell Press Inc, Fulton MO
Plus Communications, Saint Louis MO
Sayers Printing Co, St Louis MO
Schelin Printing Co, St Louis MO
John Stark Printing Co Inc, St Louis MO
Stewart Publishing & Printing Co, Marble Hill MO
John S Swift Company Inc, St Louis MO
Technical Communication Services, North Kansas City MO

TSI Graphics, St Louis MO
Ultra-Color Corporation, Saint Louis MO
Universal Printing Company, Saint Louis MO
Walsworth Publishing Co Inc, Marceline MO
Warson Graphics Inc, St Louis MO
Henry Wurst Inc, North Kansas City MO

Montana

Color World Printers, Bozeman MT

Nebraska

American Signature, Lincoln NE
Century Graphics Corp, Omaha NE
Cookbooks by Mom's Press, Kearney NE
Epsen Lithographing Co, Omaha NE
Interstate Printing Company, Omaha NE
Jacob North Printing Co Inc, Lincoln NE
Metromail Corporation, Lincoln NE
Omaha Printing Co, Omaha NE
Weber Printing Company, Lincoln NE

New Hampshire

Concord Litho Company Inc, Concord NH
Lew A Cummings Company Inc, Manchester NH
Dartmouth Printing, Hanover NH
Eastern Rainbow, Derry NH
Lighthouse Press Inc, Manchester NH
MMI Press, Harrisville NH
Odyssey Press Inc, Dover NH

New Jersey

Able Printing Co, Pennsauken NJ
Annex Printing Corp, Jersey City NJ
Barton Press Inc, West Orange NJ
Book-Mart Press Inc, North Bergen NJ
Butler Printing & Laminating Inc, Butler NJ
Catalog King, Clifton NJ
Clarkwood Corporation, Totowa NJ
Colonial Graphics, Paterson NJ
De Palma Printing Co, Ridgefield Park NJ
Edison Lithographing, North Bergen NJ
Graytor Printing, Lyndhurst NJ
Rae Horowitz Book Manufacturers, Fairfield NJ

New Mexico

New York

TSO General Corporation, Long Island
  City NY
Vail Ballou Press, Binghamton NY
Vicks Lithograph & Printing, Yorkville
  NY
Waldon Press Inc, New York NY
World Color Press Inc, New York NY
Yorktown Printing & Pennysaver,
  Yorktown Heights NY

North Carolina

Atlantic Co, Tabor City NC
Craftsman Printing Co, Charlotte NC
Delmar Printing & Publishing, Charlotte
  NC
Edwards Brothers Inc, Lillington NC
Heritage Printers, Charlotte NC
Jostens Printing & Publishing, Winston-
  Salem NC
Litho Industries Inc, Durahm NC
Parker Graphics, Fuquay-Varina NC
Printpack Inc, Hendersonville NC
Professional Press, Chapel Hill NC
Town House Press, Pittsboro NC
White Printing Inc, Charlotte NC
Winston Printing Co Inc, Winston-Salem
  NC

North Dakota

Associated Printers, Grafton ND

Ohio

Action Nicholson Color, Cleveland OH
Alden Press/Wm Feathers, Oberlin OH
Amos Press Inc, Sidney OH
Austin Printing Co Inc, Akron OH
Bedinghaus Business Forms Inc, Cincin-
  nati OH
BookMasters Inc, Ashland OH
Carey Color, Medina OH
Carpenter Lithographic Co, Springfield
  OH
Carpenter Reserve, Cleveland OH
Central Lithograph Co, Cleveland OH
Central Printing Co, Dayton OH
Champion Printing, Cincinnati OH
Communicolor, Newark OH
Complete Book Manufacturing Service,
  Ashland OH
Curless Printing Co, Blanchester OH
Danner Press Corporation, Canton OH
Direct Graphics Inc, Sidney OH

R R Donnelley & Sons Co Inc, Willard,
  Willard OH
William Exline Inc, Cleveland OH
Fisher Printing Co, Galion OH
Gowe Printing Company, Medina OH
Graphic Printing, New Carlisle OH
Gray Printing, Fostoria OH
Great Lakes Lithograph Co, Cleveland
  OH
Greenfield Printing & Publishing, Green-
  field OH
Hennegan Company, Cincinnati OH
Hiney Printing Co Inc, Akron OH
Industrial Printing Co, Toledo OH
The Johnson & Hardin Company, Cin-
  cinnati OH
Kreber Graphics, Columbus OH
The C J Krehbiel Company, Cincinnati
  OH
LBA Custom Printing Inc, Toledo OH
The Mazer Corporation, Dayton OH
National Graphics Corp, Columbus OH
Nielsen Lithographing Co, Cincinnati OH
Original Copy Centers, Cleveland OH
Penton Press Publishing, Berea OH
The Perlmuter Printing Co, Cleveland
  OH
The Press of Ohio, Kent OH
Printing Service Co, Dayton OH
S Rosenthal & Company Inc, Cincinnati
  OH
Seiple Lithograph Co, Canton OH
Shepard Poorman Communications,
  Columbus OH
Sidney Printing Works Inc, Cincinnati
  OH
Standard Publishing, Cincinnati OH
Stevenson Photo Color, Cincinnati OH
Sun Printing, Mansfield OH
John S Swift/Cincinnati, Cincinnati OH
Watkins Printing Company, Columbus
  OH
Westerman Print Co, Cincinnati OH
Younstown Lithographing Inc,
  Youngstown OH

Oklahoma

ColorGraphics Printing Corp, Tulsa OK
Gregath Publishing Co, Wyandotte OK
PennWell Printing, Tulsa OK
Pinecliffe Printers, Shawnee OK
Rodgers RF LithographCo, Tulsa OK
Shea Communications, Oklahoma City
  OK

Oregon

Commercial Documentation Services, Medford OR
Daily Journal of Commerce, Portland OR
Eagle Web Press, Salem OR
Graphic Arts Center, Portland OR
Maverick Publications Inc, Bend OR
Northwest Web, Eugene OR
Times Litho Inc, Forest Grove OR

Pennsylvania

Alcom Printing Group Inc, Bethlehem PA
Apollo Graphics Ltd, Southampton PA
Arcata Graphics/Fairfield, Fairfield PA
Art Print Co, Taylor PA
Baum Printing House Inc, Philadelphia PA
Bethel Park Printing Inc, Bethel Park PA
Braceland Brothers Inc, Philadelphia PA
Brown Printing/East Greenville Division, East Greenville PA
BSC Litho, Harrisburg PA
Chernay Printing Inc, Coopersburg PA
Dispatch Printing Inc, Erie PA
Fry Communications, Mechanicsburg PA
GRIT Commercial Printing Services, Williamsport PA
Haddon Craftsmen, Scranton PA
Havertown Printing Co, Broomall PA
Hoechstetter Printing, Pittsburgh PA
Intelligencer Printing, Lancaster PA
Interform Corporation, Bridgeville PA
International Lithographing, Philadelphia PA
Jostens Printing & Publishing, State College PA
K-B Offset Printing, State College PA
Lancaster Press, Lancaster PA
Lebanon Valley Offset Inc, Annville PA
Mack Printing Group Inc, Easton PA
Maple-Vail Book Manufacutring Group, York PA
The McFarland Press, Harrisburg PA
National Publishing Company, Philadelphia PA
Louis Neibauer Co Inc, Warminster PA
Nittany Valley Offset, State College PA
Oaks Printing Company, Bethlehem PA
Offset Paperback Manufacturers, Dallas PA
Panel Prints Inc, Old Forge PA
Pearl-Pressman-Liberty Communications Group, Philadelphia PA
Penn Colour Graphics, Huntington Valley PA
Progressive Typographers, Emigsville PA

Rapidocolor Corporation, West Chester PA
Schiff Printing, Pittsburgh PA
Science Press, Ephrata PA
The Sheridan Press, Hanover PA
SLC Graphics, Pittston PA
Smith-Edwards-Dunlap, Philadelphia PA
Sowers Printing Company, Lebanon PA
Specialty Printers of America, Scranton PA
Strine Printing Co Inc, York PA
The Veitch Printing Corp, Lancaster PA
Wickersham Printing Co Inc, Lancaster PA
Y/Z Printing Co, Elizabethtown PA

Rhode Island

Colorlith Corporation, Johnston RI
Federated Lithographers, Providence RI
United Printing Inc, Warwick RI
Universal Press, Providence RI

South Carolina

R L Bryan Company, Columbia SC
Hart Graphics, Simpsonville SC
Keys Printing Co Inc, Greenville SC
State Printing Co, Columbia SC

South Dakota

Pine Hill Press Inc, Freeman SD
Sanders Printing Company, Garretson SD
Sioux Printing Inc, Sioux Falls SD
Western Web Printing, Sioux Falls SD

Tennessee

Ambrose Printing, Nashville TN
Capitol Engraving, Nashville TN
The College Press, Collegedale TN
Donihe Graphics Inc, Kingsport TN
Fundcraft Publishing, Collierville TN
Jostens Printing & Publishing, Clarksville TN
Lakeway Printers Inc, Morristown TN
Lithograph Printing Company, Memphis TN
McQuiddy Printing Company, Nashville TN
Mercury Printing Company, Memphis TN

NEC, Nashville TN
Rich Printing Company Inc, Nashville TN
Ringier America/Dresden, Dresden TN
Starr Toof Printing, Memphis TN
Sullivan Graphics Inc, Brentwood TN
Textile Printing Co, Chattanooga TN
Triangle Booklet Company, Nashville TN
Vaughan Printing, Nashville TN
The Wimmer Companies, Memphis TN

Texas

Adventure Printing, Dallas TX
American Printing Co Inc, Galveston TX
American Signature Graphics, Dallas TX
Baker Press Inc, Dallas TX
Bennett Printing Co, Dallas TX
Best Printing Co Inc, Austin TX
Blanks Color Imaging, Dallas TX
BMP Paper & Printing Inc, Houston TX
Capital Printing Company Inc, Austin TX
Cargill's Printing & Stationery, Houston TX
Characters, Houston TX
Clarke Printing Co, San Antonio TX
Colotone Riverside Inc, Dallas TX
Dataware, Houston TX
Focus Direct, San Antonio TX
Great Impressions Printing and Graphics, Dallas TX
Gulf Printing, Houston TX
Hart Graphics, Austin TX
Henington Publishing Company, Wolfe City TX
Horticultural Printers Co, Mesquite TX
Merrill Corporation, Dallas TX
Morgan Printing, Austin TX
Motheral Printing, Fort Worth TX
Newsfoto Publishing Company, San Angelo TX
Padgett Printing Corp, Dallas TX
Printing Center, Ft Worth TX
Promotional Printing Corporation, Houston TX
Publishers Choice Reprints, The Woodlands TX
Quebecor/Dallas, Dallas TX
Taylor Publishing Company, Dallas TX
Todd Publications Inc, Smithville TX
Tracor Publications, Austin TX
Williamson Printing Corp, Dallas TX

Utah

Community Press, Provo UT
Everton Publishers Inc, Logan UT
Hawkes Publishing, Salt Lake City UT

ProLitho, Provo UT
Publishers Press, Salt Lake City UT
Sun Lithographing & Printing Co, West Valley City UT

Vermont

Academy Books, Rutland VT
Americomp, Brattleboro VT
The Book Press Inc, Brattleboro VT
Capital City Press, Montpelier VT
Daamen Printing Company, West Rutland VT
The Lane Press, Burlington VT
Northlight Studio Press, Barre VT
Offset House, Essex Junction VT
The Stinehour Press, Lunenburg VT

Virginia

3PI Prepress, Herndon VA
American Press Inc, Reston VA
Banta-Harrisonburg, Harrisonburg VA
Bassett Printing Corp, Bassett VA
Beacon Press Inc , Richmond VA
Berryville Graphics, Berryville VA
BookCrafters/Fredericksburg, Fredericksburg VA
Brunswick Publishing Corp, Lawrenceville VA
The William Byrd Press, Richmond VA
R R Donnelley & Sons Company Inc, Harrisonburg, Harrisonburg VA
Goodway Graphics of Virginia, Springfield VA
Judd's Incorporated, Strasburg VA
Kirby Lithographic Co Inc, Arlington VA
Progress Printing, Lynchburg VA
Stephenson Lithograph Inc, Alexandria VA
United Lithographic Services Inc, Falls Church VA

Washington

aBCD, Seattle WA
Alpha One Press Inc, Redmond WA
Beacon Press, Seattle WA
Consolidated Press Printing Co, Seattle WA
The Craftsman Press Inc, Seattle WA
Gorham Printing, Rochester WA
Heath Printers Inc, Seattle WA
K & H Printers & Lithographers, Everett WA
Print Northwest, Tacoma WA

Rotary Press Inc, Seattle WA
Snohomish Publishing, Snohomish WA
Trojan Lithograph Inc, Kent WA
Valco Graphics, Seattle WA

West Virginia

McClain Printing Company, Parsons WV

Wisconsin

Action Printing, Fond Du Lac WI
Arandell Corporation, Menomonee Falls
    WI
Artcraft Press, Waterloo WI
Banta Company, Menasha WI
Castle Pierce Printing Co, Oshkosh WI
Citizen Printing, Beaver Dam WI
Diamond Graphics Tech Data, Mil-
    waukee WI
Inland Printing Co Inc, La Crosse WI
La Crosse Graphics, LaCrosse WI
Litho Productions Inc, Madison WI
Marathon Communications Group,
    Wausau WI
MCP, Milwaukee WI
Moebius Printing, Milwaukee WI
OMNIPRESS, Madison WI
Outstanding Graphics, Kenosha WI
Perry Printing Corp, Waterloo WI
Perry/Baraboo, Baraboo WI
Philipp Lithographing Co, Grafton WI
Port Publications, Port Washington WI
Publishers Express Press, Ladysmith WI
Quad/Graphics, Lomira, Lomira WI
Quad/Graphics, Pewaukee, Pewaukee WI
Quad/Graphics, Sussex, Sussex WI
Quigley Printing Co Inc, Green Bay WI
Ringier America/Brookfield, Brookfield
    WI
Royle Printing, Sun Prairie WI
Schumann Printers, Fall River WI
St Croix Press Inc, New Richmond WI
Straus Printing Company, Madison WI
Times Printing, Random Lake WI
Webcrafters, Madison WI
F A Weber & Sons, Park Falls WI
Western Publishing Co Inc, Racine WI
Wisconsin Color Press Inc, Milwaukee WI
Worzalla Publishing Company, Stevens
    Point WI

Wyoming

Pioneer Printing & Stationery, Cheyenne
    WY

Canadian Printers

Alberta

Adviser Graphics, Red Deer AB

British Columbia

Ronalds Printing, Vancouver BC
Mitchell Press, Vancouver BC

Manitoba

Universal Printers Ltd, Winnipeg MB
Hignell Printing, Winnepeg MB
Friesen Printers, Altona MB

Ontario

McLaren, Morris & Todd, Mississauga
    ON
Coach House Press, Toronto ON
Best Gagne Book Mfgs, Toronto ON
Ainsworth Group Inc, Kitchener ON
St Joseph Printing Ltd, Concord ON
Network Studios Inc, Etobicoke ON
RBW Graphics, Owen Sound ON
Multiple Business Forms Inc, Missis-
    sauga ON
Dollco Printing, Ottawa ON
Tri-Graphic Printing, Ottawa ON
Quebecor Printing Inc, Richmond Hill
    ON
Webcom Ltd, Scarborough ON

Quebec

Ronalds Printing, Montreal PQ
Southam Paragon Graphics, Candiac PQ
Les Editions Marquis, Montmagny PQ
Multiple Business Forms Inc, Dorval PQ

Saskatchewan

McGraphics Inc, Saskatoon SK
Printwest, Regina SKintwest, Regina SK

Bibliography

Books on Editing, Design, and Printing

These books will help you do a better job editing, designing, and producing your books. If the titles highlighted in **bold** are not available at your local bookstore, you can order them from **Ad-Lib Publications, PO Box 1102, Fairfield, IA 52556-1102; 515-472-6617; Fax: 515-472-3186; Toll-free: 800-669-0773.** We accept **VISA, MasterCard**, and **American Express** credit cards. Add $4.50 postage and handling for the first book ordered and 50¢ for each additional book. When ordering 3 or more books, deduct 10%.

- Barker, Malcolm E., **Book Design & Production**, (San Francisco, CA: Londonborn Publications, 1990), 233 pages, softcover, **$24.95.**

  This new book is a superb step-by-step guide to designing and producing your books. Learn how and why to use a grid, how to choose the perfect typeface to match your book's content, how to design the front and back matter to make them more accessible to readers, and much more. Very good and very thorough. A wonderful book!

- Beach, Mark, **Getting It Printed: How to Work with Printers and Graphic Arts Services to Assure Quality, Stay on Schedule, and Control Costs,** (Cincinnatti OH; North Light Books, 208 pages, 1993 softcover, **$29.95.**

  Of all the books on printing and graphics, this book is a standout. It is detailed, well-designed, and easy to use. It covers everything from planning the printing job, writing specifications, requesting quotations, working with typesetters, preparing camera-ready copy, proofing, and working with your printer to get the best job. Plus it provides specific criteria for checking every step of the process. **If every printer and book publisher had this book, 90% of all printing problems would disappear.**

- Beach, Mark, and Ken Russon, <u>Papers for Printing: How to Choose the Right Paper at the Right Price for Any Printing Job</u>, (Cincinnati OH: North Light Books, (A Coast to Coast Book) 1991), 40 sample sheets, 64 pages, softcover, $39.50.
  This booklet shows how to buy and specify paper for your various printing jobs. It includes 40 printed sample sheets, a chart that compares the costs of all 40 samples, a list of 678 paper brands, a list of 591 paper merchants, and a glossary of 214 paper terms.

- Bodian, Nat G., <u>How to Choose a Winning Title</u>, (Phoenix: Oryx Press, 1989), 176 pages, softcover, $23.50.
  If you've ever agonized over the title of a book, this new guide will provide you with all the tips, insights, and real-life examples you'll need to come up with selling titles. These same tips might also help you come up with selling headlines and product names.

- Burke, Clifford, <u>Printing It: A Guide to Graphic Techniques for the Impecunious</u>, (Berkeley, CA: Wingbow Press, 1974), 127 pages, softcover, $5.95.
  A short guide that provides all the basics you need to know to prepare your own camera-ready copy for printers.

- <u>The Chicago Manual of Style</u>, (Chicago: University of Chicago Press, 1982), hardcover.
  The handbook of style for books. When you have questions about capitalization, punctuation, and usage, this book is the first place to look for answers. It will help you make the most acceptable choice when editing and copy editing your books.

- Graham, Walter B., **Complete Guide to Pasteup, Third Edition,** (Omaha, NE: Dot Pasteup Supply, 1987), 236 pages, softcover, **$19.95.**

  This is the classic book on how to prepare camera-ready copy for printing. If you are a newcomer to pasteup or have just hired a novice, this is the book to use. Very detailed and complete. Indeed, if this book has any faults, it would be the fact that it could very well overwhelm you with its details.

- International Paper Company, <u>Pocket Pal: A Graphics Arts Production Handbook</u>, (New York: International Paper Company, 1987), 216 pages, softcover, $6.95.
  A wonderful compendium of information on pre-press and printing processes. A basic reference book that still fits in your pocket!

- Judd, Karen, **Copyediting: A Practical Guide**, (Los Altos, CA: Crisp Publications, 1988), 287 pages, hardcover, **$19.95.**

  When you have questions regarding the appropriate places to use numerals versus spelling out numbers, how to use the standard punctuation marks, how to mark changes in a manuscript (with standard proofreading symbols), how to typeset equations, or how to format bibliographies and footnotes, look to this book for the answers. A very complete and easy-to-use guide. Highly recommended for your editors and copy editors.

- Lippi, Robert, <u>How to Buy Good Printing & Save Money</u>, (New York: Art Direction, 1987), 144 pages, hardcover, $15.75; softcover, $12.50.

  A good, basic, easy-to-understand guide to buying printing services. While this book is a good book, it is not nearly as attractive, comprehensive, or detailed as <u>Getting It Printed.</u>

- Middletown, Tony, <u>A Desktop Publisher's Guide to Pasteup,</u> (Colorado Springs: PLUSware, 1987), 228 pages, softcover, $15.95.

  Not just for desktop publishers. Indeed, this book is one of the most readable books on how to lay-out and paste-up books, advertisements, and other printed items. A practical guide for preparing camera-ready copy for the printer.

- Parker, Roger, **Looking Good in Print: A Guide to Basic Design for Desktop Publishing, Third Edition,** (Chapel Hill, NC: Ventana Press, 1993), 2211 pages, softcover, **$24.95.**

  Shows how to use computers to design and produce more effective brochures, newsletters, manuals, and catalogs. Also describes the common design pitfalls and how to avoid them.

- Poynter, Dan, **Publishing Short-Run Books, Fifth Edition,** Santa Barbara: Para Publishing, 1988), 142 pages, softcover, **$5.95.**

  A how-to book on preparing and reproducing small books using your local copy shop. It teaches how to set type inexpensively, pasteup camera-ready copy, print using your local copy shop, and bind your books yourself. A superb guide to producing your books and reports in short runs of 200 or less in a matter of days.

- Rice, Stanley, <u>Book Design: Systematic Aspects</u>, (New York:
  R. R. Bowker, 1978), 274 pages, hardcover, $29.95.

  A systematic and clear delineation of all the steps involved in producing a book, from editorial design decisions to printing and binding. Includes schedules, transmittal forms, press layouts, and other charts and forms to ensure that everything proceeds smoothly and efficiently. This book covers everything a book production manager needs to know to produce well-designed books on time and on budget. Highly recommended.

- Rice, Stanley, <u>Book Design: Text Format Models</u>, (New York:
  R. R. Bowker, 1978), 215 pages, hardcover, $29.95.

  An extensive collection of samples for typesetting standard text, lists, tables, poems, plays, footnotes, glossaries, bibliographies, captions, indexes, and more. These samples are a handy way to visualize the look of your book, choose the style you want, and specify the type.

- Stoughton, Mary, <u>Substance & Style</u>, (Alexandria, VA: Editorial
  Experts, 1989), 351 pages, softcover, $28.00.

  Good book design also requires careful editing. If you want your books to pass all the tests (from reviewer to retailer to reader), make sure that they contain no errors in grammar, usage, or punctuation. This book, with all its practical exercises, is the perfect tool for teaching your editors and proofreaders all the little details that must be covered when copy editing any books.

- Vandermeulen, Carl, **Photography for Student Publications**,
  (Orange City, IA: Middleburg Press, 1979), 160 pages, hardcover,
  $16.95; softcover, **$12.95.**

  This book should really be titled *Photography for Small Publications* because it is a good basic text for beginners of all kinds. It provides excellent advice on how to produce natural poses and better photos overall. It covers everything from handling cameras to setting exposures, composing photos, developing them, making prints, shooting for layouts, and much more.

- White, Jan V., <u>Editing by Design: A Guide to Effective Word-and-Picture Communication for Editors and Designers, Second Edition</u>,
  (New York: R. R. Bowker, 1982), 264 pages, softcover, $24.95.

  A graphic design guide for magazine editors and designers that shows how words and images can enhance each other and make for a more effective presentation. With the increasingly graphic nature of books today, this guide could help you to strengthen the visual impact of your books.

- White, Jan V., <u>Graphic Design for the Electronic Age</u>, (New York: Watson-Guptill, 1988), 224 pages, hardcover, $30.00; softcover, $24.95. This book has more to offer the book designer than the above book. It describes how to select a typeface and size, how to select a column style, how to handle illustrations, and how to put together and integrate a complete publication (front matter, text, and back matter).

Ad-Lib Bibliography

Books from Ad-Lib Publications

The following books, reports, newsletters, and data files are either published by **Ad-Lib Publications or Open Horizons Publishing Company** You may order any of the books from **Ad-Lib Publications, 51½ West Adams, Fairfield, IA 52556; 515- 472-6617; Fax: 515- 472-3186; Toll-free: 800-6669-0773.**

- Kiefer, Marie and Kremer, John, **Book Marketing Profitcenter, $150.00**
  These data files are continuously updated. You get the latest version on the day you order. Includes more than 3200 specialty booksellers, 140 book distributors, 217 wholesalers, 343 paperback jobbers, 164 sales representatives, 43 fulfillment services, 393 chain store buyers, 533 independent booksellers, 149 book clubs, 722 mail order catalogs, 303 mailing list sources, 231 publicity and marketing services, 220 card packs, 93 book fairs and conventions, 100 publishing associations, and much, much more! No other book marketing database offers so much for so little cost—less than 2¢ a contact. 8700 + records available in various data file formats for IBM-PCs, Macintoshes, or compatibles.

  Kiefer, Marie, **Book Promotion Hotline** *Newsletter*, **$150.00 per year.**
  This weekly newsletter features from 40 to 60 up-to-date media and book marketing contacts in each issue, including newspaper and magazine editors, book club buyers, radio and TV shows, wholesalers, distributors and more. All contacts (names, addresses, interests, etc.) are verified by phone.

- Kiefer, Marie and Kremer, John, **Book Publishing Resource Guide**, **$25.00.**
  This directory includes detailed listings for more than 7500 book marketing contacts (wholesalers, distributors, sales reps, chain stores, book clubs, and more) plus 3500 newspaper and 750 magazine editors. All the key contacts and resources you need to publish and market books and other information products. Approximately 320 pages, softcover. New Edition to be published Spring 1995.

- Kiefer, Marie and Kremer, John, **PR FLASH Profitcenter, $150.00**
  These data files provide key contact names, addresses, phone and fax numbers, subject interests, and other details for 17,019 media, including 6548 newspaper editors, 4972 magazine editors, 2761 radio shows, 1822 TV stations/shows, and 916 syndicated columns. The 17,019 records are available in a variety of file formats to use with your favorite database program on the IBM-PC, Macintosh, or compatibles. No other data-base or directory offers so much for so little cost. Updated regularly.

- Kiefer, Marie and Kremer, John, **Report: How to Sell to Mail Order Catalogs, $30.00**
  This report not only describes all the steps you need to follow to sell a book or other products to mail order catalog houses, but it also lists 722 catalogs that carry books, tapes, and other items. Current names, addresses, buyers, phone numbers, subject interests, and products carried are listed for each catalog. 57 page report. Completely updated, Spring 1995.

- Kiefer, Marie and Kremer, John, **Report: Radio Phone Interview Shows: How to Do an Interview Tour from Your Home, $40.00**
  This report lists more than 750 radio shows that feature phone interviews with authors and other experts. This reports has 68+ pages of details about each show: radio station, show name, contact person, address, phone number, subject interests, hosts, times, and much more. Radio phone interviews are one of the most cost-effective ways to promote your products and services. 94 page report. Updated regularly.

- Kiefer, Marie and Kremer, John, **The Top 200 National TV News / Talk / Magazine Shows, $30.00**

  This report features the top 200+ national TV news, talk, and magazine shows. Not only does it describe how to book guests on such shows, but it also lists the addresses and phone numbers of each show, who to contact, what subjects the show is interested in, the hosts, and other pertinent details. This report is indispensable for anyone wanting to book a guest appearance on a national TV talk show or place a story on the national news. 80 page report updated regularly.

- Kremer, John, **Book Marketing Made Easier: A Do-It-Yourself Marketing Kit for Publishers, Third Edition, $19.95**

  This book provides forms and procedures to help any book publisher prepare and carry out an effective marketing plan. Over 70 forms help you through all the steps: preparing your marketing strategy ... planning your budget ... forecasting your sales ... getting listed in the book publishing records ... researching the media ... sending out publicity releases ... obtaining reviews ... organizing author tours ... getting distribution ... setting up sales representation ... working with bookstores ... exhibiting your books ... submitting your books to catalogs ... granting subsidiary rights ... and dealing with authors. These forms take all the fuss out of book publishing. 384 pages, hardcover.

- Kremer, John, **Book Marketing Update** *Newsletter*, **$60.00 per year**

  A newsletter for anyone wanting to sell more books. Here are a few comments from our readers: "The management of our two companies have found your newsletter to be the finest newsletter about our business that we have ever seen." ... "As each issue comes, I think *There's no way you can top this* — and your next one does." ... "This is what we've been looking for since we started in publishing 3 years ago — at last someone is giving us what we need!" 16 pages, 10 issues per year.

- Kremer, John, **1001 Ways to Market Your Books — For Authors and Publishers**, softcover **$19.95.**

  The book features more than 1000 tips, techniques, and examples of how you can market your books more effectively. It covers traditional markets as well as special sales, subsidiary rights, direct mail, telemarketing, foreign sales, and much, much more. This book will help you to sell more books — and have fun doing it. This book includes two special chapters especially designed to help writers market their books and get the most out of the publication of their books. 544 pages.

• Kremer, John, **Mail Order Selling Made Easier, $19.95**

This book is an easy-to-use introduction to running a mail order operation. It includes worksheets, tables, sample letters, flow charts plus instructions to help anyone develop and carry out a successful direct response marketing program. It makes all the time-consuming details so much easier to handle. The book also features an extensive listing of resources and an annotated review of over 100 books. 288 pages, hardcover. Published in September, 1990.

"If you have any questions about the nitty-gritty of mail order, you can find the answer here," says Jeannie Spears, publisher of *The Professional Quilter*. "The material is short, concise, information-packed and easy to find. An excellent resource!"

• Kremer, John, **Mail Order Worksheet Kit, $15.00**

This kit includes full-size master copies of all the worksheets and sample letters outlined in *Mail Order Selling Made Easier*. Use these copies to organize your direct marketing programs. 60 master copies in a kit. Available for only **$10.00** when ordered with the book.

• Martin, Cliff, **Premium and Special Book Sales Report, $35.00.**

This report is a revised edition of a unique and very successful guide to selling the premium markets for publishers. This edition contains an entirely new chapter on Corporate Sponsorship; an expanded Resources section with over 120 recommended books, organizations, periodicals and key contacts; sample letters, reproducible Special Sales Contract Sheet.

• Martin, Cliff, **Talking Books: Publishing and Distributing Audiotapes For the Small Publisher, $30.00**

This new report provides easy-to-use advice on moving into the audio market and makes practical suggestions for doing your own, selling rights, or distributing other company's tapes. 59 pages

About the Author

By now many of your are aware that Marie Kiefer is the owner of Ad-Lib Publications in Fairfield, Iowa. She had worked with John Kremer, the founder of Ad-Lib Publications, for over six years before buying the company on January 1, 1991.

Two of the books she now publishes include this one: *Directory of Book Printers 1994-95* and the upcoming 1995 edition of *Book Publishing Resource Guide.*

Marie is also responsible for several special reports, including *How to Sell to Mail Order Catalogs*; *Radio Phone Interview Shows: How to Do an Author Tour from Home*; and *The Top 200 National TV News, Talk, and Magazine Shows*. In cooperation with John Kremer, Marie continues to update the *Book Marketing ProfitCenter*[TM] *Files* and the *PR ProfitCenter* [TM] *Data Files*.

Marie Kiefer is also the editor and publisher of the weekly *Book Promotion Hotline* newsletter. She has served as the Executive Director of the Mid-America Publishers Association.

After living and working on a farm which at one time included over 800 acres as well as two 150 sow herds, raising a family of four (the family now includes four + grandchildren), church activities, school functions and 4-H (both as a 4-H Leader and County 4-H Committee) kept Marie busy for over 20 years. After moving off the farm in the early '80s, Bob and Marie Kiefer started new careers. Bob to construction work and his own business Kiefer Enterprises and Marie to office positions which eventually led to being John Kremer's first full-time employee. As the saying goes *"and the rest is history"* we can now find Marie in charge of Ad-Lib Publications.